PRENTICE HALL

Trigonometry

Jerome D. Hayden
Bettye C. Hall

Consulting Authors

Jan Fair ❖ Robert Kalin

Sadie C. Bragg ❖ Mary Kay Corbitt

Mona Fabricant

Prentice Hall dedicates this
mathematics program to
all mathematics educators
and their students.

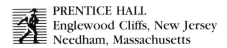
PRENTICE HALL
Englewood Cliffs, New Jersey
Needham, Massachusetts

PRENTICE HALL
Trigonometry

AUTHORS

Jerome D. Hayden
Mathematics Department Chairman, K-12
McLean County Unit District 5
Normal, Illinois

Bettye C. Hall
Director of Mathematics
Houston Independent School District
Houston, Texas

CONSULTING AUTHORS

Jan Fair
Director, CAPP Project SUCCESS
A Math/Science California
  Academic Partnership Program
Lompoc/San Luis Obispo/Santa
  Maria, California

Robert Kalin
Professor Emeritus
Florida State University
Mathematics Education Program
Tallahassee, Florida

Mona Fabricant
Professor of Mathematics
Queensborough Community
  College
Bayside, New York

Sadie C. Bragg
Professor of Mathematics and Associate Dean
  of Curriculum
Borough of Manhattan Community College
The City University of New York
New York, New York

Mary Kay Corbitt
Associate Professor of Mathematics
Valdosta State College
Valdosta, Georgia

CONSULTANTS

Herbert Hollister
Professor of Mathematics
Bowling Green State University
Bowling Green, Ohio

Glenn Miller
Mathematics Instructor
Interboro Institute
New York, New York

REVIEWERS

John F. Lamb
Professor of Mathematics
East Texas State University
Commerce, Texas

Calvin T. Long
Professor of Mathematics
Washington State University
Pullman, Washington

Francis Yu-Chaw Meng
School of Computer Science
Rochester Institute of Technology
Rochester, New York

Photo credits appear on page 472.

Cover Design: Martucci Studio

ISBN 0-13-979709-2

1 2 3 4 5 6 7 8 9 10   97 96 95 94 93

Prentice Hall
A Division of Simon & Schuster
Englewood Cliffs, New Jersey 07632

STAFF CREDITS

Editorial: Rosemary Calicchio, Enid Nagel, Debra Berger, Mary Ellen
  Cheasty, Michael Ferejohn, Tony Maksoud, John Nelson, Alan
  MacDonell, Ann Fattizzi
Design: Laura Jane Bird, Art Soares
Production: Amy Fleming, Lorraine Moffa, Suse Cioffi
Photo Research: Libby Forsyth, Emily Rose, Martha Conway
Publishing Technology: Andrew Grey Bommarito, Gwendollynn Waldron,
  Deborah Jones, Monduane Harris, Michael Colucci, Gregory Myers,
  Cleasta Wilburn
Marketing: Everett Draper, Julie Scarpa, Michelle Sergi
Pre-Press Production: Laura Sanderson, Natalia Bilash, Denise Herckenrath
Manufacturing: Rhett Conklin, Gertrude Szyferblatt
National Consultants: Susan Berk, Charlotte Mason

Contents

Trigonometric Functions

CONNECTIONS: *Meteorology • Geometry • Navigation • Construction • Engineering • Entertainment • Horology • Computer • Surveying • Routing*

Graphing Trigonometric Functions

CONNECTIONS: *Landscaping • Algebra • Architecture • Physics • Music • Oceanography • Graphing Calculator • Engineering*

Right Triangle Trigonometry and Basic Identities

CONNECTIONS: *Construction • Geometry • Architecture • Sports • Aviation • Engineering • Computer • Genetics • Physics • Metallurgy*

Oblique Triangles

CONNECTIONS: *Surveying • Engineering • Navigation • Geometry • Construction • Landscaping*

Trigonometric Identities

CONNECTIONS: *Surveying • Aviation • Mechanics • Coordinate Geometry • Indirect Measurement • Geometry*

Inverse Trigonometric Functions

CONNECTIONS: *Science • Geometry • Mechanics • Engineering • Physics • Construction*

Complex Numbers

CONNECTIONS: *Physics • Coordinate Geometry • Algebra • Electricity • Geometry • Technology*

Exponential and Logarithmic Functions

CONNECTIONS: *Biology • Archaeology • Algebra • Optics • Technology • Physics • Astronomy • Business*

Contents   **v**

9 Sequences and Series

CONNECTIONS: *Consumerism • Business • Investment • Geometry • Building Design • Sports • Number Theory • Construction • Entertainment • Finance • Physics • Statistics • Civil Engineering*

Developing Mathematical Power

Reasoning

Challenge   38, 45, 56, 77, 108, 123, 152, 167, 189, 202, 246, 268, 272, 282, 317, 330, 337, 368, 406, 416

Extra   11, 16, 22, 33, 66, 72, 87, 93, 123, 135, 139, 162, 167, 174, 213, 219, 230, 236, 254, 263, 268, 291, 297, 302, 321, 341, 358, 377, 389

Thinking Critically   9, 98, 135, 162, 167, 189, 257, 300, 305, 341, 345, 350, 357, 362

Connections

Application   7, 11, 16, 23–28, 33, 38, 45, 52–53, 65–66, 72, 77, 82–83, 86–87, 93, 99, 105, 117, 118–129, 134–135, 139, 144, 147–149, 159, 161–162, 167, 173, 179, 185, 189, 194, 196–199, 213, 219, 225, 230, 241–243, 254, 259, 264, 268, 272, 277–279, 291, 296, 302, 306–307, 312, 321, 326, 337, 341, 346, 352, 358–364, 377, 382–383, 389, 395, 401, 406, 411

Career   11, 22, 33, 38, 45, 65, 66, 72, 87, 99, 117, 124, 128, 134, 139, 147, 161, 167, 173, 179, 185, 189, 213, 242, 259, 264, 277, 302, 337, 341, 352, 358, 377, 383, 389, 390, 395, 411

Historical Note   7, 66, 99, 117, 144, 259, 278, 302, 308, 312, 401

Integrating Trigonometry   2, 12, 23, 24, 25, 33, 46, 60, 67, 72, 73, 84, 87, 100, 101, 102, 117, 118, 119, 124, 140, 148, 156, 159, 163, 168, 174, 185, 186, 190, 194, 225, 226, 230, 231, 236, 237, 250, 268, 291, 292, 301, 327, 338, 347, 352, 359, 365, 378, 383, 384, 390, 392

Preview   2, 12, 23, 46, 67, 73, 100, 118, 124, 163, 168, 186, 190, 226, 231, 237, 250, 292, 347, 359, 378, 384, 390, 396

Technology   18, 20–22, 46, 50–51, 70, 80, 84, 86–87, 89, 92–93, 98–99, 103–104, 114, 116–117, 121–123, 126–129, 133–135, 145–147, 160–162, 166–167, 172–174, 179–180, 184–185, 187–189, 219, 241, 256, 262–264, 269, 271–272, 295–297, 307, 310–312, 315–317, 325–327, 335–337, 340–341, 344–346, 353–354, 357–358, 362–365, 379, 381–383, 387–389, 393–395, 400–403, 405–406, 410–413

*T*rigonometry, like algebra and geometry, is considered a language of mathematics, and mathematics is a vital part of your life. Changes in society and in the use of technology require that you have a strong background in mathematics. The emphasis is not only on trigonometry skills, but on developing your *mathematical power*. What is mathematical power?

● The knowledge and understanding of mathematical ideas, concepts, and procedures

● The ability to solve problems within mathematics and in other subject areas

● The ability to use mathematical tools and techniques

● The ability to use mathematics in your world beyond the classroom

● The ability to use mathematics to communicate

● The ability to reason and think critically

● The understanding and appreciation of the nature and beauty of mathematics

Using your trigonometry book effectively will help you to develop your mathematical power. Let us introduce you to the parts of your trigonometry book that can help you to develop the aspects of mathematical power.

Each lesson begins with a *Preview* that connects trigonometry to a familiar topic or reviews essential skills you will need in the upcoming lesson.

Completely worked-out examples contain instruction to help you to successfully complete your assignments.

New concepts are highlighted for easy reference.

The technology logo and graphing calculator screens throughout the book identify when using technology is beneficial to and appropriate for the development of concepts. In addition, *Technology Lessons* provide you with alternative methods for exploring trigonometry. A graphing calculator is a valuable tool to use with this course. It has all the capabilities of a scientific calculator, plus you can plot and graph functions on the display screen, view formulas and solutions in their entirety, zoom in and trace graphs, and more.

Application Lessons provide a look at how trigonometry is applied and used in daily life.

Practice Exercises build a solid foundation of trigonometry skills . . .

Thinking Critically—helps improve your reasoning skills.

Applications—connects trigonometry to daily life and to other topics.

Writing in Mathematics activities help you to acquire communication skills.

Mixed Review at the end of chapters helps you to remember the skills and concepts already learned.

Test Yourself provides you with a tool for self-evaluation. Answers to these and other selected exercises can be found in the back of the book.

*R*emember . . . mathematics is not a spectator sport! Actively participating, reading carefully, and completing your assignments will contribute to your success in trigonometry and in other mathematics courses you take in the future.

1 Trigonometric Functions

Satellites in orbit are programmed to transmit data to earth from various points in their orbit. Small errors in measuring an angle may produce large errors over long distances. Trigonometric concepts play an important part in the calculations that make sure transmission signals do not miss their targets on earth.

1

Functions and the Coordinate Plane

1.1

Objectives: To define relations and functions
To graph functions in the coordinate plane

Mathematical relationships often exist between sets
of numbers. Formulas and equations are frequently
used to define such relationships.

Preview

When an object is dropped, the distance it travels
depends upon the amount of time elapsed since
its release. For a given time t, the approximate
distance d is given by the equation

$$d = 16t^2$$

where d is in feet and t is in seconds.

**Find the height of a building if an object dropped from the roof takes the
given amount of time to hit the ground.**

1. 3 s   **2.** 5 s   **3.** 0.1 min

**Find the number of seconds it takes an object to hit the ground if it is
dropped from the roof of a building with the given height.**

4. 64 ft   **5.** 256 ft   **6.** 100 ft

The solutions to the equation $d = 16t^2$ can be written as ordered pairs (t, d),
such as

$$\{(1, 16), (2, 64), (3, 144), \ldots\}$$

A set of ordered pairs of numbers is called a **relation.** The set of replacements
for the first variable is called the **domain** of the relation. The domain may be
limited to a specific set of numbers. In the equation $d = 16t^2$, it would be
reasonable to limit the time t to numbers that are greater than or equal to zero.
The domain of this relation is the set of nonnegative real numbers.

The set of replacements for the second variable is called the **range** of the
relation. In the equation $d = 16t^2$, the distance d will always be greater than
or equal to zero, because t^2 is always greater than or equal to zero. The range
of this relation is the set of nonnegative real numbers.

EXAMPLE 1 **Find the domain and range of the relation:**
   {(0, 0), (1, 2), (2, 4), (3, 6), (4, 8)}

The domain is the set of replacements for the first variable:

{0, 1, 2, 3, 4}

The range is the set of replacements for the second variable:

{0, 2, 4, 6, 8}

Each ordered pair in a relation is associated with exactly one point on the **coordinate plane.** The set of all points (x, y) associated with a relation can be graphed on a rectangular coordinate system, as shown at the right.

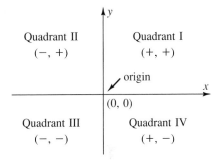

The horizontal number line is called the **x-axis** (sometimes called the axis of *abscissas*), where x represents the first number in the ordered pair. The vertical number line is called the **y-axis** (sometimes called the axis of *ordinates*), where y represents the second number in the ordered pair. The point of intersection of the axes is called the **origin.** The axes divide the plane into four regions, called **quadrants.** Every point on the plane can be represented by an ordered pair (x, y). When you plot ordered pairs, or points, it is important to label the axes. Although it is often convenient to choose the same scale for both axes, it is not necessary to do so.

EXAMPLE 2 **Find the domain and range of each relation:**

a.

b.

c.
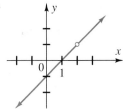

a. Domain {all real numbers}   **b.** Domain $\{x : x \geq 1\}$ **c.** Domain $\{x : x \neq 2\}$
  Range $\{y : y \geq 0\}$ Range $\{y : y$ is a real   Range $\{y : y \neq 1\}$
                                       number$\}$

In the relation given in Example 1, notice that each member x of the domain is paired with exactly one member y of the range. Because of this unique pairing, the ordered pairs represent a special type of relation called a *function*. A **function** is a relation in which each member of the domain is paired with exactly one member of the range. In this book, special types of functions called *trigonometric functions* will be studied.

EXAMPLE 3 **State whether or not each relation is a function.**

a. {(1, 2), (2, 3), (3, 4)}   **b.** {(1, 4), (2, 5), (1, 6)}   **c.** {(1, 6), (2, 7), (3, 6), (4, 7)}

a. a function   **b.** not a function   **c.** a function

A quick way to decide if a graph represents a function is to use the *vertical line test*: If any vertical line can be drawn that intersects the graph of a relation in more than one point, the graph does *not* represent a function. Likewise, if no such vertical line can be drawn, the graph *does* represent a function.

EXAMPLE 4 **State whether or not each graph represents a function.**

a.   **b.**   **c.**   **d.**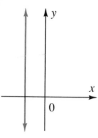

a. function   **b.** not a function   **c.** function   **d.** not a function

Special notation is used to denote a function. Symbols such as $f(x)$, $g(t)$, and $F(w)$ are used. For example, the function $d = 16t^2$ can be written

$$f(t) = 16t^2 \qquad \textit{f(t) is read f of t.}$$

Specific values of this function are written as follows:

$$f(1) = 16(1)^2 = 16(1) = 16$$
$$f(2) = 16(2)^2 = 16(4) = 64$$

EXAMPLE 5 **For the function $f(x) = 6x + 7$, find:**

a. $f(-3)$ **b.** $f(0)$ **c.** $f(3.1)$

a. $\begin{aligned} f(x) &= 6x + 7 \\ f(-3) &= 6(-3) + 7 \\ &= -18 + 7 \\ &= -11 \end{aligned}$ **b.** $\begin{aligned} f(x) &= 6x + 7 \\ f(0) &= 6(0) + 7 \\ &= 0 + 7 \\ &= 7 \end{aligned}$ **c.** $\begin{aligned} f(x) &= 6x + 7 \\ f(3.1) &= 6(3.1) + 7 \\ &= 18.6 + 7 \\ &= 25.6 \end{aligned}$

If the domain of a function is not stated, assume that it is the greatest set of real numbers for which the value of the function is also a real number. You must exclude from the domain any numbers that make a denominator equal to zero, since the value of the function would be undefined. Numbers that lead to the square root of a negative quantity must also be excluded, since the value of the function would <u>not be a</u> real number.

EXAMPLE 6 **Determine which real numbers must be excluded from the domain of each function:**

a. $f(x) = \dfrac{2}{x}$ **b.** $f(x) = \dfrac{2x}{3x - 6}$ **c.** $g(x) = \dfrac{5}{x^2 - 9}$ **d.** $g(x) = \sqrt{x}$

a. Exclude 0, since $\dfrac{2}{0}$ is undefined.   **b.** Exclude 2, since $\dfrac{4}{0}$ is undefined.

c. Exclude 3 and -3, since $\dfrac{5}{0}$ is undefined.   **d.** Exclude all real numbers less than 0.

CLASS EXERCISES

Find the domain and range of each relation, and state whether or not the relation is a function.

1. $\{(2, 4), (3, 6), (4, 8), (5, 10)\}$ **2.** $\{(-2, -1), (-1, 0), (0, 1), (1, 2)\}$

3. $\{(6, 1), (6, 2), (6, 3), (6, 4)\}$ **4.** $\{(2, 5), (3, 5), (4, 5), (5, 5), (6, 5)\}$

5. $\{(2, 4), (4, 2), (3, 6), (6, 3)\}$ **6.** $\{(0, 2), (1, 3), (1, 4), (2, 5)\}$

7. $\{(-2, -2), (1, -1), (-1, 1), (-2, 2)\}$ **8.** $\{(-2, -2), (-1, -1), (1, -1), (2, 2)\}$

For the function $f(x) = 2x + 4$**, find each value.**

9. $f(1)$ **10.** $f(10)$ **11.** $f(0)$ **12.** $f(-2)$ **13.** $f(a)$

Determine which real numbers must be excluded from the domain of each function.

14. $\dfrac{3}{x + 5}$ **15.** $\dfrac{-2x}{5x - 15}$ **16.** $\dfrac{4}{x^2 - 16}$ **17.** $\dfrac{1}{\sqrt{8x}}$

PRACTICE EXERCISES

Find the domain and range of each relation.

1. $\{(1, 0), (2, 1), (5, 3), (7, 5), (16, 9)\}$ **2.** $\{(2, 1), (3, 3), (5, 5), (10, 6)\}$

3. $\{(-2, -10), (-1, -1), (0, 8), (1, 11)\}$ **4.** $\{(-3, -2), (-2, -1), (0, 2), (8, 11)\}$

5. $\{(-3, 1), (2, 2), (4, 3), (-3, 3)\}$ **6.** $\{(-100, 100), (-100, 5), (1, -100)\}$

7. $\{(4, 25), (6, 25), (8, 25), (10, 25)\}$ **8.** $\{(7, -3), (6, 2), (5, 6), (7, -3)\}$

State whether or not each relation is a function.

9. $\{(2, 3), (2, 4), (3, 5)\}$ **10.** $\{(3, 1), (4, 1), (5, 2)\}$

11. $\{(6, 4), (7, 4), (8, 5)\}$ **12.** $\{(6, 2), (6, 3), (6, 4)\}$

13. $\{(2, 6), (3, 6), (4, 6)\}$ **14.** $\{(6, 1), (5, 1), (4, 1)\}$

State whether or not each graph represents a function.

15.   **16.**   **17.**

18.   **19.**   **20.**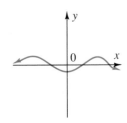

Find each value for the function $f(x) = 3x + 10$.

21. $f(5)$ **22.** $f(6)$ **23.** $f(-4)$ **24.** $f(-5)$

Find each value for the function $g(x) = 2x - 5$.

25. $g\left(\frac{1}{2}\right)$ **26.** $g\left(\frac{3}{4}\right)$ **27.** $g(1.7)$ **28.** $g(2.5)$

Find the domain and range of each relation, and state whether or not the relation is a function.

29. $\{(5, -1), (5, 3), (-1, -10), (8, 8)\}$ **30.** $\{(3, 3), (5, 3), (4, 3), (-3, 3)\}$

31. $f(x) = x^2$ **32.** $f(x) = -x^2$ **33.** $f(x) = -x$ **34.** $f(x) = x^3$

35.   **36.**   **37.**   **38.**

Find each value for the function $f(x) = x^2 - 1$

39. $f(2)$ **40.** $f(3)$ **41.** $f(-5)$ **42.** $f(-6)$ **43.** $f(\sqrt{5})$ **44.** $f(\sqrt{7})$

Determine which real numbers must be excluded from the domain of each function.

45. $f(x) = \dfrac{x}{x}$ **46.** $g(x) = \dfrac{3x}{-3x + 9}$ **47.** $f(x) = \dfrac{-x}{x^2 - 49}$ **48.** $g(x) = \dfrac{1}{\sqrt{2x}}$

Find each value for the function $f(x) = -\dfrac{1}{x^2}$.

49. $f(-3)$ **50.** $f(-4)$ **51.** $f\left(\dfrac{1}{9}\right)$ **52.** $f\left(\dfrac{1}{10}\right)$

Find each value for the functions $f(x) = \sqrt{x} - x$ and $g(x) = \sqrt{x} + 1$.

53. $f(4) + g(16)$ **54.** $f(3) + g(9)$ **55.** $f(25) - g(25)$ **56.** $f(36) - g(36)$

Applications

57. Meteorology Temperatures given in degrees Celsius can be converted to degrees Fahrenheit using the function $f(t) = \dfrac{9}{5}t + 32°$, where t is the temperature in degrees Celsius. If the temperature is 35°C, determine the temperature in degrees Fahrenheit.

58. Meteorology Temperatures given in degrees Fahrenheit can be converted to degrees Celsius by the function $g(t) = \dfrac{5}{9}(t - 32)°$, where t is the temperature in degrees Fahrenheit. If the temperature is 86°F, determine the temperature in degrees Celsius.

59. Geometry The surface area of a cube is given by the function $A = 6e^2$, where e is the length of an edge. Find the surface area of a cube whose edges each measure 3 in.

60. Geometry The volume of a cube is given by the function $V = e^3$, where e is the length of an edge. Find the length of an edge of a cube with volume 512 cm³.

HISTORICAL NOTE

Descartes and the Coordinate Plane The French mathematician and philosopher René Descartes (1596–1650) is usually considered the founder of modern coordinate geometry. It is interesting to note, however, that there are differences between the original coordinate system Descartes devised and today's *Cartesian* coordinate system, which was named in his honor. Descartes used a single horizontal axis and only positive values. At each point on the axis, he imagined the construction of a line segment whose length represented the value of the related second coordinate. These segments were not always perpendicular to the axis, but they were parallel to each other. Investigate Descartes.

1.2 The Distance Formula

Objective: To define and use the distance formula

It is important to be able to find the distance between two points in the coordinate plane. When distances are represented by irrational numbers, it is customary to simplify radicals.

Preview

To simplify a square root, factor the radicand into as many perfect squares as possible. Then take the square root of each square factor and leave the remaining factors under the square root sign.

EXAMPLE Simplify: $\sqrt{150}$

$\sqrt{150} = \sqrt{25 \cdot 6} = 5\sqrt{6}$

Simplify each expression.

1. $\sqrt{90}$ 2. $\sqrt{200}$ 3. $\sqrt{98}$ 4. $\sqrt{125}$ 5. $\sqrt{108}$ 6. $\sqrt{112}$

The distance formula is based on the **Pythagorean theorem,** which states that a and b are the lengths of the legs of a right triangle and c is the length of its hypotenuse if and only if

$$c^2 = a^2 + b^2$$

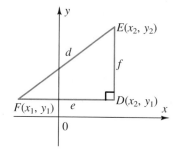

Consider two points $E(x_2, y_2)$ and $F(x_1, y_1)$. The distance between E and F is found by applying the Pythagorean theorem to the triangle at the right.

$$d^2 = e^2 + f^2$$
$$= (x_2 - x_1)^2 + (y_2 - y_1)^2 \qquad e = x_2 - x_1; f = y_2 - y_1$$

The distance formula is found by taking the square root of each side.

Distance Formula   The distance between two points $E(x_2, y_2)$ and $F(x_1, y_1)$ is

$$d = \sqrt{(x_2 - x_1)^2 + (y_2 - y_1)^2}$$

EXAMPLE 1 Find the distance between each pair of points:

a. $A(-5, 2)$ and $B(4, 2)$

b. $P(-2, -6)$ and $Q(-5, 3)$

a. $d = \sqrt{[4 - (-5)]^2 + (2 - 2)^2}$

$= \sqrt{9^2 + 0^2}$

$= \sqrt{81} = 9$

b. $d = \sqrt{[-5 - (-2)]^2 + [3 - (-6)]^2}$

$= \sqrt{(-3)^2 + 9^2}$

$= \sqrt{90} = 3\sqrt{10}$

The distance formula can be used to find the equation of a circle of radius r with center O at the origin. If $P(x, y)$ represents any point on the circle, then the distance between P and O is r.

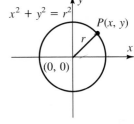

$$\sqrt{(x - 0)^2 + (y - 0)^2} = r$$

$$\sqrt{x^2 + y^2} = r$$

$$x^2 + y^2 = r^2 \qquad \textit{Square both sides.}$$

If a circle of radius r has its center at the origin, then its equation is $x^2 + y^2 = r^2$. Conversely, if the equation $x^2 + y^2 = r^2$ is given, then its graph is a circle of radius r with center at the origin.

EXAMPLE 2 Find the radius r of the circle with the equation $x^2 + y^2 = 25$.

Since $r^2 = 25, r = \sqrt{25} = 5$.

CLASS EXERCISES

1. **Thinking Critically** The formula for the distance between two points $E(x_2, y_2)$ and $F(x_1, y_1)$ is $d = \sqrt{(x_2 - x_1)^2 + (y_2 - y_1)^2}$. Would you get a different value for d if you interchanged x_1 and x_2? if you interchanged y_1 and y_2?

State whether or not each triangle is a right triangle.

2.

3.

4.

5.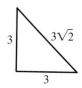

Match the equation of each circle with its description.

6. center at the origin, radius 4
7. center at the origin, radius 2
8. center at the origin, radius 7
9. center at the origin, radius $3\sqrt{6}$

a. $x^2 + y^2 = 4$
b. $x^2 + y^2 = 54$
c. $x^2 + y^2 - 49 = 0$
d. $2x^2 + 2y^2 = 32$

PRACTICE EXERCISES

Find the distance between each pair of points.

1. $(1, 1), (4, 5)$ 2. $(2, 2), (8, 10)$ 3. $(-2, 5), (5, -2)$

4. $(-7, 2), (3, -3)$ 5. $(0, 0), (6, -6)$ 6. $(0, 0), (5, -5)$

7. $(6, -5), (0, -4)$ 8. $(3, 0), (-2, 2)$ 9. $\left(\frac{1}{2}, -1\right), \left(\frac{1}{2}, 4\right)$

10. $\left(3, \frac{2}{3}\right), \left(12, \frac{2}{3}\right)$ 11. $(0.5, 1.2), (1.5, -1.8)$ 12. $(-3, 1.4), (-2, 1.4)$

Find the radius of each circle.

13. $x^2 + y^2 = 49$ 14. $x^2 + y^2 = 81$ 15. $x^2 + y^2 - 121 = 0$

16. $x^2 + y^2 - 225 = 0$ 17. $3x^2 + 3y^2 = 180$ 18. $\frac{1}{3}x^2 + \frac{1}{3}y^2 = 21$

Find the distance between each pair of points.

19. $(3.2, 5.6), (4.7, 6.1)$ 20. $(1.4, -4.3), (-0.6, -2.7)$

21. $(\sqrt{2}, -3), (2\sqrt{2}, 1)$ 22. $(5, \sqrt{2}), (9, -\sqrt{2})$

23. $(\sqrt{5}, 4\sqrt{7}), (3\sqrt{5}, 3\sqrt{7})$ 24. $(-\sqrt{2}, \sqrt{3}), (2\sqrt{2}, 2\sqrt{3})$

25. $(2a, b), (4a, -b)$ 26. $(-4r, 3s), (-r, -3s)$

Write the equation of the circle that contains the given point P and has its center at the origin.

27. $P(-3, 4)$ 28. $P(6, -8)$ 29. $P(-3, -6)$

30. $P(-9, -3)$ 31. $P(\sqrt{2}, 5)$ 32. $P(-3\sqrt{2}, -4)$

33. Find x if the distance between $(1, 3)$ and $(x, 9)$ is 10.

34. Find y if the distance between $(2, y)$ and $(7, 8)$ is 13.

35. Find the area of the rectangle with vertices $A(-5, 1), B(-3, -1), C(3, 5)$, and $D(1, 7)$.

36. Find the area of the rectangle with vertices $A(-3, 0), B(-2, -1), C(1, 2)$, and $D(0, 3)$.

37. A quadrilateral is a parallelogram if both pairs of opposite sides are equal in length. If $A(-1, 0), B(3, 8), C(3, 2)$, and D are consecutive vertices of a parallelogram, what are the coordinates of point D?

38. A quadrilateral is a rhombus if all four sides are equal in length. If $P(1, 2), Q(4, -2), R(1, -6)$, and S are consecutive vertices of a rhombus, what are the coordinates of point S?

39. **Navigation** Airplane A is located 3 mi west and 1 mi south of airport O. Airplane B is located 5 mi due north of the airport at the same altitude as airplane A. Draw a diagram using a coordinate system with origin O. How far apart are the two airplanes? Express the answer in radical form.

40. **Navigation** A rowboat is located 5 mi due east of buoy O. A sailboat is located 3 mi west and 6 mi north of the same buoy. Draw a diagram using a coordinate system with origin O. How far apart are the two boats?

41. **Construction** A stringer is a board that supports a set of stairs. If a staircase is 8 ft high and extends 10 ft horizontally, find the length of the stringer to the nearest foot.

42. **Engineering** A bridge crosses a river that is 2140 ft wide. The vertical arch support at the center of the bridge is 560 ft tall. Find the length of the longest spanning wire to the nearest foot. (Assume that the wire is taut.)

43. **Engineering** A guy wire is used to support or steady an object. A trapeze artist is standing on her landing platform, which is attached to a vertical post. A guy wire supporting the post is 87 ft long, and the base of the post is 45 ft from the wire's stake. Find the height of the platform to the nearest foot.

EXTRA

The distance between two points in space
In the coordinate system shown at the right, points in three-dimensional space are represented by ordered triples of the form (x, y, z). The following formula gives the distance d between points $P(x_2, y_2, z_2)$ and $Q(x_1, y_1, z_1)$:

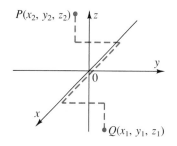

$$d = \sqrt{(x_2 - x_1)^2 + (y_2 - y_1)^2 + (z_2 - z_1)^2}$$

Use the formula above to find the distance between each pair of points.

1. $A(0, 1, 2)$, $B(5, -3, 1)$
2. $C(1, 1, 1)$, $D(1, -1, 2)$

3. Could a distance formula be written for "four-dimensional space"? If so, how? What applications might there be for this?

Angles in the Coordinate Plane

1.3

Objectives: To measure angles in rotations and in degrees
To find the measures of coterminal angles

In trigonometry, there is a distinction between the measures of angles formed by a ray rotating in a clockwise direction and the measures of angles formed by a ray rotating in a counterclockwise direction.

Preview

The Coriolis force is the inertial force caused by the earth's rotation. It affects many phenomena, including water draining from a bathtub. Above the equator, water draining from a bathtub swirls clockwise. Below the equator, it swirls counterclockwise.

1. How do you think water draining from a bathtub would behave *at* the equator?

2. If you stirred draining water with your finger, do you think you could cause it to swirl in the opposite direction?

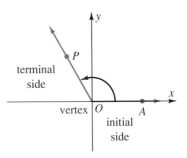

An **angle** can be defined as the union of two rays with a common endpoint, the *vertex* of the angle. In trigonometry, an angle is formed by rotating a ray about its vertex from one position, called the **initial side** of the angle, to another, called the **terminal side.**

When the vertex of an angle is the origin of the coordinate plane and its initial side coincides with the positive *x*-axis, the angle is said to be in **standard position.** Angle *AOP* (∠*AOP*) is in standard position. Its measure is the amount of rotation of $\overrightarrow{OA}$ (ray *OA*), the initial side, to the position of $\overrightarrow{OP}$, the terminal side of the angle.

If an angle is formed by a counterclockwise rotation, its measure is *positive*. If an angle is formed by a clockwise rotation, its measure is *negative*.

One unit of angle measure is the **degree.** One full counterclockwise rotation measures 360° (360 degrees), and one full clockwise rotation measures −360°. Note that the terminal and initial sides of the angle coincide under a full rotation.

EXAMPLE 1   **Find the degree measure of the angle for each rotation. Sketch the angle in standard position.**

   a. $\frac{1}{2}$ rotation, counterclockwise   **b.** $\frac{2}{3}$ rotation, clockwise

   a. $\frac{1}{2}(360°) = 180°$ **b.** $\frac{2}{3}(-360°) = -240°$

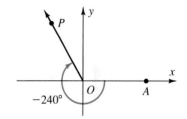

If an angle is in standard position, it is said to lie in the quadrant in which its terminal side falls. For instance, ∠AOP in Example 1b lies in quadrant II.

Coterminal angles are angles in standard position whose terminal sides coincide.

EXAMPLE 2   **Find the measure of the angle coterminal with ∠BOC under the given rotation from the position of $\overrightarrow{OC}$.**
   a. One counterclockwise rotation
   b. One clockwise rotation
   c. A counterclockwise rotation other than the one used in part a.

a.   **b.**   **c.**

60° + 360° = 420°   60° + (−360°) = −300°   60° + 2(360°) = 780°
1 counterclockwise   1 clockwise rotation   2 counterclockwise
rotation   rotations

If two angles are coterminal, their measures differ by an integral multiple of 360°.

$$420° = 1 \cdot 360° + 60° \qquad -300° = (-1) \cdot 360° + 60° \qquad 780° = 2 \cdot 360° + 60°$$

The measure of any angle coterminal with an angle of degree measure θ (the Greek letter *theta*) is found by evaluating $n \cdot 360 + \theta$, where n is an integer. That is, the measures of all angles coterminal with a given angle of measure θ may be expressed as a function of the number n of complete rotations of the terminal side

$$f(n) = (n \cdot 360 + \theta)°$$

EXAMPLE 3   **Given $f(n) = (n \cdot 360 + 48)°$, find:**

 a. $f(-1)$ **b.** $f(0)$ **c.** $f(1)$ **d.** $f(2)$

a. $f(-1) = (-1 \cdot 360 + 48)°$
$\qquad = -312°$

b. $f(0) = (0 \cdot 360 + 48)°$
$\qquad = 48°$

c. $f(1) = (1 \cdot 360 + 48)°$
$\qquad = 408°$

d. $f(2) = (2 \cdot 360 + 48)°$
$\qquad = 768°$

CLASS EXERCISES

Find the degree measure of the angle for each rotation.

1. $\frac{1}{3}$ rotation, clockwise

2. $\frac{1}{5}$ rotation, clockwise

3. $\frac{13}{12}$ rotation, counterclockwise

4. $\frac{11}{12}$ rotation, counterclockwise

In which quadrant does the terminal side of each angle lie when it is in standard position?

5. 95° **6.** −30° **7.** −40° **8.** 75°

9. 210° **10.** −340° **11.** −350° **12.** 295°

For each angle shown in standard position, name the initial and terminal sides and estimate the degree measure.

13. **14.** **15.** **16.**

Name two angles that are coterminal with the given angle.

17. 30° **18.** 150° **19.** −30° **20.** −120°

PRACTICE EXERCISES

Find the degree measure of the angle for each rotation. Then sketch each angle in standard position. Indicate its position by a curved arrow.

1. $\frac{1}{4}$ rotation, clockwise

2. $\frac{3}{4}$ rotation, clockwise

3. $\frac{1}{2}$ rotation, counterclockwise

4. $\frac{1}{8}$ rotation, counterclockwise

5. $\frac{3}{5}$ rotation, clockwise

6. $\frac{1}{3}$ rotation, counterclockwise

7. $\frac{1}{6}$ rotation, clockwise

8. $\frac{5}{6}$ rotation, clockwise

9. $\frac{4}{3}$ rotation, counterclockwise

10. $\frac{3}{10}$ rotation, clockwise

11. $\frac{5}{12}$ rotation, clockwise

12. $\frac{17}{12}$ rotation, counterclockwise

Name three angles (two positive and one negative) that are coterminal with the given angle.

13. $40°$ 14. $70°$ 15. $183°$ 16. $97°$

17. $-34°$ 18. $-62°$ 19. $151°$ 20. $209°$

Given $f(n) = (n \cdot 360 + 115)°$, find each value.

21. $f(2)$ 22. $f(1)$ 23. $f(0)$ 24. $f(4)$

25. $f(-8.4)$ 26. $f(8.4)$ 27. $f(1.8)$ 28. $f(-1.8)$

Start at the terminal side of the given angle θ in standard position. Find the measure of the resulting angle (in standard position) after the given number of rotations.

29. $\theta = 210°$, 1 clockwise

30. $\theta = 240°$, 1 clockwise

31. $\theta = 18°$, 1 counterclockwise

32. $\theta = -16°$, 1 counterclockwise

33. $\theta = -18°$, 2 counterclockwise

34. $\theta = 16°$, 2 counterclockwise

35. $\theta = 30°$, 1.5 counterclockwise

36. $\theta = 60°$, 2.5 counterclockwise

37. $\theta = 310°$, 2 clockwise

38. $\theta = -150°$, 1.5 clockwise

For each of the following angles, find a coterminal angle with measure θ such that $0° \leq \theta < 360°$.

39. $-100°$ 40. $-215°$ 41. $534°$ 42. $-386°$

43. $900°$ 44. $1250°$ 45. $-3610°$ 46. $-7201°$

The angles measuring $(n \cdot 360 + 20)°$ are coterminal for all integers n. There are two possible positions of the terminal sides for the angles $(n \cdot 180 + 20)°$ in standard position. How many different positions of the terminal sides are possible for angles in standard position given by each of the following expressions?

47. $(n \cdot 120 + 20)°$ **48.** $(n \cdot 72 + 20)°$ **49.** $(n \cdot 100 + 20)°$ **50.** $(n \cdot 150 + 20)°$

Applications

51. Entertainment Many records make $33\frac{1}{3}$ revolutions per minute. Through how many degrees does a point on the record turn during one minute?

52. Entertainment Singles are usually available on records that make 45 revolutions per minute. Through how many degrees does a point on the record turn during one minute? during 30 seconds?

EXTRA

In geometry, angles are named using capital letters of the English alphabet. For example, this angle could be called $\angle Q$ or $\angle PQR$. As mentioned previously in this lesson, Greek letters are often used in trigonometry to name angles and their measures.

Greek Alphabet

A α Alpha	B β Beta	Γ γ Gamma	Δ δ Delta
E ε Epsilon	Z ζ Zeta	H η Eta	Θ θ Theta
I ι Iota	K κ Kappa	Λ λ Lambda	M μ Mu
N ν Nu	Ξ ξ Xi	O o Omicron	Π π Pi
P ρ Rho	Σ σ Sigma	T τ Tao	Υ υ Upsilon
ϑ φ Phi	X χ Chi	Ψ ψ Psi	Ω ω Omega

It is particularly helpful to use Greek letters to name angles between 180° and 360°. To illustrate, "$\angle ABC$" could refer to *either* of the two angles shown at the right. However, the letters α and β name the angles uniquely.

Determine the degree measure of each angle labeled with a Greek letter.

1. 45° γ

2. μ 117°

3. α 95° 32°

4. σ 36°

Angle Measure in Degrees and Radians

Objectives: To measure angles and arcs in degrees and radians
To change from degree measure to radian measure and
from radian measure to degree measure

If a ray makes a complete 360° rotation about its endpoint, the path of a point on that ray is a circle.

Preview

The circumference C of a circle of radius r is given by the formula

$$C = 2\pi r$$

Find the circumference (in terms of π) of each circle of given radius r.

1. $r = 3$ **2.** $r = 6.8$ **3.** $r = 8\frac{1}{2}$ **4.** $r = 4.25$

Find the radius of each circle of given circumference C.

5. $C = 8\pi$ **6.** $C = 2\pi$ **7.** $C = 12.4\pi$ **8.** $C = 3.5\pi$

Letters of the Greek alphabet, such as θ (theta), α (alpha), and β (beta), are used to name angles as well as to represent their measures. Angle *AOP,* or angle θ, is called a *central angle.* A **central angle** of a circle is an angle whose vertex is the center of the circle. The sides of angle θ intercept arc *AP* with length s.

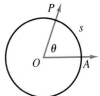

Angles and arcs can be measured in different ways. One common unit of measure is the degree. An arc of 1 degree is $\frac{1}{360}$ of the circumference of a circle. The degree measure of a central angle is equal to that of its intercepted arc. That is, a central angle that intercepts an arc of 1° also has measure 1°. A degree is subdivided into **minutes,** where $1° = 60'$ (60 min). A minute is further subdivided into **seconds,** where $1' = 60''$ (60 s).

EXAMPLE 1   **Express each angle using degrees, minutes, and seconds:**

     a. $10.5°$ **b.** $13° \, 3\frac{1}{3}'$

     a. $10.5° = 10° + (0.5°)\left(\dfrac{60'}{1°}\right)$ *Multiply by 1 in the form $\dfrac{60'}{1°}$.*

             $= 10° + 30'$ *Notice how the units of degrees in the numerator and denominator divide out.*

             $= 10°30'$

b. $13° 3\frac{1'}{3} = 13° + 3' + \left(\frac{1'}{3}\right)\left(\frac{60''}{1'}\right)$ *Multiply by 1 in the form $\frac{60''}{1'}$.*

$= 13° + 3' + 20''$ *Notice that the units of minutes in the numerator and denominator divide out.*

$= 13°3'20''$

For most applications, decimal parts of degrees are preferred to minutes and seconds. That is, 32.5° is usually used instead of 32°30′.

EXAMPLE 2   **Express 127°18′36″ in decimal degrees.**

$127°18'36'' = 127° + 18'\left(\frac{1°}{60'}\right) + 36''\left(\frac{1'}{60''}\right)\left(\frac{1°}{60'}\right)$

$= 127° + 0.3° + 0.01°$ *Units divide out.*

$= 127.31°$

Thus, 127°18′36″ equals 127.31°.

 Your calculator may have special keys for changing degrees, minutes, and seconds to decimal degrees and the reverse. If so, you can solve the first two examples by entering the given measures and pressing the appropriate keys.

Angle measures can also be given in *radians.* When a central angle of a circle intercepts an arc that has the same length as the radius of the circle, the measure of this angle is defined to be one **radian,** abbreviated 1 rad.

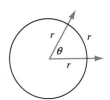

A central angle of 1 radian intercepts an arc of length r regardless of the length of the radius r of the circle. However, if the radius of one circle is longer than that of another, the intercepted arc of the larger circle is longer than that of the smaller circle.

Let r be the radius of a circle and s the length of an arc intercepted by central angle θ. The radian measure of θ can be found by dividing s by r. That is,

$$\theta = \frac{s}{r}$$

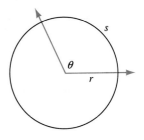

For example, if $s = 6$ and $r = 3$, then $\theta = \frac{6}{3}$, or $\theta = 2$.

Notice that no unit is given. When no unit is specified, assume that radian measure is being used. Thus, $\theta = 2$ means that the measure of angle θ is 2 radians. Similarly, $\theta = 3\pi$ means that the measure of angle θ is 3π radians.

If θ is a complete counterclockwise revolution, $s = 2\pi r$ and thus

$$\theta = \frac{2\pi r}{r} = 2\pi$$

Since 1 revolution is equal to 360° and is also equal to 2π rad,

$$2\pi \text{ rad } = 360° \qquad \text{or} \qquad \pi \text{ rad } = 180°$$

Thus,

$$1° = \frac{\pi}{180} \qquad\qquad 1° \approx 0.0174533 \text{ rad}$$

$$1 \text{ rad } = \frac{180°}{\pi} \qquad\qquad 1 \text{ rad } \approx 57.2958°$$

To express a given number of degrees in radians, multiply the number of degrees by $\frac{\pi}{180}$.

EXAMPLE 3 **Express each angle measure in radians:** **a.** 45° **b.** $-120°$

a. $45° = 45 \cdot \dfrac{\pi}{180} \text{ rad}$ **b.** $-120° = -120 \cdot \dfrac{\pi}{180} \text{ rad}$

$\qquad = \dfrac{45\pi}{180} \text{ rad}$ $\qquad\qquad\qquad = \dfrac{-120\pi}{180} \text{ rad}$

$\qquad = \dfrac{\pi}{4} \text{ rad}$ $\qquad\qquad\qquad\quad = -\dfrac{2\pi}{3} \text{ rad}$

To express a given number of radians in degrees, multiply the number of radians by $\frac{180}{\pi}$.

EXAMPLE 4 **Express each angle measure in degrees:** **a.** $\frac{\pi}{12}$ **b.** -5π

a. $\dfrac{\pi}{12} \text{ rad} = \dfrac{\pi}{12} \cdot \dfrac{180°}{\pi}$ **b.** $-5\pi \text{ rad} = -5\pi \cdot \dfrac{180°}{\pi}$

$\qquad\qquad = 15°$ $\qquad\qquad\qquad\qquad = -900°$

It is a good idea to know the following equivalent degree and radian measures. The fact that $360° = 2\pi$ radians should help you to recall the other entries in the table. For example, $120° = \frac{1}{3}(360°)$, so $120° = \frac{1}{3}(2\pi) = \frac{2\pi}{3}$ rad.

Degrees	Radians
0°	0
30°	$\dfrac{\pi}{6}$
45°	$\dfrac{\pi}{4}$
60°	$\dfrac{\pi}{3}$
90°	$\dfrac{\pi}{2}$
120°	$\dfrac{2\pi}{3}$
135°	$\dfrac{3\pi}{4}$
150°	$\dfrac{5\pi}{6}$
180°	π
270°	$\dfrac{3\pi}{2}$
360°	2π

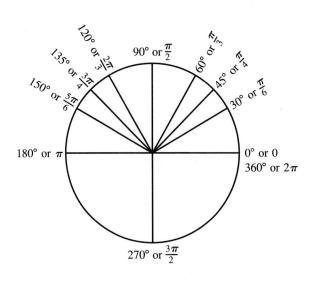

Many scientific calculators have special keys that can be used to change from degree measure to radian measure and the reverse.

EXAMPLE 5 **In which quadrant does the terminal side of each angle lie when it is in standard position?**

a. $\dfrac{2\pi}{3}$ b. $\dfrac{5\pi}{4}$ c. $\dfrac{\pi}{6}$ d. $-\dfrac{\pi}{3}$

  a. Quadrant II   b. Quadrant III   c. Quadrant I   d. Quadrant IV

Remember that for any angle θ expressed in degrees, the measure of any coterminal angle is $n \cdot 360° + \theta$, where n is an integer. Thus, for any angle θ expressed in radians, the measure of any coterminal angle is $n \cdot 2\pi + \theta$, where n is an integer.

CLASS EXERCISES

Express each angle measure in radians.

  1. 90°   2. 180°   3. 360°   4. $-720°$ 5. $-180°$ 6. 270°

Express each angle measure in degrees.

  7. 0   8. 3π 9. $-\dfrac{\pi}{2}$ 10. $-\dfrac{3\pi}{2}$ 11. $\dfrac{\pi}{4}$ 12. $\dfrac{3\pi}{4}$

In which quadrant does the terminal side of each angle lie when it is in standard position?

  13. $\dfrac{3\pi}{4}$ 14. $\dfrac{\pi}{3}$ 15. $\dfrac{5\pi}{4}$ 16. $\dfrac{11\pi}{12}$

PRACTICE EXERCISES Use appropriate technology.

In which quadrant does the terminal side of each angle lie when it is in standard position?

1. $\dfrac{4\pi}{5}$ 2. $\dfrac{2\pi}{3}$ 3. $-\dfrac{2\pi}{3}$ 4. $-\dfrac{3\pi}{4}$ 5. $-\dfrac{\pi}{12}$ 6. $-\dfrac{7\pi}{3}$

Express each angle measure in degrees and minutes.

7. $27.5°$ 8. $108.75°$ 9. $20.2°$ 10. $150.1°$ 11. $134.3°$ 12. $126.6°$

Express each angle measure in decimal degrees.

13. $135°30'$ 14. $75°30'$ 15. $220°18'$ 16. $300°24'$ 17. $25°12'$ 18. $100°36'$

Express each angle measure in radians. Give answers in terms of π.

19. $60°$ 20. $30°$ 21. $120°$ 22. $135°$ 23. $150°$ 24. $210°$

25. $54°$ 26. $28°$ 27. $335°$ 28. $-310°$ 29. $-115°$ 30. $155°$

Express each angle measure in degrees.

31. $\dfrac{4\pi}{3}$ 32. $\dfrac{2\pi}{3}$ 33. $\dfrac{5\pi}{12}$ 34. $\dfrac{7\pi}{12}$ 35. $-\dfrac{13\pi}{18}$ 36. $-\dfrac{4\pi}{9}$

37. $\dfrac{7\pi}{6}$ 38. $\dfrac{11\pi}{6}$ 39. $-\dfrac{3\pi}{4}$ 40. $-\dfrac{\pi}{4}$ 41. $\dfrac{5\pi}{3}$ 42. $\dfrac{5\pi}{4}$

Express each angle measure in degrees, minutes, and seconds, to the nearest second.

43. $21.56°$ 44. $35.54°$ 45. $160.11°$ 46. $280.99°$

47. $-313.53°$ 48. $-404.13°$ 49. $-381.87°$ 50. $-555.66°$

Express each angle measure in decimal degrees, to the nearest hundredth of a degree.

51. $18°15'15''$ 52. $25°45'45''$ 53. $315°48'03''$ 54. $333°33'33''$

55. $-102°02'02''$ 56. $-103°44'04''$ 57. $-515°15'15''$ 58. $-602°12'22''$

Express each angle measure in radians. Give answers in terms of π.

59. $450°$ 60. $630°$ 61. $-405°$ 62. $-675°$ 63. $-1035°$ 64. $-1125°$

Express each angle measure in degrees.

65. $\dfrac{21\pi}{4}$ 66. $\dfrac{19\pi}{4}$ 67. $-\dfrac{18\pi}{3}$ 68. $-\dfrac{21\pi}{3}$

69. $-\dfrac{20\pi}{3}$ 70. $-\dfrac{10\pi}{3}$ 71. $\dfrac{16\pi}{5}$ 72. $\dfrac{2\pi}{15}$

Start at the terminal side of the given angle θ in standard position. Find the radian measure of the resulting angle, in standard position, after the given number of rotations. Give answers in terms of π.

73. $\theta = \frac{\pi}{3}$, 1 clockwise

74. $\theta = \frac{\pi}{3}$, 1 counterclockwise

75. $\theta = -\frac{\pi}{3}$, 1 counterclockwise

76. $\theta = -\frac{\pi}{3}$, 1 clockwise

77. $\theta = -\frac{\pi}{6}$, $1\frac{1}{2}$ clockwise

78. $\theta = -\frac{\pi}{6}$, $1\frac{1}{2}$ counterclockwise

79. $\theta = \frac{3\pi}{2}$, 2 counterclockwise

80. $\theta = \frac{3\pi}{2}$, 2 clockwise

81. $\theta = \frac{3\pi}{4}$, $2\frac{1}{2}$ clockwise

82. $\theta = \frac{3\pi}{4}$, $2\frac{1}{2}$ counterclockwise

83. $\theta = -\frac{5\pi}{6}$, $1\frac{1}{3}$ counterclockwise

84. $\theta = -\frac{5\pi}{6}$, $1\frac{1}{3}$ clockwise

Applications

85. Horology   Through how many *degrees* does the minute hand of a clock rotate in 5 minutes?   in $1\frac{1}{2}$ hours?

86. Horology   Through how many *radians* does the minute hand of a clock rotate in 45 minutes?   in $2\frac{1}{4}$ hours?

87. Meteorology   Through how many *degrees* and how many *radians* does a weather vane rotate if a north wind changes to an east wind?

TRIGONOMETRY IN SURVEYING

Surveyors use trigonometry to find and represent boundaries of regions. An instrument called a transit is used to measure angles. The direction of a ray is sometimes given as the measure of the acute angle formed by the intersection of the ray with the true north-south line.

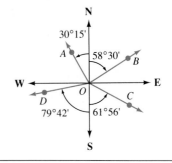

To write a direction in this way, first write N or S, then the measure of the acute angle, and finally E or W. The direction of $\overrightarrow{OA}$ is written N30°15′W, and is read 30 degrees 15 minutes west of north.

1. Write the directions for $\overrightarrow{OB}$, $\overrightarrow{OC}$, and $\overrightarrow{OD}$.

2. Find the measures of $\angle AOB$ and $\angle COD$.

Applications: Angular and Linear Velocity

1.5

Objective: To solve problems involving arc length, angular velocity, and linear velocity

In order to solve problems involving rotary motion, it is often necessary to be able to find arc length.

Preview

The first Ferris wheel built was 250 ft in diameter. Modern Ferris wheels are somewhat smaller. For example, a large Ferris wheel located in Japan is 208 ft in diameter.

1. Suppose the 250-ft Ferris wheel could make one rotation in 45 seconds and the 208-ft wheel can make one rotation in 25 seconds. On which Ferris wheel would a rider move faster?

2. Why do you think Ferris wheels are smaller today?

3. When a Ferris wheel is in motion, do all riders move at the same rate?

You have learned that the radian measure of a central angle θ can be found by dividing the length s of the intercepted arc by the radius r of the circle.

$$\theta = \frac{s}{r}$$

You can also use the formula to find s if you know θ (in radians) and r.

$$s = r \cdot \theta$$

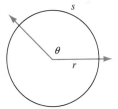

As the figure at the right shows, s is a fraction of the circumference of the circle.

EXAMPLE 1   **Find the length of the arc intercepted by a central angle of 2.5 radians in a circle of radius 10 cm.**

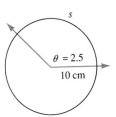

$s = r \cdot \theta$
$ = (10)(2.5)$
$ = 25$ The arc has a length of 25 cm.

Objects such as Ferris wheels, phonograph records, and pulleys turn about their axes. This is called *rotary motion*, since a point on such an object follows a circular path. Imagine point A on the edge of a record spinning about its axis O. The **angular displacement** of point A is the angle θ through which $\overrightarrow{OA}$ rotates.

EXAMPLE 2 If $\overrightarrow{OA}$ on the record pictured above makes a $\frac{4}{5}$ revolution about its axis, find the angular displacement θ of point A in radians.

$\overrightarrow{OA}$ travels through 2π radians during each complete revolution. Thus, $\theta = \frac{4}{5} \cdot 2\pi = \frac{8\pi}{5}$ rad.

The **angular velocity** of a point on a revolving ray is the angular displacement per unit of time. The Greek letter ω (omega) is usually used to represent angular velocity.

$$\omega = \frac{\theta}{t}$$

Angular velocity can be expressed in various units, such as revolutions per minute (rpm) or radians per second (rad/s). Sometimes you will have to change from one unit of measure to another.

EXAMPLE 3 Find the angular velocity ω in radians per second of a wheel turning at 25 rpm (revolutions per minute).

$\omega = \dfrac{25 \text{ revolutions}}{1 \text{ minute}} \cdot \dfrac{2\pi \text{ radians}}{1 \text{ revolution}}$ *1 revolution = 2π radians*

$= 50\pi$ rad/min

$= \dfrac{50\pi \text{ radians}}{1 \text{ minute}} \cdot \dfrac{1 \text{ minute}}{60 \text{ seconds}}$ *1 minute = 60 seconds*

$= \dfrac{5\pi}{6}$ rad/s

The **linear velocity** V of a point on a revolving ray is defined to be the linear distance traveled by the point per unit of time: $V = \frac{s}{t}$. Since $s = r \cdot \theta$ and $\omega = \frac{\theta}{t}$, the linear velocity can be expressed as the product of the angular velocity ω and the radius r.

$$V = \frac{s}{t}$$

$$V = \frac{r \cdot \theta}{t} = \frac{\theta}{t} \cdot r$$

$$V = \omega \cdot r$$

EXAMPLE 4   **The wheel of a truck is turning at 6 rps (revolutions per second). The wheel is 4 ft in diameter.**

　　a. Find the angular velocity ω of the wheel in radians per second.
　　b. Find the linear velocity V, in feet per second, of a point on the rim of the wheel.

a. $\omega = \dfrac{6 \text{ revolutions}}{1 \text{ second}} \cdot \dfrac{2\pi \text{ radians}}{1 \text{ revolution}} = 12\pi \text{ rad/s}$

b. Since the diameter is 4 ft, a point on the rim is 2 ft from the center. That is, $r = 2$ ft.

$$V = \omega \cdot r = \dfrac{12\pi \text{ radians}}{1 \text{ second}} \cdot 2 \text{ ft} = 24\pi \text{ ft/s}$$

Note that the word *radian* is not included in the answer in Example 4b. Technically, a radian is the ratio of two lengths and has no dimension. However, to avoid confusion in calculations and to indicate that angle measure is involved, the word radian is sometimes retained.

Problems dealing with pulleys often involve linear velocity and angular velocity.

EXAMPLE 5   **Two pulleys, one 6 in. and the other 2 ft in diameter, are connected by a belt. The larger pulley revolves at the rate of 60 rpm.**

　　a. Find the linear velocity of the belt in feet per minute.
　　b. Calculate the angular velocity of the smaller pulley in radians per minute.

a. The linear velocity V of the belt equals the linear velocity of a point on the rim of either pulley. You know the angular velocity of the larger pulley in revolutions per minute. Change this to radians per minute.

$$\omega = \dfrac{60 \text{ revolutions}}{1 \text{ minute}} \cdot \dfrac{2\pi \text{ radians}}{1 \text{ revolution}}$$
$$= 120\pi \text{ rad/min}$$

A point on the rim of the larger pulley is 1 ft from its center. Therefore,

$$V = \omega \cdot r$$
$$= \dfrac{120\pi \text{ radians}}{1 \text{ minute}} \cdot 1 \text{ ft} = 120\pi \text{ ft/min}$$

b. The linear velocity of the smaller pulley is 120π ft/min and the radius of that pulley is 3 in., or $\frac{1}{4}$ ft. Substitute the known information into the formula $V = \omega \cdot r$.

$$V = \omega \cdot r$$
$$120\pi = \omega \cdot \frac{1}{4}$$
$$\omega = 480\pi \text{ rad/min} \qquad \textit{Angular velocity of smaller pulley}$$

To check the answer, note that the radius of the smaller pulley is $\frac{1}{4}$ that of the larger pulley. In order to keep up, the angular velocity of the smaller pulley must be 4 times that of the larger one. Consequently, $\omega = 4(120\pi) = 480\pi$.

CLASS EXERCISES

Express answers to Exercises 1–8 in terms of π.

1. Find the length of the arc intercepted by a central angle of $\frac{5\pi}{4}$ if the radius of the circle is 20 mm.

2. Find the length of the arc intercepted by a central angle of $\frac{7\pi}{4}$ if the radius of the circle is 12 ft.

Find the angular velocity, in radians per second, of a wheel turning at each of the following numbers of revolutions per minute.

3. 200   **4.** 100   **5.** 144   **6.** 180

For Exercises 7–8, the wheel of a bicycle is turning at 5 rpm. The wheel is 24 in. in diameter.

7. Find the angular velocity of the wheel in radians per second.

8. Find the linear velocity in inches per second of a point on the rim.

9. Name at least three things in real life that have angular velocity.

PRACTICE EXERCISES

Express answers to Exercises 1–18 in terms of π.

For Exercises 1–6, find the length of the arc intercepted by the given central angle in a circle with the given radius.

1. $\theta = \frac{\pi}{2}, r = 10$ in.   **2.** $\theta = \frac{\pi}{4}, r = 8$ m   **3.** $\theta = \frac{5\pi}{6}, r = 15$ in.

4. $\theta = \frac{3\pi}{4}, r = 12$ m   **5.** $\theta = \frac{9\pi}{4}, r = 20$ in.   **6.** $\theta = \frac{7\pi}{6}, r = 25$ cm

Point *A* is on the rim of a rotating wheel. Find the angular displacement of point *A* for the given number of revolutions. Express answers in radians.

7. $\frac{1}{2}$ **8.** $\frac{1}{4}$ **9.** $\frac{1}{3}$ **10.** $\frac{2}{3}$ **11.** $\frac{3}{5}$ **12.** $\frac{5}{6}$

Find the angular velocity for each angular displacement in Exercises 7–12 for the given values of *t*. Express answers in radians per minute.

13. Exercise 7, $t = 2$ min   **14.** Exercise 8, $t = 10$ min

15. Exercise 9, $t = 5$ min   **16.** Exercise 10, $t = 3$ min

17. Exercise 11, $t = 12$ min   **18.** Exercise 12, $t = 15$ min

Calculate the linear velocity *V* of an object rotating at angular velocity ω at a distance *r* from the center. Express answers in centimeters per second.

19. $r = 10$ cm, ω $= 20\pi$ rad/s   **20.** $r = 3$ cm, ω $= 20\pi$ rad/s

21. $r = 15$ cm, ω $= 5\pi$ rad/s   **22.** $r = 25$ cm, ω $= 8\pi$ rad/s

23. Find the angular velocity in radians per second of the second hand of a clock. Express the answer in terms of π.

24. Find the angular velocity in radians per second of the minute hand of a clock. Express the answer in terms of π.

25. Calculate the angular velocity in radians per minute of a Ferris wheel 250 ft in diameter that takes 45 s to rotate once. Express the answer in terms of π.

26. If you sat on the rim of the Ferris wheel in Exercise 25, what would your linear velocity be, to the nearest foot per minute?

27. Calculate the angular velocity in radians per minute of a Ferris wheel 208 ft in diameter that takes 25 s to rotate once. Express the answer in terms of π.

28. If you sat on the rim of the Ferris wheel in Exercise 27, what would your linear velocity be, to the nearest foot per minute?

29. The Earth spins on its axis every 24 hours. If the Earth's circumference is 24,800 miles, find the linear velocity of a person standing on the equator in miles per hour.

30. A car is traveling at a speed of 45 miles per hour. Find the angular velocity of a tire in revolutions per minute (rpm), if the diameter of each rim is 13 in. with 4 in. of tire between the road surface and each rim.

31. Assume the car tire in Exercise 30 above is not properly inflated so that the measurement between the road surface and the rim is only 2 in. Find the angular velocity of the underinflated tire in rpm.

Two pulleys are connected by a belt. The smaller pulley has a diameter of 8 ft and the larger pulley has a diameter of 12 ft. The smaller pulley is revolving at the rate of 60 rpm.

32. Find the angular velocity of each pulley in radians per second. Express the answers in terms of π.

33. Find the linear velocity of a point on the rim of each pulley in feet per second. Express the answers in terms of π.

TEST YOURSELF

Give the domain and range of each relation and state whether the relation _is_ or _is not_ a function.

1. $\{(1, 3), (2, 3), (4, 5), (6, 7)\}$ **2.** $\{(2, 6), (3, 7), (3, 8), (4, 9)\}$ 1.1

For the function $f(x) = \dfrac{1}{x + 2}$, find each value if it is defined.

3. $f(5)$ **4.** $f\left(\dfrac{1}{2}\right)$ **5.** $f(-2)$

6. Find the lengths of the sides of triangle ABC with vertices $A(-4, 1)$, $B(0, 1)$, and $C(3, -2)$.   1.2

In which quadrant does the terminal side of each angle lie? Assume that the angles are in standard position.

7. $132°$ **8.** $-78°$ **9.** $400°$ 1.3

10. For a $\dfrac{13}{12}$ clockwise rotation, find the measure of the angle in degrees.

11. Express $75°$ in radians. Give the answer in terms of π.   1.4

12. Express $\dfrac{5\pi}{36}$ in degrees.

For Exercises 13–15, express answers in terms of π.   1.5

13. Find the length of the arc intercepted by a central angle of $\dfrac{2\pi}{3}$ if the radius of the circle is 18 m.

14. Find the angular velocity ω in radians per second for a pulley turning at 1800 rpm.

15. Calculate the linear velocity V of a point located 12 in. from the center of a disk rotating at 7π rad/s.

1.6 Cosine and Sine Functions

Objectives: To define the cosine and sine functions
To evaluate the sine and cosine functions of an angle given a point on its terminal side

Sometimes the Pythagorean theorem is used to find the length of a side of a right triangle in order to determine the sine and cosine of an angle.

Preview

If triangle ABC is a right triangle with hypotenuse c, the lengths of sides a, b, and c are related by the Pythagorean theorem: $c^2 = a^2 + b^2$.

EXAMPLE Find b if $a = 4.5$ and $c = 7.5$.

$$(7.5)^2 = (4.5)^2 + b^2 \qquad c^2 = a^2 + b^2$$
$$56.25 = 20.25 + b^2$$
$$36 = b^2$$
$$\sqrt{36} = b \qquad \text{Length is positive.}$$
$$6 = b$$

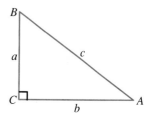

Find the length of the third side of each right triangle ABC with hypotenuse c.

1. $a = 12, b = 9, c = \underline{?}$ **2.** $b = 5, c = 5\sqrt{2}, a = \underline{?}$ **3.** $a = 1, c = 2, b = \underline{?}$

4. $a = 6, c = 15.6, b = \underline{?}$ **5.** $a = 11, b = 11, c = \underline{?}$ **6.** $b = 1.2, c = 1.3, a = \underline{?}$

The radius of a **unit circle** is 1. Consider the origin-centered unit circle shown at the right together with an angle θ in standard position drawn on the same coordinate plane. The initial side of the angle intersects the circle at $(1, 0)$ and the terminal side intersects the circle at the point $P(x, y)$.

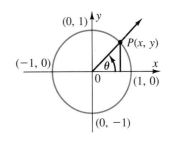

If the terminal side of θ is rotated counterclockwise about the origin, the values of x and y associated with point P will change. For each value of θ, however, there is exactly one corresponding value for x. Because of this correspondence, the set of ordered pairs (θ, x) is a function. Also, for each value of θ, there is exactly one value for y. Therefore, the set of ordered pairs (θ, y) is also a function. These two functions are called the **cosine** and the **sine** functions, respectively.

If θ is an angle in standard position whose terminal side intersects the unit circle at (x, y), then

$x =$ cosine θ *abbreviated* $x = \cos \theta$
and $y =$ sine θ *abbreviated* $y = \sin \theta$

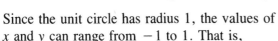

The ordered pair (x, y) can be written as $(\cos \theta, \sin \theta)$, which are the coordinates of point P in the figure.

Since the unit circle has radius 1, the values of x and y can range from -1 to 1. That is,

$$-1 \le \cos \theta \le 1 \quad \text{and} \quad -1 \le \sin \theta \le 1$$

Given an arbitrary angle θ, $\cos \theta$ and $\sin \theta$ can be found if a point (x, y) on the terminal side of θ is given, even if (x, y) is not on the unit circle. A right triangle is formed when a perpendicular is drawn from *any* point on the terminal side of θ to the x-axis. Such a triangle is called a **reference triangle.** In the figure, (x, y) and (x', y') are the coordinates of corresponding vertices of similar right triangles. The lengths of the sides are x, y, and r, and x', y', and 1, respectively. The lengths of corresponding sides of similar triangles are

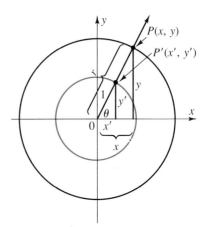

proportional, so $\dfrac{x'}{1} = \dfrac{x}{r}$ and $\dfrac{y'}{1} = \dfrac{y}{r}$.
Thus, since $x' = \cos \theta$ and $y' = \sin \theta$,

$$\cos \theta = \frac{x}{r} \quad \text{and} \quad \sin \theta = \frac{y}{r}$$

The relationship described above applies to any angle in any quadrant. Therefore, the definitions of $\cos \theta$ and $\sin \theta$ can be generalized as follows.

> If θ is an angle in standard position and if (x, y) is any point (other than the origin) on the terminal side of θ, then
>
> $$\cos \theta = \frac{x}{r} \quad \text{and} \quad \sin \theta = \frac{y}{r} \quad \text{where } r = \sqrt{x^2 + y^2}$$

Note that θ can be measured in degrees or in radians. Since r, sometimes called the **radius vector,** is the distance from point P to the origin, it is always positive. However, x and y can be positive or negative, depending upon the location of the terminal side of the angle. If x is negative and y is positive, θ is in the second quadrant; if x is negative and y is negative, θ is in the third quadrant; and if x is positive and y is negative, θ is in the fourth quadrant.

When you are asked to find *exact* values of the trigonometric functions of an angle (as in Example 1 below), leave irrational answers in radical form.

EXAMPLE 1 **Find the exact values of the sine and cosine of an angle θ in standard position if point Q is on its terminal side. Sketch the reference triangle.**
a. $Q(3, -4)$ **b.** $Q(-3, -6)$

a. $r = \sqrt{3^2 + (-4)^2}$ $r = \sqrt{x^2 + y^2}$

$\quad\quad = \sqrt{25} = 5$

Therefore, $\cos \theta = \dfrac{3}{5}$ $\cos \theta = \dfrac{x}{r}$

$\quad\quad\quad \sin \theta = \dfrac{-4}{5} = -\dfrac{4}{5}$ $\sin \theta = \dfrac{y}{r}$

b. $r = \sqrt{(-3)^2 + (-6)^2}$ $r = \sqrt{x^2 + y^2}$

$\quad\quad = \sqrt{45} = 3\sqrt{5}$

Therefore, $\cos \theta = \dfrac{-3}{3\sqrt{5}} = -\dfrac{1}{\sqrt{5}} = -\dfrac{\sqrt{5}}{5}$

$\quad\quad\quad \sin \theta = \dfrac{-6}{3\sqrt{5}} = -\dfrac{2}{\sqrt{5}} = -\dfrac{2\sqrt{5}}{5}$

If you are given either the sine or the cosine, then you can find the other function, provided you know the quadrant in which the angle terminates.

EXAMPLE 2 **Angle θ is in standard position with its terminal side in the third quadrant. Find the exact value of $\cos \theta$ if $\sin \theta = -\dfrac{3}{5}$.**

Since $\sin \theta = -\dfrac{3}{5}$, $r = 5$ and $y = -3$ *r is always positive and y is negative in the third quadrant.*

for some point (x, y) on the terminal side of θ.

$$x^2 + (-3)^2 = 5^2 \quad\quad\quad\quad x^2 + y^2 = r^2$$
$$x^2 + 9 = 25$$
$$x^2 = 16$$
$$x = 4 \quad \text{or} \quad x = -4$$

Since θ is in the third quadrant, $x < 0$. Thus, $x = -4$ and $\cos \theta = \dfrac{-4}{5} = -\dfrac{4}{5}$.

The set of real numbers is the domain for both the cosine and the sine functions. These functions can be associated with the measures of angles and they can also be associated with the lengths of arcs on the unit circle. For this reason, these functions are sometimes called **circular functions.**

CLASS EXERCISES

Let θ be an angle in standard position. For each given value of sin θ or cos θ, state the quadrants in which θ can lie.

1. $\sin \theta = \dfrac{1}{2}$ **2.** $\sin \theta = -\dfrac{\sqrt{3}}{2}$ **3.** $\cos \theta = \dfrac{3}{5}$ **4.** $\cos \theta = -\dfrac{4}{5}$

Complete the table below by writing the sign (+ or −) of cos θ and of sin θ for each quadrant in the coordinate plane.

Quadrant	I	II	III	IV
cos θ	**5.**	**6.**	**7.**	**8.**
sin θ	**9.**	**10.**	**11.**	**12.**

13. What is the minimum value of sin θ?   of cos θ?

14. What is the maximum value of sin θ?   of cos θ?

PRACTICE EXERCISES

Find the exact values of the sine and cosine of an angle θ in standard position if the point with the given coordinates is on the terminal side of the angle. Sketch the reference triangle.

1. $(3, 4)$ **2.** $(-4, 3)$ **3.** $(-5, -12)$ **4.** $(3, -6)$

5. $(-3, 0)$ **6.** $(-2, -2)$ **7.** $(-6, 3)$ **8.** $(8, 6)$

Find the exact value of sin θ if cos θ is given and θ is an angle in standard position in Quadrant I.

9. $\cos \theta = \dfrac{4}{5}$ **10.** $\cos \theta = \dfrac{3}{5}$ **11.** $\cos \theta = \dfrac{12}{13}$ **12.** $\cos \theta = \dfrac{5}{13}$

Find the exact value of cos θ if sin θ is given and θ is an angle in standard position in Quadrant II.

13. $\sin \theta = \dfrac{\sqrt{3}}{2}$ **14.** $\sin \theta = \dfrac{1}{2}$ **15.** $\sin \theta = \dfrac{3}{5}$ **16.** $\sin \theta = \dfrac{5}{13}$

Find the exact values of the sine and cosine of an angle θ in standard position if the point with the given coordinates is on the terminal side of the angle. Sketch the reference triangle.

17. $(5, 7)$ **18.** $(-2, 4)$ **19.** $(3, -3)$ **20.** $(4, 6)$ **21.** $(-6, 10)$ **22.** $(12, -18)$

Angle θ is in standard position and lies in the given quadrant. When $\sin \theta$ is given, find the exact value of $\cos \theta$. When $\cos \theta$ is given, find the exact value of $\sin \theta$.

23. $\sin \theta = \dfrac{1}{2}; \text{I}$

24. $\cos \theta = \dfrac{1}{2}; \text{I}$

25. $\sin \theta = -\dfrac{\sqrt{3}}{2}; \text{III}$

26. $\sin \theta = -\dfrac{12}{13}; \text{IV}$

27. $\cos \theta = \dfrac{5}{11}; \text{IV}$

28. $\cos \theta = -\dfrac{5}{6}; \text{II}$

Let θ be an angle in standard position. In which quadrant or quadrants can θ lie under the given conditions?

29. $\sin \theta > 0$

30. $\sin \theta < 0$ and $\cos \theta > 0$

31. $\sin \theta = \cos \theta$

32. $\sin \theta = -\cos \theta$

Applications

Computer In the program shown, the Pythagorean theorem is used to calculate specific values of the cosine and sine functions. Line 30 allows the user to enter the coordinates of a point on the terminal side of an angle. Lines 40–60 calculate the sine and cosine, and lines 70 and 80 display the results.

```
10  PRINT "ENTER THE COORDINATES OF A POINT"
20  PRINT "ON THE TERMINAL SIDE OF THE ANGLE"
30  PRINT: INPUT "ENTER X THEN Y"; X,Y
40  R = SQR(X^2 + Y^2)
50  S = Y/R
60  C = X/R
70  HOME: PRINT "THE SINE OF THE ANGLE IS"S
80  PRINT: "THE COSINE OF THE ANGLE IS"C
90  END
```

33. Why does this program give the correct sign for each cosine and sine value?

34. In the program, within what interval are degree measures of the angles assumed to be?

TRIGONOMETRY IN THE MACHINE SHOP

The *sine bar* is a precision tool used to measure angles. It is composed of two parallel cylinders of equal radii with axes 25 cm apart attached to a flat surface parallel to a plane through the axes of the cylinders. The cylinders are placed on blocks set on a horizontal table, and the bar is tilted. The sine of angle β is given by the formula

$$\sin \beta = \frac{b - a}{25}$$

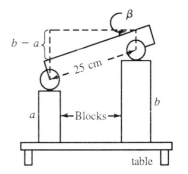

1. If $a = 1.0$ cm and $b = 3.5$ cm, find the exact value of $\sin \beta$.

2. If $a = 2.0$ cm and $b = 5.2$ cm, find $\sin \beta$ to the nearest hundredth.

The Trigonometric Functions

Objectives: To define the tangent, cotangent, secant, and cosecant functions

To evaluate the trigonometric functions of an angle

In lesson 1.6, the sine and cosine functions of an angle θ in standard position were defined to be $\sin \theta = \frac{y}{r}$ and $\cos \theta = \frac{x}{r}$, where (x, y) is a point other than the origin on the terminal side of the angle and $r = \sqrt{x^2 + y^2}$. In this lesson, the *tangent*, *cotangent*, *secant*, and *cosecant* functions will be defined.

Preview

Angle θ is in standard position. The points $(4, -2)$ and $(6, -3)$ are on the terminal side of θ.

Let $r = \sqrt{x^2 + y^2}$ and calculate the following six ratios for each of the two points.

1. $\frac{x}{r}$ 2. $\frac{y}{r}$ 3. $\frac{y}{x}$ 4. $\frac{r}{x}$ 5. $\frac{r}{y}$ 6. $\frac{x}{y}$

7. What are the values of $\sin \theta$ and $\cos \theta$? Do they differ when they are calculated using the point $(6, -3)$ instead of $(4, -2)$? Explain.

Let $P(x, y)$ be any point other than the origin on the terminal side of an angle θ in standard position. The six *trigonometric functions* are defined as follows, where $r = \sqrt{x^2 + y^2}$.

	Trigonometric Functions		
sine	$\sin \theta = \frac{y}{r}$	**cosecant**	$\csc \theta = \frac{r}{y}, y \neq 0$
cosine	$\cos \theta = \frac{x}{r}$	**secant**	$\sec \theta = \frac{r}{x}, x \neq 0$
tangent	$\tan \theta = \frac{y}{x}, x \neq 0$	**cotangent**	$\cot \theta = \frac{x}{y}, y \neq 0$

The values of these functions depend only on the size of angle θ. They do not depend on the choice of point P on the terminal side of the angle.

Notice that three pairs of reciprocal relationships exist between the six trigonometric functions. Note that no denominator may equal zero.

Reciprocal Trigonometric Functions

$$\csc \theta = \frac{1}{\sin \theta} \qquad \sec \theta = \frac{1}{\cos \theta} \qquad \cot \theta = \frac{1}{\tan \theta}$$

Remember that if two numbers are reciprocals of each other, their product is 1. Therefore, the reciprocal relationships may also be written as follows, where neither factor on the left is equal to 0:

$$\sin \theta \cdot \csc \theta = 1 \qquad \cos \theta \cdot \sec \theta = 1 \qquad \tan \theta \cdot \cot \theta = 1$$

EXAMPLE 1 **The terminal side of angle θ in standard position passes through point $P(-3, -4)$. Draw θ and find the values of the six trigonometric functions of θ.**

If P has the coordinates $(-3, -4)$, then $x = -3$, and $y = -4$, and

$$r = \sqrt{(-3)^2 + (-4)^2}$$
$$= \sqrt{9 + 16}$$
$$= \sqrt{25} = 5$$

Therefore,

$$\sin \theta = \frac{-4}{5} = -\frac{4}{5} \qquad \sin \theta = \frac{y}{r} \qquad \csc \theta = \frac{5}{-4} = -\frac{5}{4} \qquad \csc \theta = \frac{r}{y}$$

$$\cos \theta = \frac{-3}{5} = -\frac{3}{5} \qquad \cos \theta = \frac{x}{r} \qquad \sec \theta = \frac{5}{-3} = -\frac{5}{3} \qquad \sec \theta = \frac{r}{x}$$

$$\tan \theta = \frac{-4}{-3} = \frac{4}{3} \qquad \tan \theta = \frac{y}{x} \qquad \cot \theta = \frac{-3}{-4} = \frac{3}{4} \qquad \cot \theta = \frac{x}{y}$$

Recall that r is always positive. Thus, the signs of the values of the trigonometric functions of θ are determined by the signs of x and y, which depend upon the quadrant in which the terminal side of θ lies. Using the reciprocal properties, you can see that the signs of the values of $\csc \theta$ are the same as those for $\sin \theta$, the signs for $\sec \theta$ are the same as those for $\cos \theta$, and the signs for $\cot \theta$ are the same as those for $\tan \theta$. The chart at the right summarizes this information.

Signs of Trigonometric Functions

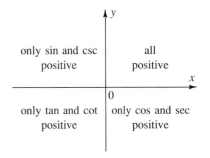

EXAMPLE 2   **In which quadrants do sin θ and tan θ have opposite signs?**

The tangent is negative and the sine is positive in quadrant II; the tangent is positive and the sine is negative in quadrant III. Therefore, sin θ and tan θ have opposite signs in quadrants II and III.

If you know the value of one of the six trigonometric functions of θ and the quadrant in which it lies, you can find the values of the other five.

EXAMPLE 3   **Let θ be an angle in standard position. Evaluate cos θ, tan θ, cot θ, sec θ, and csc θ if θ lies in quadrant IV and sin $\theta = -\dfrac{5}{13}$.**

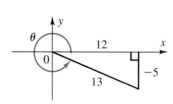

Since $\sin \theta = -\dfrac{5}{13}$ and θ is in quadrant IV,

$y = -5$ and $r = 13$. Find x.

$$x^2 + (-5)^2 = 13^2 \qquad x^2 + y^2 = r^2$$
$$x^2 + 25 = 169$$
$$x^2 = 144$$
$$x = 12 \quad \text{or} \quad x = -12$$

Since θ lies in quadrant IV, $x > 0$. Therefore, $x = 12$ and

$$\cos \theta = \frac{12}{13} \qquad \tan \theta = \frac{-5}{12} = -\frac{5}{12} \qquad \csc \theta = \frac{13}{-5} = -\frac{13}{5}$$

$$\sec \theta = \frac{13}{12} \qquad \cot \theta = \frac{12}{-5} = -\frac{12}{5}$$

CLASS EXERCISES

State the values of all six trigonometric functions of θ.

1.

2.

3.

4.

Determine in which quadrant θ lies if θ is an angle in standard position.

5. $\sin \theta > 0, \cos \theta < 0$ **6.** $\sin \theta < 0, \cos \theta > 0$ **7.** $\csc \theta < 0, \sec \theta < 0$

8. $\csc \theta > 0, \sec \theta > 0$ **9.** $\sec \theta < 0, \cot \theta > 0$ **10.** $\sec \theta < 0, \cot \theta < 0$

Find the value of each reciprocal function.

11. If $\sin \theta = \dfrac{1}{2}$, $\csc \theta = \underline{?}$

12. If $\cos \theta = -\dfrac{3}{4}$, $\sec \theta = \underline{?}$

13. If $\tan \theta = -4$, $\cot \theta = \underline{?}$

14. If $\sec \theta = 1.5$, $\cos \theta = \underline{?}$

15. If $\sec \theta = -2.5$, $\cos \theta = \underline{?}$

16. If $\cot \theta = 3$, $\tan \theta = \underline{?}$

PRACTICE EXERCISES

The terminal side of angle θ in standard position passes through point P. Draw θ and find the exact values of the six trigonometric functions of θ.

1. $P(3, 4)$ **2.** $P(6, 8)$ **3.** $P(12, -5)$ **4.** $P(-5, 12)$

5. $P(-3, -3)$ **6.** $P(2, -2)$ **7.** $P(2, -\sqrt{3})$ **8.** $P(-\sqrt{3}, -2)$

9. $P(7, -24)$ **10.** $P(24, -7)$ **11.** $P(-2, 2\sqrt{3})$ **12.** $P(3, -3\sqrt{3})$

Let θ be an angle in standard position. In which quadrant or quadrants can θ lie under the given conditions?

13. Sin θ and tan θ have the same sign.

14. Cos θ and tan θ have the same sign.

15. Sin θ is negative and cos θ is positive.

16. Sin θ, cos θ, and tan θ all have the same sign.

17. Sin θ and cos θ have opposite signs.

18. Tan θ and cot θ have the same sign.

19. Cot θ and sec θ have opposite signs.

20. Sec θ and csc θ have the same sign.

Find the exact values of the other five trigonometric functions for an angle θ in standard position lying in the given quadrant.

21. $\sin \theta = -\dfrac{4}{5}$; III

22. $\cos \theta = -\dfrac{3}{5}$; II

23. $\tan \theta = \dfrac{4}{3}$; I

24. $\tan \theta = -\dfrac{3}{4}$; IV

25. $\sec \theta = -2$; III

26. $\csc \theta = \dfrac{4}{3}$; II

27. $\cot \theta = \dfrac{24}{7}$; III

28. $\cot \theta = -\dfrac{12}{5}$; IV

A point $P(x, y)$ is on the unit circle with equation $x^2 + y^2 = 1$. **Show that the following points are on the unit circle. Then find the exact values of $\sin \theta$, $\cos \theta$, and $\tan \theta$ if P is on the terminal side of θ.**

29. $P\left(-\dfrac{3}{4}, \dfrac{\sqrt{7}}{4}\right)$

30. $P\left(\dfrac{\sqrt{2}}{2}, -\dfrac{\sqrt{2}}{2}\right)$

31. $P\left(\dfrac{1}{2}, -\dfrac{\sqrt{3}}{2}\right)$

32. $P\left(\dfrac{\sqrt{3}}{2}, -\dfrac{1}{2}\right)$

33. $P\left(-\dfrac{5}{13}, -\dfrac{12}{13}\right)$

34. $P\left(-\dfrac{3}{5}, -\dfrac{4}{5}\right)$

Determine the exact coordinates of the point at the given distance from the origin in the given quadrant, if θ is in standard position.

35. 10; II; $\sin \theta = \dfrac{3}{5}$

36. 4; IV; $\csc \theta = -2$

37. $\sqrt{2}$; IV; $\cos \theta = \dfrac{\sqrt{2}}{2}$

38. 6; III; $\cot \theta = 1$

39. 8; IV; $\tan \theta = -1$

40. $\sqrt{5}$; II; $\sin \theta = \dfrac{\sqrt{5}}{5}$

Applications

41. Surveying A surveyor needs to use the cotangent of an angle but does not have a calculator with a cot key. What function of the same angle could the surveyor use to find the cotangent?

42. Construction A 15-ft ladder forms an angle θ with the side of a building such that $\sec \theta = \dfrac{5}{3}$. The distance d from the base of the ladder to the base of the building can be found using the equation $\cos \theta = \dfrac{d}{15}$. Find the value of d.

43. Construction The wooden truss pictured at the right forms an isosceles triangle constructed in such a way that the length of its base is 5 times its height. Find the exact value of $\tan \theta$.

CHALLENGE

Geometric Interpretations of Trigonometric Functions In the unit circle figure at the right, the values of the trigonometric functions of angle θ are given by the lengths of certain line segments. For example, $\sin \theta = QP$ and $\cos \theta = OP$. Show that $\tan \theta = RS$, $\sec \theta = OS$, $\cot \theta = UT$, and $\csc \theta = OT$. (Use similar triangles and write proportions.)

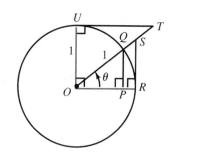

1.8 Trigonometric Functions of Special Angles

Objective: To find the values of the six trigonometric functions of special angles

Triangles with special angle-side relationships, introduced in geometry, are useful in evaluating the trigonometric functions of 30°, 45°, and 60°.

Preview

In an isosceles right triangle (45°-45°-90°), the length of the hypotenuse is $\sqrt{2}$ times the length of a leg. In a 30°-60°-90° triangle, the length of the hypotenuse is two times the length of the leg opposite the 30° angle. The length of the leg opposite the 60° angle is $\sqrt{3}$ times the length of the leg opposite the 30° angle.

EXAMPLE 1 **In the triangle at the right, $a = 5$. Find b and c.**

$$b = a = 5$$
$$c = a\sqrt{2} = 5\sqrt{2}$$

EXAMPLE 2 **In the triangle at the right, $a = 14$. Find b and c.**

$$b = a\sqrt{3} = 14\sqrt{3}$$
$$c = 2a = 2(14) = 28$$

Find the indicated lengths of the sides of a 45°-45°-90° triangle ABC with right angle C.

1. If $c = 2$, find a and b.   **2.** If $b = 3$, find a and c.   **3.** If $a = 0.5$, find b and c.

Find the indicated lengths of the sides of a 30°-60°-90° triangle ABC with right angle C and angle A equal to 30°.

4. If $a = 2$, find b and c.   **5.** If $c = 6$, find a and b.   **6.** If $b = \dfrac{3\sqrt{2}}{2}$, find a and c.

An angle in standard position whose terminal side lies on the x- or y-axis is called a **quadrantal angle.** In the figure at the right, angles POR, POD, POQ, and POP are quadrantal angles.

For an angle θ in standard position, the terminal side of θ intersects a circle of radius r at a unique point $P(x, y)$. The definitions of the trigonometric functions can be used to evaluate the trigonometric functions of $\theta = 0°$, $90°$, $180°$, $270°$, and $360°$ by setting r equal to 1.

EXAMPLE 1 **Given angle θ in standard position, evaluate sin θ, cos θ, and tan θ for each value of θ. Use $r = 1$.**
       a. $\theta = 0°$ **b.** $\theta = 90°$ **c.** $\theta = 180°$ **d.** $\theta = 270°$

a. Since $\theta = 0°$ and $r = 1$, $x = 1$ and $y = 0$.

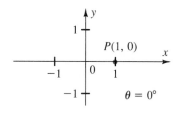

Therefore, $\sin 0° = \dfrac{0}{1} = 0$

             $\cos 0° = \dfrac{1}{1} = 1$

             $\tan 0° = \dfrac{0}{1} = 0$

b. Since $\theta = 90°$ and $r = 1$, $x = 0$ and $y = 1$.

Therefore, $\sin 90° = \dfrac{1}{1} = 1$

             $\cos 90° = \dfrac{0}{1} = 0$

             $\tan 90° = \dfrac{1}{0}$ Undefined

c. Since $\theta = 180°$ and $r = 1$, $x = -1$ and $y = 0$.

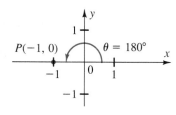

Therefore, $\sin 180° = \dfrac{0}{1} = 0$

             $\cos 180° = \dfrac{-1}{1} = -1$

             $\tan 180° = \dfrac{0}{-1} = 0$

d. Since $\theta = 270°$ and $r = 1$, $x = 0$ and $y = -1$.

Therefore, $\sin 270° = \dfrac{-1}{1} = -1$

             $\cos 270° = \dfrac{0}{1} = 0$

             $\tan 270° = \dfrac{-1}{0}$ Undefined

Because division by 0 is not defined, some values of the trigonometric functions are not defined. Example 1 illustrates this.

Choosing a convenient value for r makes it easier to find the trigonometric values of special angles such as 30°, 45°, and 60°. Since a 45°-45°-90° triangle has sides in the ratio of $1:1:\sqrt{2}$, where $\sqrt{2}$ corresponds to the hypotenuse, a convenient value for r is $\sqrt{2}$.

EXAMPLE 2 If $\theta = 45°$ and $r = \sqrt{2}$, find $\sin \theta$, $\cos \theta$, and $\tan \theta$.

Since legs OB and PB have the same length, $OB = PB = 1$. Thus, $x = 1$ and $y = 1$. Therefore,

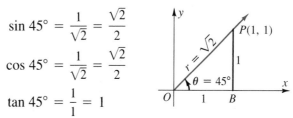

$$\sin 45° = \frac{1}{\sqrt{2}} = \frac{\sqrt{2}}{2}$$

$$\cos 45° = \frac{1}{\sqrt{2}} = \frac{\sqrt{2}}{2}$$

$$\tan 45° = \frac{1}{1} = 1$$

In a 30°-60°-90° triangle, the sides are in the ratio $1:\sqrt{3}:2$, where 2 corresponds to the hypotenuse and $\sqrt{3}$ corresponds to the side opposite the 60° angle. Thus, a convenient value for r is 2.

EXAMPLE 3 Evaluate $\sin \theta$, $\cos \theta$, $\tan \theta$, $\csc \theta$, $\sec \theta$, and $\cot \theta$ for each value of r and θ: **a.** $\theta = 60°; r = 2$ **b.** $\theta = \dfrac{\pi}{6}; r = 2$

a. $x = OB = 1$ and $y = PB = \sqrt{3}$, so

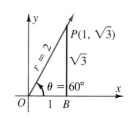

$$\sin 60° = \frac{\sqrt{3}}{2} \qquad \csc 60° = \frac{2}{\sqrt{3}} = \frac{2\sqrt{3}}{3}$$

$$\cos 60° = \frac{1}{2} \qquad \sec 60° = \frac{2}{1} = 2$$

$$\tan 60° = \frac{\sqrt{3}}{1} = \sqrt{3} \qquad \cot 60° = \frac{1}{\sqrt{3}} = \frac{\sqrt{3}}{3}$$

b. Since $\dfrac{\pi}{6}$ radians $= 30°$, triangle OPB is a 30°-60°-90° triangle. $x = OB = \sqrt{3}$ and $y = PB = 1$, so

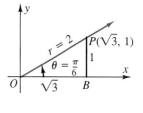

$$\sin \frac{\pi}{6} = \frac{1}{2} \qquad \csc \frac{\pi}{6} = \frac{2}{1} = 2$$

$$\cos \frac{\pi}{6} = \frac{\sqrt{3}}{2} \qquad \sec \frac{\pi}{6} = \frac{2}{\sqrt{3}} = \frac{2\sqrt{3}}{3}$$

$$\tan \frac{\pi}{6} = \frac{1}{\sqrt{3}} = \frac{\sqrt{3}}{3} \qquad \cot \frac{\pi}{6} = \frac{\sqrt{3}}{1} = \sqrt{3}$$

You can also use the methods introduced in this lesson to evaluate trigonometric functions of angles whose measures are integral multiples of 30°, 45°, and 60°. In order to do so, the concept of a *reference angle* is needed. The **reference angle** θ' of a given angle θ in standard position is the smallest positive acute angle determined by the x-axis and the terminal side of θ. Recall that an acute angle is an angle whose measure is between 0° and 90°.

The following diagrams illustrate the relationship between θ and θ' in each of the four quadrants:

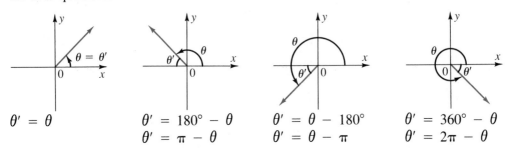

$\theta' = \theta$

$\theta' = 180° - \theta$
$\theta' = \pi - \theta$

$\theta' = \theta - 180°$
$\theta' = \theta - \pi$

$\theta' = 360° - \theta$
$\theta' = 2\pi - \theta$

To find the reference angle for an angle with negative measure or for an angle greater than 360°, first find a coterminal angle whose measure is between 0° and 360°.

EXAMPLE 4 θ **is an angle in standard position. Find the measure of its reference angle** θ'.

a. 250°   **b.** $\dfrac{5\pi}{6}$ **c.** $\dfrac{23\pi}{12}$ **d.** $-120°$ **e.** 460°

a. A 250° angle is in quadrant III.
$\theta' = 250° - 180° = 70°$

b. An angle measuring $\dfrac{5\pi}{6}$ is in quadrant II.

$\theta' = \pi - \dfrac{5\pi}{6} = \dfrac{\pi}{6}$

c. An angle measuring $\dfrac{23\pi}{12}$ is in quadrant IV.

$\theta' = 2\pi - \dfrac{23\pi}{12} = \dfrac{\pi}{12}$

d. Since $-120° + 360° = 240°$, an angle of $-120°$ is coterminal with a 240° angle, which is in quadrant III.
$\theta' = 240° - 180° = 60°$

e. Since $460° - 360° = 100°$, a 460° angle is coterminal with a 100° angle, which is in quadrant II.
$\theta' = 180° - 100° = 80°$

The possible values of the trigonometric functions of angles greater than 90° $\left(\dfrac{\pi}{2}\text{ radians}\right)$ are the same as those of angles between 0° and 90°, except for some of the signs.

EXAMPLE 5 **Evaluate sin θ, cos θ, and tan θ for each value of θ. Sketch θ and θ'.**
a. 150°   **b.** 210°   **c.** 330°

a.

b.

c.

$\sin 150° = \dfrac{1}{2}$

$\cos 150° = -\dfrac{\sqrt{3}}{2}$

$\tan 150° = -\dfrac{\sqrt{3}}{3}$

$\sin 210° = -\dfrac{1}{2}$

$\cos 210° = -\dfrac{\sqrt{3}}{2}$

$\tan 210° = \dfrac{\sqrt{3}}{3}$

$\sin 330° = -\dfrac{1}{2}$

$\cos 330° = \dfrac{\sqrt{3}}{2}$

$\tan 330° = -\dfrac{\sqrt{3}}{3}$

Thus, the value of each trigonometric function of an angle is equal to the value of the same function of its reference angle, or to the opposite of that value. For example, either $\sin \theta = \sin \theta'$ or $\sin \theta = -\sin \theta'$.

There are an infinite number of values of θ for which $\sin \theta = \dfrac{1}{2}$. For example, θ could be 30°, 150°, 390°, or $-330°$. However, if you know that θ is between 0° and 360°, then the value of θ must be either 30° or 150° for $\sin \theta = \dfrac{1}{2}$.

EXAMPLE 6 **If $0° \le \theta \le 360°$, find all values of θ for which tan $\theta = -1$.**

Tan θ is negative in quadrants II and IV. Since $\tan \theta = \dfrac{y}{x}$, let $y = 1$ and $x = -1$, or let $y = -1$ and $x = 1$. Draw the reference triangles in quadrants II and IV. The values of θ for which $\tan \theta = -1$ are 135° and 315°.

CLASS EXERCISES

Complete the following chart. Write *undefined* if a value is undefined.

	$\sin\theta$	$\cos\theta$	$\tan\theta$	$\csc\theta$	$\sec\theta$	$\cot\theta$
$\theta = 0° = 0$	1.	2.	3.	4.	5.	6.
$\theta = 90° = \dfrac{\pi}{2}$	7.	8.	9.	10.	11.	12.
$\theta = 180° = \pi$	13.	14.	15.	16.	17.	18.
$\theta = 270° = \dfrac{3\pi}{2}$	19.	20.	21.	22.	23.	24.

If $0° \le \theta \le 360°$, find the value(s) of θ that make each statement true. Express your answers in degrees.

25. $\sin\theta = 0$ **26.** $\cos\theta = 0$ **27.** $\sec\theta = 2$ **28.** $\csc\theta = -2$

PRACTICE EXERCISES

Complete the following chart.

	$\sin\theta$	$\cos\theta$	$\tan\theta$	$\csc\theta$	$\sec\theta$	$\cot\theta$
$\theta = 30° = \dfrac{\pi}{6}$	1.	2.	3.	4.	5.	6.
$\theta = 45° = \dfrac{\pi}{4}$	7.	8.	9.	10.	11.	12.
$\theta = 60° = \dfrac{\pi}{3}$	13.	14.	15.	16.	17.	18.

Give the measure of the reference angle θ' for each angle θ in standard position.

19. $171°$ **20.** $133°$ **21.** $305°$ **22.** $325°$

23. $412°$ **24.** $505°$ **25.** $-110°$ **26.** $-200°$

Find the exact values of the six trigonometric functions for each angle.

27. $150°$ **28.** $210°$ **29.** $135°$ **30.** $120°$

31. $225°$ **32.** $\dfrac{5\pi}{6}$ **33.** $\dfrac{7\pi}{6}$ **34.** $\dfrac{7\pi}{4}$

35. $390°$ **36.** $420°$ **37.** $495°$ **38.** $585°$

39. $-30°$ **40.** $-135°$ **41.** $-240°$ **42.** $1350°$

43. $-\dfrac{\pi}{3}$ **44.** $\dfrac{9\pi}{4}$ **45.** $\dfrac{10\pi}{3}$ **46.** $-\dfrac{15\pi}{4}$

If $0 \le \theta \le 2\pi$, find the values of θ that make each statement true. Express your answers in radians.

47. $\cos \theta = \dfrac{1}{2}$ **48.** $\sin \theta = \dfrac{1}{2}$ **49.** $\tan \theta = -1$ **50.** $\cot \theta = 1$

51. $\sec \theta = \sqrt{2}$ **52.** $\csc \theta = -\sqrt{2}$ **53.** $\cot \theta = -\sqrt{3}$ **54.** $\tan \theta = \sqrt{3}$

Give the measure of the reference angle θ' for each angle θ in standard position.

55. $1002°$ **56.** $1500°$ **57.** $-907°$ **58.** $-950°$

If $0° \le \theta \le 360°$, find the value of θ that makes each statement true. Express your answer in degrees.

59. $\tan \theta = 1$ and $\sin \theta \ge 0$

60. $\cos \theta = \dfrac{\sqrt{3}}{2}$ and $\tan \theta \ge 0$

61. $\sin \theta = \dfrac{1}{2}$ and $\cos \theta \le 0$

62. $\cos \theta = -\dfrac{1}{2}$ and $\sin \theta \ge 0$

63. $\sin \theta = -\dfrac{\sqrt{3}}{2}$ and $\cos \theta \le 0$

64. $\cos \theta = -\dfrac{\sqrt{3}}{2}$ and $\sin \theta \le 0$

65. $\sin \theta = -1$ and $\cos \theta = 0$

66. $\sin \theta = 0$ and $\cos \theta = -1$

67. $\sec \theta = \sqrt{2}$ and $\csc \theta = -\sqrt{2}$

68. $\cot \theta = -\sqrt{3}$ and $\sec \theta = \dfrac{2\sqrt{3}}{3}$

Applications

69. Construction   A 30-ft ladder resting against a building makes a 60° angle with the ground. The height h from the ground at which the ladder touches the building can be found using the equation $\sin 60° = \dfrac{h}{30}$. Find h to the nearest foot.

70. Routing   A square city block is 135 m on each side. Determine the distance from the NW corner to the SE corner via two of the bordering streets. Then calculate the distance between the corners if a diagonal shortcut is taken. Round your answers to the nearest meter.

CHALLENGE

A rectangular piece of paper 12 cm by 24 cm is folded along line segment AB, as shown.

Find $\csc \theta$.

Evaluating Trigonometric Functions

1.9

Objectives: To find decimal approximations for the values of the six trigonometric functions for all angles

To find the measure of an angle given the value of one of its trigonometric functions

Using the values of the trigonometric functions of angles, navigators and surveyors are able to calculate distances that they cannot measure directly.

Preview

A nautical mile is a unit of distance used on ships and aircraft. Historically, one *nautical mile* has been defined as $\frac{1}{60}$ degree of arc length on a meridian.

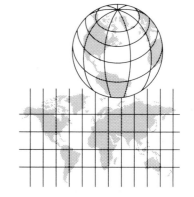

The earth is not a perfect sphere, so the actual length of a nautical mile varies with the latitude. The length can be approximated using the formula

1 nautical mile = $(6077 - 31 \cos 2\theta)$ ft

where θ is the latitude in degrees.

The latitude of Anchorage, Alaska is about 61°. If this is rounded to 60°, you can calculate the length of one nautical mile at that location, since you know the value of cos 2(60°), or cos 120°.

1. Round 61° to 60° and calculate the approximate length of a nautical mile at Anchorage. Round the answer to the nearest foot.

2. Which of the following is the best approximation of the length of a nautical mile at a location with latitude 29°?

    a. 6055 ft   **b.** 6062 ft   **c.** 6068 ft   **d.** 6075 ft

 A scientific calculator is an invaluable tool when you wish to find the values of trigonometric functions of an angle such as 61° that is not an integral multiple of a special angle. You can also use a table of trigonometric values. See pages 434–438 for the table and pages 422–423 for a discussion of its use.

A scientific calculator has at least two modes. Make sure that it is in *degree mode* when you are asked to find the values of the trigonometric functions of angles expressed in degrees, and in *radian mode* when you must find the values of the trigonometric functions of angles expressed in radians.

EXAMPLE 1 **Find each value. Round to four decimal places.**

    a. $\cos 56.2°$ **b.** $\tan 15°23'$ **c.** $\sin \dfrac{\pi}{9}$

    a. Place your calculator in degree mode. Since the angle measure is given in decimal degrees, simply enter 56.2. Use the cos key.

$$\cos 56.2° = 0.556295616$$
$$= 0.5563 \quad \text{to four decimal places}$$

    b. Change $15°23'$ to decimal degrees, unless your calculator accepts angle measures given in degrees and minutes. Use the tan key.

$$\tan 15°23' = \tan \left(15 + \frac{23}{60}\right)°$$
$$= \tan 15.38333333°$$
$$= 0.275132958$$
$$= 0.2751 \quad \text{to four decimal places}$$

    c. Place your calculator in radian mode.

$$\sin \frac{\pi}{9} = \sin 0.34906585 \quad \textit{Use the } \pi \textit{ and } \div \textit{ keys.}$$
$$= 0.342020143 \quad \textit{Use the sin key.}$$
$$= 0.3420 \quad \text{to four decimal places}$$

In Example 1, $\cos 56.2°$ is not exactly equal to 0.556295616. The exact values of $\tan 15°23'$ and $\sin \dfrac{\pi}{9}$ are not given either. Most of the values of the trigonometric functions are irrational numbers and cannot be written exactly as decimals. Nevertheless, it is customary to use the equals sign in examples of this type. For exercises in this book, values of the trigonometric functions should be rounded to four decimal places, unless otherwise indicated.

When you use a calculator to find the values of the cosecant, secant, and cotangent functions, remember that $\csc \theta$, $\sec \theta$, and $\cot \theta$ are the reciprocal functions of $\sin \theta$, $\cos \theta$, and $\tan \theta$, respectively. Therefore, use the reciprocal key, which is usually labeled $\dfrac{1}{x}$ or x^{-1}.

EXAMPLE 2 **Find sec 24.37°.**

    Make sure that your calculator is in degree mode.
    Enter 24.37. Use the cos key and the reciprocal key.

$$\sec 24.37° = 1.097815544$$
$$= 1.0978 \quad \text{to four decimal places}$$

The angles in the first two examples are all acute. You can also use a calculator to find the values of the trigonometric functions of angles whose measures are greater than 90° or less than 0°. If you make the entries correctly, the calculator will give the correct signs for the values of the trigonometric functions of these angles. However, you should check the results against your knowledge of the signs of the various functions in each quadrant.

EXAMPLE 3 **Find each value: a.** $\sin 324.72°$ **b.** $\tan \left(-\dfrac{7\pi}{10} \right)$

a. Make sure that your calculator is in degree mode.

$\sin 324.72° = -0.5776$ *Enter 324.72 and use the sin key.*

Check that the sign is correct: $270° < 324.72° < 360°$, so the terminal side of $324.72°$ is in quadrant IV where sine is negative.

b. Make sure that your calculator is in radian mode.

$\tan \left(-\dfrac{7\pi}{10} \right) = 1.3764$ *Enter* $-\dfrac{7\pi}{10}$. *Use the keys for* π, ÷, +/−, tan.

Check that the sign is correct: $-\dfrac{7\pi}{10}$ is coterminal with $\dfrac{13\pi}{10}$, whose terminal side is in quadrant III where tangent is positive.

If you know the value of a trigonometric function of an angle, a calculator will give you the measure of one angle with that value. You must then use your knowledge of the signs of the various trigonometric functions in each quadrant to decide whether there is a second angle that has the same value.

Finding the measure of an angle with a known sine and finding the sine of an angle of known measure are *inverse operations*. Therefore, you will probably have to use the inverse key (usually labeled inv) as shown in Example 4.

EXAMPLE 4 **Find the measure of each angle θ, where $0° \le \theta \le 360°$, to the nearest tenth of one degree.**
a. $\sin \theta = 0.5358$ **b.** $\cos \theta = -0.5663$ **c.** $\cot \theta = -1.3182$

a. $\sin \theta = 0.5358$ *Enter 0.5358. Make sure calculator is in degree mode.*
$\theta = 32.39818172°$ *Use the inv and the sin keys.*
$\theta = 32.4°$ to the nearest tenth

Since $\sin \theta$ is positive, there is also an angle in quadrant II whose sine is 0.5358 and whose reference angle is 32.4°.

$180° - 32.4° = 147.6°$

Check to show that $\sin 147.6° = 0.5358$.
The solutions are $\theta = 32.4°$ and $\theta = 147.6°$.

b. $\cos \theta = -0.5663$ *Enter -0.5663. Use the $+/-$ key.*
     $\theta = 124.4926144°$ *Use the inv and cos keys.*
     $\theta = 124.5°$ *to the nearest tenth*

Since $\cos \theta$ is negative, there is also an angle in
quadrant III whose cosine is -0.5663 and whose
reference angle is the same as that of $124.5°$.

$180° - 124.5° = 55.5°$ *Reference angle for 124.5°*
$180° + 55.5° = 235.5°$ *A 235.5° angle also has*
                        reference angle 55.5°.

Check to show that $\cos 235.5° = -0.5663$. (A calculator gives $\cos 235.5°$
as -0.5664, to four decimal places. The discrepancy is due to the fact that
$124.4926144°$ was rounded to $124.5°$.)
The solutions are $\theta = 124.5°$ and $\theta = 235.5°$.

c. $\cot \theta = -1.3182$ *Enter -1.3182. Use the $+/-$ key.*
     $\theta = -37.18432593°$ *Use the reciprocal, inv and tan keys.*
     $\theta = -37.2°$ *to the nearest tenth*

Since $0° \le \theta \le 360°$, angle θ cannot be negative.
Find a positive angle coterminal with $-37.2°$.

$$360° + (-37.2°) = 322.8°$$

Since $\cot \theta$ is negative, there is also an angle in
quadrant II whose cotangent is -1.3182 and
whose reference angle is the same as that of $322.8°$.

$360° - 322.8° = 37.2°$ *Reference angle for 322.8°*
$180° - 37.2° = 142.8°$ *A 142.8° angle also has reference angle 37.2°.*

Check to show that $\cot 142.8°$ is approximately equal to -1.3182.
(The calculator gives -1.3175. The discrepancy is due to the rounding
procedures used.) The solutions are $\theta = 322.8°$ and $\theta = 142.8°$.

CLASS EXERCISES

Find each value. Round to four decimal places.

  1. $\sin 24.8°$ **2.** $\cos 36.9°$ **3.** $\tan 12°18'$ **4.** $\cos(-101.7°)$

  5. $\sin(-82.5°)$ **6.** $\sec \dfrac{13\pi}{12}$ **7.** $\csc \dfrac{17\pi}{12}$ **8.** $\cot\left(-\dfrac{2\pi}{5}\right)$

Find θ, where $0° \le \theta \le 360°$, to the nearest tenth of one degree.

  9. $\sin \theta = 0.7218$ **10.** $\cos \theta = -0.5864$ **11.** $\csc \theta = -1.1111$

12. $\sec \theta = 1.1111$ **13.** $\cot \theta = -1.2222$ **14.** $\cot \theta = -2.3456$

PRACTICE EXERCISES Use appropriate technology.

Find each value. Round to four decimal places.

1. sin 18.6°   **2.** cos 21.5°   **3.** tan 83.7°   **4.** sin 85.8°

5. cos 12.7°   **6.** tan 48.4°   **7.** sin 18°32'   **8.** cos 24°25'

9. tan 87°18'   **10.** sin 73°37'   **11.** $\cos \dfrac{\pi}{5}$ **12.** $\tan \dfrac{\pi}{6}$

13. $\sin \dfrac{2\pi}{3}$ **14.** $\cos \dfrac{3\pi}{5}$ **15.** $\tan \dfrac{3\pi}{4}$ **16.** $\tan \dfrac{4\pi}{5}$

17. csc 20.7°   **18.** sec 23.9°   **19.** cot 41.4°   **20.** cot 60.6°

21. sec 145.5°   **22.** cot 160.6°   **23.** csc 98.9°   **24.** sec 175.4°

Find θ, where $0° \le \theta \le 360°$, to the nearest tenth of one degree.

25. sin θ = 0.7777   **26.** cos θ = 0.8888   **27.** tan θ = 0.6765

28. sin θ = 0.3535   **29.** cos θ = 0.6161   **30.** tan θ = 0.9808

Find each value. Round to four decimal places.

31. sin 310°16'   **32.** cos 311°55'   **33.** tan 610.7°   **34.** sin 511.1°

35. cos (− 111.8°)   **36.** tan (− 200.8°)   **37.** csc 321.2°   **38.** sec 315.6°

39. $\sin \dfrac{3\pi}{7}$ **40.** $\tan \dfrac{9\pi}{8}$ **41.** $\cot \dfrac{\pi}{11}$ **42.** $\csc \dfrac{25\pi}{8}$

43. cot (− 77°16')   **44.** $\csc \left(-\dfrac{3\pi}{7} \right)$ **45.** $\sec \left(-\dfrac{7\pi}{4} \right)$ **46.** $\cot \left(-\dfrac{6\pi}{5} \right)$

Find θ, where $0° \le \theta \le 360°$, to the nearest tenth of one degree.

47. csc θ = 2.5012   **48.** sec θ = 3.1909   **49.** cot θ = 4.4444

50. csc θ = − 1.3737   **51.** cot θ = − 6.6175   **52.** sec θ = − 1.7172

53. tan θ = 1.1213 and sin θ is positive.

54. sin θ = 0.5151 and cos θ is negative.

55. cot θ = 0.6878 and θ is in quadrant III.

56. csc θ = 2.5151 and θ is in quadrant I.

57. sin θ = − 0.5267 and 270° ≤ θ ≤ 360°

58. cos θ = − 0.0655 and 90° ≤ θ ≤ 180°

59. csc θ = 1.0222 and 90° ≤ θ ≤ 180°

60. sec θ = 1.0222 and 270° ≤ θ ≤ 360°

Computer The following program calculates and displays the values of the six trigonometric functions of an angle whose measure is given in degrees. Line 40 calculates the sine and cosine, lines 50 and 60 calculate the other functions, and lines 70–100 display the results.

```
10  PRINT "PLEASE ENTER THE MEASURE OF AN ANGLE"
20  INPUT M
30  M = M*0.01745329
40  S = SIN(M): C = COS(M)
50  T = S/C: CS = 1/S
60  SE = 1/C: CT = 1/T
70  HOME: PRINT "THE TRIG FUNCTION VALUES OF "M" DEGREES ARE"
80  PRINT: PRINT "SINE-"S, "COSINE-"C
90  PRINT "TANGENT-"T, "COSECANT-"CS
100 PRINT "SECANT-"SE, "COTANGENT-"CT
110 END
```

61. Line 30 converts the degree measure to radians. How is the constant 0.01745329 derived?

62. CT represents the cotangent. State another way that CT could be found.

TEST YOURSELF

1. Angle θ is in standard position and lies in quadrant II. If $\sin \theta = \dfrac{\sqrt{3}}{2}$, find the exact value of $\cos \theta$. **1.6**

2. Find the exact values of $\sin \theta$ and $\cos \theta$ if θ is an angle in standard position whose terminal side passes through $(-2, -2)$. Sketch the reference triangle.

Find the exact values of the other five trigonometric functions for an angle in standard position lying in the given quadrant.

3. $\sin \theta = -\dfrac{12}{13}$, III   **4.** $\cos \theta = \dfrac{4}{5}$, I   **5.** $\sec \theta = -\dfrac{12}{7}$, II **1.7**

Give the measure of the reference angle θ' for each angle θ in standard position.

6. 179°   **7.** 312°   **8.** $-\dfrac{4\pi}{3}$ **9.** $\dfrac{\pi}{4}$ **1.8**

10. Find the exact values of the six trigonometric functions of a 240° angle.

Find each value. Round to four decimal places.

11. $\sin 212.2°$ **12.** $\tan (-63.5°)$ **13.** $\csc \dfrac{\pi}{7}$ **14.** $\cot \dfrac{2\pi}{5}$ **1.9**

APPLICATION:
Space Exploration

Did you know that space scientists prefer to launch a payload, such as a communications satellite, into Earth-orbit in more than one stage? The reason for this is that while a large, heavy fuel module is required to lift a heavy rocket through the lower atmosphere, relatively little fuel is required to propel a light payload into its own trajectory through the thin atmosphere of high altitudes. As the diagram below shows, a booster rocket releases its payload at a previously selected altitude and then follows a parabolic trajectory back to Earth. Then the payload projectile starts a new trajectory that will eventually carry it into an orbit around the Earth. The trajectories of both the booster rocket and the payload projectile are curves. However, the path of the payload may be approximated by a straight line for a brief period of time at the start of its independent flight.

In the diagram, a booster rocket has reached altitude y_1 with the direction of its path forming an angle θ with an imaginary horizontal line. The value of θ has been steadily diminishing since the launch began. The space scientists on the ground now wish to fire the payload into its own trajectory, provided that θ has diminished to a certain critical value that will put the payload into Earth-orbit. This value of θ can be calculated by solving the equations

$$y = ax^2 + bx \qquad \text{\textit{Path of the booster rocket}}$$
$$\tan \theta = 2ax + b \qquad \text{\textit{Direction of the booster rocket}}$$

where x and y are the horizontal and vertical coordinates, respectively, of the position of the rocket. However, before the equations can be used, the values of constants a and b must be determined.

EXAMPLE 1 The equation of the trajectory of a booster rocket is
$y = -x^2 + 3x$. Space scientists plan to fire a payload from the
booster at a horizontal distance of 1 km from the launch site.
Find the height and direction at which the payload will be fired.

$$y = -x^2 + 3x$$
$$= -1^2 + 3(1) = 2 \qquad \text{The payload will be fired when the height is 2 km.}$$

Next, evaluate a and b in the equation for the direction.

$$\tan \theta = 2(-1)x + 3 \qquad a = -1, b = 3$$
$$= -2x + 3$$
$$= -2(1) + 3 = 1 \qquad \textit{Substitute 1 for } x_1.$$

Since $\tan 45° = 1$, the initial direction of the projectile in independent flight is
at an angle of 45° with the horizontal.

EXAMPLE 2 For the flight of Example 1, use $\sin \theta = \dfrac{h}{f}$ to find the vertical

distance h traveled by the projectile after it has traveled on an
approximately linear path of 10 km.

$$\sin 45° = \frac{h}{10} \qquad \textit{Substitute 10 for f.}$$

$$\frac{\sqrt{2}}{2} = \frac{h}{10}$$

$$h = 10 \cdot \frac{\sqrt{2}}{2} \approx 7.1 \qquad \text{The vertical distance is about 7.1 km.}$$

EXERCISES

1. Find the altitude of the payload in Example 2 after 10 km of its free flight.

2. Use $\cos \theta = \dfrac{g}{f}$ to find the horizontal distance g traveled by the payload in
Example 2.

3. In Examples 1 and 2, suppose that the payload is fired when $\theta = 58°$
rather than 45°. What will be the altitude of the rocket?

4. A booster rocket with a weather satellite payload is launched in a trajectory
with equation $y = -2x^2 + 1.9x$. The satellite is fired into its own
trajectory when the horizontal distance reaches 0.3 km. Find, to the nearest
degree, the free-flight direction of the satellite.

5. Find the maximum altitude attained by the booster rocket of Exercise 4.
Hint: What must be the value of θ at the vertex of the parabola?

Vocabulary

angle (12)	function (3)	reference triangle (30)
angular displacement (24)	initial side (12)	relation (2)
angular velocity (24)	linear velocity (24)	secant (34)
central angle (17)	minute (17)	second (17)
coordinate plane (3)	origin (3)	sine (29)
cosecant (34)	Pythagorean theorem (8)	standard position (12)
cosine (29)	quadrant (3)	tangent (34)
cotangent (34)	quadrantal angle (39)	terminal side (12)
coterminal angles (13)	radian (18)	trigonometric functions (34)
distance formula (8)	range (2)	unit circle (29)
degree (13)	reciprocal functions (35)	x-axis, y-axis (3)
domain (2)	reference angle (42)	

Functions and the Coordinate Plane   A function is a relation that pairs   **1.1**
each member of the domain with exactly one member of the range.

1. Find the domain and range of the relation $\{(1, 3), (5, 7), (5, 9)\}$. Is the relation a function?

2. Find $f(-3)$ if $f(x) = 2x - 7$.

The Distance Formula   The distance d between two points with   **1.2**
coordinates (x_2, y_2) and (x_1, y_1) is given by $d = \sqrt{(x_2 - x_1)^2 + (y_2 - y_1)^2}$.

3. Find the distance between the points $(3, 0)$ and $(-5, -6)$.

Angles in the Coordinate Plane   An angle is in standard position if its   **1.3**
vertex is the origin and its initial side lies on the positive x-axis.

4. Find the degree measure of the angle formed by a $\frac{1}{3}$ counterclockwise rotation. Sketch the angle in standard position.

5. Find the measures of two angles that are coterminal with a $110°$ angle.

Angle Measures in Degrees and Radians   To change from degrees to   **1.4**
radians, multiply the number of degrees by $\frac{\pi}{180}$. To change from radians to

degrees, multiply the number of radians by $\frac{180}{\pi}$.

6. Express $240°$ in radians. Give the answer in terms of π.

7. Express $\frac{11\pi}{4}$ in degrees.   8. Express $100°33'$ in decimal degrees.

Applications: Angular and Linear Velocity The formulas for arc length 1.5
s, angular velocity ω, and linear velocity V are:

$$s = r\theta \qquad \omega = \frac{\theta}{t} \qquad V = \frac{s}{t} = \omega r$$

9. Find the length of the arc intercepted by a central angle of $\frac{\pi}{3}$, if the radius r of the circle is 12 in. Express the answer in terms of π.

10. Find the angular velocity in radians per second and the linear velocity in inches per second of a point on the rim of a wheel with a diameter of 20 in., rotating at 150 revolutions per minute. Express the answers in terms of π.

Cosine and Sine Functions The cosine and sine functions of an angle θ
are determined by the coordinates $(x, y) = (\cos \theta, \sin \theta)$ of the intersection 1.6
of the terminal side of the angle with the unit circle. If the circle has radius r,
then $\cos \theta = \frac{x}{r}$ and $\sin \theta = \frac{y}{r}$.

11. If $\sin \theta = -\frac{15}{17}$ and θ is in quadrant III, find the exact value of $\cos \theta$.

The Trigonometric Functions The remaining functions are 1.7
$\tan \theta = \frac{y}{x} (x \neq 0)$, $\csc \theta = \frac{r}{y} (y \neq 0)$, $\sec \theta = \frac{r}{x} (x \neq 0)$, and $\cot \theta = \frac{x}{y} (y \neq 0)$.

12. Find the exact values of the six trigonometric functions of angle θ in standard position, if the point $(4, -3)$ is on its terminal side.

Trigonometric Functions of Special Angles Angles with measures 30°, 1.8
45°, and 60°, as well as quadrantal angles, are special angles in trigonometry.
Reference angles are used to find the values of trigonometric functions for
angles that are not in the first quadrant.

13. Find the exact values of the six trigonometric functions of 180°.

14. Find the exact values of the six trigonometric functions of $-30°$.

Evaluating Trigonometric Functions A scientific calculator can be used 1.9
to find values of the trigonometric functions of an angle. It can also be used to
find the measure of an angle if the value of one of its trigonometric functions
is given.

15. Find $\sin \frac{\pi}{4}$ and $\sec -215.7°$. Round to four decimal places.

16. If $0° \leq \theta \leq 360°$ and $\csc \theta = -2.2023$, find θ to the nearest tenth of one degree.

1. Write the domain and the range for the relation $\{(1, 3), (3, 8), (0, 8)\}$. Is the relation a function?

2. If $f(x) = 2x - 5$, find $f(-3)$.

3. Find the distance between the points $(-3, 4)$ and $(9, -1)$.

Find the degree measure of the angle for each rotation.

4. $\dfrac{17}{6}$ clockwise rotation

5. $\dfrac{2}{3}$ counterclockwise rotation

6. Express $225°$ in radians. Give the answer in terms of π.

7. Express $-\dfrac{5\pi}{3}$ in degrees.

Express the answers to Exercises 8–10 in terms of π.

8. Find the length of the arc intercepted by a central angle of $\dfrac{9\pi}{4}$ if the radius of the circle is 28 in.

9. Find the angular velocity in radians per minute of an object rotating at 315 revolutions per minute.

10. An object rotates at 4 revolutions per second at a distance of 15 ft from a point. Find the linear velocity of the object in feet per second.

11. If $\sin \theta = \dfrac{\sqrt{3}}{2}$ and θ is in quadrant II, find the exact value of $\cos \theta$.

12. Point P $(3, -3)$ is on the terminal side of θ, an angle in standard position. Find the exact values of the six trigonometric functions of θ.

13. Find the exact value of $\cot (-585°)$.

14. Find the value of $\cos 37.9°$. Round to four decimal places.

15. If $0° \le \theta \le 360°$ and $\tan \theta = 1.3535$, find θ to the nearest tenth of one degree.

Challenge

Find the exact coordinates of a point that separates a semicircular arc of the unit circle (from $0°$ to $180°$) into two parts, the ratio of whose lengths is $1:3$. Is there more than one such point?

Select the best choice for each question.

1. If $\triangle ABC$ is a right triangle, find the length of side x.
   A. -34 B. $\sqrt{119}$
   C. $\sqrt{34}$ D. $-\sqrt{34}$
   E. 119

2. Find $\cot \theta$ if $\sin \theta = -\frac{1}{3}$ and $\tan \theta$ is positive.

   A. $2\sqrt{2}$ B. -3 C. $\frac{2\sqrt{2}}{3}$
   D. $-2\sqrt{2}$ E. $2\sqrt{3}$

3. Solve for x.
   $$\frac{3}{x-1} + 1 = \frac{x+4}{x^2-1} - \frac{x}{1-x}$$
   A. 4 B. -2 C. $\frac{3}{13}$
   D. 2 E. No solution

4. If line m is parallel to line n, then $c - a$ equals

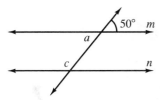

   A. $180°$ B. $130°$ C. $0°$
   D. $50°$ E. $80°$

5. Which of the following relations is not a function?
   A. $\{(3, 4), (-1, 0), (2, 2)\}$
   B. $\{(1, 2), (3, 2), (5, 2), (6, 2)\}$
   C. $\{(-1, -6), (-6, -1)\}$
   D. $\{(3, 1), (0, 1), (3, 2)\}$
   E. $\{(2, 4), (4, 6), (6, 8)\}$

6. $\left(\dfrac{x-2}{x+4} \div \dfrac{x+1}{x+4}\right) = \dfrac{x-2}{x+1}$ where
   A. $x \neq 2$ B. $x \neq 0$ C. $x \neq -1$
   D. $x \neq -1, x \neq -4$ E. $x \neq -4$

7. The graphs of $4x + y = 4$ and $4x - 3y = 2$ are
   A. equivalent B. parallel
   C. perpendicular D. intersecting
   E. not straight

8. If $\cos \theta = \frac{\sqrt{3}}{2}$ and θ is in quadrant IV, then $\sin \theta$ equals
   A. $\frac{1}{2}$ B. 2 C. $-\frac{1}{2}$
   D. $\frac{\sqrt{3}}{3}$ E. $-\sqrt{3}$

9. For what value(s) of θ does $\sin 2\theta = -1$, where $0 \leq \theta \leq 2\pi$?
   I. $\frac{\pi}{2}$ II. $\frac{\pi}{4}$ III. $\frac{3\pi}{4}$
   A. I only B. II only C. III only
   D. I and III only E. II and III only

10. If f and g are inverse functions and $g(x) = 3x - 1$, then $f(x)$ equals
    A. $3x - 1$ B. $\frac{1}{3}x + \frac{1}{3}$ C. $x + \frac{1}{3}$
    D. $x - 3$ E. $\frac{1}{3}x + 1$

11. Which of the following is the smallest?
    A. $(4x^2)^0$ B. $\left(\frac{2}{3}\right)^{-2}$ C. $\frac{\sqrt{2}}{2}$
    D. $\left(\frac{1}{2}\right)^3$ E. $\sin 30°$

12. Which of the following is (are) true?
    I. $\frac{\pi}{2}$ is a quadrantal angle.
    II. If $\theta = \frac{\pi}{3}$, then $0° \leq \theta \leq 45°$.
    III. $\frac{3\pi}{4} = 135°$
    A. I only B. I and II only
    C. I and III only D. II and III only
    E. I, II, and III

Solve for x.

Example $\dfrac{x - 5}{10} = \dfrac{2x + 6}{15}$

$(x - 5)(15) = 10(2x + 6)$ If $\dfrac{a}{b} = \dfrac{c}{d}$, then $ad = bc$.

$15x - 75 = 20x + 60$ Distributive property

$-135 = 5x$ Subtract 60 and 15x from both sides.

$-27 = x$

1. $-\dfrac{3}{5} = \dfrac{6}{x}$

2. $\dfrac{x + 2}{5} = \dfrac{x - 3}{7}$

3. $-\dfrac{x}{7} = \dfrac{x + 3}{5}$

4. $\dfrac{x + 3}{x - 4} = \dfrac{x + 5}{x - 7}$

5. $\dfrac{x + 2}{x + 7} = \dfrac{x + 3}{x - 2}$

6. $\dfrac{2}{x + 3} = \dfrac{2x - 4}{x^2 + 3x + 2}$

Factor each polynomial.

Example $4x^2 + 12x + 9 = (2x + 3)^2$ $a^2 + 2ab + b^2 = (a + b)^2$

7. $4x^2 - 9$

8. $x^2 - 9y^2$

9. $8 - x^3$

10. $9x^2 - 24x + 16$

11. $x^3 - 27y^3$

12. $x^8 - 16$

Solve each equation.

Example $0 = x^3 - 4x$

$0 = x(x^2 - 4)$

$0 = x(x + 2)(x - 2)$ $a^2 - b^2 = (a + b)(a - b)$

$x = 0, -2, 2$ The solutions are 0, -2, and 2.

13. $0 = 2x^3 - 4x$

14. $0 = -x^3 + x$

15. $0 = x^3 - x^2 - 2x$

16. $x^3 = 4x$

17. $-x^3 + 4x^2 = -5x$

18. $x^3 = x^2 + 6x$

Graph each equation.

Example $y = x^2 - 1$

Make a table of values.

x	-2	-1	0	1	2
y	3	0	-1	0	3

19. $y = -2x + 3$

20. $y = 5x$

21. $y = x^2 + 2$

22. $y = x^2 - 3$

23. $y = -x^2 + 1$

24. $y = -x^2 - 2$

2 Graphing Trigonometric Functions

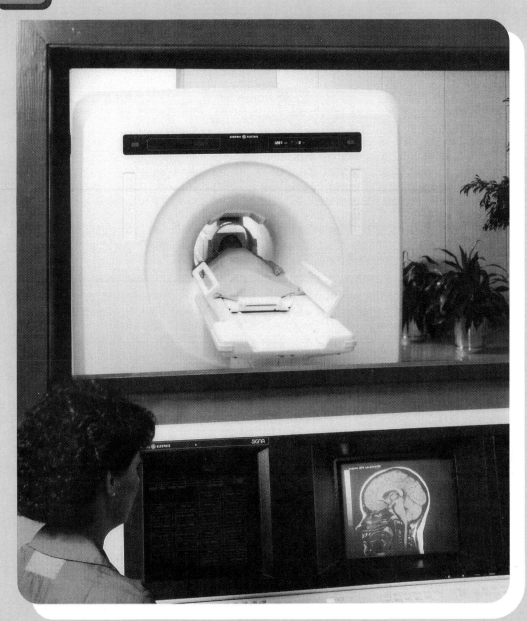

Magnetic Resonance Imaging (MRI) has been heralded as the most significant advance in medical technology since the advent of the x-ray. It enables physicians to take images of body organs from any angle on any plane without moving the patient.

Periodic Functions and Symmetry

2.1

Objectives: To define periodic functions and odd and even functions
To determine the symmetry of a graph

Many functions are *periodic*; that is, their graphs repeat in cycles. Also, many functions have graphs that are symmetric with respect to one of the axes or to the origin.

Preview

If a figure can be "folded" along a line so that the two parts coincide, the figure is said to be *symmetric* with respect to that line. For example, square *WXYZ* is symmetric with respect to either diagonal *WY* or diagonal *XZ*.

If a figure appears to be in exactly the same position after a 180° rotation about a point, it is said to be *symmetric* with respect to that point. For example, square *WXYZ* is symmetric with respect to point *M*.

E, F, G, and H are the midpoints of the sides of rectangle ABCD, and AB > BC. Tell whether or not each figure is symmetric with respect to the given line or point.

1. Line segment *AB*, to line *FH*

2. Line segment *AD*, to point *E*

3. Rectangle *ABCD*, to line *FH*

4. Rectangle *ABCD*, to point *I*

5. Rectangle *ABGE*, to line *AG*

6. Triangle *AGD*, to line *EG*

7. Triangle *AGD*, to point *I*

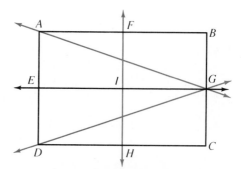

A periodic function is a function that repeats itself. That is, a function $f(x)$ is **periodic** if there exists a smallest positive number p such that

$$f(x + p) = f(x)$$

whenever both $f(x + p)$ and $f(x)$ are defined. The number p is said to be the **period** of the function, and one **cycle** of the graph is completed in each period.

The graphs of two periodic functions are shown below. The arrowheads indicate that the graphs continue the same way, without bound, to the left and to the right. The period of $f(x)$ is 2 and the period of $g(x)$ is 4.

$f(x + 2) = f(x)$

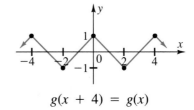

$g(x + 4) = g(x)$

Recall that a graph represents a function if no vertical line can be drawn that intersects the graph in more than one point. Given the graph of a relation, you can determine if the graph is a function, if it is periodic, and what the period is.

EXAMPLE 1 **Examine the graph at the right.**
   a. Does it represent a function?
   b. Is it periodic?
   c. If it is periodic, what is the period?

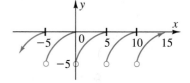

   a. The vertical line test shows that the graph represents a function.

   b. The graph repeats in cycles, so the function is periodic.

   c. The graph repeats every 5 units; that is, $f(x + 5) = f(x)$. The period is 5.

If you know the pattern of a periodic function, as well as its period, you can extend the graph.

EXAMPLE 2 **The figure at the left below shows one cycle of the graph of a function with period 3. Extend the graph over the interval $-3 < x \le 9$.**

   To extend the graph, repeat the pattern every 3 units.

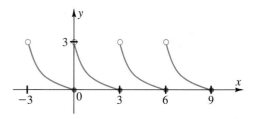

If $f(-x) = f(x)$ for every value in its domain, then the function is said to be an **even function.** If $f(-x) = -f(x)$ for every value in its domain, then it is said to be an **odd function.**

EXAMPLE 3 **Determine whether each function is even, odd, or neither.**
  a. $f(x) = x^2 + 2$ **b.** $f(x) = x^3$ **c.** $f(x) = x^2 + x$

a. For any value of x,

$$f(x) = x^2 + 2 \quad \text{and} \quad f(-x) = (-x)^2 + 2$$
$$= x^2 + 2$$

Since $f(-x) = f(x)$, the function is even.

b. For any value of x,

$$f(x) = x^3 \quad \text{and} \quad f(-x) = (-x)^3$$
$$= -x^3$$

Since $f(-x) = -f(x)$, the function is odd.

c. For any value of x,

$$f(x) = x^2 + x \quad \text{and} \quad f(-x) = (-x)^2 + (-x)$$
$$= x^2 - x$$

Since $f(-x) \neq f(x)$ and $f(-x) \neq -f(x)$, the function is neither even nor odd.

A graph may be symmetric with respect to the x-axis, the y-axis, or the origin. Knowing that the graph of a relation is symmetric with respect to one of the axes or to the origin makes it easier to sketch.

For every point (a, b) on the graph of a relation, if the point $(-a, b)$ is also on the graph, then the graph is **symmetric with respect to the y-axis.** An *even* function is symmetric with respect to the y-axis.

For every point (a, b) on the graph of a relation, if the point $(a, -b)$ is also on the graph, then the graph is **symmetric with respect to the x-axis.**

For every point (a, b) on the graph of a relation, if the point $(-a, -b)$ is also on the graph, then the graph is **symmetric with respect to the origin.** An *odd* function is symmetric with respect to the origin.

The figures below provide examples of each type of symmetry.

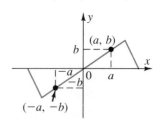

y-axis symmetry   x-axis symmetry   origin symmetry

EXAMPLE 4 Determine whether each graph is symmetric with respect to the *x*-axis, the *y*-axis, the origin, or to none of these.

a.

b.

c.
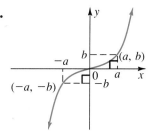

a. For any point (a, b) on the graph, $(a, -b)$ is also on the graph. The graph is symmetric with respect to the *x*-axis.

b. For any point (a, b) on the graph, $(-a, b)$ is also on the graph. The graph is symmetric with respect to the *y*-axis.

c. For any point (a, b) on the graph, $(-a, -b)$ is also on the graph. The graph is symmetric with respect to the origin.

CLASS EXERCISES

Which of these graphs represent periodic functions?

1.

2.

3.

4.
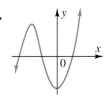

Determine whether each graph is symmetric with respect to the *x*-axis, the *y*-axis, the origin, or to none of these.

5.

6.

7.

8.

Determine whether each function is even, odd, or neither.

9.

10.

11.

12.
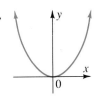

PRACTICE EXERCISES

For each of the graphs in Exercises 1–6, answer the following questions:

a. Does the graph represent a function?

b. Is the graph periodic?

c. If the graph is periodic, what is the period?

1.

2.

3.

4.

5.

6.

Given one cycle of the graph of $f(x)$ and the period of the function, sketch the graph of the function over the specified domain.

7. period 2; $0 \leq x \leq 8$

8. period 3; $0 \leq x \leq 6$

9. period 1; $-2 \leq x \leq 3$

10. period 2; $-1 \leq x \leq 5$

11. period 2; $-4 \leq x \leq 6$

12. period 4; $4 \leq x \leq 8$

Determine whether each function is even, odd, or neither.

13. $f(x) = 2x^2$

14. $f(x) = 5x^3$

15. $f(x) = x$

16. $f(x) = 3x^2 - 7$

17. $f(x) = 3x^2 + 5$

18. $f(x) = 2x + 1$

19. $f(x) = x^3 + x$

20. $f(x) = x^5$

21. $f(x) = -x^2$

22. $f(x) = -2x^2 + 1$ **23.** $f(x) = -x^3$ **24.** $f(x) = -3x^3 - x$

Determine whether each graph is symmetric to the x-axis, the y-axis, the origin, or to none of these.

25.

26.

27.

28.

29.

30.
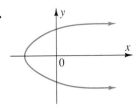

Make three copies of each figure below. Extend the first graph so that it is symmetric to the x-axis. Extend the second graph so that it is symmetric to the y-axis. Extend the third graph so that it is symmetric to the origin.

31.

32.

33.

34.

35.

36.
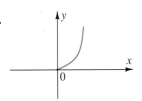

37. Which of the graphs constructed in Exercises 31–36 represent functions?

Applications

38. Landscaping If a rectangular garden is plotted on a coordinate system so that its boundary is symmetric with respect to both the x-axis and the y-axis, and the area of the portion in the third quadrant is 187 square units, what is the area of the garden?

39. Algebra Graph $x^2 + y^2 = 36$. Describe the symmetry of the graph.

40. Algebra   Graph $\dfrac{x^2}{9} + \dfrac{y^2}{16} = 1$. Describe the symmetry of the graph.

41. Architecture   The front of a building is to be symmetric with respect to a vertical line through the center of its base. The left half of a diagram of the front of the building is shown at the right. Copy the figure and complete the drawing of the front of the building.

HISTORICAL NOTE

Trigonometric concepts were developed in ancient times, often for use in the study of astronomy. However, it was not until the 17th century, when a satisfactory system of algebraic symbols had been devised, that trigonometry was thought of as a unique branch of mathematics. Listed below are some of the major contributions to the development of trigonometry.

Century	Mathematicians	Contributions
2nd, BC	Hipparchus	Related certain functions of angle measures to astronomy
	Claudius Ptolemy	Summarized in the *Almagest* the theorems known to Hipparchus Computed entries for a table of chords of circles that was roughly equivalent to a sine table
15th	Regiomontanus (Johann Müller)	Consolidated known material dealing with trigonometry Did much to establish trigonometry as a science independent of astronomy
16th	François Viète	Computed entries for more extensive trigonometric tables Obtained many of the trigonometric identities algebraically Used all six trigonometric functions to develop methods of solving plane and spherical triangles
18th	Joseph Fourier	Developed Fourier series, used to approximate values of any function

Find information on Ptolemy's *table of chords*. Explain why such a table is similar to a table of sines.

Graphs of the Sine and Cosine Functions

Objectives: To develop the properties of the sine and cosine functions
To graph the sine and cosine functions

The sine and cosine functions are defined for all real numbers, and these functions have many real-world applications.

Preview

An oscilloscope is an electronic instrument used to display changing electrical signals. It displays electrical or sound waves as patterns on a fluorescent screen. Many of the graphs shown on an oscilloscope are sine or cosine waves.

1. What is the period of the graph shown if the side of each square is one?

2. Is the graph symmetric with respect to one of the axes? If so, which one?

3. Is it symmetric with respect to the origin?

4. What is the maximum value of the function?   the minimum value?

5. How do you think an oscilloscope can be used by engineers or physicians?

Two acute angles are **complementary** if the sum of their measures is 90°. For example, 40° and 50° are complementary angles. The sine, tangent, and secant functions are associated with the **cofunctions** cosine, cotangent, and cosecant, respectively. The value of any defined trigonometric function of an acute angle is equal to the value of the cofunction of its complement.

Cofunctions

sine and cosine	tangent and cotangent	secant and cosecant
$\sin x = \cos (90° - x)$ $\cos x = \sin (90° - x)$	$\tan x = \cot (90° - x)$ $\cot x = \tan (90° - x)$	$\sec x = \csc (90° - x)$ $\csc x = \sec (90° - x)$

Thus, $\sin 30° = \cos (90° - 30°) = \cos 60° = \dfrac{1}{2}$

$\cos 30° = \sin (90° - 30°) = \sin 60° = \dfrac{\sqrt{3}}{2}$

$$\tan 30° = \cot (90° - 30°) = \cot 60° = \frac{\sqrt{3}}{3}$$

$$\cot 30° = \tan (90° - 30°) = \tan 60° = \sqrt{3}$$

$$\sec 30° = \csc (90° - 30°) = \csc 60° = \frac{2\sqrt{3}}{3}$$

$$\csc 30° = \sec (90° - 30°) = \sec 60° = 2$$

EXAMPLE 1 **Express each function in terms of its cofunction:**
  a. $\sin 47°$ **b.** $\cot 78°$

  a. $\sin 47° = \cos (90° - 47°)$ **b.** $\cot 78° = \tan (90° - 78°)$
  $\qquad\quad = \cos 43°$ $\qquad\qquad = \tan 12°$

The sine and cosine functions exhibit a number of the properties discussed in the previous lesson. They are periodic. It can be shown that one is odd and the other is even, and that one is symmetric with respect to the origin, and the other is symmetric with respect to the y-axis. Knowledge of these properties makes it easier to sketch the graphs of $y = \sin x$ and $y = \cos x$.

Recall that a function is odd if $f(-x) = -f(x)$, and even if $f(-x) = f(x)$. To determine whether the sine function is odd, even, or neither, check the values of $\sin x$ for $x = \pm\frac{\pi}{6}$ and a few other values of x.

$$\sin \frac{\pi}{6} = \frac{1}{2} \qquad \sin \left(-\frac{\pi}{6}\right) = -\frac{1}{2} \qquad \sin \frac{2\pi}{3} = \frac{\sqrt{3}}{2} \qquad \sin \left(-\frac{2\pi}{3}\right) = -\frac{\sqrt{3}}{2}$$

$$\sin \frac{4\pi}{3} = -\frac{\sqrt{3}}{2} \qquad \sin \left(-\frac{4\pi}{3}\right) = \frac{\sqrt{3}}{2} \qquad \sin \frac{11\pi}{6} = -\frac{1}{2} \qquad \sin \left(-\frac{11\pi}{6}\right) = \frac{1}{2}$$

It appears that $\sin (-\theta) = -\sin \theta$. The figure below demonstrates that $\sin \theta = \frac{y}{r}$ and $\sin (-\theta) = -\frac{y}{r} = -\sin \theta$. Thus, $\sin (-\theta) = -\sin \theta$.
The sine is an odd function, and its graph is symmetric with respect to the origin.

The figure also shows that $\cos \theta = \frac{x}{r}$ and $\cos (-\theta) = \frac{x}{r}$. Thus, $\cos (-\theta) = \cos \theta$.
The cosine is an even function, and its graph is symmetric with respect to the y-axis.

EXAMPLE 2 **Use the fact that sine is an odd function to find $\sin (-46°)$ if $\sin 46° = 0.7193$.**

$$\sin (-46°) = -\sin 46° = -0.7193$$

After one complete clockwise or counterclockwise rotation, the values of the sine and cosine function repeat, since angles with measures less than 0 or greater than 2π are coterminal with angles with measures between 0 and 2π. Thus, the sine and cosine functions are periodic and appear to have period 2π. How could you verify that 2π is the smallest value of p such that $\sin(\theta + p) = \sin\theta$ and $\cos(\theta + p) = \cos\theta$?

EXAMPLE 3 **Use the fact that the sine is a periodic function to find $\sin \dfrac{13\pi}{6}$.**

The sine function is periodic with period 2π, so $\sin(\theta + 2\pi) = \sin\theta$.

$$\sin\frac{13\pi}{6} = \sin\left(\frac{\pi}{6} + 2\pi\right) = \sin\frac{\pi}{6} = 0.5$$

Before $y = \sin x$ and $y = \cos x$ are graphed for the first time, it is helpful to construct a table of values for $0 \le x \le 2\pi$. The properties of symmetry, along with the fact that $\sin x$ and $\cos x$ have periods of 2π, can be used to extend the graphs in either direction.

When $y = \sin x$ is graphed on the traditional xy-coordinate plane, each point on the graph has coordinates of the form $(x, \sin x)$, where x is given in radians. In other words, x is used in place of θ.

EXAMPLE 4 **Sketch the graph of $y = \sin x$ over the interval $-2\pi \le x \le 2\pi$.**

To graph $y = \sin x$ over the interval $0 \le x \le 2\pi$, use radian measure and make a table of values. Round the values of $\sin x$ to the nearest hundredth.

x	0	$\dfrac{\pi}{6}$	$\dfrac{\pi}{4}$	$\dfrac{\pi}{3}$	$\dfrac{\pi}{2}$	$\dfrac{2\pi}{3}$	$\dfrac{3\pi}{4}$	$\dfrac{5\pi}{6}$	
$\sin x$	0	0.5	0.71	0.87	1	0.87	0.71	0.5	

x	π	$\dfrac{7\pi}{6}$	$\dfrac{5\pi}{4}$	$\dfrac{4\pi}{3}$	$\dfrac{3\pi}{2}$	$\dfrac{5\pi}{3}$	$\dfrac{7\pi}{4}$	$\dfrac{11\pi}{6}$	2π
$\sin x$	0	-0.5	-0.71	-0.87	-1	-0.87	-0.71	-0.5	0

Plot the points and sketch the curve over the interval $0 \le x \le 2\pi$. To sketch the curve over the interval $-2\pi \le x \le 0$, use the facts that $y = \sin x$ has a period of 2π and that the graph is symmetric with respect to the origin.

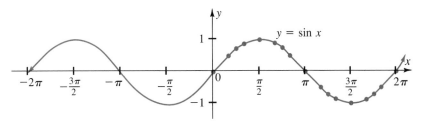

The graph of $y = \cos x$ can be constructed in a similar manner.

EXAMPLE 5   **Sketch the graph of $y = \cos x$ over the interval $-2\pi \leq x \leq 2\pi$.**

Begin by constructing a table of values for the interval $0 \leq x \leq 2\pi$.

x	0	$\dfrac{\pi}{6}$	$\dfrac{\pi}{4}$	$\dfrac{\pi}{3}$	$\dfrac{\pi}{2}$	$\dfrac{2\pi}{3}$	$\dfrac{3\pi}{4}$	$\dfrac{5\pi}{6}$	
$\cos x$	1	0.87	0.71	0.5	0	-0.5	-0.71	-0.87	

x	π	$\dfrac{7\pi}{6}$	$\dfrac{5\pi}{4}$	$\dfrac{4\pi}{3}$	$\dfrac{3\pi}{2}$	$\dfrac{5\pi}{3}$	$\dfrac{7\pi}{4}$	$\dfrac{11\pi}{6}$	2π
$\cos x$	-1	-0.87	-0.71	-0.5	0	0.5	0.71	0.87	1

Plot the points and sketch the curve over the interval $0 \leq x \leq 2\pi$. To extend the curve over the interval $-2\pi \leq x \leq 0$, use the facts that $y = \cos x$ has a period of 2π and that the graph is symmetric with respect to the y-axis.

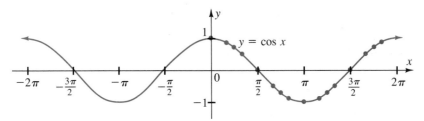

A scientific calculator can be used to find specified values of the functions being graphed. Graphing calculators or computers can be used to graph trigonometric functions. Their displays are helpful when you are checking the accuracy of graphs that you have sketched.

CLASS EXERCISES

Use properties of the sine and cosine functions to match each value in the column on the left with one on the right.

1. $\cos 65°$

2. $\cos (-25°)$

3. $\cos 75°$

4. $\sin 35°$

5. $\cos (-35°)$

6. $-\cos (-15°)$

7. $\sin (-25°)$

8. $-\sin 55°$

a. $\cos 25°$

b. $\sin (-55°)$

c. $\sin (-75°)$

d. $\sin 25°$

e. $\sin 15°$

f. $\sin 55°$

g. $\cos 55°$

h. $-\cos 65°$

PRACTICE EXERCISES

Express each function in terms of its cofunction.

1. $\sin 28°$ **2.** $\tan 32°$ **3.** $\csc 85°$ **4.** $\cos 14°$

5. $\sec 27°$ **6.** $\sec 10°$ **7.** $\cot 53°$ **8.** $\csc 42°$

Use the fact that the sine is an odd function and the cosine is an even function to determine the following values.

9. If $\sin 32° = 0.5299$, find $\sin(-32°)$.

10. If $\sin 129° = 0.7771$, find $\sin(-129°)$.

11. If $\cos 48° = 0.6691$, find $\cos(-48°)$.

12. If $\cos 117° = -0.4540$, find $\cos(-117°)$.

13. If $\sin 238° = -0.8480$, find $\sin(-238°)$.

14. If $\sin 306° = -0.8090$, find $\sin(-306°)$.

Use the fact that the sine and cosine are periodic functions to determine the following values.

15. If $\sin 54° = 0.8090$, find $\sin 414°$.

16. If $\sin 134° = 0.7193$, find $\sin 494°$.

17. If $\cos 62° = 0.4695$, find $\cos 422°$.

18. If $\cos 153° = -0.8910$, find $\cos 513°$.

19. If $\sin 222° = -0.6691$, find $\sin 582°$.

20. If $\sin 312° = -0.7431$, find $\sin 672°$.

Graph each function over the given interval.

21. $y = \sin x,\ 2\pi \leq x \leq 4\pi$ **22.** $y = \cos x,\ 2\pi \leq x \leq 4\pi$

23. $y = \cos x,\ -6\pi \leq x \leq 0$ **24.** $y = \sin x,\ -6\pi \leq x \leq 0$

Use the fact that the sine and cosine are periodic functions, and your knowledge of cofunctions, to determine the following values.

25. If $\cos 68° = 0.3746$, find $\sin 22°$.

26. If $\cos 54° = 0.5878$, find $\sin 36°$.

27. If $\sin 27° = 0.4540$, find $\cos 63°$.

28. If $\cos 74° = 0.2756$, find $\sin 16°$.

29. If $\sin 141° = 0.6293$, find $\sin 861°$.

30. If $\cos 123° = -0.5446$, find $\cos 843°$.

31. If $\sin 227° = -0.7314$, find $\sin 947°$.

32. If $\sin 34° = 0.5592$, find $\sin (-326°)$.

33. If $\sin 139° = 0.6561$, find $\sin (-221°)$.

34. If $\cos 162° = -0.9511$, find $\cos (-198°)$.

Graph each function over the interval $-2\pi \le x \le 2\pi$.

35. $y = \sin x + 2$ **36.** $y = \cos x - 3$ **37.** $y = 2 \cos x$

38. $y = 3 \sin x$ **39.** $y = \sin 2x$ **40.** $y = \cos 2x$

Applications

41. Algebra Sketch the graphs of $y = x$ and $y = \sin x$ on the same coordinate plane. Find approximate solutions to $x = \sin x$, where x is in radians.

42. Algebra Sketch the graphs of $y = x$ and $y = \cos x$ on the same coordinate plane. Find approximate solutions to $x = \cos x$, where x is in radians.

CAREER

Air traffic control specialists are guardians of the airways. They keep track of planes in the airspace surrounding an airport, monitor them to ensure a safe and orderly flow of traffic, and provide pilots with take off and landing directions as well as weather information.

Individuals must have the ability to analyze what is shown on the radar screen, make predictions, and be prepared to make split-second decisions. Thus, they must exercise common sense, patience, and emotional control, and be able to enunciate clearly. In addition, they must pass a physical exam and possess 20/20 vision or vision that can be corrected to 20/20.

Educational requirements include a degree in air traffic control, at least 3 years of college, 2 years of college combined with work experience, or military experience. They must also train for $2\frac{1}{2}$ months in Oklahoma City.

Amplitude and Period

2.3

Objectives: To find the amplitude and period of a trigonometric
function from its equation
To graph sine and cosine functions with various amplitudes
and periods

Amplitude and frequency are important concepts that are essential to the
understanding of many fields, including music.

Preview

The loudness, or *amplitude*, of a musical sound is a comparative measure of
the strength of the sound that is heard. Loudness is related to the distance from
the source and the energy of the vibration.

Pitch is determined by the number of vibrations per second, or the *frequency* of
the vibrations that yield a sound wave. Frequency is the reciprocal of the
period. For example, if the period of a sound wave is 0.01 second, then the
frequency is $\frac{1}{0.01}$, or 100 vibrations per second.

1. The period of a sound wave is 0.002 seconds. What is the frequency?

2. The lowest frequency of the human voice is about 60 vibrations per second.
   What is the period of a sound wave with this frequency?

3. The highest frequency of the human voice is about 1300 vibrations per
   second. What is the period of a sound wave with this frequency?

4. The periods of sound waves produced by a piano range from about
   0.0002416 to 0.03676 seconds. What is the range of the frequencies?

Several observations can be made about the basic sine and cosine functions,
$y = \sin x$ and $y = \cos x$.

- Their graphs are *continuous*; that is, they have no breaks.
- They each have a period of 2π; that is, $\sin (x + 2\pi) = \sin x$ and
  $\cos (x + 2\pi) = \cos x$ for all real values of x.
- Their domains are the set of all real numbers.
- For all values of x, $-1 \leq \sin x \leq 1$ and $-1 \leq \cos x \leq 1$. Thus, the range
  of both functions is the set of all real numbers from -1 through 1.

The maximum value of each function is 1, and the minimum value is -1. The
functions are said to be *bounded* by these values. The **amplitude** of a periodic
function is defined to be half the difference of the maximum value M and the

minimum value m. Thus, the amplitude is equal to $\frac{1}{2}(M - m)$. The amplitude is related to the height of the graph.

EXAMPLE 1 **Sketch the graphs of $y = \cos x$, $y = -2 \cos x$, and $y = \frac{1}{2} \cos x$ on the same coordinate plane. Then determine their amplitudes.**

x	0	$\dfrac{\pi}{6}$	$\dfrac{\pi}{3}$	$\dfrac{\pi}{2}$	$\dfrac{2\pi}{3}$	$\dfrac{5\pi}{6}$	π	$\dfrac{3\pi}{2}$	2π
$\cos x$	1	0.87	0.5	0	-0.5	-0.87	-1	0	1
$-2 \cos x$	-2	-1.73	-1	0	1	1.73	2	0	-2
$\frac{1}{2} \cos x$	0.5	0.43	0.25	0	-0.25	-0.43	-0.5	0	0.5

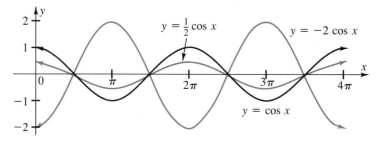

For $y = \cos x$, $M = 1$ and $m = -1$, so the amplitude is $\frac{1}{2}[1 - (-1)] = 1$.

For $y = -2 \cos x$, $M = 2$ and $m = -2$, so the amplitude is $\frac{1}{2}[2 - (-2)] = 2$.

For $y = \frac{1}{2} \cos x$, $M = \frac{1}{2}$ and $m = -\frac{1}{2}$, so the amplitude is $\frac{1}{2}\left[\frac{1}{2} - \left(-\frac{1}{2}\right)\right] = \frac{1}{2}$.

The amplitude of each of the functions in Example 1 is equal to the absolute value of a, the coefficient of $\cos x$. This is, in fact, the case for all functions of the form $y = a \sin bx$ or $y = a \cos bx$. Since the maximum value of the sine or cosine is 1 and the minimum value is -1, the maximum and minimum values of $y = a \sin bx$ and $y = a \cos bx$ are $|a|$ and $-|a|$, respectively. Thus, the amplitude of $y = a \sin bx$ and $y = a \cos bx$ is $\frac{1}{2}[|a| - (-|a|)] = \frac{1}{2}(2|a|) = |a|$.

Note that the graph of $y = -a \cos x$ is the reflection of the graph of $y = a \cos x$ about the x-axis. This is illustrated at the right for $y = -\cos x$ and $y = \cos x$. Similarly, the graph of $y = -a \sin x$ is the reflection of the graph of $y = a \sin x$ about the x-axis.

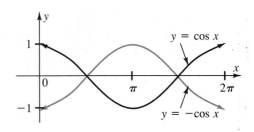

Sometimes it is convenient to use different scales on the x- and y-axes of a trigonometric graph. This is illustrated in Example 1 and in later examples.

EXAMPLE 2 **Sketch the graphs of $y = \sin x$ and $y = \sin 2x$ on the same coordinate plane.**

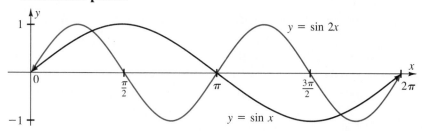

Example 2 shows that $y = \sin x$ has a period of 2π, but $y = \sin 2x$ has a period of π. That is, the period of $y = \sin 2x$ is half that of $y = \sin x$. In general, the period of a function of the form $y = a \sin bx$ or $y = a \cos bx$, where $b \neq 0$, is $\dfrac{2\pi}{|b|}$.

EXAMPLE 3 **Find an equation of the graph shown below.**

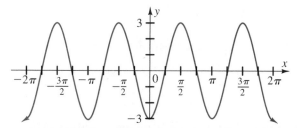

First determine whether the graph represents a sine or cosine function. Note that the curve has y-axis symmetry and when $x = 0$, $y \neq 0$. These two facts are true for cosine graphs but not for sine graphs. Therefore, use $y = a \cos bx$.

Find the period: $\dfrac{2\pi}{|b|} = \pi$, so $|b| = 2$ and $b = \pm 2$. Choose $b = 2$, so $y = a \cos 2x$.

Find the amplitude: $|a| = 3$, so $a = \pm 3$. From the graph, if $x = 0$, $y = -3$. Therefore, $-3 = a \cos 2(0) = a \cos 0 = a \cdot 1 = a$.

Thus, an equation of the graph is $y = -3 \cos 2x$.

For $y = a \sin bx$ and $y = a \cos bx$

The amplitude is $|a|$.   The period is $\dfrac{2\pi}{|b|}$.

CLASS EXERCISES

Determine the amplitude and the period of each function.

1. $y = -\frac{1}{2}\cos x$

2. $y = -\sin x$

3. $y = 2\cos 6x$

4. $y = \frac{3}{2}\cos\frac{1}{2}x$

5. $y = -6\sin(-4x)$

6. $y = -\frac{1}{5}\cos x$

7. $y = \sin 3x$

8. $y = \cos 2x$

9. $y = \cos\left(-\frac{1}{2}x\right)$

10. $y = 3\sin\frac{3}{2}x$

11. $y = -\cos x$

12. $y = -4\sin\frac{1}{4}x$

PRACTICE EXERCISES

Determine the amplitude and period of each function.

1. $y = \sin 4x$

2. $y = \cos 5x$

3. $y = \sin x$

4. $y = \cos 6x$

5. $y = 3\sin x$

6. $y = 2\cos x$

7. $y = 4\cos x$

8. $y = -2\sin x$

9. $y = \sin(-2x)$

10. $y = 2\sin(-4x)$

11. $y = -4\cos 5x$

12. $y = 3\cos(-2x)$

13. $y = 4\cos\frac{1}{2}x$

14. $y = 3\sin\frac{2}{3}x$

15. $y = 2\cos\frac{3}{2}x$

16. $y = 5\sin\frac{5}{3}x$

17. $y = -2\cos\frac{5}{4}x$

18. $y = -3\sin\frac{3}{5}x$

Give the amplitude and period of each function graphed below. Then write an equation of each graph.

19.

20.

21.

22.

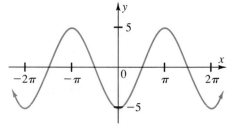

Give the amplitude and period of each function. Then sketch the graph of the function over the interval $-2\pi \le x \le 2\pi$.

23. $y = 3 \sin x$ **24.** $y = 2 \cos x$ **25.** $y = \sin 2x$

26. $y = \cos 2x$ **27.** $y = 3 \sin 2x$ **28.** $y = 4 \cos 2x$

29. $y = 3 \cos \frac{1}{2} x$ **30.** $y = 2 \sin \frac{1}{3} x$ **31.** $y = 2 \cos \frac{3}{2} x$

32. $y = 3 \sin \frac{2}{3} x$ **33.** $y = -3 \cos x$ **34.** $y = -2 \sin x$

35. $y = \cos (-x)$ **36.** $y = \sin (-x)$

37. $y = -2 \sin (-2x)$ **38.** $y = -3 \cos (-3x)$

39. Write four equations of the form $y = a \cos bx$ with amplitude 5 and period 3.

40. Write four equations of the form $y = a \sin bx$ with amplitude $\frac{2}{3}$ and period $\frac{3}{2} \pi$.

Applications

Physics   The *frequency* of a periodic function is defined to be the reciprocal of the period.

41. Find the frequency of the function $y = \cos 2x$.

42. Find the frequency of the function $y = -4 \sin \frac{1}{4} x$.

43. The voltage V in a particular electrical circuit is given by the formula $V = 6 \cos 90\pi t$, where t is measured in seconds. Find the amplitude, the period, and the frequency of this function.

CHALLENGE

In how many ways can *TRIGONOMETRY* be spelled using adjacent letters in this triangle? One possible solution is shown in red.
Hint: Look for a pattern.

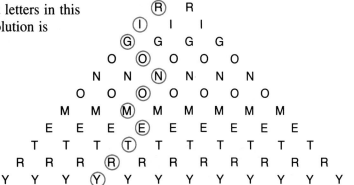

Phase Shift and Vertical Shift

2.4

Objectives: To find the phase shift and vertical shift of sine and cosine functions from their equations
To graph sine and cosine functions with various phase shifts and vertical shifts

Before shifts in the graphs of the sine and cosine functions are introduced, it will be helpful to review shifts in the graphs of quadratic functions.

Preview

The graph of an equation of the form $y = (x + a)^2 + b$ is the same shape as that of $y = x^2$. However, it is shifted a units to the left if $a > 0$, $|a|$ units to the right if $a < 0$, b units up if $b > 0$, and $|b|$ units down if $b < 0$.

To illustrate, the graph of $y = (x - 1)^2 + 3$ is 1 unit to the right of the graph of $y = x^2$ and 3 units above it.

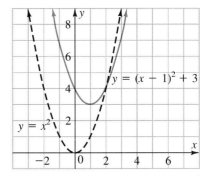

State how the graph of each equation is shifted from the graph of $y = x^2$. Check by graphing.

1. $y = (x + 3)^2$ **2.** $y = x^2 - 4$ **3.** $y = (x + 2)^2 - 1$ **4.** $y = (x - 4)^2 + 5$

Consider the graphs of the sine and cosine shown on the same coordinate plane. If the graph of the cosine were shifted to the right $\frac{\pi}{2}$ units, it would coincide with the graph of the sine. This relationship implies that $\sin x = \cos\left(x - \frac{\pi}{2}\right)$. The distance that one graph must be shifted to the right or to the left to coincide with another graph is called the **phase shift**.

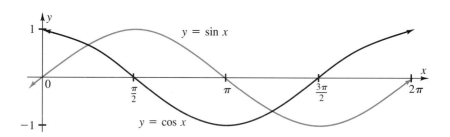

EXAMPLE 1   **Sketch the graph of $y = \sin\left(x - \dfrac{\pi}{4}\right)$.**

Construct a table of values, rounding to the nearest hundredth. Then graph the points and join them with a smooth curve.

x	0	$\dfrac{\pi}{4}$	$\dfrac{\pi}{2}$	$\dfrac{3\pi}{4}$	π	$\dfrac{5\pi}{4}$	$\dfrac{3\pi}{2}$	$\dfrac{7\pi}{4}$	2π	$\dfrac{9\pi}{4}$
$x - \dfrac{\pi}{4}$	$-\dfrac{\pi}{4}$	0	$\dfrac{\pi}{4}$	$\dfrac{\pi}{2}$	$\dfrac{3\pi}{4}$	π	$\dfrac{5\pi}{4}$	$\dfrac{3\pi}{2}$	$\dfrac{7\pi}{4}$	2π
$\sin\left(x - \dfrac{\pi}{4}\right)$	-0.71	0	0.71	1	0.71	0	-0.71	-1	-0.71	0

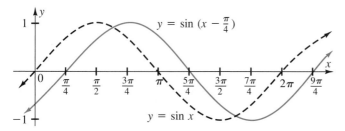

The graph of $y = \sin\left(x - \dfrac{\pi}{4}\right)$ is the same as that of $y = \sin x$, except that

it is shifted $\dfrac{\pi}{4}$ units to the right. The phase shift of $y = \sin\left(x - \dfrac{\pi}{4}\right)$ is $\dfrac{\pi}{4}$, the

period is 2π, and the amplitude is 1.

In general, the graph of a function of the form $y = a \sin b(x + c)$ or $y = a \cos b(x + c)$ is the same as that of $y = a \sin bx$ or $y = a \cos bx$, but it is shifted c units to the left if $c > 0$, and $|c|$ units to the right if $c < 0$. The period of a function of the form $y = a \sin b(x + c)$ or $y = a \cos b(x + c)$ is $\dfrac{2\pi}{|b|}$, and the amplitude is $|a|$.

EXAMPLE 2   **Determine the amplitude, period, and phase shift of**
$$y = 3 \cos\left(x + \frac{\pi}{6}\right).$$ **Then sketch its graph.**

Amplitude $= |3| = 3$ *Amplitude* $= |a|$

Period $= \dfrac{2\pi}{|1|} = 2\pi$ *Period* $= \dfrac{2\pi}{|b|}$

Phase shift $= \left|\dfrac{\pi}{6}\right|$

$= \dfrac{\pi}{6}$ units to the left, since $c > 0$

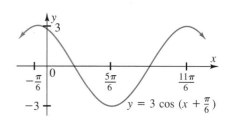

In Example 3, the equation must be factored before the phase shift can be determined.

EXAMPLE 3   **Determine the amplitude, period, and phase shift of $y = \sin(2x + \pi)$. Then sketch its graph.**

$y = \sin(2x + \pi)$

$y = \sin 2\left(x + \dfrac{\pi}{2}\right)$ *Factor.*

Amplitude $= |1| = 1$ *Amplitude $= |a|$*

Period $= \dfrac{2\pi}{|2|} = \pi$ *Period $= \dfrac{2\pi}{|b|}$*

Phase shift $= \left|\dfrac{\pi}{2}\right| = \dfrac{\pi}{2}$ units to the left, since $c > 0$

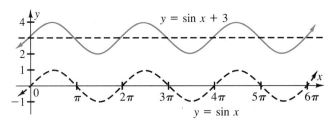

Phase shifts are horizontal shifts. However, if a constant term d is added to the right side of an equation of the form $y = a \sin b(x + c)$ or $y = a \cos b(x + c)$, a **vertical shift** results. You can verify graphs using a graphing calculator or computer.

EXAMPLE 4   **Sketch each graph: a. $y = \sin x + 3$ b. $y = \cos x - 2$**

a. The graph of $y = \sin x$ is shifted up 3 units.

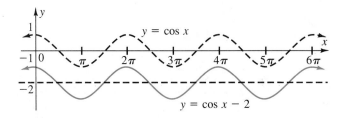

b. The graph of $y = \cos x$ is shifted down 2 units.

In general, the graphs of $y = a \sin b(x + c) + d$ and $y = a \cos b(x + c) + d$ are shifted d units upward from those of $y = a \sin bx$ and $y = a \cos bx$ if $d > 0$, and $|d|$ units downward if $d < 0$. Combinations of phase shifts and vertical shifts also occur, as illustrated in Example 5.

EXAMPLE 5 Determine the amplitude, period, phase shift, and vertical shift for the graph of $y = \sin\left(x - \dfrac{\pi}{4}\right) + 2$. Then sketch the graph.

Amplitude $= |1| = 1$ Amplitude $= |a|$

Period $= \dfrac{2\pi}{|1|} = 2\pi$ Period $= \dfrac{2\pi}{|b|}$

Phase shift $= \left|-\dfrac{\pi}{4}\right| = \dfrac{\pi}{4}$ units to the right

Vertical shift $= |2| = 2$ units up

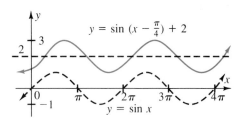

For $y = a \sin b(x + c) + d$ and $y = a \cos b(x + c) + d$

The amplitude is $|a|$.   The period is $\dfrac{2\pi}{|b|}$.

The phase shift from $y = a \sin bx$ or $y = \cos bx$ is c units to the left if $c > 0$, and $|c|$ units to the right if $c < 0$.
The vertical shift from $y = a \sin bx$ or $y = a \cos bx$ is d units upward if $d > 0$, and $|d|$ units downward if $d < 0$.

EXAMPLE 6 Sketch the graph of $y = 3 \cos\left(\dfrac{1}{2}x - \dfrac{\pi}{2}\right) - 2$.

$y = 3 \cos\left(\dfrac{1}{2}x - \dfrac{\pi}{2}\right) - 2$

$y = 3 \cos \dfrac{1}{2}(x - \pi) - 2$

Amplitude $= |3| = 3$

Period $= \dfrac{2\pi}{\left|\dfrac{1}{2}\right|} = 4\pi$

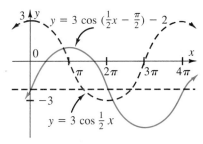

Phase shift $= |-\pi| = \pi$ units to the right
Vertical shift $= |-2| = 2$ units down

CLASS EXERCISES

Determine the amplitude, period, phase shift, and vertical shift for each.

1. $y = 3 \sin 2x$

2. $y = \cos(x - \pi)$

3. $y = 2 \sin 3x - 5$

4. $y = \cos 2x + 2$

5. $y = 4 \sin(2x + 2\pi)$

6. $y = 2 \cos \dfrac{\pi}{2}x - 6$

7. $y = 4 \sin\left(2x - \dfrac{\pi}{2}\right) - 3$

8. $y = -2 \sin\left(3x - \dfrac{3\pi}{4}\right) + 2$

PRACTICE EXERCISES

Determine the amplitude, period, phase shift, and vertical shift for each.

1. $y = 2 \sin 3x$

2. $y = \sin (x - \pi)$

3. $y = 3 \cos 4x$

4. $y = 3 \sin 6x - 3$

5. $y = \cos 2x - 5$

6. $y = 2 \sin (3x + 3\pi)$

7. $y = \frac{1}{4} \sin 2x$

8. $y = 3 \cos \frac{1}{2}x + 4$

9. $y = 4 \sin \left(3x - \frac{\pi}{3} \right) + 2$

10. $y = -2 \cos \left(3x - \frac{4\pi}{3} \right) - 3$

Sketch the graph of each function over the interval $0 \le x \le 4\pi$.

11. $y = \sin 2x + 3$

12. $y = \cos (x - \pi)$

13. $y = \cos 2x - 1$

14. $y = 3 \sin x - 1$

15. $y = \sin (x - \pi)$

16. $y = \sin (x - \pi) - 1$

17. $y = \cos 3x - 5$

18. $y = 2 \sin (2x + 2\pi)$

19. $y = \cos 2(x - \pi)$

20. $y = \sin \frac{1}{2} (x - \pi)$

21. $y = \frac{1}{2} \sin 2x$

22. $y = 2 \cos \frac{1}{2}x + 4$

23. $y = 3 \sin \left(2x - \frac{\pi}{2} \right) + 1$

24. $y = -2 \cos \left(3x - \frac{4\pi}{3} \right) - 2$

25. Find an equation for a sine function which has an amplitude of 4, a period of 180°, and a y-intercept of -3.

26. Find an equation for a sine function which has an amplitude of 3, a period of 90°, and a y-intercept of 2.

27. Find an equation for a cosine function which has an amplitude of $\frac{3}{5}$, a period of 270°, and y-intercept of 5.

28. Find an equation for a cosine function which has an amplitude of $\frac{3}{2}$, a period of 90°, and a y-intercept of -4.

Applications

Music   Sound waves can be represented by sine or cosine functions.

29. A tuning fork for G (below middle C) has a frequency of 196 vibrations per second $\left(\text{frequency} = \dfrac{1}{\text{period}} \right)$. If the amplitude is 0.04 mm, find an equation for the resulting sound wave.

30. The vibrations may not be exactly in phase when two instruments are played together. Suppose the sound wave from one instrument can be represented by $y_1 = 3 \sin (3\pi x)$, and the sound wave from another instrument can be represented by $y_2 = 3 \sin \left(3\pi x + \dfrac{3\pi}{2} \right)$. Sketch the graphs of both.

31. A musical sound wave is represented by $y = 0.004 \sin (300\pi x + 150\pi^2)$. Determine its amplitude, period, and phase shift. Sketch the graph of the sound wave for $0 \le x \le 0.01$, letting x increase in steps of 0.0005 radians.

TEST YOURSELF

Determine whether each function is even, odd, or neither.

1. $f(x) = 2x^2$ **2.** $f(x) = 3x^3 + 1$ **3.** $f(x) = x^7$ 2.1

Determine whether each graph is symmetric with respect to the x-axis, the y-axis, the origin, or to none of these.

4.   **5.**   **6.**

Use the fact that the sine is an odd function and the cosine is an even function to determine the following values.

7. If sine $121° = 0.8572$, find $\sin (-121°)$. 2.2

8. If $\cos 121° = -0.5150$, find $\cos (-121°)$.

Determine the amplitude and period of each function.

9. $y = -5 \cos x$ **10.** $y = \sin \frac{3}{4} x$ 2.3

11. $y = 4 \cos \frac{1}{3} x$ **12.** $y = \frac{1}{2} \sin (-2x)$

Determine the amplitude, period, phase shift, and vertical shift of each.

13. $y = 2 \cos \left(x - \dfrac{\pi}{2} \right) + 5$ **14.** $y = \dfrac{1}{2} \cos \left(x + \dfrac{\pi}{2} \right) - 7$

15. $y = \sin (2x + 3\pi) - 1$ **16.** $y = \sin (3x + \pi) + 2$

Sketch the graph of each function over the interval $0 \le x \le 4\pi$.

17. $y = 2 \sin \left(x - \dfrac{\pi}{2} \right) + 1$ **18.** $y = 3 \cos \left(x + \dfrac{\pi}{4} \right) - 2$

2.5 Graphing by Addition of Ordinates

Objectives: To graph functions that are the sums or differences of two sine functions, two cosine functions, or a sine function and a cosine function

The function $y = 2x + 4$ is actually the sum of two functions, $y = 2x$ and $y = 4$. For every value in the domain, there is a corresponding value in the range so that the point $(x, 2x + 4)$ is on the graph of $y = 2x + 4$. The sums of trigonometric functions can also be graphed.

Preview

If two strings of a piano have the same frequency, they are in tune with each other. If they are off by even a small number of vibrations per second, the sound has *beats*. The figure at the right shows the graphs of $y_1 = \sin(\pi x)$, $y_2 = \sin(2\pi x)$, and their sum,

$$y = \sin(\pi x) + \sin(2\pi x)$$

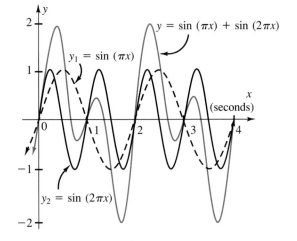

1. How many cycles does y_1 complete in four seconds?

2. How many cycles does y_2 complete in four seconds?

3. The graphs of y_1 and y_2 coincide at $x = 0$ and $x = 2$. Then both graphs rise for a fraction of a second. What number does the amplitude of y approach during this time?

On page 80, it was shown that the graph of $y = \sin x + 3$ is the same as the graph of $y = \sin x$ shifted up 3 units. In other words, each value of the function $y = \sin x + 3$ is the sum of the values $y = \sin x$ and $y = 3$. Similarly, the graph of $y = \cos x - 2$ is the same as the graph of $y = \cos x$ shifted down 2 units. That is, each value of the function $y = \cos x - 2$ is the sum of the values $y = \cos x$ and $y = -2$.

 A function that is the sum of the two trigonometric functions can be sketched by first graphing the two functions and then adding the two y-values for each value of x. A graphing calculator is useful for checking the results.

EXAMPLE 1 **Graph $y = \sin x + \cos x$ over the interval $0 \le x \le 4\pi$.**

First, sketch the graphs of $y = \sin x$ and $y = \cos x$.

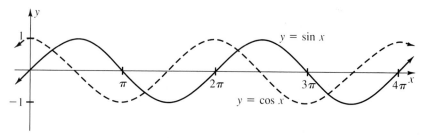

For each value of x, add the two values of y and sketch their sum.

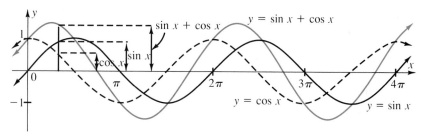

The y-coordinates of a point are sometimes called *ordinates*, and the method illustrated in Example 1 is called graphing by **addition of ordinates.** Check the graph using a graphing calculator or computer.

EXAMPLE 2 **Graph $y = 2 \sin x + \sin \frac{1}{2} x$ over the interval $0 \le x \le 4\pi$.**

First, sketch the graphs of $y = 2 \sin x$ and $y = \sin \frac{1}{2} x$.
Then, for each value of x, add the two values of y to sketch
$y = 2 \sin x + \sin \frac{1}{2} x$.

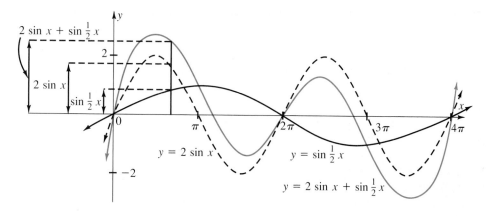

CLASS EXERCISES

1. Complete the table below for the function $y = \sin x + 2 \sin x$.

x	0	$\frac{\pi}{6}$	$\frac{\pi}{4}$	$\frac{\pi}{3}$	$\frac{\pi}{2}$	$\frac{2\pi}{3}$	$\frac{3\pi}{4}$	$\frac{5\pi}{6}$	π
$\sin x$									
$2 \sin x$									
$\sin x + 2 \sin x$									

2. Complete the table below for the function $y = \cos x + \cos 2x$.

x	0	$\frac{\pi}{6}$	$\frac{\pi}{4}$	$\frac{\pi}{3}$	$\frac{\pi}{2}$	$\frac{2\pi}{3}$	$\frac{3\pi}{4}$	$\frac{5\pi}{6}$	π
$\cos x$									
$2x$									
$\cos 2x$									
$\cos x + \cos 2x$									

PRACTICE EXERCISES   Use appropriate technology.

Graph each by addition of ordinates over at least a one-period interval.

1. $y = 2 \sin x + \sin x$
2. $y = 3 \sin x + \sin x$
3. $y = \sin x + 2 \cos x$
4. $y = 2 \sin x + \cos x$
5. $y = 3 \sin x + \sin 2x$
6. $y = 4 \sin x + \sin 2x$
7. $y = 2 \sin x - \cos x$
8. $y = 2 \cos x - \sin x$
9. $y = \sin 2x + \cos 2x$
10. $y = \sin 3x + \cos 3x$
11. $y = \sin 2x - \cos 2x$
12. $y = \cos 2x - \sin 2x$
13. $y = \sin 3x + \cos 2x$
14. $y = \sin x + \cos 2x$
15. $y = -2 \sin x - \cos x$
16. $y = -\sin x - 2 \cos x$
17. $y = \cos \left(x - \frac{\pi}{4}\right) + \sin \left(x - \frac{\pi}{4}\right)$
18. $y = \sin \left(x + \frac{\pi}{3}\right) + \cos \left(x + \frac{\pi}{2}\right)$
19. $y = \sin x + \cos x + \sin 2x$
20. $y = 2 \sin x + \sin \frac{1}{2}x + \sin 2x$

Applications

21. **Music** Two strings of a piano are out of tune. Their sound waves can be represented by $y_1 = \sin 2\pi x$ and $y_2 = \sin 3\pi x$. Graph $y_1 = \sin 2\pi x$ and $y_2 = \sin 3\pi x$. Then graph $y = \sin 2\pi x + \sin 3\pi x$ on the same plane.

22. Oceanography The graph of a tide function can be represented by the equation $y = 2 \sin\left(x - \dfrac{\pi}{4}\right) + 5 \sin x$. Graph the function.

TRIGONOMETRY AND THE CALENDAR

The table below lists times of sunrise and sunset at Lancaster, Pennsylvania, which is located at approximately 40°N latitude, on various dates during a year. The number of hours of daylight is shown in the last column.

Date	Sunrise (AM)	Sunset (PM)	Hours of Daylight	Date	Sunrise (AM)	Sunset (PM)	Hours of Daylight
Jan 1	7:22	4:44	9.37	Jul 13	4:42	7:29	14.78
15	7:20	4:58	9.63	30	4:57	7:16	14.32
30	7:11	5:16	10.08	Aug 15	5:12	6:56	13.73
Feb 15	6:54	5:35	10.68	29	5:25	6:36	13.18
28	6:36	5:50	11.23	Sep 15	5:41	6:09	12.47
Mar 15	6:11	6:07	11.93	30	5:55	5:44	11.28
30	5:48	6:22	12.57	Oct 15	6:10	5:20	11.17
Apr 14	5:24	6:38	13.23	30	6:27	5:06	10.65
29	5:03	6:53	13.83	Nov 15	6:45	4:43	9.97
May 14	4:46	7:08	14.37	30	7:02	4:36	9.57
29	4:34	7:20	14.77	Dec 15	7:15	4:36	9.35
Jun 15	4:30	7:31	15.02	30	7:21	4:44	9.38
30	4:34	7:33	14.98				

A graph of the hours of daylight is shown in red below. The graph is periodic and approximates a sine curve.

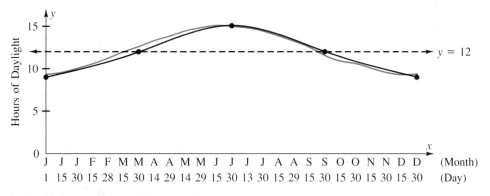

1. Find the equation of the sine function shown by the black curve.
2. Find the times of sunrise and sunset for the area in which you live. (You must know the approximate latitude of your area to find the times.) Make a table and a graph similar to those shown above.

<table>
<tr><td>**2.6**</td><td># Graphs of the Tangent and Cotangent Functions</td></tr>
</table>

Objectives: To determine the properties of the tangent and cotangent functions

To graph the tangent and cotangent functions

The tangent and cotangent functions are not defined for all real numbers, and their graphs are not like the graphs of the sine and cosine functions.

Preview

The graph of a hyperbola is discontinuous, and it has two **asymptotes,** which are lines that the graph approaches but never crosses as it gets infinitely far from the origin. The asymptotes of a hyperbola with an equation of the form

$$\frac{x^2}{a^2} - \frac{y^2}{b^2} = 1 \text{ have equations of the form}$$

$$y = \frac{b}{a}x \quad \text{and} \quad y = -\frac{b}{a}x$$

Similarly, the asymptotes of a hyperbola with an equation of the form $\dfrac{y^2}{a^2} - \dfrac{x^2}{b^2} = 1$

have equations of the form

$$y = \frac{a}{b}x \quad \text{and} \quad y = -\frac{a}{b}x$$

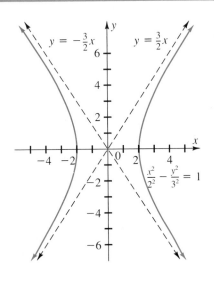

Find the equations of the asymptotes of each hyperbola. Then sketch the graph of the hyperbola.

1. $\dfrac{x^2}{5^2} - \dfrac{y^2}{2^2} = 1$

2. $\dfrac{y^2}{3^2} - \dfrac{x^2}{4^2} = 1$

3. $x^2 - \dfrac{y^2}{36} = 1$

In order to graph the tangent and cotangent functions, it will be useful to construct a table of values. Recall that the tangent is not defined for $\pm\dfrac{\pi}{2}, \pm\dfrac{3\pi}{2}, \ldots$. Near these values of x, the value of tan x increases or decreases without bound. Consider the values of tan x as x approaches $\dfrac{\pi}{2}$ from 0. Round answers to the nearest hundredth.

x	0	$\dfrac{\pi}{6}$	$\dfrac{\pi}{4}$	$\dfrac{\pi}{3}$	$\dfrac{\pi}{2.1}$	$\dfrac{\pi}{2.01}$	$\dfrac{\pi}{2.001}$	$\dfrac{\pi}{2}$
$\tan x$	0	0.58	1	1.73	13.34	127.96	1273.88	undefined

The value of $\tan x$ increases without bound as x approaches $\dfrac{\pi}{2}$ from the left. The tangent of $\dfrac{\pi}{2}$ is not defined. Similarly, as x decreases from π to $\dfrac{\pi}{2}$, $\tan x$ decreases without bound. This can be seen by reading the table below from right to left.

x	$\dfrac{\pi}{2}$	$\dfrac{\pi}{1.999}$	$\dfrac{\pi}{1.99}$	$\dfrac{\pi}{1.9}$	$\dfrac{2\pi}{3}$	$\dfrac{3\pi}{4}$	$\dfrac{5\pi}{6}$	π
$\tan x$	undefined	-1272.6	-126.68	-12.07	-1.73	-1.00	-0.58	0

After x increases from 0 to π, the values of $\tan x$ begin to repeat. That is, $\tan (x + \pi) = \tan x$. Thus, the period of the tangent function is π.

EXAMPLE 1 **Graph $y = \tan x$ over the interval $-\pi \leq x \leq 2\pi$.**

Use the tables of values shown above to find the coordinates of the points over the interval $0 \leq x \leq \pi$, and sketch the curve over that interval. Then use the fact that the period of the function is π to extend the graph in both directions.

 You can check your results with a graphing utility.

Tan x is undefined for some values of x. Therefore, the tangent function is said to be **discontinuous,** since its graph has breaks. *Asymptotes* occur at these breaks, and the corresponding x-values are not in the domain of the function.

The asymptotes are lines with equations of the form $x = \left(\dfrac{2n + 1}{2}\right)\pi$, where n is an integer. The range of $y = \tan x$ is the set of all real numbers. The amplitude is not defined because $\tan x$ increases or decreases without bound as x approaches the asymptotes.

Since the tangent and cotangent functions are reciprocal functions, the value of $\cot x$ is undefined when $\tan x = 0$, and it is zero when $\tan x$ is undefined.

EXAMPLE 2   Graph $y = \cot x$ over the interval $0 < x < 2\pi$.

First, make a table of values for the interval $0 \le x \le \pi$.

x	0	$\dfrac{\pi}{6}$	$\dfrac{\pi}{4}$	$\dfrac{\pi}{3}$	$\dfrac{\pi}{2}$	$\dfrac{2\pi}{3}$	$\dfrac{3\pi}{4}$	$\dfrac{5\pi}{6}$	π
$\cot x$	undefined	1.73	1	0.58	0	-0.58	-1	-1.73	undefined

The asymptotes for $y = \cot x$ have equations of the form $x = n\pi$, where n is an integer. It is helpful to sketch the asymptotes before drawing the graph. Then use the table of values to plot points and sketch the curve over the interval $0 < x < \pi$. Use the fact that the period is π to extend the graph. The range of $y = \cot x$ is the set of all real numbers, and the amplitude is not defined.

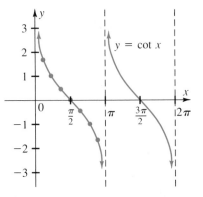

Even though the tangent and cotangent are cofunctions, the graph of the cotangent function is not a phase shift of the tangent function. The functions are reciprocals, and this is reflected in their graphs. When the absolute value of the tangent is very small, the value of the cotangent is very large. The reverse is also true.

Constants affect the graphs of the tangent and cotangent functions in many of the same ways that they affect the sine and cosine functions.

For $\quad y = a \tan b(x + c) + d \quad$ and $\quad y = a \cot b(x + c) + d$

The constant a vertically stretches or shrinks the graph compared to the graph of $y = \tan x$ or $y = \cot x$.

The period is $\dfrac{\pi}{|b|}$.

The phase shift from $y = a \tan bx$ or $y = a \cot bx$ is c units to the left if $c > 0$, and $|c|$ units to the right if $c < 0$.

The vertical shift from $y = a \tan bx$ or $y = a \cot bx$ is d units upward if $d > 0$, and $|d|$ units downward if $d < 0$.

Asymptotes have these equations:

For tangent functions, $b(x + c) = \left(\dfrac{2n + 1}{2}\right)\pi$, if n is an integer.

For cotangent functions, $b(x + c) = n\pi$, if n is an integer.

EXAMPLE 3 **Determine the period, the phase shift, and the vertical shift of**
$y = 3 \tan 2\left(x + \dfrac{\pi}{4}\right) - 5$**. Describe any stretching or shrinking.**

Period: $\dfrac{\pi}{|2|} = \dfrac{\pi}{2}$ $Period = \dfrac{\pi}{|b|}$

Phase shift: $\left|\dfrac{\pi}{4}\right| = \dfrac{\pi}{4}$ units to the left, since $c > 0$

Vertical shift: $|-5| = 5$ units down, since $d < 0$

Since $a = 3$, the graph is stretched vertically by a factor of 3.

You can use your knowledge of the period, the phase shift, and the vertical shift to graph the tangent and cotangent functions.

EXAMPLE 4 **Sketch each graph over the given interval:**

a. $y = \tan 3x$; $-\pi \le x \le \pi$ **b.** $y = \cot\left(x - \dfrac{\pi}{4}\right)$; $-\dfrac{3\pi}{4} \le x \le \dfrac{17\pi}{4}$

a. Period: $\dfrac{\pi}{|3|} = \dfrac{\pi}{3}$ $b = 3$

Phase shift: none $c = 0$

Vertical shift: none $d = 0$

Asymptotes: $3x = \left(\dfrac{2n+1}{2}\right)\pi$, where

n is an integer. For $-\pi < x < \pi$, the

equations are $x = -\dfrac{5\pi}{6}$, $x = -\dfrac{\pi}{2}$, $x = -\dfrac{\pi}{6}$,

$x = \dfrac{\pi}{6}$, $x = \dfrac{\pi}{2}$, and $x = \dfrac{5\pi}{6}$.

b. Period: $\dfrac{\pi}{|1|} = \pi$ $b = 1$

Phase shift: $\left|-\dfrac{\pi}{4}\right| = \dfrac{\pi}{4}$ units

to the right, since $c < 0$

Vertical shift: none $d = 0$

Asymptotes: $x - \dfrac{\pi}{4} = n\pi$, where n is an

integer. For $-\dfrac{3\pi}{4} < x < \dfrac{17\pi}{4}$, the

equations are $x = -\dfrac{3\pi}{4}$, $x = \dfrac{\pi}{4}$,

$x = \dfrac{5\pi}{4}$, $x = \dfrac{9\pi}{4}$, $x = \dfrac{13\pi}{4}$, and $x = \dfrac{17\pi}{4}$.

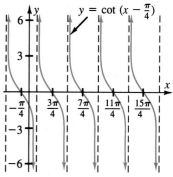

CLASS EXERCISES

Determine the period of each function.

1. $y = \tan \frac{1}{4}x$ **2.** $y = \tan \frac{1}{2}x$ **3.** $y = \cot \frac{2}{3}x$ **4.** $y = \cot \frac{3}{4}x$

Determine the phase shift of each function.

5. $y = \tan \left(x - \dfrac{\pi}{4} \right)$ **6.** $y = \tan \left(x + \dfrac{\pi}{3} \right)$

7. $y = \cot \left(x - \dfrac{2\pi}{3} \right)$ **8.** $y = \cot (2x + \pi)$

PRACTICE EXERCISES   Use appropriate technology.

Determine the period, the phase shift, and the vertical shift, if any, of each function. Describe any vertical stretching or shrinking.

1. $y = \tan 3(x - \pi)$ **2.** $y = \cot 2(x + \pi)$

3. $y = \tan \left(x - \dfrac{\pi}{4} \right)$ **4.** $y = \tan \left(x + \dfrac{\pi}{2} \right)$

5. $y = \cot \left(x + \dfrac{2\pi}{3} \right)$ **6.** $y = \cot \left(x - \dfrac{3\pi}{4} \right)$

7. $y = \tan 2(x + \pi) + 3$ **8.** $y = \cot 3(x - \pi) - 5$

9. $y = \tan \left(x - \dfrac{4\pi}{3} \right) - 1$ **10.** $y = \cot \left(x - \dfrac{5\pi}{6} \right) + 4$

11. $y = 5 \cot \left(x + \dfrac{2\pi}{3} \right) - 2$ **12.** $y = 4 \tan \left(x - \dfrac{3\pi}{4} \right) + 3$

13. $y = 3 \tan \left(x - \dfrac{\pi}{6} \right) + 2$ **14.** $y = 3 \cot \left(x + \dfrac{\pi}{6} \right) + 1$

Graph each function over a two-period interval.

15. $y = \cot 2x$ **16.** $y = \tan 2x$

17. $y = 3 \cot 2x$ **18.** $y = 4 \cot 2x$

19. $y = \tan (x - \pi)$ **20.** $y = \cot (x + \pi)$

21. $y = \cot 3 \left(x + \dfrac{2\pi}{3} \right)$ **22.** $y = \cot 2 \left(x - \dfrac{3\pi}{4} \right)$

23. $y = \tan \left(x - \dfrac{2\pi}{3} \right) + 2$ **24.** $y = \cot \left(x + \dfrac{\pi}{6} \right) - 2$

25. $y = \sin x + \tan x$ **26.** $y = \cos x + \tan x$

27. $y = \sin x + \cot x$ **28.** $y = \cos x + \cot x$

Applications

Graphing Calculator Before graphing a trigonometric function, determine whether your calculator is set for radians or degrees and set range values for the x- and y-axes accordingly. Some calculators have automatic trigonometry settings, which will save you considerable time.

Most graphing utilities draw functions as continuous curves or lines; thus, in this graph the asymptotes of $y = \tan\left(x - \dfrac{\pi}{4}\right)$ appear as part of the graph. Some graphing calculators have noncontinuous modes that avoid this problem.

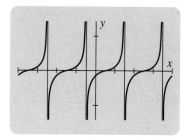

29. In the graph above, what is the distance between the tickmarks along the x-axis?

EXTRA

The values of x, $\sin x$, and $\tan x$ are very nearly the same for small positive values of x.

1. Graph $y = x$ and $y = \sin x$ carefully on the same coordinate plane over the interval $0 \le x \le \dfrac{\pi}{3}$. Label the vertical axis 0, 0.1, 0.2, . . . , 0.8, 0.9, and 1, as shown. Over approximately what interval do the nonnegative values of x and $\sin x$ differ by less than 0.05?

2. Without graphing, give the approximate interval over which the negative values of x and $\sin x$ differ by less than 0.05.

3. Graph $y = x$ and $y = \tan x$ carefully on the same coordinate plane over the interval $0 \le x \le \dfrac{\pi}{4}$. Over approximately what interval do the nonnegative values of x and $\tan x$ differ by less than 0.05?

4. Without graphing, give the approximate interval over which the negative values of x and $\tan x$ differ by less than 0.05.

Graphs of the Secant and Cosecant Functions

Objectives: To determine the properties of the secant and cosecant functions

To graph the secant and cosecant functions

The sine and cosecant are reciprocal functions, as are the cosine and secant. Use these relationships to evaluate specific values of the cosecant and secant functions.

Preview

EXAMPLE Evaluate: $\sec \dfrac{\pi}{3}$

$$\sec \frac{\pi}{3} = \frac{1}{\cos \dfrac{\pi}{3}} = \frac{1}{\dfrac{1}{2}} = 2$$

Find each value. Write *undefined*, where appropriate.

1. $\csc \dfrac{\pi}{3}$ 2. $\sec \dfrac{\pi}{2}$ 3. $\csc \pi$ 4. $\sec \dfrac{2\pi}{3}$

5. $\sec \dfrac{\pi}{4}$ 6. $\csc \dfrac{7\pi}{4}$ 7. $\sec \pi$ 8. $\csc 2\pi$

Since the secant and cosecant functions are not defined when the values of the cosine and sine functions, respectively, are equal to zero, their graphs are discontinuous and have asymptotes. Consider the values of $\sec x$ as x approaches $\dfrac{\pi}{2}$.

x	0	$\dfrac{\pi}{6}$	$\dfrac{\pi}{4}$	$\dfrac{\pi}{3}$	$\dfrac{\pi}{2.1}$	$\dfrac{\pi}{2.01}$	$\dfrac{\pi}{2}$
$\sec x$	1	1.15	1.41	2	13.38	127.96	undefined

x	$\dfrac{\pi}{2}$	$\dfrac{\pi}{1.99}$	$\dfrac{\pi}{1.9}$	$\dfrac{2\pi}{3}$	$\dfrac{3\pi}{4}$	$\dfrac{5\pi}{6}$	π
$\sec x$	undefined	-126.69	-12.11	-2	-1.41	-1.15	-1

The value of $\sec x$ increases without bound as x approaches $\dfrac{\pi}{2}$ from the left. The secant of $\dfrac{\pi}{2}$ is not defined. As x decreases from π to $\dfrac{\pi}{2}$, $\sec x$ decreases without bound. Thus, $x = \dfrac{\pi}{2}$ is an asymptote.

x	π	$\dfrac{7\pi}{6}$	$\dfrac{5\pi}{4}$	$\dfrac{4\pi}{3}$	$\dfrac{3\pi}{2.1}$	$\dfrac{3\pi}{2.01}$	$\dfrac{3\pi}{2.001}$	$\dfrac{3\pi}{2}$
$\sec x$	-1	-1.15	-1.41	-2	-4.49	-42.66	-424.63	undefined

x	$\dfrac{3\pi}{2}$	$\dfrac{3\pi}{1.999}$	$\dfrac{3\pi}{1.99}$	$\dfrac{3\pi}{1.9}$	$\dfrac{5\pi}{3}$	$\dfrac{7\pi}{4}$	$\dfrac{11\pi}{6}$	2π
$\sec x$	undefined	424.20	42.23	4.07	2	1.41	1.15	1

As x increases from π to $\dfrac{3\pi}{2}$, $\sec x$ decreases without bound. As x decreases from 2π to $\dfrac{3\pi}{2}$, $\sec x$ increases without bound. Thus, $x = \dfrac{3\pi}{2}$ is an asymptote. Other asymptotes occur at $x = \left(\dfrac{2n+1}{2}\right)\pi$, where n is an integer.

To graph $y = \sec x$, first sketch the asymptotes. It is also helpful to sketch the graph of $y = \cos x$. Since the period of $y = \cos x$ is 2π, the period of its reciprocal function, $y = \sec x$, is also 2π. The graphs of $y = \cos x$ and $y = \sec x$ intersect at the points where both $\cos x$ and $\sec x$ equal 1, and at the points where both $\cos x$ and $\sec x$ equal -1.

EXAMPLE 1 Graph $y = \sec x$ over the interval $-\dfrac{\pi}{2} < x < \dfrac{7\pi}{2}$.

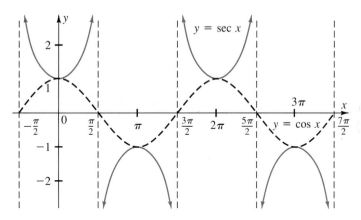

The amplitude is not defined because $y = \sec x$ increases or decreases without bound as x approaches the asymptotes. The range of $y = \sec x$ is the set of all real numbers y such that $|y| \geq 1$.

The graph of $y = \csc x$ can be constructed in a similar manner. Asymptotes occur when $\sin x = 0$, since $\csc x = \dfrac{1}{\sin x}$. Since the period of $y = \sin x$ is 2π, the period of $y = \csc x$ is also 2π. The asymptotes for $y = \csc x$ occur when $x = n\pi$, where n is any integer.

EXAMPLE 2   **Graph $y = \csc x$ over the interval $0 < x < 4\pi$.**

Sketch the asymptotes and the graph of $y = \sin x$. Then graph the cosecant function.

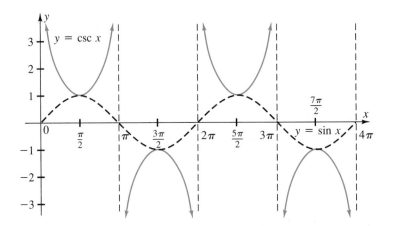

The amplitude is not defined because $y = \csc x$ increases or decreases without bound as x approaches the asymptotes. The range of $y = \csc x$ is the set of all real numbers y such that the $|y| \geq 1$. The graph of the cosecant function is that of the secant function shifted $\dfrac{\pi}{2}$ units to the right.

Constants affect the graphs of the secant and cosecant functions in many of the same ways that they affect the graphs of the sine and cosine functions.

> For $y = a \sec b(x + c) + d$ and $y = a \csc b(x + c) + d$
>
> The constant a vertically stretches or shrinks the graph compared to the graph of $y = \sec x$ or $y = \csc x$.
>
> The period is $\dfrac{2\pi}{|b|}$.
>
> The phase shift from $y = a \sec bx$ or $y = a \csc bx$ is c units to the left if $c > 0$, and $|c|$ units to the right if $c < 0$.
>
> The vertical shift from $y = a \sec bx$ or $y = a \csc bx$ is d units upward if $d > 0$, and $|d|$ units downward if $d < 0$.
>
> Asymptotes have these equations:
>
> For secant functions, $b(x + c) = \left(\dfrac{2n + 1}{2}\right)\pi$, if n is an integer.
>
> For cosecant functions, $b(x + c) = n\pi$, if n is an integer.

EXAMPLE 3 **Determine the period, the phase shift, and the vertical shift of**
$$y = 4 \sec 3\left(x + \frac{\pi}{3}\right) - 6.$$ **Describe any stretching or shrinking.**

The period is $\dfrac{2\pi}{|3|} = \dfrac{2\pi}{3}$. The phase shift is $\dfrac{\pi}{|3|} = \dfrac{\pi}{3}$ units to the left.

The vertical shift is $|-6| = 6$ units down, and the graph will be stretched vertically by a factor of 4.

The properties are also useful when you graph a secant or cosecant function.

EXAMPLE 4 **Sketch each graph over the given interval:**

a. $y = \sec \dfrac{4}{3}x; \quad -\dfrac{9\pi}{8} \le x \le \dfrac{15\pi}{8}$ **b.** $y = \csc\left(x - \dfrac{\pi}{4}\right); \quad -\dfrac{3\pi}{4} \le x \le \dfrac{13\pi}{4}$

a. $y = \sec \dfrac{4}{3}x$

Period: $\dfrac{2\pi}{\left|\dfrac{4}{3}\right|} = \dfrac{3}{2}\pi.$

Phase shift: none
Vertical shift: none
No vertical stretching or shrinking

Asymptotes: $\dfrac{4}{3}x = \left(\dfrac{2n+1}{2}\right)\pi$

$\qquad\qquad x = \dfrac{3}{4}\left(\dfrac{2n+1}{2}\right)\pi, = \left(\dfrac{6n+3}{8}\right)\pi$

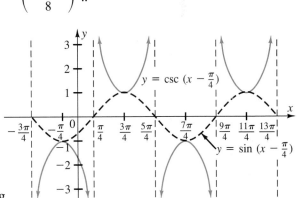

b. $y = \csc\left(x - \dfrac{\pi}{4}\right)$

Period: $\dfrac{2\pi}{|1|} = 2\pi$

Phase shift: $\left|-\dfrac{\pi}{4}\right| =$

$\dfrac{\pi}{4}$ units to the right

Vertical shift: none
No vertical stretching or shrinking
Asymptotes: $x - \dfrac{\pi}{4} = n\pi$

$\qquad\qquad x = \dfrac{\pi}{4} + n\pi$

CLASS EXERCISES

1. **Thinking Critically**   Why are there no values of sec x and csc x between -1 and 1?

Determine the period of each function.

2. $y = \sec \dfrac{1}{4}x$ 3. $y = \sec \dfrac{1}{2}x$ 4. $y = \csc \dfrac{2}{3}x$ 5. $y = \csc \dfrac{3}{4}x$

Find the phase shift of each function.

6. $y = \sec \left(x - \dfrac{\pi}{2} \right)$ 7. $y = \sec \left(x + \dfrac{\pi}{4} \right)$

8. $y = \csc (x + \pi)$ 9. $y = \csc \left(x - \dfrac{2\pi}{3} \right)$

PRACTICE EXERCISES   Use appropriate technology.

Determine the period, the phase shift, and the vertical shift, if any, of each function. Describe any vertical stretching or shrinking.

1. $y = \sec 3(x - \pi)$ 2. $y = \csc 2(x + \pi)$

3. $y = \sec \left(x - \dfrac{\pi}{4} \right)$ 4. $y = \sec \left(x + \dfrac{\pi}{2} \right)$

5. $y = \csc \left(x + \dfrac{2\pi}{3} \right)$ 6. $y = \csc \left(x - \dfrac{3\pi}{4} \right)$

7. $y = \sec 2(x + \pi) + 3$ 8. $y = \csc 3(x - \pi) - 5$

9. $y = \sec \left(x - \dfrac{4\pi}{3} \right) - 1$ 10. $y = \csc \left(x + \dfrac{5\pi}{6} \right) + 4$

11. $y = \csc \left(3x + \dfrac{2\pi}{3} \right) - 2$ 12. $y = \sec \left(4x - \dfrac{3\pi}{4} \right) + 3$

13. $y = \sec \left(2x + \dfrac{\pi}{2} \right) - 1$ 14. $y = \csc \left(2x - \dfrac{3\pi}{2} \right) + 1$

Graph each function over a two-period interval.

15. $y = \csc 2x$ 16. $y = \sec 2x$

17. $y = 3 \csc 2x$ 18. $y = 4 \csc 2x$

19. $y = \sec (x - \pi)$ 20. $y = \csc (x + \pi)$

21. $y = \csc 2 \left(x + \dfrac{2\pi}{3} \right)$ 22. $y = \csc 3 \left(x - \dfrac{3\pi}{4} \right)$

23. $y = \sec(2x - \pi)$ **24.** $y = \sec(2x + \pi)$

25. $y = \sin x + \sec x$ **26.** $y = \cos x + \sec x$

27. $y = \sin x + \csc x$ **28.** $y = \cos x + \csc x$

Applications

29. Engineering   The tension at a given point T in a cable supporting a distributed load can be represented by the equation $T = T_h \sec \theta$, where T_h is the tension at which the cable is horizontal, and θ is the angle between the cable and the horizontal at any point. Sketch the graph when $T_h = 400$ lb, letting θ vary from $0°$ to $90°$ in intervals of $10°$.

BIOGRAPHY: Grace Chisholm Young

Grace Chisholm received the first doctorate that was ever granted to a woman in Germany. Her dissertation was entitled "The algebraic groups of spherical trigonometry." Spherical trigonometry is the study of the relations between the sides and angles of triangles drawn on spheres.

She was born in England in 1868 when Great Britain was the center of technological advance. It was also a time of educational reform. It was only at the end of the 19th century that girls were given the opportunity to attend day and boarding schools which would prepare them to qualify for university entrance exams. Grace received an informal education at home until she was ten.

At seventeen, she passed the Cambridge Senior Exam. Because she was a woman, she was not allowed to apply for university entrance. Her family wanted her to be a social worker to pass her time until she married. However, she was determined to continue her studies. Her mother forbade her to study medicine, her first choice, so she chose mathematics. She won a scholarship to Girton College, a new women's college at Cambridge. Since women were not admitted to graduate school in England, she decided to go to Germany to continue her studies after receiving her degree from Cambridge. At the University of Gottingen in Germany, she became a student of Felix Klein, a leading mathematician.

A year after her graduation, Grace Chisholm married Will Young, a former tutor of hers at Girton College. They collaborated on over 220 articles and books, and Grace Chisholm spent her life raising six children and writing mathematics. In 1906 she and her husband published a book entitled *The Theory of Sets of Points*. This was the first textbook on set theory. She also spoke six languages and wrote two children's books on biology. Investigate Grace Chisholm Young.

Application: Harmonic Motion

2.8

Objective: To solve problems involving simple harmonic motion

Many motions, including the oscillations of a pendulum, the orbit of a satellite or a space shuttle, and the vibrations of a tuning fork, are periodic in nature. Such motions can often be modeled by sine or cosine curves and are examples of **simple harmonic motion.** Because the graphs of these motions are similar to the graph of the sine, they are said to be **sinusoidal.**

Preview

An oscilloscope can convert sound waves into electrical impulses and display them on a fluorescent screen. The sound of a tuning fork produces a sine-like wave on the screen when used with a microphone.

microphone

oscilloscope

tuning fork

Suppose that the equation for the instantaneous movement of a particular tuning fork is $y = 2 \sin (400\pi t)$, where t is the elapsed time in seconds.

1. What is the amplitude of $y = 2 \sin (400\pi t)$?

2. What is the period?

3. What is the phase shift?

4. What is the vertical shift?

The equations representing sinusoidal graphs can be expressed in the form $y = a \sin b(t + c) + d$ or $y = a \cos b(t + c) + d$, where a, b, c, and d are constants and t represents time. The period of each function is $\dfrac{2\pi}{|b|}$, the amplitude is $|a|$, the horizontal shift is $|c|$, and the vertical shift is $|d|$. The **frequency** is the reciprocal of the period and is therefore equal to $\dfrac{|b|}{2\pi}$.

Simple harmonic motion is the motion that occurs when an object is shifted from its rest position and the force attempting to return it to rest is directly proportional to the shift. One example of simple harmonic motion is the vertical oscillation of a weight suspended from a spring. If the weight is pulled down a certain distance, it will oscillate up and down. The maximum distance the weight is pulled from its rest position is called the **displacement,** *a*. The time the weight takes to reach a given position is represented by *t*.

EXAMPLE 1 **A weight is at rest hanging from a spring. It is then pulled down 6 cm and released. The weight bounces back and forth through its rest position, making one full cycle from −6 to 6 and then back to −6 every 3 s. This motion can be described by a sine function. Find an equation for the function. Then sketch the graph.**

The weight moves up and down between a minimum of −6 and a maximum of 6. So, its amplitude is 6. Its period is 3, since a complete cycle takes 3 s. There is no vertical shift, since the rest position is at 0; that is, $d = 0$.

The motion can be modeled by either the sine or the cosine curve, but the sine has been selected. Thus, $y = a \sin b(t + c)$. The period is $\frac{2\pi}{|b|}$, so $3 = \frac{2\pi}{|b|}$. Therefore, letting *b* be positive, $b = \frac{2\pi}{3}$.

Substituting $a = 6$, $b = \frac{2\pi}{3}$, and $d = 0$ into $y = a \sin b(t + c) + d$ gives

$$y = 6 \sin \frac{2\pi}{3} (t + c)$$

Remember that $y = -6$ when $t = 0$, since the spring was pulled down 6 cm to start a cycle. Therefore,

$$-6 = 6 \sin \frac{2\pi}{3} (0 + c)$$

$$-6 = 6 \sin \frac{2\pi}{3} c$$

$$-1 = \sin \frac{2\pi}{3} c$$

$$\frac{2\pi}{3} c = -\frac{\pi}{2} \qquad \textit{If } \sin \frac{2\pi}{3} c = -1\textit{, one possible value for } \frac{2\pi}{3} c \textit{ is } -\frac{\pi}{2}.$$

$$c = -\frac{3}{4}$$

Note that other possible values for *c* would also result in a correct equation. Substituting $-\frac{3}{4}$ for *c* yields $y = 6 \sin \frac{2\pi}{3} \left(t - \frac{3}{4} \right)$.

Before sketching the graph, note that the phase shift is $\frac{3}{4}$ units to the right.

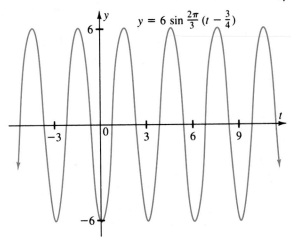

$$y = 6 \sin \frac{2\pi}{3}\left(t - \frac{3}{4}\right)$$

The motion of the tides in the Bay of Fundy is well known around the world. It has been shown that such motion is sinusoidal.

EXAMPLE 2   **At high tide in the Bay of Fundy, the water level is approximately 8 m above the level at low tide, and the time between high tides is 12.4 h. Write an equation for the motion of these tides.**

Since the time between high tides is 12.4 h, the period is 12.4. Therefore,

$$12.4 = \frac{2\pi}{|b|}, \text{ and } b = \frac{2\pi}{12.4} = \frac{\pi}{6.2}.$$

The water level at high tide is approximately 8 m above the level at low tide, so $a = \frac{8}{2} = 4$. Also, there is no vertical shift, so $d = 0$.

$$y = a \sin b(t + c) + d$$

$$y = 4 \sin \frac{\pi}{6.2}(t + c) \qquad \textit{Substitute } a = 4, b = \frac{\pi}{6.2}, \textit{ and } d = 0.$$

Assume that a period starts at high tide. Then $y = 4$ when $t = 0$.

$$4 = 4 \sin \frac{\pi}{6.2}(0 + c)$$

$$\sin \frac{\pi}{6.2}c = 1$$

$$\frac{\pi}{6.2}c = \frac{\pi}{2}$$

$$c = 3.1$$

Note that other possible values of c would also result in a correct equation.

$$y = 4 \sin \frac{\pi}{6.2}(t + 3.1) \qquad \textit{Substitute 3.1 for c.}$$

Use the information given in Examples 1 and 2 to answer each of the following questions.

1. What is the distance of the weight from its rest position after 1 s?

2. When is the first time that the weight will be 6 cm above its rest position?

3. How many meters has the water level dropped 2 hours after a high tide in the Bay of Fundy?

4. How many hours after a high tide does the water level in the Bay of Fundy first reach a level halfway between that of high tide and low tide?

PRACTICE EXERCISES Use appropriate technology.

A weight is hanging at rest from a spring. At $t = 0$, the weight is pulled down q units and released. It makes r full cycles every second. Write an equation using each of the following values of q and r, and then graph the equation.

1. $q = 3\,\text{cm}, r = 1$ 2. $q = 6\,\text{cm}, r = 1$

3. $q = 5\,\text{cm}, r = \frac{1}{2}$ 4. $q = 8\,\text{cm}, r = \frac{1}{2}$

5. $q = 5\,\text{in.}, r = 2$ 6. $q = 8\,\text{in.}, r = 2$

7. $q = 20\,\text{in.}, r = \frac{1}{4}$ 8. $q = 16\,\text{in.}, r = \frac{1}{4}$

9. $q = 20\,\text{in.}, r = 4$ 10. $q = 16\,\text{in.}, r = 4$

For Exercises 11–15, assume that the equations are of the form $y = a \sin b(t + c) + d$.

11. If the amplitude of the sound wave produced by middle C is 1 and the frequency is 264 cycles per second, write an equation for the graph seen on an oscilloscope when middle C is sounded.

12. If the amplitude of the wave produced by A below middle C is 1 and the frequency is 220 cycles per second, what is an equation for the sound wave?

13. Although the tides in Fishfin are not as famous as those in the Bay of Fundy, the high tide of 3.2 m is quite impressive. If the tides occur every 12.4 h in Fishfin, what is an equation for the changes in the tides?

14. The pendulum of a clock swings so that the distance that its tip travels from its vertical position as a function of time is sinusoidal. If the displacement of the pendulum on a grandfather clock is 5 in. and the pendulum makes one complete cycle in 2 s, write an equation for its motion.

15. A point A is moving at a constant speed in a clockwise direction around the circumference of a circle. The circle is located on the coordinate plane as shown. The projection of A on the x-axis is the point B. B moves back and forth between the points (6, 0) and (−6, 0). Assume that the position x of B is a sinusoidal function of the time t. If A makes one complete rotation in 4 s and starts at (0,6) when $t = 0$, what is an equation for the position of B?

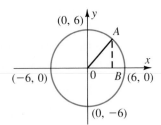

16. The graph of a simple harmonic motion is shown at the right. Write an equation for the motion.

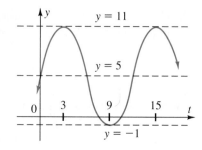

TEST YOURSELF

Graph each by addition of ordinates over at least a one-period interval.

1. $y = \cos x + \cos 3x$

2. $y = \sin x + \sin \frac{1}{2} x$

2.5

Determine the period, phase shift, and vertical shift, if any, of each.

3. $y = \tan 4(x + \pi)$

4. $y = \tan 2(x - \pi)$

2.6

5. $y = \cot \left(x + \frac{\pi}{4} \right) - 1$

6. $y = \cot \left(x - \frac{\pi}{4} \right) + 2$

7. $y = \csc \left(x + \frac{\pi}{2} \right) + 1$

8. $y = \csc \left(x - \frac{\pi}{2} \right) + 3$

2.7

9. $y = \sec 2\left(x + \frac{\pi}{4} \right) - 3$

10. $y = \sec 3\left(x - \frac{\pi}{4} \right) - 2$

Graph each function over a two-period interval.

11. $y = \tan \frac{1}{2} x$

12. $y = \cot \left(x + \frac{\pi}{4} \right)$

2.6

13. $y = \sec \frac{1}{2} x$

14. $y = \csc \left(x - \frac{\pi}{2} \right)$

2.7

15. A weight is hanging at rest from a spring. It is pulled down 8 in. and released when $t = 0$. The weight oscillates up and down from its rest position and makes one full cycle every 4 s. Write an equation for the function and then sketch its graph.

2.8

104　Chapter 2 Graphing Trigonometric Functions

APPLICATION:
Magnetic Resonance Imaging

Did you know that Magnetic Resonance Imaging (MRI) is a technique used to view tissues in the human body? This is done by using radio signals to effect temporary electromagnetic changes in hydrogen (or other) atoms. These weak changes are detected by magnetic fields. Using computer technology, these magnetic impressions are converted into visual images.

An MRI system has three magnetic *gradient coils*. Their axes (x, y, and z) are mutually perpendicular. If the plane that is to contain the image (the *slice plane*) is perpendicular to one of the coil axes, then the magnetic field, measured in gauss units, will be needed from only the gradient coil for that axis. Otherwise, contributions will be needed from more than one coil.

EXAMPLE   A slice plane forms a 30° angle with the axis of the y-gradient coil. The field strength of the slice plane is composed of the contributions of the y-gradient coil and the z-gradient coil. Find the magnetic field components from each coil.

The two components are added as *vectors*, not as real numbers (see Lesson 4.7). The fields are related by the equations

$G_y = G \cos \theta$ $G_z = G \sin \theta$
$G_y = 8000 \cos 30° = 6928$ gauss $G_z = 8000 \sin 30° = 4000$ gauss

EXERCISES

1. For a vertebral disk image, a slice plane must form a 20° angle with the y-coil and requires a magnetic field of 10,000 gauss. Find the contributions of the y- and z-coils.

2. For a slice plane image of a shoulder bone, an angle of 15° with the z-axis is required. The contribution of the z-coil is to be 12,000 gauss. Find the magnetic field in the slice plane and the contribution from the y-coil.

Vocabulary

addition of ordinates (85)
amplitude (73)
asymptotes (88)
cofunctions (67)
complementary angles (67)
cycle (60)
discontinuous (89)
displacement (101)
even function (61)
frequency (100)

odd function (61)
period (60)
periodic (60)
phase shift (78)
simple harmonic motion (100)
sinusoidal (100)
symmetric with respect to the origin (62)
symmetric with respect to the x-axis (62)
symmetric with respect to the y-axis (62)
vertical shift (80)

Periodic Functions and Symmetry   A function $f(x)$ is periodic (with   **2.1**
period p) if there exists a smallest positive number p such that $f(x + p) = f(x)$
when both $f(x + p)$ and $f(x)$ are defined. If $f(-x) = f(x)$ for every value x in
the domain of a function, $f(x)$ is an even function. If $f(-x) = -f(x)$ for
every value of x, $f(x)$ is an odd function. Even functions are symmetric with
respect to the y-axis; odd functions are symmetric with respect to the origin.

1. Determine if the graph at the right
   represents a periodic function. If so, state
   the period. Also state whether the graph is
   symmetric with respect to the y-axis, the
   x-axis, the origin, or to none of these.

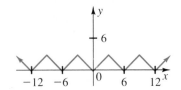

Graphs of the Sine and Cosine Functions   The sine and cosine functions   **2.2**
are periodic with period 2π. Sine is an odd function. Cosine is an even function.

2. Express cos 26° in terms of its cofunction.

3. Use the fact that the cosine function is even to find cos $(-505°)$, if
   cos 505° $= -0.8192$.

4. Graph $y = \sin x$ over the interval $2\pi \le x \le 6\pi$.

Amplitude and Period   For $y = a \sin bx$ and $y = a \cos bx$, the amplitude   **2.3**
is $|a|$ and the period is $\dfrac{2\pi}{|b|}$, $b \ne 0$.

5. Determine the amplitude and period of $y = -4 \cos 3x$.

6. Graph $y = 3 \sin 2x$ over the interval $-2\pi \le x \le 2\pi$.

Phase Shift and Vertical Shift   For $y = a \sin b(x + c) + d$ and $\quad$ 2.4
$y = a \cos b(x + c) + d$, the amplitude is $|a|$, the period is $\dfrac{2\pi}{|b|}$, the phase shift
is $|c|$ (to the left if $c > 0$, to the right if $c < 0$), and the vertical shift
is $|d|$ (up if $d > 0$, down if $d < 0$).

Determine the amplitude, period, phase shift, and vertical shift for each. Then sketch the graph over a one-period interval.

   7. $y = 3 \cos 2(x + \pi) - 4$ $\qquad\qquad$ **8.** $y = 4 \sin 3(x - \pi) + 2$

Addition of Ordinates   To graph a function that is the sum of two $\quad$ 2.5
functions, graph the two functions. Then add the two y-values.

Graph each by addition of ordinates over a one-period interval.

   9. $y = 3 \cos x + \cos x$ $\qquad\qquad$ **10.** $y = \sin 2x + \cos x$

Graphs of the Tangent and Cotangent Functions   The tangent and $\quad$ 2.6
cotangent functions are discontinuous and have vertical asymptotes. For
$y = a \tan b(x + c) + d$ and $y = a \cot b(x + c) + d$, the period is $\dfrac{\pi}{|b|}$, and the
phase shift and vertical shift are determined by the constants c and d.

Determine the period, the phase shift, and the vertical shift, if any, for each function. Then graph the function over a two-period interval.

   11. $y = \tan 3x$ $\qquad\qquad\qquad$ **12.** $y = 2 \cot \left(x + \dfrac{\pi}{2}\right)$

Graphs of the Secant and Cosecant Functions   The secant and $\quad$ 2.7
cosecant are discontinuous functions and have vertical asymptotes. For
$y = a \csc b(x + c) + d$ and $y = a \sec b(x + c) + d$, the period is $\dfrac{2\pi}{|b|}$, and the
phase shift and vertical shift are determined by c and d.

Determine the period, the phase shift, and the vertical shift, if any, for each function. Then graph the function over a two-period interval.

   13. $y = 2 \sec 3x$ $\qquad\qquad\qquad$ **14.** $y = \csc (x - \pi) + 1$

Harmonic Motion   Simple harmonic motion can be modeled by $\quad$ 2.8
equations of the form $y = a \sin b(t + c) + d$ and $y = a \cos b(t + c) + d$.

   15. The amplitude of the sound wave for C below middle C is 1, and the
      frequency is 132 cycles per second. Find an equation for the graph.

Summary and Review **107**

1. State whether this graph represents a periodic function. If it does, give the period.

2. State whether this graph is symmetric to the x-axis, the y-axis, or the origin.

Express each function in terms of its cofunction.

3. $\cos 88°$

4. $\csc 54°$

5. $\tan 8°$

6. Use the properties of the cosine function to find $\cos 123°$, given that $\cos (-123°) = -0.5446$.

Determine the amplitude, the period, the phase shift, and the vertical shift of each function. Then sketch the graph over the interval $0 \le x \le 2\pi$.

7. $y = 4 \sin x$

8. $y = \cos 3x$

9. $y = \sin \left(x - \dfrac{\pi}{2} \right) + 1$

10. Graph the function $y = 2 \sin x + 2 \cos x$ by addition of ordinates over the interval $0 \le x \le 2\pi$.

Determine the period, the phase shift, and the vertical shift of each.

11. $y = \tan 4x$

12. $y = \cot \left(x + \dfrac{\pi}{4} \right)$

13. $y = 2 \sec \dfrac{1}{2} x$

14. $y = \csc \left(2x - \dfrac{2\pi}{3} \right) - 2$

15. A weight hanging from a spring is pulled down 10 cm and released at time $t = 0$. The weight oscillates up and down from its rest position, making one full cycle every 2 seconds. Write an equation to describe its motion.

Challenge

The graph of the function $y = \sin x \cos x$ is a sine curve. What is its period?

Select the best choice for each question.

1. For $0 \le x \le 2\pi$, the following represents the graph of

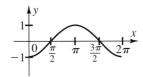

A. $y = \cos x$ **B.** $y = \sec x$
C. $y = \sin x$ **D.** $y = -\cos x$
E. $y = -\sin x$

2. The amplitude of $y = \sin 2\theta$ is
A. 2π **B.** 1 **C.** 2 **D.** $\frac{1}{2}$ **E.** π

3. Evaluate $4x^3 - 3x^2 + 2x - 1$ when $x = 1$.
A. -2 **B.** -10 **C.** -4 **D.** 2
E. -12

4. In the figure, $\sin \theta \cdot \tan \theta$ equals

A. $\frac{x}{r}$ **B.** $\frac{y}{r}$ **C.** $\frac{x^2}{ry}$

D. $\frac{y^2}{rx}$ **E.** $\frac{xy}{r^2}$

5. $y = -2 \sin 4x$ has
A. amplitude -2 and period 4
B. amplitude -2 and period π
C. amplitude $\frac{1}{2}$ and period $\frac{\pi}{2}$
D. amplitude -8 and period 4
E. amplitude 2 and period $\frac{\pi}{2}$

6. The length of a rectangle is $2x + 3$ and the width is $x + 5$. The area of the rectangle expressed in terms of x is
A. $3x + 8$ **B.** $2x^2 + 2x + 15$
C. $3x^2 + 8x + 8$ **D.** $2x^2 + 2x + 8$
E. $2x^2 + 13x + 15$

7. $\dfrac{3}{4 - \sqrt{5}}$ equals

A. $\dfrac{3(4 + \sqrt{5})}{-1}$ **B.** $\dfrac{11}{4 + \sqrt{5}}$

C. $\dfrac{12 + \sqrt{5}}{11}$ **D.** $\dfrac{12 + 3\sqrt{5}}{11}$

E. $\dfrac{12 - 3\sqrt{5}}{11}$

8. In $\triangle ABC$, x equals

A. 10 **B.** 5 **C.** 90 **D.** 45 **E.** 20

9. The graph of $y = \sin x$
   I. rises from $x = 0$ to $x = \dfrac{\pi}{2}$
   II. falls from $x = \dfrac{\pi}{2}$ to $x = \pi$
   III. rises from $x = \pi$ to $x = 2\pi$
A. I only   **B.** II only   **C.** III only
D. II and III only   **E.** I and II only

10. If $2p + q = 4$, then $6p + 3q$ equals
A. $-\frac{2}{3}$ **B.** 12 **C.** 24 **D.** 8 **E.** 2

11. Twelve times the first of two consecutive even integers equals eight times the second. The sum of the integers is
A. 22 **B.** 26 **C.** 10 **D.** 14 **E.** 3

12. $\sin 90°$ equals
A. 1 **B.** 0 **C.** -1
D. $\frac{1}{2}$ **E.** undefined

13. If $MN = NP$, then

A. $MO - NP = MN$
B. $MO = MO + NP$
C. $MN + NO = MN + OP$
D. $MN - NO = NP - NO$
E. $MN - OP = MO$

1. Sketch the graph of $y = \sin\left(x - \frac{\pi}{3}\right)$ over the interval $0 \le x \le 4\pi$.

2. Angle θ is in standard position with its terminal side in the second quadrant. Find the exact value of $\sin\theta$ if $\cos\theta = -\frac{1}{2}$.

3. Determine whether the function $f(x) = x^2 + 3$ is odd, even, or neither.

State whether or not each relation is a function.

4. $\{(2, 3), (3, 6), (4, 6)\}$

5. $\{(1, 4), (4, 1), (1, 1)\}$

6. Determine the amplitude, period, phase shift, and vertical shift for the function $y = 4\cos(2x + \pi)$.

7. A weight is hanging at rest from a spring. It is pulled down 6 cm and released. The weight oscillates up and down from its rest position and makes one full cycle every 2 s. Write an equation for the function.

8. Find the exact values of the six trigonometric functions of $\theta = -150°$.

9. Express the function $\cos 49°$ in terms of its cofunction.

10. Name two angles that are coterminal with an angle of $135°$.

11. Graph $y = \sin x - \cos x$ over the interval $0 \le x \le 4\pi$.

12. If $0° < \theta < 360°$, find all values of θ for which $\cos\theta = -\frac{\sqrt{3}}{2}$.

13. The wheel of a car is turning at 8 rps. It is 3 ft in diameter. Find the linear velocity, in feet per second, of a point on the rim of the wheel.

14. Graph $y = \sec\left(2x - \frac{\pi}{2}\right)$ over a two-period interval.

15. Express $315°$ in radians.

16. In which quadrants do $\sin\theta$ and $\cos\theta$ have opposite signs?

17. Express $\frac{11\pi}{6}$ in degrees.

18. Find the radius r of a circle with the equation $2x^2 + 2y^2 = 36$.

Find θ, where $0° \le \theta \le 360°$, to the nearest tenth of a degree.

19. $\cos\theta = 0.6587$

20. $\cot\theta = 1.4019$

3 Right Triangle Trigonometry and Basic Identities

A laser (an acronym for light amplification by stimulated emission of radiation) is a device that produces a nearly perfect plane wave of light. Lasers are used as surgical tools, to treat cancer, to align surveying equipment, and in compact disc players.

Solving Right Triangles

Objectives: To solve right triangles, given the measures of one angle and one side or the measures of two sides

In this chapter, trigonometric functions will sometimes be referred to as trigonometric *ratios*, since they can be thought of as the ratios of lengths of sides of right triangles.

Preview

Recall that a calculator can be used to find both a value of a trigonometric function and an angle measure for a given trigonometric-function value.

EXAMPLE 1 Find $\cot \dfrac{\pi}{5}$ to four decimal places.

$\cot \dfrac{\pi}{5} = 1.3764$ *Use radian mode. Enter $\dfrac{\pi}{5}$ and use the tan key and the reciprocal key.*

EXAMPLE 2 Find θ to the nearest tenth of one degree if $\sin \theta = 0.7083$ and $0° < \theta < 90°$.

$\theta = 45.1°$ *Use degree mode. Enter 0.7083 and use the inverse key and the sine key.*

Find each value to four decimal places.

1. $\sec 203.8°$ **2.** $\tan \dfrac{\pi}{7}$ **3.** $\sin \dfrac{7\pi}{4}$

Find θ to the nearest tenth of one degree, where $0° < \theta < 90°$.

4. $\cos \theta = 0.6128$ **5.** $\cot \theta = 1.2341$ **6.** $\csc \theta = 3.1215$

The table on page 113 compares trigonometric *functions* and trigonometric *ratios*. Right triangle *ABC* can be used to express the values of the trigonometric ratios in terms of the sides of the triangle, where

a = length of leg opposite θ
b = length of leg adjacent to θ
$c = \sqrt{a^2 + b^2}$ = hypotenuse

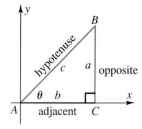

Trigonometric Functions	Right Triangle Ratios ($0° < \theta < 90°$)
$\sin \theta = \dfrac{y}{r}$	$\sin \theta = \dfrac{a}{c} = \dfrac{\text{length of leg opposite } \theta}{\text{length of hypotenuse}}$
$\cos \theta = \dfrac{x}{r}$	$\cos \theta = \dfrac{b}{c} = \dfrac{\text{length of leg adjacent to } \theta}{\text{length of hypotenuse}}$
$\tan \theta = \dfrac{y}{x}$	$\tan \theta = \dfrac{a}{b} = \dfrac{\text{length of leg opposite } \theta}{\text{length of leg adjacent to } \theta}$
$\csc \theta = \dfrac{r}{y}$	$\csc \theta = \dfrac{c}{a} = \dfrac{\text{length of hypotenuse}}{\text{length of leg opposite } \theta}$
$\sec \theta = \dfrac{r}{x}$	$\sec \theta = \dfrac{c}{b} = \dfrac{\text{length of hypotenuse}}{\text{length of leg adjacent to } \theta}$
$\cot \theta = \dfrac{x}{y}$	$\cot \theta = \dfrac{b}{a} = \dfrac{\text{length of leg adjacent to } \theta}{\text{length of leg opposite } \theta}$

Capital letters are usually used to represent the angles of triangles, or their measures. Lowercase letters refer to the sides opposite their respective angles, or to their measures. The right triangle ratios can be used to *solve* a right triangle, that is, to find the unknown measures of the sides and angles.

EXAMPLE 1 **Solve right triangle ABC if $b = 32$, $\angle A = 25°$, and $\angle C = 90°$. Find a and c to the nearest unit.**

Since you know θ and b, use $\tan 25°$ to find a.

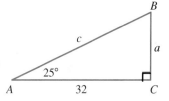

$$\tan 25° = \frac{a}{32} \qquad \tan A = \frac{a}{b}$$
$$a = 32 \tan 25° \qquad \text{Calculation-ready form}$$
$$a = 15 \qquad\quad \text{To the nearest unit}$$

Since you know θ and b, use $\cos 25°$ to find c.

$$\cos 25° = \frac{32}{c} \qquad\qquad \cos A = \frac{b}{c}$$

$$c = \frac{32}{\cos 25°} = 35 \qquad \text{To the nearest unit}$$

Since angles A and B are complementary, $\angle B = 90° - 25° = 65°$.

A **significant digit** is any nonzero digit or any zero that serves a purpose other than to locate the decimal point. Consider the following examples:

0.00*304*	Three significant digits
29.40	Four significant digits
301,450	Five significant digits

In general, the greater the number of significant digits in a measurement, the greater the accuracy. Use the table below as a guide to decide how accurate your answers should be.

Angle measures accurate to nearest . . .	1°	10' or 0.1°	1' or 0.01°
Significant digits in lengths of sides	2	3	4

To illustrate, if angle measures are given to the nearest degree, compute the lengths of the sides to two significant digits. If lengths are given to three significant digits, compute angle measures to the nearest 0.1°.

EXAMPLE 2   **Solve right triangle ABC if $\angle C = 90°$, $c = 7.25$, and $b = 4.37$.**

To find the value of $\angle A$, use the cosine function.

$$\cos A = \frac{4.37}{7.25}$$

$A = 52.9°$ *Since b and c are given to 3 significant digits, round to the nearest 0.1°.*

Since angles A and B are complementary, $\angle B = 90° - 52.9° = 37.1°$.

To find the value of a, use $\sin 52.9°$.

$$\sin 52.9° = \frac{a}{7.25}$$

$$a = 7.25 \sin 52.9°$$

$$a = 5.78 \qquad \textit{Round to 3 significant digits.}$$

   In Example 3, the angle measures are expressed in degrees and minutes. If your calculator has a key that converts from degrees and minutes to decimal degrees, you can omit the next-to-last steps shown in the calculation of c and in the calculation of b.

EXAMPLE 3   **Solve triangle ABC if $a = 22.14$, $\angle B = 32°12'$, and $\angle C = 90°$.**

To find c, use $\sec 32°12'$.

$$\sec 32°12' = \frac{c}{22.14}$$

$$c = 22.14 \sec 32°12'$$

$$c = 22.14 \sec \left(32 + \tfrac{12}{60}\right)^{\circ}$$

$$c = 26.16 \qquad \textit{Round to 4 significant digits.}$$

To find b, use $\tan 32°12'$.

$$\tan 32°12' = \frac{b}{22.14}$$

$$b = 22.14 \tan 32°12'$$

$$b = 22.14 \tan \left(32 + \frac{12}{60}\right)°$$

$$b = 13.94 \qquad \textit{Round to 4 significant digits.}$$

Since angles A and B are complementary, $\angle A = 90° - 32°12' = 57°48'$.

In order to solve a right triangle, it is helpful to draw a reasonably accurate figure and label the parts of known measure. You can usually choose which side or angle measure to determine first, and more than one trigonometric function may be appropriately applied. The Pythagorean theorem can also be used when the measures of the two sides are known.

To illustrate, it is possible to solve first for c, b, or $\angle A$ in Example 3. In order to find b first, $\tan B$ or $\cot B$ can be used. After b is found, $\sin B$, $\cos B$, $\sec B$, $\csc B$, or the Pythagorean theorem can be used to determine c. Remember that calculations with rounded values may yield slightly different answers.

CLASS EXERCISES

For triangle ABC at the right, express the given trigonometric function in terms of a, b, and c.

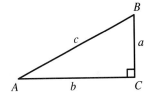

1. $\sin B$ **2.** $\sin A$ **3.** $\cot A$

4. $\sec A$ **5.** $\cos B$ **6.** $\tan B$

State two equations, using reciprocal functions, that could be used to find x or θ.

7.

8.

9.

10.

11.

12.

PRACTICE EXERCISES   Use appropriate technology.

Solve each right triangle *ABC*.

1.

2.

3.

4.

5.

6.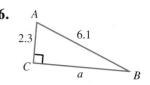

Solve each right triangle *ABC* ($\angle C = 90°$) given the measure of one angle and the length of one side.

7. $\angle A = 58°$, $c = 27$

8. $\angle B = 29°$, $c = 14$

9. $\angle B = 15.1°$, $c = 10.7$

10. $\angle A = 17.2°$, $c = 29.4$

11. $\angle A = 63°$, $a = 11$

12. $\angle B = 24°$, $b = 36$

13. $\angle B = 42.5°$, $a = 188$

14. $\angle A = 70.5°$, $b = 276$

Solve each right triangle *ABC* ($\angle C = 90°$) given the measure of one angle and the length of one side.

15. $\angle A = 36°41'$, $a = 19.32$

16. $\angle B = 42°35'$, $a = 71.22$

17. $\angle B = 72°28'$, $a = 84.84$

18. $\angle A = 80°12'$, $a = 36.22$

Solve each right triangle *ABC* ($\angle C = 90°$) given the lengths of two sides.

19. $b = 17.62$, $c = 23.91$

20. $b = 13.42$, $c = 26.31$

21. $a = 18.65$, $b = 14.22$

22. $a = 7.613$, $c = 14.05$

23. $a = 1632$, $c = 2015$

24. $a = 1503$, $b = 1635$

Use the given information for triangle *ABC* ($\angle C = 90°$) to express the other five trigonometric functions of $\angle A$ in terms of *t*.

25. $\sin A = \dfrac{3}{t}$

26. $\tan A = \dfrac{1}{t}$

27. $\sec A = t$

28. For any right triangle *DEF* ($\angle F = 90°$), find the numerical value of $(\cos D)^2 + (\cos E)^2 + (\cos F)^2$.

29. Construction A ladder rests against a building at a point that is 63 ft from the ground. If the ladder makes a 47° angle with the ground, what is the length of the ladder?

30. Geometry The base of an isosceles triangle is 14 cm in length and the angle opposite the base measures 86°. Find the length of each of the congruent sides.

31. Architecture The Leaning Tower of Pisa is 179 ft high, making a 5.1° angle with the vertical. Find *x*.

32. Sports A ski slope has a 42° incline and is 170 yd long. Find the vertical drop *d*.

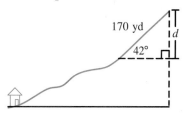

HISTORICAL NOTE

The Pythagorean theorem There is evidence that the Pythagorean theorem was known and used long before the time of the Greek mathematician Pythagoras, who lived in the sixth century B.C. In fact, the Babylonians knew of the theorem more than 1000 years earlier. The ancient Egyptians may have been familiar with the concept also. It has been speculated that Egyptian surveyors used Pythagorean triples to construct right angles. To do this, a rope divided by knots into segments in the ratio 3:4:5 was stretched out into a triangle. The desired angle was opposite the longest segment.

Since the time of Pythagoras, this theorem has been proved in many ways. Locate two proofs of the theorem and write them in your own words.

Application: Angles of Elevation and Depression

Objective: To define and use angles of elevation and depression

Solar heating panels are placed on roofs facing due south, if possible, in order to absorb maximum energy. Trigonometry can be used to determine where trees should be planted so that they do not block sunlight striking the panels.

Preview

The angle of the sun at noon varies, with the smallest angle occurring on December 21, the winter solstice. Richard lives in Winnipeg, where the angle of the sun at noon on December 21 is 14°. He has solar panels on his roof with the bottom edge of the panels 25 ft above the ground. Richard wants to plant a tree 110 ft from the panels. What is the maximum height the tree can attain if it is not to block sunlight to the panels?

$$\tan 14° = \frac{x}{110}$$

$$x = 110 \tan 14°$$

$$x = 27 \text{ ft} \qquad \textit{Two significant digits}$$

maximum height = 25 ft + 27 ft = 52 ft

1. In Philadelphia, the angle of the sun on December 21 is 24°. Solar panels are placed on a roof 42 ft above the ground. How far from the house should a tree be planted if the tree is to attain a maximum height of 75 ft?

2. On December 21, the angle of the sun at noon in Toronto is 18°. How far from the ground should solar panels be placed so that sunlight will not be blocked by a fully grown, 33-ft tree 68 ft from the house?

If an observer sights an object, the angle formed between a horizontal line and his or her *line of sight* is called the **angle of elevation** if the line of sight is above the horizontal and the **angle of depression** if the line of sight is below the horizontal. In the figure at the right, lines n and p are parallel. Therefore, the angle of elevation from the person on the ground to the airplane is equal in measure to the angle of depression from the plane to the person on the ground.

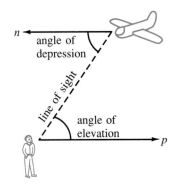

EXAMPLE 1 A 98-ft extension ladder rests on top of
a hook and ladder truck with its base
11 ft from the ground. When the angle
of elevation of the ladder is 73°, how
high up the building will it reach?

$$\sin 73° = \frac{h}{98}$$
$$h = 98 \sin 73°$$
$$h = 94 \text{ ft} \qquad \textit{Two significant digits}$$

The vertical reach is about 11 ft + 94 ft, or 105 ft.

The altimeter of an airplane records the height at which the plane is flying.
The altimeter reading, along with the angle of depression from the plane to a
point on the ground, can be used to find the distance to that point from a point
directly below the plane.

EXAMPLE 2 The altimeter of a jet airplane approaching O'Hare Airport in Chicago
records 5900 ft as it passes over the Sears Tower. At the same time,
the angle of depression from the plane to the near end of the runway
is 5°. How far is it from the base of the Sears Tower to the end of
the runway?

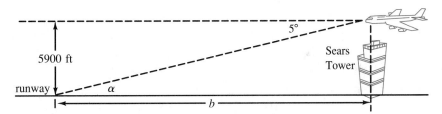

Since the angle of depression from the plane to the runway is equal
to the angle of elevation from the runway to the plane, $\alpha = 5°$.
The required distance is b.

$$\cot 5° = \frac{b}{5900}$$
$$b = 5900 \cot 5°$$
$$b = 67,000 \text{ ft} \qquad \textit{Two significant digits}$$

The base of the Sears Tower is about 67,000 ft from the runway.
Since 5280 ft = 1 mi, the distance is about 13 mi.

In the next example, two known distances are used to find an angle of depression.

EXAMPLE 3   **A passenger in a helicopter shines
a light on a car stranded 45 ft from
a point just below the helicopter. If
the helicopter is hovering at 85 ft,
what is the angle of depression from
the light source to the car?**

The angle of depression from the helicopter to the car is equal to θ,
the angle of elevation from the car to the helicopter.

$$\tan \theta = \frac{85}{45}$$

$$\theta = 62° \qquad \textit{To the nearest degree}$$

The angle of depression is approximately 62°.

CLASS EXERCISES

For each figure, give two trigonometric equations that could be used to find x.

1.

2.

3.

4.

5.

6.

7. If a guy wire for a tree is 14 ft long,
making a 41° angle with the ground,
how far is the base of the tree
from the stake anchoring the wire?

8. From a height of 38 m above sea
level, two ships are sighted due west.
The angles of depression are 53° and
28°. How far apart are the ships?

1. The angle of depression is measured from the top of a 43-ft tower to a reference point on the ground. Its value is found to be 63°. How far is the base of the tower from the point on the ground?

2. The angle of depression from a searchlight to its target is 58°. How long is the beam of light, if the searchlight is 26 ft above ground?

3. A NASA recovery helicopter hovers 75 ft above a space capsule. If the angle of depression to the recovery ship is 40°, how far is the ship from the space capsule?

4. The entrance of the old town library is 2.3 ft above ground level. A ramp from the ground level to the library entrance is scheduled to be built. The angle of elevation from the base of the ramp to its top is to be 15°. Find the length of the ramp.

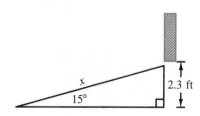

5. A closed-circuit television camera is mounted on a wall 7.4 ft above a security desk in an office building. It is used to view an entrance door 9.3 ft from the desk. Find the angle of depression from the camera lens to the entrance door.

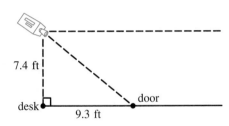

6. To illuminate the entrance of an apartment building, a night light is mounted on a 6.6-m pole. If the base of the pole is 24 m from the entrance, find the angle of depression from the light.

7. The angle of elevation from the bottom of the world's longest slide, located in Peru, Vermont, is approximately 10.3°. The slide has a vertical drop of 821 ft. Find the length of the slide.

8. The largest doors in the world are located in the Vehicle Assembly Building near Cape Canaveral, Florida. If the angle of elevation from a point on the ground 199 ft from the base of the doors is 66.6°, how high are the doors?

9. The world's longest escalator is at the Leningrad Underground in Lenin Square. The escalator has an angle of elevation of 10.36° and a vertical rise of 195.8 ft. Find the length of the escalator.

10. The world's tallest fountain is in Fountain Hills, Arizona. If the angle of elevation to the top of the fountain from a point 755 ft from its base is 36.5°, find the height of the fountain.

11. The extension ladder on top of a 6.0-ft high hook and ladder truck is 150 ft long. If the angle of elevation of the ladder is 70°, to what height on a building will the ladder reach?

12. A child holds the end of a kite string 36 in. above the ground. The string is taut and it makes a 68° angle with the horizontal. How high off the ground is the kite if 540 in. of string are out?

13. The Statue of Liberty stands on a 150-ft pedestal. From a point 280 ft from the base of the pedestal, the angle of elevation to the top of Liberty's torch is 47°. Find the height of the statue.

14. A television antenna stands on the edge of the top of a 52-story building. From a point 320 ft from the base of the building, the angle of elevation to the top of the antenna is 64°. If each story is 12 ft high, find the height of the antenna.

15. A ranger's tower is located 44 m from a tall tree. From the top of the tower, the angle of elevation to the top of the tree is 28°, and the angle of depression to the base of the tree is 36°. How tall is the tree?

16. A building is 16.3 m from a television tower. From the top of the building, the angle of depression to the base of the tower is 43.5°, and the angle of elevation to the top of the tower is 23.8°. Find the height of the tower.

17. A jet took off at a rate of 260 ft/s and climbed in a straight path for 3.2 min. What was the angle of elevation of its path if its final altitude was 12,000 ft?

18. A jet took off at a rate of 280 ft/s and climbed in a straight path for 5.1 min. What was the angle of elevation of its path if its final altitude was 15,000 ft?

19. An engineer determines that the angle of elevation from her position to the top of a tower is 52°. She measures the angle of elevation again from a point 47 m farther from the tower and finds it to be 31°. Both positions are due east of the town. Find the height of the tower.

20. From one point on the ground, the angle of elevation of the peak of a mountain is 10.38°, and from a point 15,860 ft closer to the mountain, the angle of elevation is 14.67°. Both points are due south of the mountain. Find the height of the mountain.

CHALLENGE

1. The area of a parallelogram can be given in terms of the lengths x and y of two adjacent sides and a trigonometric function of the angle formed by those two sides. Which of the following formulas gives the area A of this parallelogram?

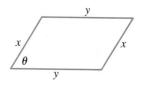

  a. $A = xy \cos \theta$ **b.** $A = xy \tan \theta$ **c.** $A = xy \sin \theta$ **d.** $A = xy \cot \theta$

2. Would the answer to Exercise 1 be the same if θ were one of the other three angles of the parallelogram?

3. The area of a rectangle can be given in terms of the length p of a diagonal and two trigonometric functions of an angle formed by the diagonal and a side. Which of the following formulas gives the area A of this rectangle?

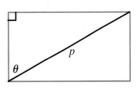

  a. $A = p^2 \sec \theta$ **b.** $A = p^2 \cot \theta$ **c.** $A = p^2 \sin \theta \tan \theta$ **d.** $A = p^2 \sin \theta \cos \theta$

4. Would the answer to Exercise 3 be the same if θ were one of the other three angles formed by the diagonal and a side of the rectangle?

Applications

Objective: To solve real-world problems using trigonometry

Right triangle trigonometry can be used to solve problems in navigation, surveying, and construction.

Preview

In air and sea navigation, the angle measured clockwise from north to the line of travel is the **course** of the plane or ship. The clockwise angle from north to the line of sight to a point of reference is called the **bearing** of the point.

Directions are also written by referring to north or south, using an acute angle. In the figure at the right below, the direction of $\vec{OA}$ is N 60°E and that of $\vec{OB}$ is S15°E.

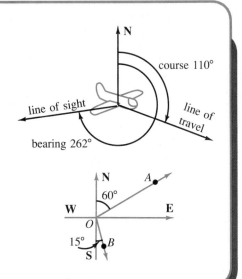

Write the bearing for each direction.

1. N15°E   **2.** N 20°W   **3.** S18°E

4. S 45°W   **5.** N 72°W   **6.** S 59°E

In order to solve navigation problems, it is often necessary to find or to interpret bearings.

EXAMPLE 1   **The pilot of a plane traveling on a course of 30° sights the San Francisco International Airport. The pilot's line of sight forms a right angle with the plane's line of travel, as shown at the right. Find the bearing of the airport.**

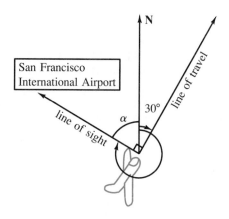

$$\alpha = 90° - 30° = 60°$$
$$\text{bearing} = 360° - \alpha$$
$$= 360° - 60° = 300°$$

EXAMPLE 2 After traveling 65 mi on the same course, the new bearing of the airport from the plane in Example 1 is 238°. Find the distance from the plane to San Francisco International Airport.

$$\beta = \text{bearing} - 180° - 30°$$
$$= 238° - 210° = 28°$$

Find the distance using the secant ratio.

$$\sec 28° = \frac{d}{65}$$
$$d = 65 \sec 28° \quad \textit{Calculation-ready form}$$
$$d = 74 \quad \textit{Two significant digits}$$

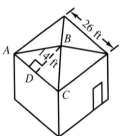

The plane is approximately 74 mi from the airport.

Small errors in angle measurement can cause large errors in computed distances. To minimize such errors, surveyors measure angles using an extremely accurate instrument called a *transit*.

EXAMPLE 3 A bridge is being built across a river, from B to C. A surveyor using a transit determines that $\angle A = 43°20'$. It is also known that the distance from A to C is 465 m. Find BC, the distance across the river.

$$\tan 43°20' = \frac{BC}{465}$$
$$BC = 465 \tan 43°20'$$
$$BC = 439 \quad \textit{Three significant digits}$$

The distance across the river is 439 m.

Trigonometric ratios can also be used to solve construction problems.

EXAMPLE 4 This hip roof is in the form of four congruent isosceles triangles with a common vertex that forms the peak of the roof. If each triangle has a base of 26 ft and a height of 14 ft, find the measure of each base angle.

Since side AC is bisected by the altitude, $AD = 13$ ft.

$$\tan \alpha = \frac{14}{13}$$
$$\alpha = 47° \quad \textit{To the nearest degree}$$

Each base angle measures approximately 47°.

Give the course of each ship or airplane and its bearing to the lighthouse or tower.

1.

2.

Tower

3.

4. To find the distance *AC* across the western corner of Ocean Pond, a surveyor determined the measurements shown. Find *AC*.

5. A hip roof is in the form of four congruent isosceles triangles. The base of each triangle is 24 ft and the height is 11 ft. Find the measure of each base angle.

6. Each of the four congruent isosceles triangles that form a hip roof has a height of 12 ft and base angles of 36°. What is the length of the base of each triangle?

PRACTICE EXERCISES   Use appropriate technology.

1. A tugboat is 36 km due north of lighthouse *A*. Lighthouse *B* has a bearing of 90° from lighthouse *A*. The lighthouses are 53 km apart. Find the bearing of lighthouse *B* from the tugboat and the distance from lighthouse *B* to the tugboat.

2. A boat is 23 mi due west of lighthouse *A*. Lighthouse *B* is 14 mi due north of lighthouse *A*. Find the bearing of lighthouse *B* from the boat and the distance from lighthouse *B* to the boat.

3. Two ships leave the same port at 7 AM. The first ship sails towards Europe on a 54° course at a constant rate of 36 mi/h. The second ship, with a tropical destination, sails on a 144° course at a constant speed of 42 mi/h. Find the distance between the ships at 11 AM.

4. A jet flew 140 mi on a course of 196° and then 120 mi on a course of 106°. Then the jet returned to its starting point via the shortest route possible. Find the total distance that the jet traveled.

5. To find the length of a diagonal of a rectangular city block, a surveyor determined the measurements shown below. How long is the diagonal?

750 ft
35°

6. To find the distance *d* across a pond, a surveyor determined the measurements shown below. How far is it across the pond?

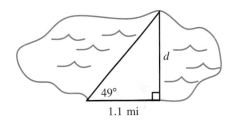

d
49°
1.1 mi

7. To find the distance *AC* across a canyon, a hiker walks 29 yd along one side from *A* to *B*. If segments *AB* and *AC* are perpendicular and $\angle B = 73°$, find *AC*.

A   *C*
29 yd
73°
B

8. To find the width of a river between points *P* and *R*, a surveyor locates a point *Q* such that segments *PQ* and *PR* are perpendicular and *PQ* = 420 yd. Using a transit, the surveyor finds that $\angle Q = 66°$. Find *PR*, the width of the river.

P   *R*
420 yd
66°
Q

9. A hip roof is in the form of four congruent isosceles triangles with a common vertex that forms the peak of the roof. If each triangle has an 18-ft base and a height of 16 ft, find the measure of each base angle.

10. A hip roof is in the form of four congruent isosceles triangles with base angles of 30° and heights of 15 ft. Find the length of the base of each triangle.

11. The bearing of a buoy from a ship 8.7 mi away is 64°. The ship is headed due north, and the navigator plans to change course when the buoy has a bearing of 154°. How much farther will the ship travel before a change of course is needed?

12. A pilot of a San Antonio-to-Houston express plane traveling on a course of 79° sights the Austin airport. His line of sight forms a right angle with the plane's line of travel, as shown below. Find the bearing of the Austin airport.

13. After the plane in Exercise 12 travels 45 min at 180 mi/h along the same course, the airport has a new bearing of 280°. How far is the plane from the airport then?

14. The navigator of a ship on a 44° course sights a buoy with bearing 134°. After the ship sails 15 km along the same course, the navigator sights the same buoy with bearing 168°. Find the distance between the ship and the buoy at the time of each sighting.

15. The two sections of the Sault Sainte Marie railroad bridge in Michigan are each 210 ft in length. Suppose that the maximum angle of elevation of each section is 75°. When the bridge is closed, the water level is normally 13 ft below the bridge. When the bridge is fully opened, what is the distance from the water to point A on the upper left corner of the right section?

16. When the Sault Sainte Marie bridge is fully opened, find the distance between the separated ends of the sections (see Exercise 15).

17. A space-shuttle pilot flying toward the Suez Canal finds that the angle of depression to one end of the canal is 38.25° and the angle of depression to the other end is 52.75°. If the canal is 100.6 mi long, find the altitude of the space shuttle.

18. A helicopter is hovering near the Fort Pitt Tunnel in Pittsburgh, Pennsylvania. The tunnel is 3560 ft long and the angles of depression to the two ends of the tunnel are 22.2° and 68.5°. Find the altitude of the helicopter.

TEST YOURSELF

Solve each right triangle *ABC* ($\angle C = 90°$) using the given information.

1. $\angle A = 43°$, $c = 19$ **2.** $\angle B = 21°$, $a = 7.6$ **3.1**

3. $a = 15$, $b = 19$ **4.** $a = 23.2$, $c = 31.6$

5. A 22-ft flagpole casts a 62-ft shadow. Calculate the angle of elevation to the sun.   **3.2**

6. The angle of depression from a helicopter to its landing port is 64°. If the altitude of the helicopter is 1600 m, find the direct distance from the helicopter to the landing port.

7. An ocean liner is 19 mi due west of lighthouse *A*. Lighthouse *B* is 32 mi due south of lighthouse *A*. Find the bearing of lighthouse *B* from the ocean liner and the distance from lighthouse *B* to the liner.   **3.3**

8. Two planes take off from an airport at 9:15 A.M. One flies on a course of 86° at a constant rate of 190 mi/h. The second one flies on a course of 176° at 160 mi/h. How far apart are they at 10:45 A.M.?

9. To find the distance across a pond, a surveyor measured 72 m along a line perpendicular to the line of sight across the pond, and also measured an angle of 52°, as shown below. What is the distance *d* across the pond?

10. The sloping edges *AB* and *AC* of a symmetrical roof each measure 27 ft and form 35° angles with the horizontal. Find the distance *BC* between the edges of the roof.

Fundamental Identities

Objective: To introduce and prove the reciprocal, ratio, Pythagorean, and odd-even identities

Many fundamental relationships between trigonometric functions can be derived directly from definitions. These relationships are useful in simplifying more complicated trigonometric expressions.

Preview

In both algebra and trigonometry, sentences in one variable may be true for all values of the variable, false for all values, or true for just some values.

True for *all* values	False for *all* values	True for *some* values
$ab = ba$	$3(x + 3) = 3x + 5$	$7x + 2 = 5$
$2x^2 + 4x^2 = 6x^2$	$x + 3 = 2 + x$	$3(x + 2) = 16$
$-1 \leq \sin \theta \leq 1$	$\sin \theta = 3\sqrt{3}$	$\cos \theta = 1$

Classify each sentence as *always true*, *always false*, or *sometimes true*.

1. $-x = -(x)$

2. $x - 5 = x + (-5)$

3. $(x + 2) - 4 = x + 2$

4. $\dfrac{x^2}{2x} = \dfrac{1}{2}$

5. $\dfrac{1}{x^2} = \dfrac{1}{x}$

6. $\dfrac{1}{2}(2x) = 3x\left(\dfrac{1}{3}\right)$

A **trigonometric identity** is a trigonometric equation that is true for all values in the domain of its variable. Examples are the **reciprocal identities** (Chapter 1).

$$\csc \theta = \frac{1}{\sin \theta} \qquad \sec \theta = \frac{1}{\cos \theta} \qquad \cos \theta = \frac{1}{\tan \theta}$$

An identity may be proved using the basic definitions from Chapter 1.

EXAMPLE 1 Use the definitions $\sin \theta = \dfrac{y}{r}$ and $\csc \theta = \dfrac{r}{y}$ to prove $\csc \theta = \dfrac{1}{\sin \theta}$.

$$\frac{1}{\sin \theta} = \frac{1}{\dfrac{y}{r}} = \frac{1 \cdot r}{\dfrac{y}{r} \cdot r} = \frac{r}{y} = \csc \theta$$

Therefore, $\csc \theta = \dfrac{1}{\sin \theta}$.

In Example 1, csc θ must have no value of θ for which sin $\theta = 0$. Similar restrictions exist for sec θ (cos $\theta \neq 0$) and cot θ (tan $\theta \neq 0$). Assume that a denominator in a trigonometric expression is not zero, even if not explicitly stated.

A proof of an identity may be written using a vertical line to separate the expressions to be proved equivalent. Use known algebraic and trigonometric identities to replace one of the expressions with an equivalent expression. Repeat the process until an expression is obtained that is the same as that on the other side of the line. The procedure is illustrated in Example 2.

The **ratio identities** can also be proved using definitions.

$$\tan \theta = \frac{\sin \theta}{\cos \theta}, \quad \cos \theta \neq 0 \qquad \cot \theta = \frac{\cos \theta}{\sin \theta}, \quad \sin \theta \neq 0$$

EXAMPLE 2   Prove: $\tan \theta = \dfrac{\sin \theta}{\cos \theta}$, $\cos \theta \neq 0$

$\tan \theta$	$\dfrac{\sin \theta}{\cos \theta}$	
	$\dfrac{\frac{y}{r}}{\frac{x}{r}}$	*Definitions of sin θ and cos θ* *($x \neq 0$, since cos $\theta \neq 0$.)*
	$\dfrac{y}{x}$	*Multiply numerator and* *denominator by r.*
$= \tan \theta$		*Therefore, $\tan \theta = \dfrac{\sin \theta}{\cos \theta}$, $\cos \theta \neq 0$.*

The expression $(\sin x)^2$ is usually written as $\sin^2 x$, and similarly for the other trigonometric functions. Such expressions are found in the three **Pythagorean identities,** so called because they can be derived from the Pythagorean theorem.

$$\sin^2 \theta + \cos^2 \theta = 1 \qquad 1 + \cot^2 \theta = \csc^2 \theta \qquad 1 + \tan^2 \theta = \sec^2 \theta$$

EXAMPLE 3   Prove: $\sin^2 \theta + \cos^2 \theta = 1$

$\sin^2 \theta + \cos^2 \theta$	1	
$\left(\dfrac{y}{r}\right)^2 + \left(\dfrac{x}{r}\right)^2$		*Definitions of sin θ and cos θ*
$\dfrac{y^2}{r^2} + \dfrac{x^2}{r^2}$		
$\dfrac{x^2 + y^2}{r^2}$		
$\dfrac{r^2}{r^2}$		$x^2 + y^2 = r^2$
1	$=$	Thus, $\sin^2 \theta + \cos^2 \theta = 1$.

Recall that a function $f(x)$ is *even* if $f(-x) = f(x)$ for all real values of x in its domain and *odd* if $f(-x) = -f(x)$ for all real values of x in its domain. The last two identities of this lesson will be the **odd-even identities.**

$$\sin(-\theta) = -\sin\theta \qquad \cos(-\theta) = \cos\theta \qquad \tan(-\theta) = -\tan\theta$$

EXAMPLE 4 **Prove: $\sin(-\theta) = -\sin\theta$**

Let θ be an angle that intersects a circle of radius r at a point (x, y). Then $-\theta$ will intersect the circle at the point $(x, -y)$.

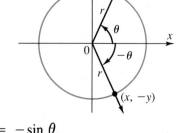

$$
\begin{array}{c|c}
\sin(-\theta) & -\sin\theta \\[4pt]
\dfrac{-y}{r} & \\[8pt]
-\dfrac{y}{r} & \\[8pt]
-\sin\theta = & \text{Therefore, } \sin(-\theta) = -\sin\theta.
\end{array}
$$

Any identity can be *verified* for a specific value of the variable. Keep in mind that verification for a specific value does not constitute a proof of the identity.

EXAMPLE 5 **Verify the following identities for $\theta = 30°$.**

    a. $\tan\theta = \dfrac{\sin\theta}{\cos\theta}$ **b.** $\sin^2\theta + \cos^2\theta = 1$

a. $\tan 30°$ $\qquad \begin{array}{c} \dfrac{\sin 30°}{\cos 30°} \\[10pt] \dfrac{\frac{1}{2}}{\frac{\sqrt{3}}{2}} \\[12pt] \dfrac{1}{\sqrt{3}} \\[10pt] \dfrac{\sqrt{3}}{3} \\[8pt] = \ \tan 30° \end{array}$

b. $\sin^2(30°) + \cos^2(30°)$ $\qquad\Big|\qquad 1$

$$\left(\frac{1}{2}\right)^2 + \left(\frac{\sqrt{3}}{2}\right)^2$$

$$\frac{1}{4} + \frac{3}{4}$$

$$1 \qquad\qquad =$$

Thus, $\sin^2(30°) + \cos^2(30°) = 1$.

Therefore, $\tan 30° = \dfrac{\sin 30°}{\cos 30°}$.

Identities may assume alternate forms. For example, $\sin\theta = \cos\theta\tan\theta$ is an alternate form of $\tan\theta = \dfrac{\sin\theta}{\cos\theta}$.

Here is a summary of the **fundamental identities.**

Reciprocal Identities

$$\csc \theta = \frac{1}{\sin \theta}, \quad \sin \theta \neq 0$$

$$\sec \theta = \frac{1}{\cos \theta}, \quad \cos \theta \neq 0$$

$$\cot \theta = \frac{1}{\tan \theta}, \quad \tan \theta \neq 0$$

Ratio Identities

$$\tan \theta = \frac{\sin \theta}{\cos \theta}, \quad \cos \theta \neq 0$$

$$\cot \theta = \frac{\cos \theta}{\sin \theta}, \quad \sin \theta \neq 0$$

Pythagorean Identities

$$\sin^2 \theta + \cos^2 \theta = 1$$

$$1 + \cot^2 \theta = \csc^2 \theta$$

$$1 + \tan^2 \theta = \sec^2 \theta$$

Odd-Even Identities

$$\sin(-\theta) = -\sin \theta$$

$$\cos(-\theta) = \cos \theta$$

$$\tan(-\theta) = -\tan \theta$$

CLASS EXERCISES

Complete by writing 1 or −1 to make each sentence an identity.

1. $\sin \theta \csc \theta =$ _?_

2. $\sec^2 \theta - \tan^2 \theta =$ _?_

3. $\dfrac{\sin(-\theta)}{\sin \theta} =$ _?_ $(\sin \theta \neq 0)$

4. $\cot \theta \tan \theta =$ _?_

5. $\tan^2 \theta - \sec^2 \theta =$ _?_

6. $\cos^2 \theta + \sin^2 \theta =$ _?_

7. $\cot^2 \theta - \csc^2 \theta =$ _?_

8. $\sec \theta \cos \theta =$ _?_

9. $\dfrac{\cos \theta}{\cos(-\theta)} =$ _?_ $(\cos(-\theta) \neq 0)$

10. $\dfrac{\sin \theta}{\cos \theta \tan \theta} =$ _?_ $(\cos \theta \tan \theta \neq 0)$

PRACTICE EXERCISES Use appropriate technology.

Write two equivalent ways to express each identity.

1. $\csc \theta = \dfrac{1}{\sin \theta}, \quad \sin \theta \neq 0$

2. $\sec \theta = \dfrac{1}{\cos \theta}, \quad \cos \theta \neq 0$

3. $\tan \theta = \dfrac{\sin \theta}{\cos \theta}, \quad \cos \theta \neq 0$

4. $\cot \theta = \dfrac{\cos \theta}{\sin \theta}, \quad \sin \theta \neq 0$

5. $\sin^2 \theta + \cos^2 \theta = 1$

6. $1 + \cot^2 \theta = \csc^2 \theta$

7. $1 + \tan^2 \theta = \sec^2 \theta$

8. Use the two ratio identities to show that $(\tan \theta)(\cot \theta) = 1$.

Verify each identity for the given angle measure.

9. $1 + \cot^2 30° = \csc^2 30°$

10. $1 + \tan^2 30° = \sec^2 30°$

11. $\sin^2 45° + \cos^2 45° = 1$

12. $\cot 45° = \dfrac{\cos 45°}{\sin 45°}$

13. $\tan(-60°) = \dfrac{\sin(-60°)}{\cos(-60°)}$

14. $\sec(-60°) = \dfrac{1}{\cos(-60°)}$

15. $\sin(-120°) = -\sin 120°$

16. $\cos(-120°) = \cos 120°$

17. $1 - \sin^2\left(\dfrac{3\pi}{4}\right) = \cos^2\left(\dfrac{3\pi}{4}\right)$

18. $\tan^2\left(\dfrac{3\pi}{4}\right) = \sec^2\left(\dfrac{3\pi}{4}\right) - 1$

Use the definitions of the trigonometric functions to prove each identity.

19. $\sec\theta = \dfrac{1}{\cos\theta}, \quad \cos\theta \neq 0$

20. $\cot\theta = \dfrac{1}{\tan\theta}, \quad \tan\theta \neq 0$

21. $\cot\theta = \dfrac{\cos\theta}{\sin\theta}, \quad \sin\theta \neq 0$

22. $1 + \cot^2\theta = \csc^2\theta$

23. $1 + \tan^2\theta = \sec^2\theta$

24. $\cos(-\theta) = \cos\theta$

25. $\sin\theta \csc\theta = 1$

26. $\sin\theta = \cos\theta \tan\theta$

27. $\cos^2\theta = 1 - \sin^2\theta$

28. $\sin^2\theta = 1 - \cos^2\theta$

Verify each identity for the given angle measure.

29. $\sin 45° \cot 45° = \cos 45°$

30. $\tan 60° = \sin 60° \sec 60°$

31. $\sin 135° \cos 135° \tan 135° = 1 - \cos^2 135°$

32. $\cos 30° \cot 30° = \csc 30° - \sin 30°$

In triangle *ABC*, sides *a*, *b*, and *c* measure 9, 40, and 41, respectively.

33. Show that triangle *ABC* is a right triangle.

34. Verify the identity $\cos A = \dfrac{\sin A}{\tan A}$ for triangle *ABC*.

35. Verify the identity $\csc B \cos B \tan B = 1$ for triangle *ABC*.
36. Verify the identity $\sec B \sin B \cot B = 1$ for triangle *ABC*.

Applications

37. Engineering   A 20.2-m walkway between two buildings is higher at one end than at the other. The sine of the angle of elevation is 0.1386. Use a Pythagorean identity to find the cosine of that angle. Then find the distance between the buildings.

38. Aviation A plane is flying 8100 m above the ground. The secant of the angle of depression to the end of the runway is 5.7588. Find the tangent of the angle of elevation from the end of the runway to the plane. Then find the ground distance from the plane to the runway.

Graphing Calculator A graphing calculator or computer graphics utility can be used to provide a striking visual demonstration of the Pythagorean Identities.

Graph the following on one screen:

$$y = \sin^2 x$$
$$y = \cos^2 x$$
$$y = \sin^2 x + \cos^2 x$$

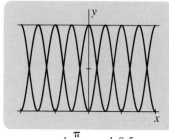

The graph demonstrates clearly that $\sin^2 x + \cos^2 x = 1$ for all values of x.

x scl: $\dfrac{\pi}{2}$ y scl: 0.5

39. What equations should be graphed to provide an illustration of the identity $1 + \cot^2 \theta = \csc^2 \theta$?

40. What equations should be graphed to provide an illustration of the identity $1 + \tan^2 \theta = \sec^2 \theta$?

LOGICAL REASONING

When you prove a trigonometric identity, or any other statement in mathematics, every step must be mathematically sound. Here is an algebraic "proof" that $1 = 2$. Can you find the error in the reasoning?

Given: s and t are positive real numbers, and $s = t$.

Proof:

1.	$s = t$	Given.
2.	$st = t^2$	Multiply both sides by t.
3.	$st - s^2 = t^2 - s^2$	Subtract s^2 from both sides.
4.	$s(t - s) = (t + s)(t - s)$	Factor both sides.
5.	$s = t + s$	Divide both sides by $t - s$.
6.	$t = t + t$	Substitute t for s.
7.	$t = 2t$	$t + t = t(1 + 1) = 2t$
8.	$1 = 2$	Divide both sides by t.
		($t \neq 0$, since it was given that t is a positive real number.)

3.5 Equivalent Trigonometric Expressions

Objective: To use the fundamental identities to write equivalent trigonometric expressions

The skills used to manipulate and simplify algebraic expressions can also be used with trigonometric expressions.

Preview

Recall the following patterns that are used to factor polynomials:

$$am + bm + cm = m(a + b + c) \qquad a^2 - b^2 = (a + b)(a - b)$$
$$a^2 + 2ab + b^2 = (a + b)(a + b) \qquad a^3 + b^3 = (a + b)(a^2 - ab + b^2)$$
$$a^2 - 2ab + b^2 = (a - b)(a - b) \qquad a^3 - b^3 = (a - b)(a^2 + ab + b^2)$$

EXAMPLE Simplify: $\dfrac{x^2 - 25}{x^2 + 10x + 25}$

$$\frac{x^2 - 25}{x^2 + 10x + 25} = \frac{(x + 5)(x - 5)}{(x + 5)(x + 5)} = \frac{x - 5}{x + 5}$$

Simplify.

1. $\dfrac{9x^2 - 4}{9x^2 + 12x + 4}$ 2. $\dfrac{x^2 - 36}{x^2 - 12x + 36}$ 3. $\dfrac{27x^3 - 8}{9x^2 - 4}$ 4. $\dfrac{x^2 - 1}{x^3 + 1}$

The patterns for factoring polynomials can be used along with the fundamental identities to convert one trigonometric expression to another.

EXAMPLE 1 Convert the first trigonometric expression to the second.

   a. $(1 - \cos \theta)(1 + \cos \theta);\quad \sin^2 \theta$

   b. $\dfrac{9 \sin^2 \theta - 24 \sin \theta + 16}{9 \sin^2 \theta - 16};\quad \dfrac{3 \sin \theta - 4}{3 \sin \theta + 4}$

   a. $(1 - \cos \theta)(1 + \cos \theta) = 1 - \cos^2 \theta$

$$= \sin^2 \theta \qquad\qquad \sin^2 \theta + \cos^2 \theta = 1,$$
$$\text{so } 1 - \cos^2 \theta = \sin^2 \theta.$$

   b. $\dfrac{9 \sin^2 \theta - 24 \sin \theta + 16}{9 \sin^2 \theta - 16} = \dfrac{(3 \sin \theta - 4)(3 \sin \theta - 4)}{(3 \sin \theta + 4)(3 \sin \theta - 4)}$

$$= \frac{3 \sin \theta - 4}{3 \sin \theta + 4}$$

Algebraic manipulations can be used together with the fundamental trigonometric identities to write a trigonometric expression in terms of a single function.

EXAMPLE 2 **Simplify each expression by writing it in terms of cosine only.**

$$\textbf{a. } \sin \theta \cot \theta \qquad \textbf{b. } (\csc^2 \theta - 1)(\sin^2 \theta) \qquad \textbf{c. } \frac{1 + \sin \theta}{\cos \theta} + \frac{\cos \theta}{1 + \sin \theta}$$

a. $\sin \theta \cot \theta = (\sin \theta)\left(\dfrac{\cos \theta}{\sin \theta}\right) = \cos \theta$

b. $(\csc^2 \theta - 1)(\sin^2 \theta) = \cot^2 \theta \sin^2 \theta = \left(\dfrac{\cos^2 \theta}{\sin^2 \theta}\right)(\sin^2 \theta) = \cos^2 \theta$

c. $\dfrac{1 + \sin \theta}{\cos \theta} + \dfrac{\cos \theta}{1 + \sin \theta} = \dfrac{(1 + \sin \theta)(1 + \sin \theta)}{(\cos \theta)(1 + \sin \theta)} + \dfrac{\cos \theta \cos \theta}{(\cos \theta)(1 + \sin \theta)}$

$$= \frac{1 + 2 \sin \theta + \sin^2 \theta}{(\cos \theta)(1 + \sin \theta)} + \frac{\cos^2 \theta}{(\cos \theta)(1 + \sin \theta)}$$

$$= \frac{1 + 2 \sin \theta + (\sin^2 \theta + \cos^2 \theta)}{(\cos \theta)(1 + \sin \theta)}$$

$$= \frac{1 + 2 \sin \theta + 1}{(\cos \theta)(1 + \sin \theta)} \qquad \textit{sin}^2\, \theta + \textit{cos}^2\, \theta = 1$$

$$= \frac{2 + 2 \sin \theta}{(\cos \theta)(1 + \sin \theta)} = \frac{2(1 + \sin \theta)}{(\cos \theta)(1 + \sin \theta)} = \frac{2}{\cos \theta}$$

CLASS EXERCISES

Write each expression in terms of sine or cosine only.

1. $\tan \theta \cos \theta$ **2.** $\cot \theta \cos \theta$ **3.** $\dfrac{\cot \theta}{\cos \theta}$ **4.** $\tan \theta \sin \theta$

5. $\dfrac{\sec \theta}{\cos \theta}$ **6.** $\dfrac{\csc \theta}{\sin \theta}$ **7.** $\dfrac{\cot \theta}{\sin \theta}$ **8.** $\dfrac{\tan \theta}{\cos \theta}$

PRACTICE EXERCISES

Convert the first expression to the second expression.

1. $(1 - \sin \theta)(1 + \sin \theta); \cos^2 \theta$

2. $(\csc \alpha + 1)(\csc \alpha - 1); \cot^2 \alpha$

3. $(\sec \beta + 1)(\sec \beta - 1); \tan^2 \beta$

4. $(1 + \tan \alpha)(1 - \cot \alpha); \tan \alpha - \cot \alpha$

5. $\dfrac{\sin^2 \alpha - 25}{\sin^2 \alpha + 10 \sin \alpha + 25}; \dfrac{\sin \alpha - 5}{\sin \alpha + 5}$

6. $\dfrac{\sin^2 \alpha - 36}{\sin^2 \alpha + 12 \sin \alpha + 36}; \dfrac{\sin \alpha - 6}{\sin \alpha + 6}$

7. $\dfrac{\cos^2 \alpha - 14 \cos \alpha + 49}{\cos^2 \alpha - 49}; \dfrac{\cos \alpha - 7}{\cos \alpha + 7}$

8. $\dfrac{\cos^2 \alpha + 16 \cos \alpha + 64}{\cos^2 \alpha - 64}; \dfrac{\cos \alpha + 8}{\cos \alpha - 8}$

9. $\dfrac{\tan^3 \theta - 27}{\tan^2 \theta - 9} ; \dfrac{\tan^2 \theta + 3 \tan \theta + 9}{\tan \theta + 3}$

10. $\dfrac{\tan^3 \theta - 64}{\tan^2 \theta - 16} ; \dfrac{\tan^2 \theta + 4 \tan \theta + 16}{\tan \theta + 4}$

11. $\dfrac{\cot^3 \theta + 8}{\cot^2 \theta - 4} ; \dfrac{\cot^2 \theta - 2 \cot \theta + 4}{\cot \theta - 2}$

12. $\dfrac{\cot^3 \theta + 125}{\cot^2 \theta - 25} ; \dfrac{\cot^2 \theta - 5 \cot \theta + 25}{\cot \theta - 5}$

13. $\dfrac{1}{\sin^2 \theta} - \dfrac{1}{\tan^2 \theta} ; 1$

14. $\dfrac{1}{\cos^2 \theta} - \dfrac{1}{\cot^2 \theta} ; 1$

15. $\dfrac{1 - \cos^2 \theta}{\cos^2 \theta} ; \tan^2 \theta$

16. $\dfrac{1 - \sin^2 \theta}{\sin^2 \theta} ; \cot^2 \theta$

Write each expression in terms of sin θ.

17. $\tan \theta \sec \theta$

18. $(\tan^2 \theta)\left(\dfrac{1}{\sec^2 \theta}\right) + \dfrac{1}{\sin \theta}$

19. $\dfrac{1 + \cos \theta}{\sin \theta} + \dfrac{\sin \theta}{1 + \cos \theta}$

20. $\dfrac{1}{1 - \cos \theta} + \dfrac{1}{1 + \cos \theta}$

Write each expression in terms of cos θ.

21. $\dfrac{\sin^2 \theta}{\sec^2 \theta - 1}$

22. $\dfrac{\sin^2 \theta}{\cos \theta} + \cos \theta$

23. $\dfrac{1}{\sec \theta - \tan \theta} + \dfrac{1}{\sec \theta + \tan \theta}$

24. $(1 - \tan^2 \theta)(\sec^2 \theta)$

Write each expression in terms of tan θ.

25. $\dfrac{\sec \theta}{\csc \theta}$

26. $\dfrac{1 + \tan^2 \theta}{\csc^2 \theta}$

27. $\dfrac{\sec \theta}{\sin \theta} - \dfrac{\sec \theta}{\csc \theta}$

28. $\dfrac{\sec \theta - \cos \theta}{\sin \theta}$

29. Convert $\dfrac{\sin^3 \theta - \cos^3 \theta}{\sin \theta - \cos \theta}$ to $\sin \theta \cos \theta + 1$.

30. Convert $\dfrac{\sin^3 \theta + \cos^3 \theta}{\sin \theta + \cos \theta}$ to $1 - \sin \theta \cos \theta$.

31. If $s = a \sec \theta - b \tan \theta$ and $t = b \sec \theta - a \tan \theta$, find $s^2 - t^2$.

32. If $\sin \alpha = \dfrac{(a + b)^2}{a - b}$ and $\cos \alpha = \dfrac{a^2 + b^2}{a^2 - b^2}$, find $\tan \alpha$.

33. Genetics In an analysis of heredity, a genetic engineer was led to the expression $\dfrac{1}{1 + \sin \theta} + \dfrac{1}{1 - \sin \theta}$. Show that this expression is equivalent to $\dfrac{2}{\cos^2 \theta}$.

Computer The program at the right evaluates the expression $\dfrac{\tan^2 \theta - \sin^2 \theta}{\tan^2 \theta \sin^2 \theta}$ for any angle measure (in degrees) entered by the user. Line 20 changes the angle measure to radians. Lines 40 and 50 evaluate the expression.

```
1   HOME
10  INPUT"ENTER AN ANGLE
    MEASURE IN DEGREES";A
20  R = A*0.0174533
30  S = SIN(R):T = TAN(R)
40  S = S^2:T = T^2
50  E = (T - S)/(T * S)
60  PRINT E
70  END
```

34. Run the program several times, using a different angle measure each time.

35. The same result should be displayed each time the program is run. Explain.

EXTRA

Recall that $x!$ (x factorial) is equal to $x(x - 1)(x - 2)(x - 3) \ldots (3)(2)(1)$.

$$5! = 5(4)(3)(2)(1) = 120 \qquad 8! = 8(7)(6)(5)(4)(3)(2)(1) = 40{,}320$$

In calculus, $\sin x$ and $\cos x$ are each written as the sum of an infinite number of terms using the following formulas, where x is in radians:

$$\sin x = x - \frac{x^3}{3!} + \frac{x^5}{5!} - \frac{x^7}{7!} + \frac{x^9}{9!} - \cdots$$

$$\cos x = 1 - \frac{x^2}{2!} + \frac{x^4}{4!} - \frac{x^6}{6!} + \frac{x^8}{8!} - \cdots$$

EXAMPLE **Find sin 2, using the first five terms of the above formula.**

$$\sin 2 = 2 - \frac{2^3}{3!} + \frac{2^5}{5!} - \frac{2^7}{7!} + \frac{2^9}{9!}$$

$$= 2 - \frac{8}{6} + \frac{32}{120} - \frac{128}{5040} + \frac{512}{362880} = 0.9093$$

Check that this result is correct to the nearest ten-thousandth.

1. Find cos 2 using the first five terms of the formula for cos x. How many more terms are needed to obtain a result correct to the nearest ten-thousandth?

2. Find sin 1 and cos 1 using the first four terms of the formulas.

3. Use the results from Exercise 2 and a fundamental identity to approximate the value of tan 1.

Proving Identities

Objective: To use the fundamental identities to prove other identities

Recall that a *complex rational expression* has a rational expression in its numerator or denominator, or in both. Complex expressions occur in trigonometry as well as in algebra.

Preview

One method used to simplify complex rational expressions is to multiply the numerator and the denominator by the lowest common denominator (LCD) of all the rational expressions in the numerator and the denominator.

EXAMPLE

Simplify: a. $\dfrac{\dfrac{1}{x} + \dfrac{1}{y}}{\dfrac{1}{x} - \dfrac{1}{y}}$ b. $\dfrac{\sin x - 1}{1 - \dfrac{1}{\sin x}}$

a. $\dfrac{\dfrac{1}{x} + \dfrac{1}{y}}{\dfrac{1}{x} - \dfrac{1}{y}} = \dfrac{\left(\dfrac{1}{x} + \dfrac{1}{y}\right)(xy)}{\left(\dfrac{1}{x} - \dfrac{1}{y}\right)(xy)} = \dfrac{y + x}{y - x}$ $LCD = xy$

b. $\dfrac{\sin x - 1}{1 - \dfrac{1}{\sin x}} = \dfrac{(\sin x - 1)(\sin x)}{\left(1 - \dfrac{1}{\sin x}\right)(\sin x)} = \dfrac{(\sin x - 1)(\sin x)}{\sin x - 1} = \sin x$ $LCD = \sin x$

Simplify.

1. $\dfrac{1 - \dfrac{1}{x}}{\dfrac{1}{y}}$

2. $\dfrac{1 + \dfrac{x^2}{y^2}}{\dfrac{1}{x^2}}$

3. $\dfrac{3 + \dfrac{1}{\sin^2 A}}{\dfrac{1}{\sin^2 A}}$

4. $\dfrac{1 + \dfrac{1}{\cos A}}{2 - \dfrac{1}{\sin A}}$

To prove that a trigonometric equation is an identity means to show that the equation is true for any permissible replacement of the variable. Recall that one way to write a proof of an identity is to continue to write equivalent expressions for one side of the equation until you arrive at an expression identical to the one on the other side of the equation.

EXAMPLE 1   Prove: cot θ = cos θ csc θ

$$
\begin{array}{c|l}
\cot \theta & \cos \theta \csc \theta \\
& (\cos \theta)\left(\dfrac{1}{\sin \theta}\right) \\
& \dfrac{\cos \theta}{\sin \theta} \\
= & \cot \theta \qquad\qquad \text{Therefore, } \cot \theta = \cos \theta \csc \theta.
\end{array}
$$

Another way to write a proof of an identity is to replace expressions on *both* sides of the given equation with equivalent expressions.

EXAMPLE 2   Prove: cos θ sec θ = tan θ cot θ

$$
\begin{array}{c|c}
\cos \theta \sec \theta & \tan \theta \cot \theta \\
(\cos \theta)\left(\dfrac{1}{\cos \theta}\right) & (\tan \theta)\left(\dfrac{1}{\tan \theta}\right) \\
1 \qquad = & 1 \qquad \text{Therefore, } \cos \theta \sec \theta = \tan \theta \cot \theta.
\end{array}
$$

There are several strategies to use when you prove identities.

1. Know the fundamental identities and look for ways to apply them.

2. Write all expressions in terms of sines and cosines.

3. If you choose to work with only one side of an identity, continuously refer back to the other side to see what you are trying to obtain.

4. When one side contains only one trigonometric function, attempt to rewrite all the functions on the other side in terms of that function.

5. Use the Pythagorean identities to substitute for expressions equal to 1.

6. Perform algebraic operations.

   a. Factor.

   b. Simplify complex rational expressions.

   c. Find the LCD and combine fractions.

   d. Combine like terms.

   e. Multiply both numerator and denominator by the same expression to obtain an equivalent fraction.

   f. Replace a binomial with a monomial.

Proving an identity is not the same as solving an equation. The vertical line separating the left side from the right side is used to show that each side is done independently. Thus, such equation-solving techniques as operating on both sides are not to be used in the proof of an identity.

EXAMPLE 3   **Prove:** $2 \sec^2 \theta = \dfrac{1}{1 - \sin \theta} + \dfrac{1}{1 + \sin \theta}$

$2 \sec^2 \theta$ $\Big|$ $\dfrac{1}{1 - \sin \theta} + \dfrac{1}{1 + \sin \theta}$

$\dfrac{(1 + \sin \theta) + (1 - \sin \theta)}{(1 - \sin \theta)(1 + \sin \theta)}$ *LCD* $= (1 - \sin \theta)(1 + \sin \theta)$

$\dfrac{2}{1 - \sin^2 \theta}$

$\dfrac{2}{\cos^2 \theta}$ $1 - \sin^2 \theta = \cos^2 \theta$

$= 2 \sec^2 \theta$ Therefore, $2 \sec^2 \theta = \dfrac{1}{1 - \sin \theta} + \dfrac{1}{1 + \sin \theta}$.

In Example 4, a Pythagorean identity is used.

EXAMPLE 4   **Prove:** $\dfrac{\cos^2 \theta}{1 + 2 \sin \theta + \sin^2 \theta} = \dfrac{\sec \theta - \tan \theta}{\sec \theta + \tan \theta}$

$\dfrac{\cos^2 \theta}{1 + 2 \sin \theta + \sin^2 \theta}$ $\Big|$ $\dfrac{\sec \theta - \tan \theta}{\sec \theta + \tan \theta}$

$\cos^2 \theta = 1 - \sin^2 \theta$ $\dfrac{1 - \sin^2 \theta}{(1 + \sin \theta)(1 + \sin \theta)}$ $\dfrac{\dfrac{1}{\cos \theta} - \dfrac{\sin \theta}{\cos \theta}}{\dfrac{1}{\cos \theta} + \dfrac{\sin \theta}{\cos \theta}}$

Factor $1 - \sin^2 \theta.$ $\dfrac{(1 + \sin \theta)(1 - \sin \theta)}{(1 + \sin \theta)(1 + \sin \theta)}$ $\dfrac{\left(\dfrac{1}{\cos \theta} - \dfrac{\sin \theta}{\cos \theta}\right)(\cos \theta)}{\left(\dfrac{1}{\cos \theta} + \dfrac{\sin \theta}{\cos \theta}\right)(\cos \theta)}$

$\dfrac{1 - \sin \theta}{1 + \sin \theta}$ $=$ $\dfrac{1 - \sin \theta}{1 + \sin \theta}$

Therefore, $\dfrac{\cos^2 \theta}{1 + 2 \sin \theta + \sin^2 \theta} = \dfrac{\sec \theta - \tan \theta}{\sec \theta + \tan \theta}$.

CLASS EXERCISES

Factor each expression.

1. $\cot^2 \theta + 2 \cot \theta \csc \theta + \csc^2 \theta$ **2.** $9 - \sec^2 \theta$

3. $1 - \sin^2 \alpha$ **4.** $\sin^2 \alpha + 2 \sin \alpha \cos \alpha + \cos^2 \alpha$

5. $\cos^4 \theta - \sin^4 \theta$ **6.** $9 \cot^4 \theta + 6 \cot^2 \theta + 1$

7. $\sec^4 \beta + 6 \sec^2 \beta + 5$ **8.** $\csc^3 \beta + \sec^3 \beta$

PRACTICE EXERCISES

Prove each identity.

1. $\dfrac{1 - \cos^2 \theta}{\cos^2 \theta} = \left(\dfrac{\sin \theta}{\cos \theta}\right)^2$

2. $\dfrac{\sin^2 \theta + \cos^2 \theta}{\cos^2 \theta} = \left(\dfrac{1}{\cos \theta}\right)^2$

3. $\dfrac{\cos \alpha}{1 - \sin^2 \alpha} = \dfrac{1}{\cos \alpha}$

4. $\dfrac{1 - \cos^2 \lambda}{\sin \lambda} = \sin \lambda$

5. $\dfrac{1}{\tan \alpha} = \dfrac{\cot \alpha}{\sin^2 \alpha + \cos^2 \alpha}$

6. $\dfrac{1 + \cot^2 \alpha}{\csc^2 \alpha} = 1$

7. $\dfrac{1 + \cos \mu}{\cos \mu} = \sec \mu + 1$

8. $\dfrac{\csc^2 \mu - 1}{\csc^2 \mu} = \cos^2 \mu$

9. $\dfrac{\sin^2 \alpha}{1 - \sin^2 \alpha} = \sec^2 \alpha - 1$

10. $\cot \alpha - 1 = \dfrac{\csc \alpha - \sec \alpha}{\sec \alpha}$

11. $\sin^2 \theta \cos \theta \sec \theta = 1 - \cos^2 \theta$

12. $\dfrac{1}{\sin \phi} = \sin \phi + \cos \phi \cot \phi$

13. $\sin^2 \beta = 1 - \dfrac{1}{\sec^2 \beta}$

14. $\dfrac{\cos \beta}{\sec \beta} - \dfrac{\cot \beta}{\tan \beta} = -\cos^2 \beta \cot^2 \beta$

15. $\dfrac{\cos \lambda}{\sec \lambda + \tan \lambda} = 1 - \sin \lambda$

16. $\dfrac{1 + \cot^2 \alpha}{\csc \alpha} = \dfrac{1}{\sin \alpha}$

17. $\csc^2 \beta = \cos^2 \beta + \cot^2 \beta + \sin^2 \beta$

18. $\dfrac{1}{\tan \beta + \cot \beta} = \sin \beta \cos \beta$

19. $2 \cos^2 \alpha - 1 = \cos^4 \alpha - \sin^4 \alpha$

20. $(\sin \lambda - \tan \lambda)^2 = (\tan^2 \lambda)(\cos \lambda - 1)^2$

21. $\dfrac{\cos \lambda}{1 + \sin \lambda} + \dfrac{\cos \lambda}{1 - \sin \lambda} = \dfrac{2}{\cos \lambda}$

22. $\dfrac{\sec \lambda}{\cos \lambda} - 1 = \dfrac{\sin^2 \lambda}{\cos^2 \lambda}$

23. $\dfrac{\sin^4 \theta - 1}{\cos^2 \theta} = \cos^2 \theta - 2$

24. $\dfrac{\sin \theta - 1}{1 - \sin^2 \theta} = \dfrac{\csc \theta}{-\csc \theta - 1}$

25. $\tan^2 \alpha + \cos^2 \alpha - 1 = \tan^2 \alpha - \sin^2 \alpha$

26. $2 \csc \lambda = \dfrac{1}{\csc \lambda - \cot \lambda} + \dfrac{1}{\csc \lambda + \cot \lambda}$

27. $\dfrac{1 - \sin \alpha}{1 + \sin \alpha} = (\tan \alpha - \sec \alpha)^2$

28. $(\sin \theta + \cos \theta)^2 \tan \theta = \tan \theta + 2 \sin^2 \theta$

29. $(1 + \sin \theta + \cos \theta)(1 - \sin \theta - \cos \theta) = -2 \sin \theta \cos \theta$

30. $(\sec \theta)(\sin \theta \sec \theta + 1) = \dfrac{\sec^2 \theta - \tan^2 \theta + \tan \theta}{\cos \theta}$

31. $\dfrac{\tan^2 \alpha}{1 - \cos^2 \alpha} + \dfrac{\sin \alpha}{\sec^2 \alpha - 1} = (\cos \alpha)(\sec^3 \alpha + \cot \alpha)$

32. $\dfrac{\cot \alpha}{1 - \sin^2 \alpha} + \dfrac{\cos \alpha}{\csc^2 \alpha - 1} = (\sec \alpha)(\sin^2 \alpha + \csc \alpha)$

33. $\sqrt{(5 \cos \theta + 12 \sin \theta)^2 + (12 \cos \theta - 5 \sin \theta)^2} = 13$

34. $\dfrac{-2 \sin \alpha \cos \alpha}{1 - \sin \alpha - \cos \alpha} = 1 + \sin \alpha + \cos \alpha$

35. $\dfrac{2 \sin \theta \cos \theta}{\sin \theta + \cos \theta - 1} = \sin \theta + \cos \theta + 1$

Hint: Multiply the numerator and denominator on the left side by $(\sin \theta + \cos \theta + 1)$.

36. $\dfrac{\sin \alpha - \cos \alpha - 1}{\sin \alpha + \cos \alpha - 1} = -\dfrac{\cos \alpha + 1}{\sin \alpha}$

Hint: Multiply the numerator and denominator on the left side by $(\sin \theta + \cos \theta + 1)$.

Applications

Computer This program evaluates $\cos x + \sin x \tan x$ for $0° \le x \le 360°$. The value of the expression is displayed.

37. Run the program for several values of x.

38. For what angle measures from $0°$ through $360°$ will an error message be displayed? Why?

39. Write $\cos x + \sin x \tan x$ in terms of a single function of x.

```
10  HOME
20  INPUT"ENTER AN ANGLE MEASURE
    FROM 0 TO 360 DEGREES";A
30  IF A < 0 OR A > 360 THEN 20
40  R = A*0.0174533
50  E = COS(R) + SIN(R)*TAN(R)
60  PRINT E
70  END
```

HISTORICAL NOTE

Milestones in the History of Computing Devices Human beings have been interested in building work-saving calculating devices since ancient times. Around 300 BC the first known calculating device, the abacus, was created by the Babylonians and/or the Egyptians. In the seventeenth century, mechanical devices were developed in France by Blaise Pascal and in Germany by Gottfried Leibniz.

During the next two centuries, mechanical, rotary, and electrical calculating devices continued to be invented and refined. Each refinement generally upgraded the capability and features of the devices. Through the 1960s, the devices were rather large. This changed with the invention of the electronic microchip, and the hand-held calculator became a reality in the early 1970s.

Find information on the mechanical calculating device invented by Pascal or Leibniz. Write a report on your findings.

3.7 Graphical Representation of Identities

Objectives: To check identities by graphing both sides of the equation

You should be familiar with the graphs of the trigonometric functions when you use a computer or a graphing calculator to check identities. If you know the general characteristics of a graph, you are more likely to catch errors made while entering the equations in the computer or calculator.

Preview

Graph each function.

1. $y = \sin x$

2. $y = \tan x$

3. $y = \sec x$

4. $y = 2 \sin x$

5. $y = \cos 2x$

6. $y = \sin x + 2$

EXAMPLE Graph: $y = \cos x$

 To check that a trigonometric equation is an identity, first graph the left side of the equation using a graphing calculator or a computer. Then, without clearing the screen, graph (overwrite) the right side. If the graphs coincide, you can be fairly sure that the original equation is an identity. Note that graphs are inexact, so this method of verification does not constitute a proof.

EXAMPLE 1 **Verify graphically that $\tan x = \dfrac{\sin x}{\cos x}$ is an identity.**

Graph the left side. $y = \tan x$

Without clearing the screen,

graph the right side. $y = \dfrac{\sin x}{\cos x}$

The graphs coincide. This suggests

that $\tan x = \dfrac{\sin x}{\cos x}$ is an identity.

x scl: $\dfrac{\pi}{2}$ y scl: 10

If the graphs are not identical and the equations have been graphed correctly, then the given equation is *not* an identity.

EXAMPLE 2 **Check graphically to see whether $\sec x \sin^2 x = \cos^2 x$ is an identity.**

Since the graphs of the two sides are not the same, this is *not* an identity.

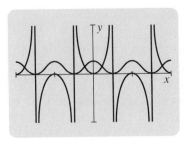

x scl: $\dfrac{\pi}{2}$ y scl: 3

CLASS EXERCISES

Simplify each expression. Then check your result by graphing the given expression and your answer.

1. $\dfrac{1}{\sec \theta}$

2. $\dfrac{1}{\cot x}$

3. $\dfrac{1}{\tan B}$

4. $\dfrac{1}{\csc x}$

5. $\cot x \tan x$

6. $\sin A \csc A$

PRACTICE EXERCISES   Use appropriate technology.

Check graphically to see whether or not each equation is an identity.

1. $\dfrac{\sin A}{\csc A} = \sin^2 A$

2. $\dfrac{\cos A}{\sec A} = \cos^2 A$

3. $\dfrac{\tan x}{\sin x} = \dfrac{1}{\cos x}$

4. $\dfrac{\cos x}{\sin x} = \cot x$

5. $\cos B \csc B = \cot B$

6. $\cos \beta \sec \beta = \tan \beta \cot \beta$

7. $\sec A \cot A = \csc A$

8. $\sin \alpha \cot \alpha = \cos \alpha$

9. $\dfrac{\sin x \csc x}{\tan x} = \cot x$

10. $\dfrac{\cot x \tan x}{\sin x} = \csc x$

11. $\cos x \sin x \cot x = 2 \cos x$

12. $\sin x + \cos x \tan x = 2 \sin x$

13. $\dfrac{1 - \cos^2 \beta + \sin \beta}{1 + \sin \beta} = \sin \beta$

14. $\dfrac{\sec \theta + 1}{\sin \theta} = \dfrac{1 + \cos \theta}{\sin \theta \cos \theta}$

15. $(1 - \sin \theta)^2 = (\sec \theta - \tan \theta)^2$

16. $(\sec^2 \theta)(\tan^2 \theta - 1) = (\tan^2 \theta)(\tan^2 \theta + 1)$

17. $\tan^2 \beta \cos^2 \beta = 1 - \cot^2 \beta \sin^2 \beta$ **18.** $\sin^2 \beta + \dfrac{\cos^2 \beta}{\sin^2 \beta} + \cos^2 \beta = \dfrac{1}{\cos^2 \beta}$

19. $\dfrac{\sin^2 A + \cos^2 A}{\sin A \cos A} = \sec A \csc A$ **20.** $\dfrac{\sec \lambda \sin \lambda + \tan \lambda}{\cos \lambda \cot \lambda} = \dfrac{2 \sin^2 \lambda}{\cos^3 \lambda}$

21. $\dfrac{\sin \alpha}{\cos \alpha - 1} + \dfrac{\sin \alpha}{\cos \alpha + 1} = \dfrac{2 \cos \alpha}{-\sin \alpha}$ **22.** $\dfrac{\cot \theta}{\cos \theta} - \dfrac{\csc^2 \theta}{\sec \theta} = \cot \theta$

23. $\dfrac{\sin \mu}{\sec \mu - 1} = \dfrac{\sin \mu + 1}{\tan \mu}$ **24.** $\dfrac{2 + 2 \cos^2 \mu}{1 - \cos^2 \mu} = 4 \cot \mu \csc \mu$

25. $\dfrac{2 \sin^2 x - 1}{\sin x \cos x} = \tan x - \cot x$ **26.** $\dfrac{\tan \lambda + \sin \lambda}{\sin \lambda + \sec \lambda} = 1 + \sin \lambda$

27. $\dfrac{2 \cos^2 \beta + \cot \beta}{\tan \beta} = \dfrac{2 \sin \beta \cos^3 \beta + \cos^2 \beta}{\sin \beta}$

28. $\dfrac{\sin \alpha - \sin^3 \alpha + \cos^2 \alpha}{\cos^2 \alpha} = \sin \alpha + 1$

Applications

29. Physics A scientist working with the expression $4 \cos \theta + 3 \sin \theta$ guessed that it was about equivalent to $5 \cos (\theta - 36°52')$. Verify this graphically.

30. Metallurgy A metallurgist has to find the values of θ between $0°$ and $360°$ for which $\sec \theta + \csc \theta = 0$. Find the solutions graphically.

TEST YOURSELF

Verify each trigonometric identity.

                                                                3.4

1. $\cot^2 \left(\dfrac{2\pi}{3} \right) = \csc^2 \left(\dfrac{2\pi}{3} \right) - 1$ **2.** $\cos 210° \cot 210° = \csc 210° - \sin 210°$

Convert the first expression to the second expression.

                                                                3.5

3. $\dfrac{\cot \theta}{\sin \theta} (\sec \theta - \cos \theta); \ 1$ **4.** $(\cos \alpha + 1)(\cos \alpha - 1) - \cos^2 \alpha; \ -1$

Prove each identity.

5. $\dfrac{\cot \theta}{\sec \theta + 1} = \dfrac{\sec \theta - 1}{\tan^3 \theta}$ **6.** $\dfrac{\tan \beta}{\sec \beta} + \dfrac{\cot \beta}{\csc \beta} = \sin \beta + \cos \beta$ 3.6

Check graphically to see whether or not each equation is an identity.   3.7

7. $(\cot x + \csc x)^2 = \dfrac{\sec x + 1}{\sec x - 1}$ **8.** $2 \sin x = \dfrac{\sin x}{1 + \cos x} + \dfrac{1 + \cos x}{\sin x}$

APPLICATION:
Compact Discs

Did you know that a revolution in recorded music began with the introduction of the compact disc (CD) in 1983? Compact discs have captured approximately one-quarter of the audio market. It is likely that CDs will one day make conventional long-playing vinyl records obsolete. What is a CD, how is sound stored on one, and how does a CD player retrieve that sound?

A CD is a plastic coated metallic disc that is less than five inches in diameter. Despite its small size, it is capable of providing approximately one hour of uninterrupted playing time. Its tracks are so fine that over 300 of them could fit next to one another in a single groove of a standard vinyl record! There are no grooves on the surface of a CD, and the CD player has no stylus. CDs are scanned by a beam of laser light in the CD player that is focused deep into the disc. Because there is no physical contact between the CD and the sensing device of the player, there is virtually no wear on the CD as there is on a conventional record with a stylus. Thus, if handled with reasonable care, a CD will not wear out nor deteriorate with age.

As do other music reproduction systems, a CD system stores and retrieves sound waves. The sound waves are translated by a microphone into an alternating electrical current. The waveform of this current represents the original sound wave input and is called an *analog*. The electrical analog is then used to produce a permanent physical equivalent of the sound wave, which can then be stored in the grooves of a conventional vinyl record.

Unlike analog systems such as vinyl records, CDs are *digitally* encoded using data from audio waveforms. They do not reproduce and store the actual shape of the waveform. Instead, a digital system records frequent measurements of the height of the graph. Indeed, the measurements are made 44,100 times each second. The CD player uses a binary code to handle the numbers rapidly, using just the symbols 0 and 1. Electrically, 1 corresponds to "on" and 0 to "off." A compact disc is encoded using 16 binary digits. Thus, the audio waveforms are sampled $2^{16} = 65,536$ times, which leads to very precise plotting.

The following diagram demonstrates how information is recorded on a CD and how it is played back. The original sound wave is sampled, and the wave height is recorded as a binary number (for purposes of this example, only 3 binary digits are used). When a digitally recorded compact disc is played, the sampled points are recreated, giving the signal a stair-step appearance. If the points are connected with a smooth line, the curve is shaped like the graphs of the sine or cosine functions.

EXERCISES

1. Draw the graph of a sound wave with sample points 7, 6, 6, 5, 3, 2, 1, 0, 1, 2, 3, 5, 6, 6, 7.

2. Convert the sample points from Exercise 1 to binary notation.

3. Referring to the sound wave shown below, find the voltage of selected sample points and then convert them to binary notation.

Vocabulary

angle of depression (118)
angle of elevation (118)
bearing (124)
course (124)
fundamental identities (133)

odd-even identities (132)
Pythagorean identities (131)
ratio identities (131)
reciprocal identities (133)
significant digit (113)
trigonometric identity (130)

Solving Right Triangles Right triangle trigonometry can be used to find   3.1
lengths of missing sides and measures of missing angles.

Solve each triangle ABC ($\angle C = 90°$).

1. $\angle B = 21°$, $a = 170$

2. $a = 11.3$, $b = 20.5$

Angles of Elevation and Depression Angles of elevation and depression   3.2
are used to find distances that cannot be measured directly.

3. A girl who is 54 in. tall casts a 140-in. shadow. Find the angle of elevation
from the tip of the shadow to the sun.

4. The angle of depression from an airplane to an airport is 25°. If the plane
is flying at an altitude of 1200 m and is directly over a football stadium,
how far is it from the stadium to the airport?

Applications In navigation, the *course* is the angle measured clockwise   3.3
from north to the line of travel, and the *bearing* is the angle measured
clockwise from north to the line of sight to a point of reference.

5. An airplane is 24.3 km due east of radar station A, and radar station B
is 37.4 km due south of radar station A. Find the bearing of radar station B
from the plane and the distance between radar station B and the plane.

6. To find the distance AB across a lake, a surveyor
made the following measurements:

$$AC = 420 \text{ yd} \qquad \angle C = 57°$$

Find AB.

7. A surveyor determined that the diagonal of a city block formed an angle of
28° with one of the streets forming the block. The side of the block on that
street was 600 ft long. How long was the diagonal?

Fundamental Identities   The ratio, reciprocal, Pythagorean, and odd-even   **3.4**
identities are considered the fundamental trigonometric identities.

Reciprocal Identities

$$\csc \theta = \frac{1}{\sin \theta}, \quad \sin \theta \neq 0$$

$$\sec \theta = \frac{1}{\cos \theta}, \quad \cos \theta \neq 0$$

$$\cot \theta = \frac{1}{\tan \theta}, \quad \tan \theta \neq 0$$

Ratio Identities

$$\tan \theta = \frac{\sin \theta}{\cos \theta}, \quad \cos \theta \neq 0$$

$$\cot \theta = \frac{\cos \theta}{\sin \theta}, \quad \sin \theta \neq 0$$

Pythagorean Identities

$$\sin^2 \theta + \cos^2 \theta = 1$$

$$1 + \cot^2 \theta = \csc^2 \theta$$

$$1 + \tan^2 \theta = \sec^2 \theta$$

Odd-Even Identities

$$\sin(-\theta) = -\sin \theta$$

$$\cos(-\theta) = \cos \theta$$

$$\tan(-\theta) = -\tan \theta$$

Verify each identity for the given angle measure.

8. $1 + \cot^2 150° = \csc^2 150°$

9. $\tan \dfrac{7\pi}{6} = \dfrac{\sin \dfrac{7\pi}{6}}{\cos \dfrac{7\pi}{6}}$

Equivalent Trigonometric Expressions   The fundamental identities,   **3.5**
along with algebraic manipulations, can be used to write a trigonometric
expression equivalent to a given expression.

Convert the first trigonometric expression to the second expression.

10. $\dfrac{\csc^2 \theta - 10 \csc \theta + 25}{\csc^2 \theta - 25}; \dfrac{\csc \theta - 5}{\csc \theta + 5}$

11. $\dfrac{\cos^2 \theta}{1 - \cos^2 \theta}; \cot^2 \theta$

Proving Identities   Trigonometric identities can be proved using algebraic   **3.6**
manipulations and the fundamental identities.

Prove each identity.

12. $\sin \theta \sec \theta + \cot \theta = \dfrac{1}{\sin \theta \cos \theta}$

13. $\dfrac{\sec^2 \beta - 1}{\sec \beta - 1} = 1 + \dfrac{1}{\cos \beta}$

Graphical Representation of Identities   If the graphs of the two   **3.7**
sides of an equation are identical, it is probable that the equation is an
identity. If the graphs are not the same, the equation is not an identity.

Check graphically to see whether or not each equation is an identity.

14. $\dfrac{1}{\cos x} - \dfrac{\cos x}{1 + \sin x} = \tan x$

15. $\dfrac{\csc^2 B - \cot B}{\sin^2 B - \tan B} = \tan^2 B$

Solve each triangle ABC ($\angle C = 90°$).

1. $\angle A = 56°$, $b = 14$ **2.** $a = 19.2$, $b = 42.1$

3. The angle of elevation of a 16-ft ladder leaning against a wall is 36°. How far is the base of the ladder from the wall?

4. From an airplane flying over a bridge at an altitude of 640 m, the angle of depression to the airport is 26°. How far is it from the bridge to the airport?

5. A boat is 21 m due west of buoy A. Buoy B is 38 m due south of buoy A. Find the bearing of buoy B from the boat and the distance between buoy B and the boat.

Verify each identity for the given angle measure.

6. $\dfrac{\cot 60°}{\sin^2 60° + \cos^2 60°} = \cot 60°$

7. $\dfrac{\sec^2 135°}{1 + \tan^2 135°} = 1$

Convert the first trigonometric expression to the second expression.

8. $\dfrac{\cos^2 \theta - 81}{\cos^2 \theta + 18 \cos \theta + 81}; \dfrac{\cos \theta - 9}{\cos \theta + 9}$

9. $\dfrac{\tan^2 \theta - 100}{\tan^2 \theta - 20 \tan \theta + 100}; \dfrac{\tan \theta + 10}{\tan \theta - 10}$

10. $\dfrac{\sin^2 \alpha}{\cos^2 \alpha} + \sin \alpha \csc \alpha; \sec^2 \alpha$

11. $\dfrac{\sec^2 \alpha - 1}{\csc^2 \alpha - 1}; \tan^4 \alpha$

Prove each identity.

12. $\sec \theta = \cos \theta + \tan \theta \sin \theta$

13. $\dfrac{\cos A}{\sec A} + \dfrac{\sin A}{\csc A} = \sec^2 A - \tan^2 A$

14. $\sin \beta \cos \beta + \sin \beta = \dfrac{\sin^3 \beta}{1 - \cos \beta}$

Check graphically to see whether or not each equation is an identity.

15. $\tan 2x = \dfrac{2 \sin x \cos x}{\sin^2 x - \cos^2 x}$

16. $\cos x + \sin x = \sqrt{2} \sin \left(x + \dfrac{\pi}{4} \right)$

Challenge

Show that the area of any isosceles triangle with legs of lengths a and base angles of measure θ is given by the expression $a^2 \cos \theta \sin \theta$.

Solve. Grid in your response on the answer sheet.

1. Evaluate $9^0 + 9^{-\frac{1}{2}} + 16^{\frac{1}{4}}$.

2. Find the value of $-4 \sin 270$.

3. The diagonal of a square is $7\sqrt{2}$ cm. Find the length of the side of the square in cm.

Use the diagram for Exercises 4 and 5.

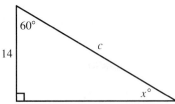

4. What is the value of x?

5. Find the length of c.

6. In her closet Andrea has 16 items of clothing, all of which are blouses or skirts. If 75 percent of the items are blouses, how many of the items are skirts?

7. In the figure below, if $\overline{AB}$ is a line segment, then what is the value of x?

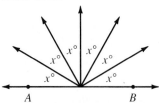

8. Eileen scored 87, 93, 100, and 85 on four tests. What must she score on her fifth test to have an average of 90?

9. Evaluate: $\dfrac{1 + \tan^2 135°}{\sec^2 135°}$

10. An observer in a lighthouse 225 ft above sea level spots a ship in the harbor. The angle of depression from the observer to the ship is 45°. How far in feet is the ship from the shore?

11. The length of the hypotenuse of a right triangle is 26 in. One leg is 14 in. longer than the other. Find the length in inches of the shorter leg.

Use the diagram for Exercise 12.

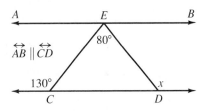

12. What is the value in degrees of $200° - x$?

13. If $\sin x = -\dfrac{\sqrt{3}}{2}$ and x is in quadrant IV, find the exact value of $\cos x$.

14. For a science project, 27 students were divided into two groups. One group had 3 more than twice the number of students in the other. How many students were in the smaller group?

Simplify each expression.

Example $\dfrac{x}{r} - \dfrac{t}{y} = \dfrac{x(y)}{r(y)} - \dfrac{t(r)}{y(r)} = \dfrac{xy - tr}{ry}$

1. $\dfrac{2}{q} + \dfrac{1}{3}$

2. $\dfrac{2}{x} + \dfrac{3}{2y}$

3. $\dfrac{2}{p} - \dfrac{p + q}{q}$

4. $\dfrac{2}{5} - 4x$

5. $\dfrac{x}{x - 3} - \dfrac{3}{x - 3}$

6. $\dfrac{x - 3}{x^2 - 4} + \dfrac{x}{x + 2}$

Solve for x.

Example

$$\dfrac{1}{2} + \dfrac{2x}{x + 4} = 3$$

$$2(x + 4)\left(\dfrac{1}{2} + \dfrac{2x}{x + 4)}\right) = 3 \cdot 2\,(x + 4)$$ *Multiply both sides by LCD.*

$$x + 4 + 4x = 6x + 24$$
$$5x + 4 = 6x + 24$$
$$-20 = x$$

7. $\dfrac{2}{3}x - \dfrac{4}{5} = R$

8. $\dfrac{x}{a} + \dfrac{x}{b} = 2$

9. $\dfrac{2x}{a} + \dfrac{c}{d} = b$

10. $\dfrac{ax}{b} + c = d$

11. $\dfrac{2x}{5y} = \dfrac{R - S}{4}$

12. $\dfrac{x}{x - 3} + \dfrac{4}{7} = -2$

Solve for x.

Examples $8 - 2x > 16$ $14 + \dfrac{x}{3} < 9$

$$-2x > 8 \qquad\qquad \dfrac{x}{3} < -5$$
$$x < -4 \qquad\qquad\; x < -15$$

13. $-2x > 7x$

14. $\dfrac{2}{3} - 5x > 7$

15. $5 + x > -2x + 7$

16. $-3x - 5 < -7x - 10$ **17.** $2x < 3x + 7$

18. $\dfrac{x}{2} < \dfrac{2}{3} + 8$

19. Two angles are supplementary. The measure of one angle is $\dfrac{4}{5}$ the measure of the other angle. Find the measure of the smaller angle.

20. The measure of an angle is 46° more than the measure of its complement. Find the measure of the larger angle.

4 Oblique Triangles

Robotics is the use of mechanical devices to perform tasks which are normally performed by humans. Robots are programmed to use linear and rotary (angular) motion to conduct routine, heavy, dangerous, or repetitive tasks. They can even be programmed to play chess!

The Law of Sines

Objectives: To introduce and prove the law of sines
To use the law of sines to solve triangles when the
measures of two angles and one side are known

In Chapter 3, you learned to solve *right* triangles. In this chapter, you will find
the unknown sides and angles of *oblique* (nonright) triangles.

Preview

A triangle is called *scalene* if no two of its sides have the same length. Draw a
scalene triangle with no right angles on a full sheet of paper. Label the angles
A, B, and C and the sides opposite these angles a, b, and c, respectively.

1. Use a ruler to measure a, b, and c and a protractor to measure $\angle A$, $\angle B$,
and $\angle C$.

2. Use a calculator to compute the following three ratios: $\dfrac{\sin A}{a}$, $\dfrac{\sin B}{b}$, $\dfrac{\sin C}{c}$.

3. What appears to be true about the three ratios you calculated in Exercise 2?

4. Repeat the activity with a second triangle. What can you conclude?

The activity in the Preview suggests a relationship called the *law of sines*.

Law of Sines

For any $\triangle ABC$ in which a, b, and c are the lengths of the sides opposite
the angles with measures A, B, and C, respectively,

$$\frac{\sin A}{a} = \frac{\sin B}{b} = \frac{\sin C}{c}$$

The law of sines can also be expressed in the following alternate form:

$$\frac{a}{\sin A} = \frac{b}{\sin B} = \frac{c}{\sin C}$$

To prove the law of sines, consider two cases, one in which all angles are
acute, and one in which one angle is *obtuse* (has a measure between 90° and
180°). Let A, B, and C be the angles of any scalene triangle, and let a, b, and
c be the sides opposite these angles. Draw the altitude (height) h_1 from C to
side AB or to the extension of AB. Draw the altitude h_2 from A to side BC.

156 Chapter 4 Oblique Triangles

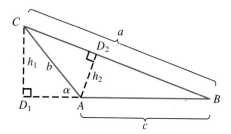

All three angles are acute (case 1).　　　$\angle A$ is obtuse (case 2).

In the acute triangle (case 1), $\sin A = \dfrac{h_1}{b}$ and $\sin B = \dfrac{h_1}{a}$.

Thus, $h_1 = b \sin A$ and $h_1 = a \sin B$.

$b \sin A = a \sin B$　　*Substitution*

$\dfrac{\sin A}{a} = \dfrac{\sin B}{b}$　　*Divide both sides by ab.*

Similarly, $\sin B = \dfrac{h_2}{c}$ and $\sin C = \dfrac{h_2}{b}$. Thus, $h_2 = c \sin B$ and $h_2 = b \sin C$.

$c \sin B = b \sin C$　　*Substitution*

$\dfrac{\sin B}{b} = \dfrac{\sin C}{c}$　　*Divide both sides by bc.*

Therefore, $\dfrac{\sin A}{a} = \dfrac{\sin B}{b} = \dfrac{\sin C}{c}$.

In the obtuse triangle (case 2), $\sin B = \dfrac{h_1}{a}$ and $\sin \alpha = \sin (180° - A) = \dfrac{h_1}{b}$.

Thus, $h_1 = a \sin B$ and $h_1 = b \sin (180° - A)$.

$b \sin (180° - A) = a \sin B$　　*Substitution*

$b \sin A = a \sin B$　　*sin (180° − A) = sin A (See top of page 158.)*

$\dfrac{\sin A}{a} = \dfrac{\sin B}{b}$　　*Divide both sides by ab.*

Similarly, $\sin B = \dfrac{h_2}{c}$ and $\sin C = \dfrac{h_2}{b}$. Thus, $h_2 = c \sin B$ and $h_2 = b \sin C$.

$c \sin B = b \sin C$　　*Substitution*

$\dfrac{\sin B}{b} = \dfrac{\sin C}{c}$　　*Divide both sides by bc.*

Therefore, $\dfrac{\sin A}{a} = \dfrac{\sin B}{b} = \dfrac{\sin C}{c}$.

You will be asked to prove case 3 (the right triangle) in the exercises.

The proof of the law of sines uses the following trigonometric identity:

$$\sin(180° - \theta) = \sin \theta$$

To see why this is an identity, refer to the figure at the right. Angle θ intersects a circle of radius r at point (x, y). Similarly, $180° - \theta$ intersects the circle at point $(-x, y)$. Since $\sin \theta$ and $\sin(180° - \theta)$ are each equal to $\frac{y}{r}$, $\sin(180° - \theta) = \sin \theta$.

The figure can also be used to prove the following identity, which will be used later in this chapter.

$$\cos(180° - \theta) = -\cos \theta$$

You will be asked to prove this identity in the exercises.

The law of sines can be used to find the unknown sides of a triangle when the measures of two angles and an included side are known.

EXAMPLE 1   **Solve $\triangle ABC$, pictured at the right, if $\angle A = 63°$, $\angle B = 49°$, and $c = 78$. Give the lengths of the sides to two significant digits and the angle measure to the nearest degree.**

$\angle C = 180° - (63° + 49°) = 68°$ *Find $\angle C$.*

$$\frac{\sin A}{a} = \frac{\sin C}{c} \quad \text{and} \quad \frac{\sin B}{b} = \frac{\sin C}{c} \qquad \textit{Use the law of sines.}$$

$$\frac{\sin 63°}{a} = \frac{\sin 68°}{78} \qquad\qquad \frac{\sin 49°}{b} = \frac{\sin 68°}{78}$$

$$78 \sin 63° = a \sin 68° \qquad\qquad 78 \sin 49° = b \sin 68°$$

$$\frac{78 \sin 63°}{\sin 68°} = a \qquad\qquad \frac{78 \sin 49°}{\sin 68°} = b \qquad \textit{Calculation-ready form}$$

$$75 = a \qquad\qquad\qquad 63 = b$$

Therefore, $\angle A = 63°$, $\angle B = 49°$, $\angle C = 68°$, $a = 75$, $b = 63$, and $c = 78$.

The law of sines can also be used when the measures of two angles and the side opposite an angle are given.

EXAMPLE 2   **Find the unknown sides for the triangle pictured at the right. Give the lengths of the sides to three significant digits and the angle measure to the nearest tenth of a degree.**

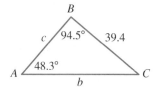

158   Chapter 4 Oblique Triangles

$$\angle C = 180° - (94.5° + 48.3°) = 37.2° \qquad \textit{Find } \angle C.$$

$$\frac{\sin A}{a} = \frac{\sin B}{b} \qquad \text{and} \qquad \frac{\sin A}{a} = \frac{\sin C}{c}$$

$$\frac{\sin 48.3°}{39.4} = \frac{\sin 94.5°}{b} \qquad\qquad \frac{\sin 48.3°}{39.4} = \frac{\sin 37.2°}{c}$$

$$b \sin 48.3° = 39.4 \sin 94.5° \qquad\qquad c \sin 48.3° = 39.4 \sin 37.2°$$

$$b = \frac{39.4 \sin 94.5°}{\sin 48.3°} \qquad\qquad c = \frac{39.4 \sin 37.2°}{\sin 48.3°}$$

$$b = 52.6 \qquad\qquad\qquad c = 31.9$$

Therefore, $\angle A = 48.3°$, $\angle B = 94.5°$, $\angle C = 37.2°$, $a = 39.4$, $b = 52.6$, and $c = 31.9$.

The law of sines can be combined with techniques from earlier chapters to solve real-world problems.

EXAMPLE 3 **From a point A, the angle of elevation to the top of a tree is 38°. From a point B 25 ft closer to the tree, the angle of elevation to the top is 48°. How far is it from point B to T? How tall is the tree? Express your answers to the nearest foot.**

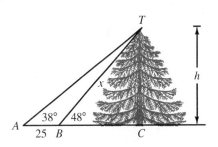

$$\angle ABT = 180° - 48° = 132° \qquad\qquad \textit{$\angle ABT$ and $\angle CBT$ are supplementary.}$$
$$\angle ATB = 180° - (132° + 38°) = 10°$$

$$\frac{\sin 10°}{25} = \frac{\sin 38°}{x} \qquad\qquad \textit{Use the law of sines.}$$

$$x \sin 10° = 25 \sin 38°$$

$$x = \frac{25 \sin 38°}{\sin 10°} \qquad\qquad \textit{Calculation-ready form}$$

$$= 88.6363 \qquad\qquad \textit{Do not round to the nearest whole number yet, since the value of x is used in calculating h.}$$

$$\sin 48° = \frac{h}{x} \qquad\qquad \textit{Definition of sine}$$

$$h = x \sin 48°$$

$$= 88.6363(\sin 48°) \qquad\qquad \textit{On a calculator, simply retain the displayed}$$
$$= 65.8696 \qquad\qquad\qquad \textit{value of x (88.63632829) and multiply by}$$
$$\qquad\qquad\qquad\qquad\qquad\qquad \textit{sin 48°.}$$

The tree is approximately 66 ft high, and it is approximately 89 ft from point B to the top of the tree.

CLASS EXERCISES

Find the measures of the unknown angle or angles in each triangle.

1.

2.

3.

4.

Solve each triangle for the indicated side. Leave the answer in calculation-ready form.

5.

6.

7.

8.

9. Niels, a geometry student, recently learned that "If two angles of a triangle have unequal measures, then the sides opposite those angles also have unequal measures in the same order." In symbols this means that if $A > B$, then $a > b$. (See figure.) From this, Niels thought he could conclude the following proportion: $\dfrac{A}{B} = \dfrac{a}{b}$. Alicia, a trigonometry student, disagreed with Niels. Who is right? Why?

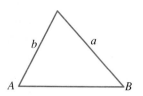

PRACTICE EXERCISES   Use appropriate technology.

Solve each triangle for the indicated side. Express your answer to two significant digits.

1.

2.

3.

4.

5. $\angle A = 39°$, $\angle B = 42°$, $c = 47$; find a.

6. $\angle A = 41°$, $\angle B = 57°$, $c = 52$; find b.

7. $\angle B = 72°$, $\angle C = 31°$, $a = 103$; find b.

8. $\angle B = 34°$, $\angle C = 71°$, $a = 115$; find b.

9. $\angle A = 48°$, $\angle B = 38°$, $b = 49$; find c.

10. $\angle A = 35°$, $\angle B = 56°$, $a = 51$; find c.

11. $\angle A = 128°$, $\angle C = 19°$, $a = 47$; find c.

12. $\angle B = 119°$, $\angle C = 21°$, $b = 59$; find c.

13. $\angle A = 49.7°$, $\angle B = 48.6°$, $a = 31$; find b.

14. $\angle A = 48.7°$, $\angle B = 40.2°$, $b = 29$; find a.

Solve each triangle *PQR*. Give angle measures to the nearest degree and lengths to two significant digits.

15. $p = 18$, $\angle Q = 46°$, $\angle R = 39°$ 16. $p = 24$, $\angle Q = 51°$, $\angle R = 38°$

17. $q = 48$, $\angle P = 63°$, $\angle R = 51°$ 18. $q = 75$, $\angle P = 42°$, $\angle R = 20°$

Solve each triangle *PQR*. Express the lengths of the sides to three significant digits and angle measures to the nearest tenth of a degree.

19. $\angle P = 32.6°$, $\angle R = 46.9°$, $r = 115$ 20. $\angle P = 45.8°$, $\angle R = 32.6°$, $p = 113$

21. $\angle P = 54.2°$, $\angle Q = 45.9°$, $r = 76.1$ 22. $\angle P = 76.7°$, $\angle Q = 29.3°$, $r = 87.0$

23. $\angle Q = 113.4°$, $\angle P = 27.5°$, $p = 56.3$ 24. $\angle Q = 129.7°$, $\angle P = 23.8°$, $p = 112$

25. $\angle R = 54.9°$, $\angle Q = 110.3°$, $q = 73.2$ 26. $\angle R = 21.7°$, $\angle Q = 97.5°$, $q = 85.3$

27. $\angle P = 47.3°$, $\angle Q = 65.2°$, $p = 96.4$ 28. $\angle P = 55.9°$, $\angle Q = 73.8°$, $p = 73.4$

29. $\angle Q = 132.7°$, $\angle P = 28.1°$, $p = 67.4$ 30. $\angle Q = 31.7°$, $\angle P = 42.9°$, $p = 87.6$

31. Is it true that for any $\triangle ABC$, $\dfrac{a}{b} = \dfrac{\sin A}{\sin B}$? Justify your answer.

32. If $\sin A = \sin B$ in $\triangle ABC$, when will $A = B$? Justify your answer.

33. Prove that the law of sines holds for a right triangle.

34. Prove that for any $\triangle ABC$, $\dfrac{a - b}{a + b} = \dfrac{\sin A - \sin B}{\sin A + \sin B}$.

35. Prove that $\cos(180° - \theta) = -\cos \theta$ is an identity.

Applications

Surveying In Exercises 36–40, give angle measures to the nearest degree and lengths to two significant digits, unless otherwise specified.

36. From two points P and Q that are 140 ft apart, the lines of sight to a flagpole across a river make angles of 79° and 58°, respectively, with the line joining P and Q. What are the distances from P and Q to the flagpole?

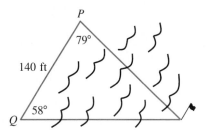

37. Suppose that a parcel of land is triangular, with vertices A and B on the roadway and the third vertex marked at point C. A surveyor measures the distance from A to B and finds that it is 245.8 ft. The lines of sight from A and B to C make angles of 79.46° and 51.67°, respectively, with the line from A to B. Find the measure of angle C and the lengths of sides AC and BC. Give the angle measure to the nearest hundredth of a degree and the lengths to the nearest tenth of a foot.

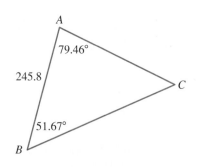

38. As you walk on a straight level path toward a mountain, the measure of the angle of elevation to the peak from one point is 33°. From a point 1000 ft closer, the angle of elevation is 35°. How high is the mountain?

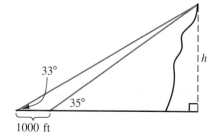

39. A flagpole stands on the edge of the bank of a river. From a point on the opposite bank directly across from the flagpole, the measure of the angle of elevation to the top of the pole is 25°. From a point 200 ft further away and in line with the pole and the first point, the measure of the angle of elevation to the top of the pole is 21°. Draw a diagram. Then find the distance across the river.

40. You measure the angle of elevation to an airplane as 46.3°. At the same time your friend, who is 600 ft closer to a point directly beneath the plane, measures the angle of elevation as 47.5°. Draw a diagram. Then find the altitude of the plane.

MATH CLUB ACTIVITY

The measure of $\angle A$ is 30°. What is the measure of $\angle M$, the angle formed by the bisectors of angles B and C? (The triangle shown appears to be isosceles. Is this necessarily so? Explain your reasoning.)

Law of Sines: The Ambiguous Case

Objective: To solve a triangle when the measures of two sides and an angle opposite one of them are given

For some triangles, the law of sines must be used with care.

Preview

The arm of a power shovel is attached to the shovel's beam at a point 60 ft from the base of the power shovel. The distance from the beam to the bottom of the shovel is 18 ft. A triangle is sometimes formed by the beam, the shovel arm, and the ground. If the angle α between the beam and the ground is too large, the shovel will not touch the ground and there is no triangle (Figure 1). If the shovel arm can touch the ground at only one point, then only one triangle is determined (Figure 2). If the shovel can touch the ground at two points, then two triangles are determined and earth can be picked up as the shovel moves from one point to the other (Figure 3).

Figure 1   Figure 2   Figure 3

1. Refer to the power shovel above. Given sides of 60 ft and 18 ft and an angle α of 15°, how many triangles can be constructed?

2. How many triangles can be constructed if α = 50°? *Hint*: Recall that −1 ≤ sin α ≤ 1.

3. If the bottom of the shovel touches the ground at a right angle, what is the measure of angle α, to the nearest hundredth of a degree?

4. If you are given two sides of a triangle and an angle opposite one side, is a unique triangle determined?

When the measures of *two sides and the angle opposite one of them* are given, you may be able to use the law of sines to find the measure of another angle of the triangle. Two angle measures, one, or none may satisfy the given value of the sine ratio. When two angle measures satisfy the given sine ratio, a unique triangle cannot be determined; this is called the *ambiguous case*.

EXAMPLE 1 Solve $\triangle ABC$, where $a = 9.1$, $b = 12$, and $\angle A = 35°$. **Round lengths to two significant digits and angle measures to the nearest degree.**

$$\frac{\sin 35°}{9.1} = \frac{\sin B}{12} \qquad \textit{Use the law of sines.}$$

$$\frac{12 \sin 35°}{9.1} = \sin B \qquad \textit{Calculation-ready form}$$

$$0.7564 = \sin B$$

$$\angle B = 49° \quad \text{or} \quad \angle B = 180° - 49° = 131°$$

Since there are two possible solutions to $\triangle ABC$, this is an example of the ambiguous case. Both solutions are shown below.

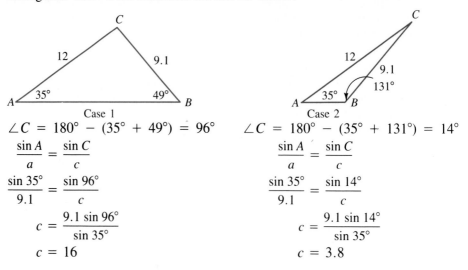

Case 1

$$\angle C = 180° - (35° + 49°) = 96°$$

$$\frac{\sin A}{a} = \frac{\sin C}{c}$$

$$\frac{\sin 35°}{9.1} = \frac{\sin 96°}{c}$$

$$c = \frac{9.1 \sin 96°}{\sin 35°}$$

$$c = 16$$

Case 2

$$\angle C = 180° - (35° + 131°) = 14°$$

$$\frac{\sin A}{a} = \frac{\sin C}{c}$$

$$\frac{\sin 35°}{9.1} = \frac{\sin 14°}{c}$$

$$c = \frac{9.1 \sin 14°}{\sin 35°}$$

$$c = 3.8$$

One solution is $\angle A = 35°$, $\angle B = 49°$, $\angle C = 96°$, $a = 9.1$, $b = 12$, and $c = 16$. Another solution is $\angle A = 35°$, $\angle B = 131°$, $\angle C = 14°$, $a = 9.1$, $b = 12$, and $c = 3.8$.

Sometimes a triangle has no solution.

EXAMPLE 2 Solve $\triangle ABC$, where $a = 25$, $b = 46$, and $\angle A = 37°$.

$$\frac{\sin 37°}{25} = \frac{\sin B}{46} \qquad \textit{Use the law of sines.}$$

$$\frac{46 \sin 37°}{25} = \sin B \qquad \textit{Calculation-ready form}$$

$$1.1073 = \sin B$$

There is no angle whose sine is equal to 1.1073 because $-1 \le \sin \theta \le 1$. Therefore, there is no triangle that satisfies the given conditions.

Below is a summary of the possible solutions of a triangle, given two sides and the angle opposite one of them.

1. If $\angle A$ is an acute angle and $a < b$, there are three possibilities.

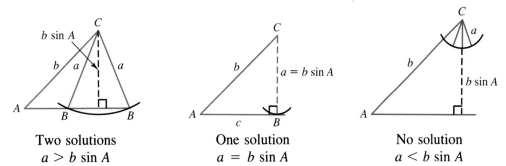

Two solutions	One solution	No solution
$a > b \sin A$	$a = b \sin A$	$a < b \sin A$

2. If $\angle A$ is an acute angle and $a \geq b$, then there is exactly one solution.

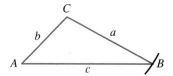

3. If $\angle A$ is an obtuse or right angle, there are two possibilities.

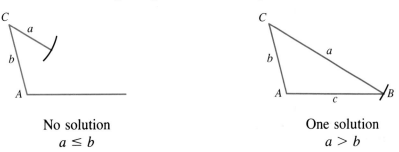

No solution	One solution
$a \leq b$	$a > b$

EXAMPLE 3 **How many solutions are there in each case?**

   a. $\angle C = 47°$,   **b.** $\angle C = 97°$,   **c.** $\angle A = 114°$,
      $c = 20$, $c = 45$, $a = 21$,
      $a = 12$ $a = 39$ $b = 32$

a.

b.

c.

One solution, since C is acute and $c \geq a$.	One solution; C is obtuse and $c > a$.	No solution, since A is obtuse and $a \leq b$.

CLASS EXERCISES

Find the number of solutions in each case. When in doubt, carefully draw a sketch.

1. $\angle A = 94°, a = 25, b = 15$ 2. $\angle A = 84°, a = 25, b = 15$

3. $\angle A = 80°, a = 20, b = 20$ 4. $\angle A = 40°, a = 19, b = 20$

5. $\angle A = 30°, a = 10, b = 20$ 6. $\angle A = 65°, a = 10, b = 20$

7. $\angle A = 40°, a = 10, b = 20$ 8. $\angle A = 150°, a = 10, b = 20$

9. $\angle A = 45°, a = 10, b = 10\sqrt{2}$ 10. $\angle A = 60°, a = 10\sqrt{3}, b = 20$

PRACTICE EXERCISES   Use appropriate technology.

Determine how many solutions exist. When either one or two solutions exist, solve the triangle or triangles. In Exercises 1–10, round all lengths to two significant digits and all angle measures to the nearest degree. In Exercises 11–28, round all lengths to three significant digits and all angle measures to the nearest tenth of a degree.

1. $\angle A = 67°, a = 18, b = 20$ 2. $\angle A = 71°, a = 37, b = 40$

3. $\angle A = 32°, a = 7, b = 10$ 4. $\angle A = 29°, a = 15, b = 19$

5. $\angle A = 87°, a = 47, b = 50$ 6. $\angle A = 79°, a = 52, b = 55$

7. $\angle A = 113°, a = 49, b = 54$ 8. $\angle A = 110°, a = 76, b = 85$

9. $\angle A = 37°, a = 49, b = 54$ 10. $\angle A = 40°, a = 75, b = 85$

11. $\angle A = 59.8°, a = 80.8, b = 73.9$ 12. $\angle A = 69.8°, a = 74.5, b = 21.3$

13. $\angle A = 31.9°, a = 30.6, b = 37.9$ 14. $\angle A = 29.8°, a = 28.6, b = 35.8$

15. $\angle A = 85.8°, a = 23.9, b = 26.4$ 16. $\angle A = 76.4°, a = 27.3, b = 29.0$

17. $\angle C = 47.1°, b = 15.3, c = 11.9$ 18. $\angle C = 51.6°, b = 32.4, c = 28.0$

19. $\angle B = 36.3°, b = 46.3, c = 51.2$ 20. $\angle B = 41.2°, b = 83.2, c = 76.2$

21. $\angle B = 54.3°, a = 62.5, b = 29.6$ 22. $\angle B = 58.7°, a = 118.6, b = 62.4$

23. $\angle C = 108.7°, a = 51.2, c = 54.3$ 24. $\angle C = 103.4°, a = 89.4, c = 98.4$

25. $\angle C\ 123°, b = 106.9, c = 104.3$ 26. $\angle C = 115°, b = 54.8, c = 53.4$

27. $\angle C\ 23.47°, a = 26.49, c = 20.5$ 28. $\angle C = 29.7°, a = 78.92, c = 58.9$

In Exercises 29–33, give angle measures to the nearest degree.

29. One angle of a triangle measures 51°, an adjacent side is 49 cm, and the opposite side is 46 cm. What are the measures of the other two angles?

30. One of the angles of a triangle is to be obtuse. Another angle is to measure 73°, with an adjacent side of 54 cm and an opposite side of 100 cm. Does such a triangle exist? If so, what are the measures of the other two angles?

31. One of the angles of a triangle measures 31°. An adjacent side measures 40 cm, and the opposite side is twice as long as the other adjacent side. What are the measures of the other two angles?

32. One of the angles of a triangle measures 42°. An adjacent side measures 40 cm, and the opposite side is three times as long as the other adjacent side. What are the measures of the other two angles?

33. If $\angle A$ is acute and the measures of sides a and b are known, show that two triangles will satisfy these conditions if and only if $b \sin A < a < b$.

34. Thinking Critically The ambiguous case involves two sides and an angle opposite one of the sides. In terms you used in geometry, this would be expressed as SSA. Explain the conditions under which SSA could be used as a method for proving two triangles congruent.

35. Thinking Critically Given three side lengths d, e, and f, what conditions are necessary for a triangle to be formed?

Applications

In Exercises 36–37, give angle measures to the nearest degree.

36. Engineering If a pole has one 62-ft guy wire that makes an angle of 39° with the ground, and a second 50-ft guy wire is available for the opposite side of the pole, what angle measure will the second wire make with the ground?

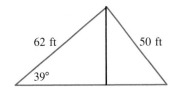

37. Surveying Two roads intersect at an obtuse angle. Two points are chosen: A, 0.15 mi from the intersection, and B, 0.23 mi from the intersection. If the line of sight from A to B makes an angle of 42° with one road at point A, what is the measure of the obtuse angle between the roads?

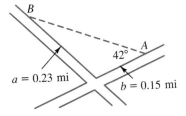

The Law of Cosines

Objectives: To introduce and prove the law of cosines
To use the law of cosines to solve triangles when the measures of two sides and the included angle or the measures of three sides are given

The law of sines cannot be used to solve every triangle. If the measures of *two sides and the included angle* or the measures of *all three sides* are given, another formula called the *law of cosines* is used.

Preview

Mrs. McDonald, Mr. Irizarry, and Ms. Rothstein have cottages on a lake, as shown in the diagram. Suppose you have measured the distances from the McDonald cottage to the Irizarry cottage and from the Irizarry cottage to the Rothstein cottage, as shown.

1. Describe how you could use a scale drawing to find the distance from the McDonald cottage to the Rothstein cottage.

2. To use the law of sines, what additional information would you need?

3. If the measure of ∠*I* were 90° instead of 123°, would you need additional information? Explain your answer.

If you know the lengths of sides *a* and *b* in a triangle *ABC*, and if the included angle is a right angle, you can find side *c* using the Pythagorean theorem, $c^2 = a^2 + b^2$. If the included angle is not a right angle, you can use a generalization of this theorem called the *law of cosines*.

Law of Cosines

For any triangle *ABC*, where *a*, *b*, and *c* are the lengths of the sides opposite the angles with measures *A*, *B*, and *C*, respectively,

$$a^2 = b^2 + c^2 - 2bc \cos A$$
$$b^2 = a^2 + c^2 - 2ac \cos B$$
$$c^2 = a^2 + b^2 - 2ab \cos C$$

To prove the law of cosines it is necessary to consider two cases, one in which the included angle is acute and another in which the included angle is obtuse.

In the figure, $\angle A$, $\angle B$, and $\angle C$ represent the angles of any acute triangle, and a, b, and c, respectively, represent the lengths of the sides opposite these angles. Let h be the length of an altitude that separates $\triangle ABC$ into two right triangles with common side AD. Using the Pythagorean theorem and the cosine ratio, you can derive a relationship among a, b, c, and the measure of $\angle C$. Note that since x is the measure of segment DC, $a - x$ is the measure of segment BD.

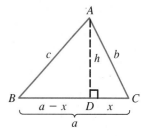

$$c^2 = (a - x)^2 + h^2 \qquad \textit{Apply the Pythagorean theorem to } \triangle ABD.$$
$$= a^2 - 2ax + x^2 + h^2$$
$$= a^2 - 2ax + b^2 \qquad \textit{In } \triangle ADC, b^2 = h^2 + x^2, \textit{ so substitute.}$$
$$= a^2 - 2a(b \cos C) + b^2 \qquad \cos C = \frac{x}{b}, \textit{ so } x = b \cos C.$$
$$= a^2 + b^2 - 2ab \cos C$$

Therefore, $c^2 = a^2 + b^2 - 2ab \cos C$.

You can derive similar expressions for a^2 and b^2. In the exercises, you will be asked to prove the case for an included angle that is obtuse.

EXAMPLE 1 **Find the length of side c for the given triangle.**

$$\angle C = 60°$$
$$a = 10$$
$$b = 14$$

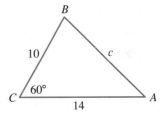

$$c^2 = a^2 + b^2 - 2ab \cos C$$
$$= 10^2 + 14^2 - 2\,(10)(14) \cos 60° \qquad \textit{Calculation-ready form}$$
$$= 156$$
$$c = 12 \qquad \textit{To two significant digits}$$

Therefore, the length of side c is approximately 12 units.

In most cases, you will need a calculator or table to find the cosine of the given angle. Once you have used the law of cosines to find the missing side, you can use the law of sines to complete the solution.

EXAMPLE 2 Solve $\triangle ABC$ if $\angle B = 98.1°$, $a = 17.2$, and $c = 21.5$. Compute the length of the side to three significant digits and the angle measures to the nearest tenth of a degree.

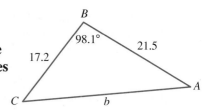

$$b^2 = a^2 + c^2 - 2ac \cos B$$
$$= (17.2)^2 + (21.5)^2 - 2(17.2)(21.5) \cos 98.1°$$
$$= 862.3006$$
$$b = 29.3650$$

$$\frac{\sin C}{c} = \frac{\sin B}{b} \qquad \textit{Use the law of sines.}$$

$$\frac{\sin C}{21.5} = \frac{\sin 98.1°}{29.3650}$$

$$\sin C = \frac{21.5 \sin 98.1°}{29.3650} \qquad \textit{On a calculator, retain the}$$
$$\textit{displayed value of b to use}$$
$$= 0.7249 \qquad \textit{in the calculation of sin C.}$$
$$\angle C = 46.5°$$

$$\angle A = 180° - (98.1° + 46.5°) = 35.4°$$

Thus, $\angle A = 35.4°$, $\angle C = 46.5°$, and $b = 29.4$.

If the third form of the law of cosines, $c^2 = a^2 + b^2 - 2ab \cos C$, is solved for $\cos C$, it yields an expression for $\cos C$ in terms of a, b, and c, the three sides of the triangle.

$$c^2 = a^2 + b^2 - 2ab \cos C$$
$$2ab \cos C = a^2 + b^2 - c^2$$
$$\cos C = \frac{a^2 + b^2 - c^2}{2ab}$$

Similarly, you can show that $\cos A = \dfrac{b^2 + c^2 - a^2}{2bc}$ and $\cos B = \dfrac{a^2 + c^2 - b^2}{2ac}$.

If the lengths of all three sides of a triangle are given, you can find two of the missing angle measures by applying this form of the law of cosines twice. The third angle measure can be found using the fact that the sum of the measures of the angles of a triangle is 180°.

EXAMPLE 3 Solve $\triangle ABC$ if $a = 43$, $b = 39$, and $c = 51$. Round angle measures to the nearest degree.

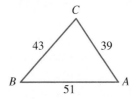

$$\cos A = \frac{b^2 + c^2 - a^2}{2bc}$$

$$= \frac{39^2 + 51^2 - 43^2}{2(39)(51)} \qquad \text{\textit{Calculation-ready form}}$$

$$= 0.5714$$

$$\angle A = 55°$$

$$\cos B = \frac{a^2 + c^2 - b^2}{2ac}$$

$$= \frac{43^2 + 51^2 - 39^2}{2(43)(51)} \qquad \text{\textit{Calculation-ready form}}$$

$$= 0.6678$$

$$\angle B = 48°$$

$$\angle C = 180° - (\angle A + \angle B) \qquad \textit{A + B + C = 180°}$$
$$= 180° - (55° + 48°) \qquad \text{\textit{Calculation-ready form}}$$
$$= 77°$$

Therefore, $\angle A = 55°$, $\angle B = 48°$, and $\angle C = 77°$, to the nearest degree.

The law of cosines can be used to solve real-world problems.

EXAMPLE 4 **A field is triangular in shape with sides 473, 512 and 734.**
To the nearest tenth of a degree, what is the measure of the angle between the sides measuring 512 and 734?

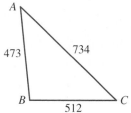

$$\cos C = \frac{734^2 + 512^2 - 473^2}{2(734)(512)}$$

$$= 0.7679$$

$$\angle C = 39.8° \qquad \text{\textit{To the nearest tenth of a degree}}$$

The following table indicates which law to use when solving triangles.

Given Information	Appropriate Law
Three sides	law of cosines
Two sides and the included angle	law of cosines
Two sides and an angle opposite one side (ambiguous case)	law of sines
One side and two angles	law of sines

CLASS EXERCISES

Write an equation to solve for the indicated side. Do not solve the equation.

1.

2.

3.

4.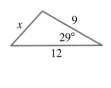

Write an expression to find the cosine of the unknown angle. Do not complete the computation.

5.

6.

7.

8.

PRACTICE EXERCISES   Use appropriate technology.

Solve for the length of the missing side of each triangle. Round your answer to two significant digits.

1.

2.

3.

4.

5. $\angle C = 115°$, $a = 11$, $b = 21$

6. $\angle C = 113°$, $a = 13$, $b = 23$

7. $\angle A = 32°$, $b = 23$, $c = 47$

8. $\angle A = 34°$, $b = 24$, $c = 46$

9. $\angle B = 31°$, $a = 17$, $c = 14$

10. $\angle B = 32°$, $a = 15$, $c = 18$

Solve each triangle for the specified angle measure. Round your answer to the nearest degree.

11. $a = 11$, $b = 14$, $c = 17$; $\angle A$

12. $a = 12$, $b = 16$, $c = 19$; $\angle A$

13. $a = 23$, $b = 43$, $c = 31$; $\angle B$

14. $a = 21$, $b = 42$, $c = 31$; $\angle B$

15. $a = 12$, $b = 12$, $c = 17$; $\angle C$

16. $a = 17$, $b = 17$, $c = 24$; $\angle C$

Solve each triangle *PQR*. Unless otherwise directed, round lengths to two significant digits and angle measures to the nearest degree.

17. $\angle R = 30°, p = 18, q = 16$

18. $\angle R = 45°, p = 13, q = 19$

19. $\angle P = 83°, r = 43, q = 51$

20. $\angle P = 77°, r = 76, q = 49$

21. $\angle Q = 113°, p = 27, r = 43$

22. $\angle Q = 129°, p = 45, r = 71$

23. $p = 15, q = 19, r = 23$

24. $p = 27, q = 33, r = 41$

25. $p = 310, q = 250, r = 160$

26. $p = 200, q = 410, r = 280$

27. $p = 104.3, q = 135.7, r = 154.6$
(Give angle measures to the nearest hundredth of a degree.)

28. $p = 65.5, q = 92.7, r = 114$
(Give angle measures to the nearest tenth of a degree.)

29. Prove that the Pythagorean theorem is a special case of the law of cosines.

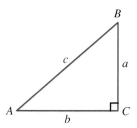

30. Given quadrilateral *ABCD* with $AB = 47, BC = 31, CD = 61,$ $\angle B = 117°,$ and $\angle C = 76°,$ determine *AD*. *Hint*: This is a three-step process.

31. A parallelogram has adjacent sides of measure *a* and *b* and diagonals of measure *p* and *q*. Use the law of cosines to show that $2a^2 + 2b^2 = p^2 + q^2$.

32. Derive the law of cosines for the case in which $\angle C$ is an obtuse angle. *Hint*: Recall that $\cos(180° - C) = -\cos C$.

33. Show that in $\triangle ABC$, if $\angle C$ is acute, $c < \sqrt{a^2 + b^2}$, and if $\angle C$ is obtuse, $c > \sqrt{a^2 + b^2}$.

Applications

Surveying

34. A triangular field is 452 ft on one side and 572 ft on another. The sides meet in an angle of 67.1°. Find the length of the third side, to the nearest foot.

35. If a triangular parcel of land has sides of lengths 541 ft, 429 ft, and 395 ft, what are the measures of the angles between the sides, to the nearest tenth of a degree?

36. The east end of Silver Pond is 5.6 mi from the intersection of Hennick Road and Nutt Road, and the west end is 3.2 mi from the intersection. If the two roads are straight and intersect at an angle of 74°, what is the distance from the west end to the east end of the pond, to the nearest tenth of a mile?

37. A parcel of land has sides measuring 175 ft, 234 ft, 295 ft, and 415 ft, and the angle between the sides of lengths 234 ft and 295 ft has measure 137.1°. What is the measure of the angle opposite this angle, to the nearest tenth of a degree?

Navigation

38. Two roads intersect at an angle of 102.1°. Your friend's mailbox is 476 ft from the intersection. Your mailbox is on the other road and is 615 ft from the intersection. How far is it from your mailbox to your friend's, to the nearest foot?

39. A plane leaves the airport and travels due east at 540 mi/h. Another plane leaves the airport $\frac{1}{4}$ h later and travels 20° east of north at the rate of 575 mi/h. To the nearest ten miles, how far apart are they $\frac{1}{2}$ h after the second plane leaves?

TRIGONOMETRY IN THREE DIMENSIONS

The law of cosines was proved on page 169 for two dimensions. Given the following information and figure, prove a three-dimensional version of the law of cosines.

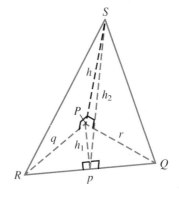

Given: *PQRS* is a right tetrahedron such that *PS* is perpendicular to plane *PQR*, *PS* = *h*, the area of △*PQR* = *C*, and the measure of ∠*RPQ* = γ. Let the areas of the lateral faces △*PSQ*, △*PSR*, and △*RSQ* be *A*, *B*, and *D*, respectively.

Prove: $D^2 = A^2 + B^2 + C^2 - 2AB \cos \gamma$

The Law of Tangents

Objectives: To introduce the law of tangents

To use the law of tangents to solve triangles when the measures of two sides and the included angle are known

If you are given two sides of a triangle and the included angle, you can use the law of cosines to find the third side and then use it again to find the other angles. In this lesson, you will learn another way to compute the angles.

Preview

To solve a system of two linear equations in two variables, add multiples of one equation to another to obtain one equation in one variable. Solve this equation, substitute the known value into one of the original equations, and then solve to find the second value.

EXAMPLE **Solve the following system for x and y:**

$$\begin{cases} 2x + y = 17 \\ x - y = 4 \end{cases}$$

$$3x = 21 \qquad \text{\textit{Add the two equations.}}$$
$$x = 7$$

$$7 - y = 4$$
$$-y = 4 - 7$$
$$y = 3 \qquad \text{\textit{Substitute 7 for x in one of the equations.}}$$

Solve each system of equations.

1. $\begin{cases} x - y = 10 \\ x + y = 4 \end{cases}$ **2.** $\begin{cases} 2x - y = 3 \\ x - y = 7 \end{cases}$ **3.** $\begin{cases} x - 2y = 3 \\ x + y = 15 \end{cases}$ **4.** $\begin{cases} 2x + 3y = 16 \\ 3x + 4y = 7 \end{cases}$

Another law that can be used to solve triangles is the *law of tangents*.

Law of Tangents

For any $\triangle ABC$ in which a, b, and c are the lengths of the sides opposite angles with measures A, B, and C, respectively,

$$\frac{a - b}{a + b} = \frac{\tan \frac{1}{2}(A - B)}{\tan \frac{1}{2}(A + B)} \qquad \frac{c - a}{c + a} = \frac{\tan \frac{1}{2}(C - A)}{\tan \frac{1}{2}(C + A)} \qquad \frac{b - c}{b + c} = \frac{\tan \frac{1}{2}(B - C)}{\tan \frac{1}{2}(B + C)}$$

The law of tangents is used to determine the other angles when two sides and an included angle are known. It generates a system of two equations in two variables, which are solved to give the missing angles. The derivation of the law involves certain identities that are developed in a later chapter. Therefore, the law is presented here without proof.

EXAMPLE 1 **Find, to the nearest degree, the measures of the other two angles of a triangle in which $a = 13$, $b = 10$, and $\angle C = 58°$. Then find the measure of side c to two significant digits.**

Since $\angle C = 58°$, $A + B = 180° - 58° = 122°$.

$$\frac{13 - 10}{13 + 10} = \frac{\tan \frac{1}{2}(A - B)}{\tan \frac{1}{2}(122°)} \qquad \textit{Use the law of tangents.}$$

$$\frac{3}{23} = \frac{\tan \frac{1}{2}(A - B)}{\tan 61°}$$

$$\tan \frac{1}{2}(A - B) = \frac{3 \tan 61°}{23} \qquad \textit{Calculation-ready form}$$

$$= 0.2353$$

$$\frac{1}{2}(A - B) = 13.24°$$

$$A - B = 26.48°$$

Thus, $A + B = 122°$ and $A - B = 26.48°$.

$$A - B = 26.48°$$
$$\underline{A + B = 122°}$$
$$2A = 148.48° \qquad \textit{Add the two equations.}$$
$$A = 74.24°$$
$$B = 122° - 74.24° = 47.76°$$

$$\frac{\sin 74.24°}{13} = \frac{\sin 58°}{c} \qquad \textit{Use the law of sines.}$$

$$c = \frac{13 \sin 58°}{\sin 74.24°}$$

$$c = 11.46$$

Therefore, $\angle A = 74°$, $\angle B = 48°$, and $c = 11$.

If the measures of two angles and the sum or difference of the lengths of the sides opposite them are known, the law of tangents can be used to determine the measures of the sides.

EXAMPLE 2 In $\triangle ABC$, the sum of sides a and b is 142, $\angle A = 48°$, and $\angle B = 32°$. Find sides a and b. Then find $\angle C$ and side c. Express lengths to two significant digits and the angle measure to the nearest degree.

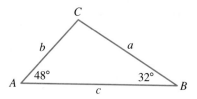

You know that $a + b = 142$.

$$\frac{a - b}{142} = \frac{\tan \frac{1}{2}(48° - 32°)}{\tan \frac{1}{2}(48° + 32°)}$$ *Use the law of tangents.*

$$\frac{a - b}{142} = \frac{\tan 8°}{\tan 40°}$$

$$a - b = \frac{142 \tan 8°}{\tan 40°}$$ *Calculation-ready form*

$$= 23.7836$$

Thus, $a + b = 142$ and $a - b = 23.7836$.

$$a - b = 23.7836$$
$$\underline{a + b = 142}$$
$$2a = 165.7836$$ *Add the two equations.*
$$a = 82.8918$$
$$b = 142 - 82.8918 = 59.1082$$

$$\angle C = 180° - (48° + 32°) = 100°$$
$$\frac{\sin 48°}{82.8918} = \frac{\sin 100°}{c}$$ *Use the law of sines.*
$$c = \frac{82.8918 \sin 100°}{\sin 48°}$$
$$c = 109.8473$$

Therefore, $\angle C = 100°$, $a = 83$, $b = 59$, and $c = 110$.

The law of tangents can be used to solve surveying problems if two sides and the included angle of a triangular parcel have been measured.

EXAMPLE 3 A triangular lot has two sides of length 248 ft and 379 ft, and the angle between them is 73.5°. The third side contains thick brush so that a line of sight along it cannot be established. What are the measures of the other two angles of the parcel, to the nearest tenth of a degree?

$$Q + P = 180° - 73.5° = 106.5°$$

$$\frac{379 - 248}{379 + 248} = \frac{\tan \frac{1}{2}(Q - P)}{\tan \frac{1}{2}(106.5°)} \qquad \textit{Use the law of tangents.}$$

$$\frac{131}{627} = \frac{\tan \frac{1}{2}(Q - P)}{\tan 53.25°}$$

$$\tan \frac{1}{2}(Q - P) = \frac{131 \tan 53.25°}{627} \qquad \textit{Calculation-ready form}$$

$$= 0.2798$$

$$\frac{1}{2}(Q - P) = 15.63°$$
$$Q - P = 31.26°$$

Thus, $Q + P = 106.5°$ and $Q - P = 31.26°$.

$$Q - P = 31.26°$$
$$\underline{Q + P = 106.5°}$$
$$2Q = 137.76° \qquad \textit{Add the two equations.}$$
$$Q = 68.88°$$
$$P = 106.5° - 68.88° = 37.62°$$

Thus, $\angle P = 37.6°$ and $\angle Q = 68.9°$.

CLASS EXERCISES

1. If $\angle A = 72°$ and $\angle B = 29°$, find $\tan \frac{1}{2}(A - B)$ and $\tan \frac{1}{2}(A + B)$.

2. If $\angle A = 34°$ and $\angle B = 43°$, find $\tan \frac{1}{2}(A - B)$ and $\tan \frac{1}{2}(A + B)$.

3. If $\angle A = 100°$ and $\angle C = 12°$, find $\tan \frac{1}{2}(A - C)$ and $\tan \frac{1}{2}(A + C)$.

4. If $\angle A = 97°$ and $\angle C = 15°$, find $\tan \frac{1}{2}(A - C)$ and $\tan \frac{1}{2}(A + C)$.

5. If $\angle B = 88°$ and $\angle C = 66°$, find $\tan \frac{1}{2}(B - C)$ and $\tan \frac{1}{2}(B + C)$.

6. If $\angle B = 77°$ and $\angle C = 43°$, find $\tan \frac{1}{2}(B - C)$ and $\tan \frac{1}{2}(B + C)$.

PRACTICE EXERCISES Use appropriate technology.

Use the law of tangents to find the measures of the unknown angles of
△*ABC*. Then use the law of sines to find the length of the unknown side.
Express angle measures to the nearest degree and lengths to two
significant digits.

1. $\angle C = 47°, a = 46, b = 19$
2. $\angle C = 51°, a = 71, b = 27$
3. $\angle C = 104°, a = 27, b = 15$
4. $\angle C = 97°, a = 36, b = 19$
5. $\angle A = 43°, b = 61, c = 47$
6. $\angle A = 51°, b = 93, c = 87$
7. $\angle A = 123°, b = 15, c = 13$
8. $\angle A = 114°, b = 28, c = 25$
9. $\angle B = 26°, a = 45, c = 34$
10. $\angle B = 56°, a = 52, c = 32$
11. $\angle B = 111°, a = 87, c = 54$
12. $\angle B = 154°, a = 96, c = 82$

Solve △*ABC*. In Exercises 13–16, express angle measures to the nearest
degree and lengths to two significant digits. In Exercises 17–20, express
angle measures to the nearest tenth of a degree and lengths to three
significant digits.

13. $\angle A = 47°, \angle B = 29°, a + b = 49$
14. $\angle A = 37°, \angle B = 62°, a + b = 76$
15. $\angle A = 94°, \angle C = 22°, a + b = 114$
16. $\angle A = 42°, \angle C = 101°, a + b = 213$
17. $\angle A = 43.6°, \angle B = 21.7°, a + b = 92.3$
18. $\angle A = 92.7°, \angle B = 11.6°, a + b = 27.4$
19. $\angle B = 36.7°, \angle C = 48.9°, b + c = 104.3$
20. $\angle B = 103.4°, \angle C = 21.8°, b + c = 67.5$

21. Use the law of tangents to show that in △*ABC*, if $A > B$, then $a > b$.

22. Use the law of tangents to show that if two angles of a triangle are
congruent, then the triangle is isosceles.

Applications

23. **Surveying** A surveyor wishes to calculate the
distance between two signs. The distance from
the surveyor to the first sign is 70 m, and the
distance to the second sign is 100 m. The angle
between the two lines of sight measures 56°.
Find the distance between the two signs, to the
nearest meter.

24. Geometry The sides of a parallelogram are 12 ft and 8 ft, and each of the larger angles measures 125°. Find the length of the shorter diagonal of the parallelogram, to the nearest foot.

TEST YOURSELF

In all exercises, give angle measures to the nearest degree and lengths to two significant digits.

Solve each triangle.

1. $\angle A = 33°$, $\angle B = 56°$, $a = 15$ 2. $\angle A = 62°$, $\angle B = 48°$, $b = 21$ **4.1**

3. $\angle A = 43°$, $\angle C = 50°$, $c = 12$ 4. $\angle B = 65°$, $\angle C = 50°$, $c = 14$

5. How many triangles are there with angle A equal to 95°, adjacent side 23, and opposite side 31? If there are any, solve them.   **4.2**

6. How many triangles are there with angle A equal to 47°, adjacent side 14, and opposite side 12? If there are any, solve them.

Solve each triangle.

7. $\angle C = 84°$, $a = 42$, $b = 50$ 8. $\angle A = 78°$, $b = 75$, $c = 48$ **4.3**

9. $a = 32$, $b = 25$, $c = 17$ 10. $a = 19$, $b = 40$, $c = 27$

11. How many triangles are there with an angle of 51°, adjacent side 27, and opposite side 11? If there are any, solve them.

12. If a triangular plot of land has sides of 476 m, 512 m, and 544 m, what are the measures of the three angles?

Use the law of tangents to solve $\triangle ABC$.

13. $\angle C = 120°$, $a = 20$, $b = 35$ 14. $\angle B = 118°$, $a = 10$, $c = 17$ **4.4**

Solve $\triangle ABC$.

15. $\angle A = 11°$, $\angle B = 118°$, $a + c = 39$ 16. $\angle A = 95°$, $\angle C = 21°$, $a + b = 110$

17. The lengths of two sides of a parallelogram are 8 cm and 6 cm, and each of the smaller angles measures 78°. Find the length of the longer diagonal of the parallelogram, to the nearest centimeter.

The Area of a Triangle

4.5

Objectives: To find the area of a triangle when the measures of two sides and the included angle are known

To use the law of sines to find the area of a triangle when the measures of one side and two angles are known

You have learned how to find the missing parts of a triangle when you know three sides, two sides and the included angle, or two angles and any side. Since a triangle is completely determined in these cases, its area is also determined.

Preview

The area K of a triangle is given by the formula $K = \frac{1}{2} bh$, where b is the length of one side and h is the length of the altitude to that side.

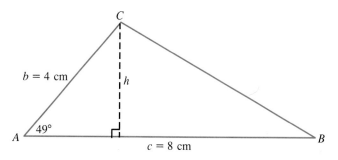

1. Refer to △ *ABC* above. Use a centimeter ruler to measure the length of the altitude to side *AB*. Then compute the area of the triangle.

2. Use trigonometry to find the length of the altitude from *C* to side *AB*. Then compute the area again.

For each triangle, use the methods given in Exercises 1 and 2 to find *h* and the area *K*.

3.

4.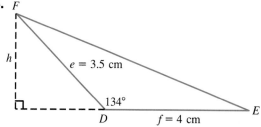

The formula $K = \frac{1}{2}bh$ presented in the Preview can be transformed into other area formulas. Assume that sides a and b are given, and that h is the altitude to side AC. Consider the case in which $\angle A$ is acute.

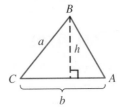

Then, $K = \frac{1}{2}bh$

But, $h = a \sin C$ $\sin C = \frac{h}{a}$

Therefore, $K = \frac{1}{2}ab \sin C$ *Substitute.*

You can write two other expressions for area involving the sine of $\angle A$ or $\angle B$. The results would be the same if an angle of the triangle is obtuse. The proof of the case in which one angle is obtuse is left as an exercise.

> The area K of any triangle ABC is given by any one of these formulas:
>
> $$K = \frac{1}{2}bc \sin A \qquad K = \frac{1}{2}ac \sin B \qquad K = \frac{1}{2}ab \sin C$$

EXAMPLE 1   **In $\triangle ABC$, $\angle C = 57°$, $a = 31$ cm, and $b = 42$ cm. Find the area of the triangle to the nearest ten square centimeters.**

$K = \frac{1}{2}ab \sin C$

$\quad = \frac{1}{2}(31)(42) \sin 57°$ *Calculation-ready form*

$\quad = 546$

The area is approximately 550 cm^2.

The preceding formula can be used in conjunction with the law of sines to derive another area formula that can be used when at least two angles and one side are known.

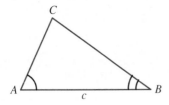

Assume that $\angle A$, $\angle B$, and c are known. Then $\angle C = 180° - (\angle A + \angle B)$.

$$\frac{b}{\sin B} = \frac{c}{\sin C}$$ *Use the law of sines to find b.*

$$b = \frac{c \sin B}{\sin C}$$

$$K = \left(\frac{1}{2}\right)\left(\frac{c \sin B}{\sin C}\right)(c)(\sin A)$$ *Substitute in the area formula $K = \frac{1}{2}bc \sin A$.*

$$K = \frac{c^2 \sin A \sin B}{2 \sin C}$$

It can also be shown that $K = \dfrac{a^2 \sin B \sin C}{2 \sin A}$ and $K = \dfrac{b^2 \sin A \sin C}{2 \sin B}$ are area formulas for $\triangle ABC$.

EXAMPLE 2   **Given $\angle A = 61°$, $\angle B = 78°$, and $c = 37$ in., compute the area of the triangle to the nearest ten square inches.**

$$\angle C = 180° - (61° + 78°) = 41°$$

$$K = \frac{c^2 \sin A \sin B}{2 \sin C}$$

$$K = \frac{(37)^2 (\sin 61°)(\sin 78°)}{2 \sin 41°}$$ *Calculation-ready form*

$$K = 893 \text{ in.}^2$$ The area is approximately 890 in.^{2}

CLASS EXERCISES

Substitute values into the appropriate area formula, but do not solve.

1.

2.

3.

4.

Find the area of the given triangle to the nearest ten square units.

5.

6.

7.

8.

Use appropriate technology.

Find the area of the given triangle to the nearest ten square units.

1.

2.

3.

4.

5. $\angle A = 92°$, $\angle B = 61°$, $b = 15$

6. $\angle B = 61°$, $\angle C = 93°$, $b = 25$

7. $\angle A = 110°$, $\angle B = 28°$, $c = 47$

8. $\angle B = 27°$, $\angle C = 108°$, $a = 53$

9. $\angle A = 34°$, $b = 11$, $c = 11$

10. $\angle C = 35°$, $a = 17$, $c = 17$

11. $\angle A = 71°$, $b = 21$, $c = 87$

12. $\angle A = 73°$, $b = 161$, $c = 41$

13. $\angle B = 90°$, $a = 23$, $c = 47$

14. $\angle A = 90°$, $b = 28$, $c = 15$

15. $\angle B = 60°$, $\angle C = 60°$, $a = 10$

16. $\angle B = 60°$, $\angle A = 60°$, $c = 100$

Find the area of the given triangle to the nearest square unit.

17. $\angle A = 87.3°$, $b = 11.3$, $c = 19.2$

18. $\angle B = 86.4°$, $b = 37.7$, $c = 21.4$

19. $\angle A = 41.3°$, $\angle B = 112.5°$, $c = 46.7$

20. $\angle B = 113.9°$, $\angle C = 43.9°$, $a = 57.9$

21. $\angle R = 62.4°$, $q = 23.6$, $p = 19.2$

22. $\angle R = 57.3°$, $q = 17.6$, $p = 19.8$

23. $\angle Q = 83.9°$, $\angle R = 46.7°$, $p = 46.7$

24. $\angle Q = 67.3°$, $\angle R = 51.2°$, $p = 31.3$

25. $\angle P = 108.4°$, $\angle Q = 27.3°$, $r = 19.7$

26. $\angle P = 97.8°$, $\angle Q = 34.9°$, $r = 17.3$

27. $\angle P = 67.3°$, $\angle Q = 76.4°$, $p = 19.6$

28. $\angle Q = 53.7°$, $\angle R = 61.9°$, $q = 23.1$

29. Prove that $K = \frac{1}{2} ab \sin C$ is an area formula for $\triangle ABC$ when $\angle A$ is an obtuse angle (that is, when the altitude from B does not intersect segment AC).

30. Use the results of this lesson to show that the area of a parallelogram is given by $K = ab \sin \theta$, where a and b are the lengths of the adjacent sides and θ is the measure of the included angle.

31. Compute the area of trapezoid *ABCD*, if it has base angles of 70° and 45° and bases of 110 cm and 80 cm. Round your answer to the nearest hundred square centimeters.

Applications

Surveying

32. To the nearest square foot, what is the area of a triangular parcel of land, if two adjacent sides measure 184 ft and 246 ft and the angle between them measures 76.0°?

33. To the nearest square foot, what is the area of a triangular pond, if two adjacent sides measure 112 ft and 96 ft and the angle between them measures 65.4°?

34. What is the area, to the nearest 0.1 acre, of a triangular field that is 476 ft on one side and 723 ft on another, if the angle between these sides measures 71.3°? (1 acre = 43,560 ft^2)

35. What is the area, to the nearest 0.1 acre, of a triangular field that is 529 ft on one side and 849 ft on another, if the angle between these sides measures 102.7°?

CAREER

CAD (Computer-Aided Design) Technician

CAD technicians create layouts, designs, and line drawings for any step in the manufacturing process, including product development, testing, and modification. They also provide assistance to architects and engineers. They use light pens to move, rotate, or zoom in on a portion of a computer display, and they edit and modify computer drawings.

Requirements include skill in mechanical drawing and drafting, the ability to use a computer program, proficiency in mathematics, an eye for detail, and the ability to visualize spatial relationships.

CAD technicians need at least a high school education. Technical school or college study in technical graphics, architectural drafting, and product design is useful. Some companies provide on-the-job training. A bachelor's degree, experience, training, and serious effort can lead to excellent job prospects in this field and related fields like robotics.

Heron's Formula

4.6

Objectives: To use Heron's formula to find the area of a triangle when the lengths of the three sides are known

To use Heron's formula to find the length of an altitude of a triangle when the lengths of the three sides are known

You know how to find the area of a triangle when the measure of at least one angle is known. It is also possible to find the area of a triangle when only the lengths of the three sides are known. Indeed, in real-life settings it is often easier to measure the sides of a triangle than it is to measure altitudes or angles.

Preview

A city block is triangular in shape. The frontage along High Street is 423 ft, that along Conover Street is 764 ft, and that along First Street is 524 ft. A developer wishes to find the area of the plot.

1. What additional information would he need to find the area using one of the formulas from the previous lesson?

2. What law or laws could he use to find that information?

3. Find, to three significant digits, the area of the triangle using a formula from the previous lesson.

If you know the measures of three sides of a triangle, you can find the measure of an angle using the law of cosines. You can then apply a formula from the previous lesson to find the area of the triangle. However, *Heron's formula* often involves fewer calculations.

Heron's Formula

If a, b, and c are measures of the sides of a triangle, then the area K of the triangle is given by

$$K = \sqrt{s(s-a)(s-b)(s-c)}, \quad \text{where } s = \frac{a+b+c}{2}$$

The quantity s is called the *semiperimeter* of a triangle.

To prove Heron's formula, start with the area formula $K = \frac{1}{2} bc \sin A$.

$K = \frac{1}{2} bc \sin A$

$K^2 = \frac{1}{4} b^2 c^2 \sin^2 A$ *Square both sides.*

$K^2 = \frac{1}{4} b^2 c^2 (1 - \cos^2 A)$ *Replace $\sin^2 A$ with $1 - \cos^2 A$.*

$K^2 = \left[\frac{bc}{2}(1 + \cos A)\right]\left[\frac{bc}{2}(1 - \cos A)\right]$ *Factor.*

Using the law of cosines, replace $\cos A$ with $\dfrac{b^2 + c^2 - a^2}{2bc}$.

$K^2 = \left[\frac{bc}{2}\left(1 + \frac{b^2 + c^2 - a^2}{2bc}\right)\right]\left[\frac{bc}{2}\left(1 - \frac{b^2 + c^2 - a^2}{2bc}\right)\right]$

$K^2 = \left[\frac{2bc + b^2 + c^2 - a^2}{4}\right]\left[\frac{2bc - b^2 - c^2 + a^2}{4}\right]$ *Multiply and simplify.*

$K^2 = \left[\frac{(b^2 + 2bc + c^2) - a^2}{4}\right]\left[\frac{a^2 - (b^2 - 2bc + c^2)}{4}\right]$ *Regroup terms.*

$K^2 = \left[\frac{(b + c)^2 - a^2}{4}\right]\left[\frac{a^2 - (b - c)^2}{4}\right]$ $\begin{aligned} b^2 + 2bc + c^2 &= (b + c)^2 \\ b^2 - 2bc + c^2 &= (b - c)^2 \end{aligned}$

$K^2 = \left[\frac{b + c + a}{2}\right]\left[\frac{b + c - a}{2}\right]\left[\frac{a - b + c}{2}\right]\left[\frac{a + b - c}{2}\right]$ *Factor.*

Let $s = \dfrac{a + b + c}{2}$. The formula then simplifies to

$K^2 = s(s - a)(s - b)(s - c)$

Thus, the area of $\triangle ABC$ is given by

$K = \sqrt{s(s - a)(s - b)(s - c)}$, where $s = \dfrac{a + b + c}{2}$

Before the development of calculators, Heron's formula required a great deal of time-consuming computation. Calculators make the use of the formula more practical.

EXAMPLE 1 **Find the area of a triangle with sides 23 cm, 26 cm, and 31 cm, to the nearest ten square centimeters.**

23 cm   26 cm

31 cm

$s = \dfrac{23 + 26 + 31}{2} = 40$ *Find s, where $s = \dfrac{a + b + c}{2}$.*

$K = \sqrt{40(40 - 23)(40 - 26)(40 - 31)}$ *Use $K = \sqrt{s(s - a)(s - b)(s - c)}$.*

$K = \sqrt{40(17)(14)(9)}$

$K = 293$ The area is approximately 290 cm^2.

Heron's formula can also be used to find the length of the altitude of a triangle, if the measures of three sides are known.

EXAMPLE 2 **Find the length of the altitude from A to side BC, if $a = 24$ ft, $b = 14$ ft, and $c = 18$ ft. Round the answers to the nearest foot.**

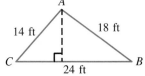

$$s = \frac{24 + 14 + 18}{2} = 28$$

$$K = \sqrt{28\,(28 - 24)\,(28 - 14)\,(28 - 18)}$$ *Use Heron's formula.*
$$K = \sqrt{28\,(4)\,(14)\,(10)}$$ *Calculation-ready form*
$$K = 125.2$$

Next, use $K = \frac{1}{2}\,ah$, where $K = 125.2$ and the base, a, is 24.

$$125.2 = \frac{1}{2}\,(24)\,(h) \qquad K = \frac{1}{2}\,ah$$
$$h = 10.4$$

The altitude from A to side BC is approximately 10 ft long.

CLASS EXERCISES

Substitute values into the appropriate area formula, but do not solve.

1. **2.** **3.** **4.**

5. **6.** **7.** **8.**

PRACTICE EXERCISES   Use appropriate technology.

Find the area of the given triangle to the nearest whole square unit.

1.

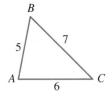

2.

Find the area of the given triangle to the nearest ten square units.

3.

4.

5. $a = 31, b = 23, c = 14$

6. $a = 61, b = 23, c = 47$

7. $a = 22, b = 25, c = 30$

8. $a = 40, b = 45, c = 50$

9. $a = 12, b = 12, c = 12$

10. $a = 20, b = 20, c = 20$

11. $a = 18, b = 24, c = 30$

12. $a = 10, b = 24, c = 26$

Find the area of the given triangle to the nearest hundred square units.

13. $a = 127, b = 246, c = 291$

14. $a = 221, b = 119, c = 246$

15. $p = 176, q = 187, r = 176$

16. $p = 267, q = 283, r = 296$

Find the length of the altitude from vertex A to side BC for each triangle to the nearest unit.

17. $a = 10, b = 12, c = 16$

18. $a = 15, b = 22, c = 18$

19. $a = 48, b = 26, c = 50$

20. $a = 38, b = 42, c = 16$

Find the area of the given quadrilateral to the nearest hundred square units.

21.

22.

23.

24.

25. Thinking Critically   If you knew the lengths of the sides of polygon $ABCD$, with $m\angle C = 90$, how could you find the area? (*Hint*: Draw an auxiliary line.)
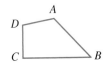

26. Construction   Find the area of a triangular patio that has sides 18 ft, 20 ft, and 22 ft, to the nearest ten square feet.

27. Landscaping   Find the area of a triangular flower garden with sides 11 ft, 19 ft, and 18 ft, to the nearest ten square feet.

Vectors in the Plane

4.7

Objectives: To find the magnitudes of the *x*- and *y*-components
of a vector
To find the sum of two vectors
To find the measure of the angle between two vectors

Many quantities can be described by their magnitude and direction.

Preview

The velocity of an airplane traveling at 620 mi/h due east has both magnitude
(speed) and direction. A 20-N force exerted at the end of a plank in order to
move a large rock has both magnitude and direction.

1. Name some quantities other than
   velocity and force that have both
   magnitude and direction.

2. Name some quantities that have
   only magnitude.

3. Name some quantities that have
   only direction.

4. Name some quantities that have
   neither magnitude nor direction.

Any quantity that has both magnitude and direction is called a **vector
quantity.** A vector quantity may be represented by a directed line segment, or
vector. The length of any vector is proportional to the magnitude of the
corresponding vector quantity. The direction of the vector represents the
direction of the vector quantity.

In the figure at the right, vector *AB*,
written $\overrightarrow{AB}$, has the same magnitude and
direction as $\overrightarrow{CD}$. The vectors are said to be
equivalent. This fact can be denoted by
$\overrightarrow{AB} = \overrightarrow{CD}$. *A* and *C* are called the **initial
points** of the vectors, and *B* and *D* are
called the **terminal points.**

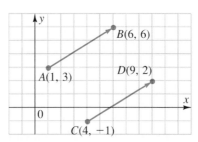

A vector can also be represented by a single letter such as **v.** The norm, or
magnitude, of a vector **v** is represented as $\|\mathbf{v}\|$ and is found by using the
distance formula.

190 Chapter 4 Oblique Triangles

EXAMPLE 1 **Draw an arrow representing the vector from $A(-1, 3)$ to $B(3, 6)$, and find its magnitude.**

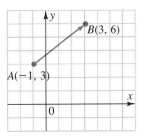

$$\|\overrightarrow{AB}\| = \sqrt{(x_2 - x_1)^2 + (y_2 - y_1)^2}$$

$$\|\overrightarrow{AB}\| = \sqrt{[3 - (-1)]^2 + (6 - 3)^2}$$

$\|\overrightarrow{AB}\| = 5$ The magnitude of $\overrightarrow{AB}$ is 5.

If a vector is placed on a standard coordinate plane with its tail, or initial point, at the origin, the vector is said to be in **standard position.** Assume that the endpoint of $\overrightarrow{OC}$ has coordinates (a, b). By the distance formula, $\|\overrightarrow{OC}\| = \|\mathbf{v}\| = \sqrt{a^2 + b^2}$.

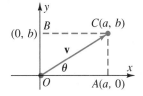

Furthermore, if θ is the measure of the angle between the vector and the positive x-axis, called the **direction angle** of $\overrightarrow{OC}$, then $\dfrac{a}{\|\mathbf{v}\|} = \cos \theta$ and $\dfrac{b}{\|\mathbf{v}\|} = \sin \theta$.

Vectors $\overrightarrow{OA}$ and $\overrightarrow{OB}$ are the **x-** and **y-components,** respectively, of $\overrightarrow{OC}$. If you know the magnitude and direction angle of a vector, you can determine the magnitude of its x-component and of its y-component.

EXAMPLE 2 **Find the magnitudes of the x- and y-components of $\overrightarrow{OC}$, given that the measure of the direction angle is 60° and $\|\overrightarrow{OC}\| = 10$.**

$a = 10 \cos 60° = 5$ The magnitude of the x-component is 5.
$b = 10 \sin 60° = 5\sqrt{3}$ The magnitude of the y-component is $5\sqrt{3}$.

The sum, or **resultant,** of two vectors is also a vector. Vectors **u** and **v** can be added using *parallelogram* addition or *tail-to-head* (triangular) addition. Each method yields the same resultant vector **w,** as illustrated below.

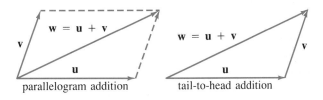

parallelogram addition   tail-to-head addition

Note that in parallelogram addition, vectors **u** and **v** have the same initial point. In tail-to-head addition, the head (terminal point) of **u** is the same as the tail of **v.**

EXAMPLE 3   Find each of the following:

  a. $\overrightarrow{AB} + \overrightarrow{AC}$, by parallelogram addition   **b. u + v,** by tail-to-head addition

a.

b.

In the figure at the right, both vectors have the same magnitude and are parallel. They have different directions, however, and therefore are not equivalent. $\overrightarrow{EF}$ would be equivalent to a vector with the magnitude of $\overrightarrow{AB}$ but the opposite direction. Such a vector is said to be the **negative** of $\overrightarrow{AB}$.

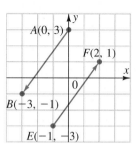

The negative of **v** is written as $-\mathbf{v}$. It has the same length as **v** and the opposite direction. Negative vectors are especially useful in vector subtraction. Given the vectors **u** and **v,** the vector difference $\mathbf{u} - \mathbf{v}$ is defined as follows:

$$\mathbf{u} - \mathbf{v} = \mathbf{u} + (-\mathbf{v})$$

It is also possible to multiply a real number, or **scalar,** by a vector. To do so, multiply the scalar by the magnitude of the vector.

EXAMPLE 4   Use the vectors shown to draw each of the following.

  a. u − v   b. 2v   c. −2u

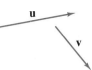

  a. Draw $-\mathbf{v}$, then $\mathbf{u} + (-\mathbf{v})$.   **b.**   **c.** $-2\mathbf{u} = 2(-\mathbf{u})$

The magnitude and direction of the sum or difference of two vectors can be determined from the magnitudes and directions of the vectors. Let **u** and **v** be vectors, and let θ be the measure of the angle between them. (If you know the directions of both vectors, you can easily compute θ.) By the law of cosines,

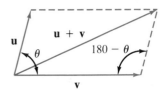

$$\|\mathbf{u} + \mathbf{v}\|^2 = \|\mathbf{u}\|^2 + \|\mathbf{v}\|^2 - 2\|\mathbf{u}\|\,\|\mathbf{v}\|\cos\,(180° - \theta)$$
$$\|\mathbf{u} + \mathbf{v}\|^2 = \|\mathbf{u}\|^2 + \|\mathbf{v}\|^2 - 2\|\mathbf{u}\|\,\|\mathbf{v}\|\cos\,\theta \qquad \text{cos } (180° - \theta) = -\cos \theta$$

EXAMPLE 5 **Find the magnitude and direction of the sum w of two vectors u and v of magnitudes 6.9 and 13, respectively, if the angle between them measures 48°. Round the magnitude to the nearest whole number and the direction to the nearest degree.**

$$\|\mathbf{u} + \mathbf{v}\|^2 = \|\mathbf{u}\|^2 + \|\mathbf{v}\|^2 + 2\|\mathbf{u}\|\,\|\mathbf{v}\|\cos\,\theta \qquad \textit{Use the above equation.}$$
$$\|\mathbf{w}\|^2 = 6.9^2 + 13^2 + 2(6.9)(13)\cos 48°$$
$$\|\mathbf{w}\|^2 = 336.6520$$
$$\|\mathbf{w}\| = 18.3481 \qquad \text{The magnitude of } \mathbf{w} \text{ is } 18.$$

$$\cos \alpha = \frac{\|\mathbf{w}\|^2 + \|\mathbf{v}\|^2 - \|\mathbf{u}\|^2}{2\|\mathbf{w}\|\,\|\mathbf{v}\|} \qquad \textit{Use the formula derived from the law of cosines to find } \alpha.$$

$$\cos \alpha = \frac{18.3481^2 + 13^2 - 6.9^2}{2(18.3481)(13)} = 0.9602$$

$$\alpha = 16.22° \qquad \text{The angle between } \mathbf{v} \text{ and } \mathbf{w} \text{ is } 16°.$$

The figures below represent the flight of an airplane in still air and in the presence of wind from the south. In both cases the plane's engines provide *air velocity*, which is a vector quantity. The magnitude and direction of the air velocity are called the *air speed* and *heading*, respectively. Notice that the heading is measured clockwise from the north. The *ground velocity* is the vector sum of the air velocity and wind velocity.

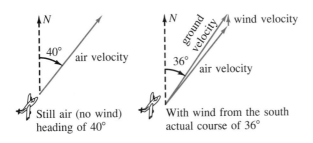

Still air (no wind) heading of 40°

With wind from the south actual course of 36°

EXAMPLE 6   An airplane has an air speed of 450 mi/h and a heading of 60 degrees. A wind is blowing from the north at 42 mi/h. Find the plane's ground velocity, including the *ground speed* (magnitude) to two significant digits and the *course* (direction) to the nearest degree. Draw a diagram, showing the sum of the air velocity and wind velocity as **v**, the ground velocity.

$$a^2 = b^2 + c^2 - 2bc \cos A \qquad \textit{Law of cosines}$$

$$\|v\|^2 = 450^2 + 42^2 - 2(450)(42) \cos 60°$$

$$\|v\|^2 = 185{,}364$$

$$\|v\|^2 = 430.5392 \qquad \textit{The ground speed is 430 mi/h.}$$

$$\frac{\sin A}{a} = \frac{\sin C}{c} \qquad \textit{Use the law of sines.}$$

$$\frac{\sin 60°}{430.5392} = \frac{\sin \alpha}{42}$$

$$\sin \alpha = \frac{42 \sin 60°}{430.5392}$$

$$\alpha = 4.85° \qquad \textit{To the nearest degree } \alpha = 5°.$$

The course is approximately $60° + 5° = 65°$, and the ground speed is approximately 430 mi/h.

CLASS EXERCISES

Draw each vector $\overrightarrow{OC}$. Then use a ruler and protractor to find the vector's magnitude and the measure of its direction angle.

1. $O(0, 0), C(7, 2)$ **2.** $O(0, 0), C(-3, 5)$ **3.** $O(0, 0), C(6, -2)$

Use a ruler and a protractor to draw the following vectors in standard position. Draw their sum. Measure the angle between them.

4. $\|u\| = 4$; **u** has a direction angle of 55°.
   $\|v\| = 7$; **v** has a direction angle of 20°.

5. $\|u\| = 9$; **u** has a direction angle of 45°.
   $\|v\| = 5$; **v** has a direction angle of $-30°$.

6. $\|w\| = 3.5$; **w** has a direction angle of 110°.
   $\|z\| = 5.2$; **z** has a direction angle of $-15°$.

7. $\|w\| = 9.2$; **w** has a direction angle of 245°.
   $\|z\| = 7.4$; **z** has a direction angle of 15°.

PRACTICE EXERCISES

Draw each vector $\overrightarrow{AB}$ and find its magnitude.

1. $A(2, 5), B(7, 17)$ **2.** $A(1, 3), B(5, 6)$

3. $A(11, 5), B(-2, -6)$ **4.** $A(-5, 7), B(2, -4)$

Find the magnitudes of the *x*- and *y*-components, to the nearest whole number.

5. $\|\mathbf{v}\| = 15$, and $\mathbf{v}$ has a direction angle of $30°$.

6. $\|\mathbf{v}\| = 29$, and $\mathbf{v}$ has a direction angle of $45°$.

7. $\|\mathbf{v}\| = 18$, and $\mathbf{v}$ has a direction angle of $25°$.

8. $\|\mathbf{v}\| = 25$, and $\mathbf{v}$ has a direction angle of $43°$.

9. $\|\mathbf{v}\| = 12$, and $\mathbf{v}$ has a direction angle of $112°$.

Use the vectors shown to draw each of the following.

10. s + t   **11.** $\overrightarrow{AB} + \overrightarrow{CD}$ **12.** u + v

13. $\overrightarrow{CD} + \overrightarrow{AB}$ **14.** s + $\overrightarrow{CD}$ **15.** v − t

16. 2u   **17.** 3t   **18.** − 2s

Find the magnitude and direction of the sum w of the following vectors. Express the direction as a measure of the angle between v and w. In Exercises 19–24, round magnitudes to two significant digits and angle measures to the nearest degree. In Exercises 25–28, round magnitudes to three significant digits and angle measures to the nearest tenth of a degree. Assume that v lies on the positive *x*-axis.

19. $\|\mathbf{u}\| = 18$, $\|\mathbf{v}\| = 7$, and the angle between them measures $60°$.

20. $\|\mathbf{u}\| = 10$, $\|\mathbf{v}\| = 15$, and the angle between them measures $60°$.

21. $\|\mathbf{u}\| = 14$, $\|\mathbf{v}\| = 19$, and the angle between them measures $45°$.

22. $\|\mathbf{u}\| = 21$, $\|\mathbf{v}\| = 26$, and the angle between them measures $45°$.

23. $\|\mathbf{u}\| = 6.5$, $\|\mathbf{v}\| = 7.2$, and the angle between them measures $150°$.

24. $\|\mathbf{u}\| = 9.2$, $\|\mathbf{v}\| = 10.6$, and the angle between them measures $150°$.

25. $\|\mathbf{u}\| = 12.4$, $\|\mathbf{v}\| = 19.3$, and the angle between them measures $48°$.

26. $\|\mathbf{u}\| = 15.7$, $\|\mathbf{v}\| = 18.9$, and the angle between them measures $54°$.

27. $\|\mathbf{u}\| = 32.5$, $\|\mathbf{v}\| = 29.4$, and the angle between them measures $114°$.

28. $\|\mathbf{u}\| = 23.7$, $\|\mathbf{v}\| = 39.4$, and the angle between them measures $127°$.

Use the vectors from Exercises 10–18 to draw each of the following.

29. s + (t + v)   **30.** (s + t) + v   **31.** 2(u + v)   **32.** $\frac{1}{2}$(s − t)

In Exercises 33 and 34, an airplane has the indicated air velocity and is affected by the given wind velocity. Find the ground speed to two significant digits and the course to the nearest degree.

In Exercises 35–40, round magnitudes to two significant digits and angle measures to the nearest degree.

35. The air speed of an airplane is 250 mi/h, and its heading is 45 degrees. The wind is blowing from the north at 30 mi/h. Find the ground speed and course of the plane.

36. The air speed of an airplane is 520 mi/h, and its heading is 110 degrees. The wind is blowing from the south at 15 mi/h. Find the ground speed and course of the plane.

37. A plane has a heading of 26 degrees and an air speed of 575 mi/h. The wind is blowing in a direction of 67 degrees east of north at a speed of 23 mi/h. Find the ground speed and course of the plane.

38. The air speed of a plane is 280 mi/h, its heading is 190 degrees, and its course is 195°. If the wind is blowing at 23 mi/h and the ground speed of the plane is 275 mi/h, in what direction is the wind blowing?

39. A pilot wants to maintain a course of 40 degrees and a ground speed of 300 mi/h against a 45-mi/h wind from 20 degrees west of north. What should be his heading and air speed?

40. Two planes left an airport at the same time and in still air. One flew east at 450 mi/h, and the other flew at 380 mi/h. In what direction did the second pilot fly if the two planes were 800 mi apart after 1 h?

41. The magnitudes of the *x*- and *y*-components of **u** are *a* and *b*, respectively. The magnitudes of the *x*- and *y*-components of **v** are *c* and *d*, respectively. Use principles of coordinate geometry to show that the magnitudes of the *x*- and *y*-coordinates of **u** + **v** are *a* + *c* and *b* + *d*, respectively.

42. Use the law of cosines to show that $\|\mathbf{u}\|^2 + \|\mathbf{v}\|^2 = \|\mathbf{u} + \mathbf{v}\|^2$ if and only if the angle between **u** and **v** is 90°.

43. The magnitudes of the *x*- and *y*-components of **u** are *a* and *b*, respectively. The magnitudes of the *x*- and *y*-components of **v** are *c* and *d*, respectively. Show that the angle between **u** and **v** is 90° if and only if $ac + bd = 0$.

In Exercises 44–45, give magnitudes to two significant digits and directions to the nearest degree.

Physics

44. Forces of 8.0 lb and 15 lb act on a body at right angles. Find the magnitude of the resultant force.

45. Find the measure of the angle between forces of 80 lb and 70 lb, if the magnitude of the resultant force is 98 lb.

46. Two forces are pushing an ice shanty along the ice. One force has a magnitude of 330 lb in a direction due east. The other force has a magnitude of 110 lb in a direction 54° east of north. What are the magnitude and direction of the resultant force? Give the magnitude to the nearest ten pounds and the direction to the nearest degree.

TEST YOURSELF

Find the area of each triangle to the nearest ten square units.

1. $\angle C = 55°, a = 32, b = 43$

2. $\angle B = 38°, a = 10, c = 11$ **4.5**

3. $\angle A = 112°, \angle B = 23°, c = 37$

4. $\angle B = 43°, \angle C = 29°, a = 73$

Use Heron's formula to find the area of each triangle to the nearest ten square units.

5. $a = 30, b = 24, c = 13$

6. $a = 62, b = 24, c = 45$ **4.6**

7. $a = 12, b = 24, c = 27$

8. $a = 31, b = 35, c = 40$

9. Given $A(3, 4)$ and $B(-2, 7)$, what is the magnitude of $\overrightarrow{AB}$? **4.7**

10. Find the magnitude of the *x*- and *y*-components of **v**, if $\|\mathbf{v}\| = 10$ and **v** has a direction angle of 45°.

11. If $\|\mathbf{u}\| = 9$, $\|\mathbf{v}\| = 11$, and the angle between **u** and **v** is 39°, find the magnitude and direction of **w**, the sum of **u** and **v**. Express the direction as a measure of the angle between **v** and **w**.

12. An airplane has an air speed of 500 mi/h and a heading of 120°. A wind is blowing from the north at 30 mi/h. Find, to two significant digits, the plane's ground speed and course.

APPLICATION:
Industrial Robots

Did you know that robots are used in industry to perform routine, heavy, dangerous, or repetitive tasks? As you observe them in operation, they appear to think, but this impression is an illusion; their every move has been programmed beforehand by a human being. To understand how a robot is controlled, visualize three-dimensional space.

To perform its tasks, a robot may use both linear and rotary (angular) motion. Two types of robot configurations will be considered. *Cartesian motion* acts in straight lines for forward-backward, right-left, and up-down directions. *Cylindrical motion* acts in straight lines for forward-backward and up-down direction. Horizontal motion is angular.

Illustrated at the right is a three-dimensional *rectangular coordinate system*. The coordinates of a point $P(x, y, z)$ are ordered triples. To program robotic motion, a programmer needs to know the coordinates of two different positions. The difference between the two positions is then calculated.

EXAMPLE 1   Suppose for a robot operating in a rectangular system, the coordinates of the first position of the end of its arm are P_1 (7, -2, 1), and the coordinates of a second position are P_2 (2, 4, -6). Calculate the distance the end of the robot's arm moves along each axis.

	P_1	P_2	Difference	
x	7	2	-5	*moved backward 5 units*
y	-2	4	6	*moved right 6 units*
z	1	-6	-7	*moved downward 7 units*

In the *cylindrical coordinate system*, a point is represented as an ordered triple (r, θ, z). In the diagram, point P is r units from the z-axis and z units directly above a point P' in the xy-plane. Angle θ is the measure of the counterclockwise acute angle between OP' and the positive x-axis.

EXAMPLE 2   A robot using cylindrical coordinates has its arm rotating about the z-axis. The end of the arm is at P_1 so that θ has a value of 45° with $x_1 = 5$ and $z_1 = 8$. The arm moves to P_2 by rotating to $\theta = -30°$ with $x_2 = 4$ and $z_2 = 6$. Find the distance the robot moves along the x- and z-axes, the length of the arm at P_1 and P_2, the amount the arm length changed from P_1 to P_2, and the y-coordinate at P_2.

	P_1	P_2	Difference	
x	5	4	-1	*moved backward 1 unit*
θ	45°	$-30°$	$-75°$	*moved counterclockwise 75°*
z	8	6	-2	*moved downward 2 units*

arm length at P_1: $\cos 45° = \dfrac{5}{\text{arm length}}$

arm length at $P_1 = 7.07$ units

arm length at P_2: $\cos(-30°) = \dfrac{4}{\text{arm length}}$

arm length $= 4.62$ units

$\tan(-30°) = \dfrac{y_2}{4}$

$y_2 = 4\tan(-30°) = -2.31$

(top view)

The difference in the length of the arm between P_1 and P_2 is -2.45 units. (The arm retracted, or "shrank.") The y-coordinate of P_2 is -2.31.

Robot engineers are trying to design a robot that will direct some of its own movements. However, this goal is several years away.

EXERCISES

Suppose a robot operating in a rectangular system has the end of its arm located at $P_1(4, 6, 3)$. The arm then moves to $P_2(7, 5, 9)$.

1. Draw a diagram for the robot at P_1 and at P_2.

2. Calculate the distance the end of the robot's arm moves along each axis.

3. A robot using cylindrical coordinates has the end of its arm located at $P_1(r_1, 48°, -2)$ with $x_1 = 9$. The arm then rotates to $P_2(r_2, 76°, 8)$ with $x_2 = 5$. Find the distance the end of the robot's arm moves along the x- and z-axes, the length of the arm at P_1 and P_2, the amount the arm changed from P_1 to P_2, and the y-coordinate at P_2.

Vocabulary

ambiguous case (163)
components of a vector (191)
direction of a vector (191)
direction angle of a vector (191)
equivalent vector (190)
initial point of a vector (190)
Heron's formula (186)
law of cosines (168)
law of sines (156)
magnitude of a vector (190)

law of tangents (175)
negative of a vector (192)
parallelogram addition (191)
resultant of a vector (191)
scalar (192)
standard position of a vector (191)
tail-to-head addition (191)
terminal point of a vector (190)
vector (190)

In all exercises, round lengths to two significant digits and angle measures to the nearest degree, unless otherwise stated.

Law of Sines   The law of sines can be used to solve triangles when   4.1
two angles and any side are given.

$$\frac{\sin A}{a} = \frac{\sin B}{b} = \frac{\sin C}{c}$$

1. In $\triangle ABC$, find b if $c = 26$, $\angle B = 49°$, and $\angle C = 53°$.

2. Solve $\triangle ABC$ if $a = 41$, $\angle B = 28°$, and $\angle C = 120°$.

Law of Sines: The Ambiguous Case   When the measures of two sides   4.2
and an angle opposite one of them are given, the law of sines must be used with care,
since *two* triangles, *one* triangle, or *no* triangles may be determined.

Solve $\triangle ABC$.

3. $\angle A = 56°, b = 12, a = 11$ **4.** $\angle B = 125°, b = 13, c = 20$

5. $\angle A = 99°, a = 25, b = 15$ **6.** $\angle C = 105°, a = 11, c = 14$

Law of Cosines   The law of cosines can be used to solve a triangle when   4.3
two sides and the included angle or three sides are given.

$$a^2 = b^2 + c^2 - 2bc \cos A \qquad \cos A = \frac{b^2 + c^2 - a^2}{2bc}$$

$$b^2 = a^2 + c^2 - 2ac \cos B \qquad \cos B = \frac{a^2 + c^2 - b^2}{2ac}$$

$$c^2 = a^2 + b^2 - 2ab \cos C \qquad \cos C = \frac{a^2 + b^2 - c^2}{2ab}$$

Solve $\triangle ABC$.

7. $\angle C = 58°, a = 12, b = 16$ **8.** $a = 25, b = 33, c = 28$

Law of Tangents   When two sides and the included angle are given, the law   **4.4**
of tangents may allow you to solve the triangle with fewer calculations.

$$\frac{a - b}{a + b} = \frac{\tan\frac{1}{2}(A - B)}{\tan\frac{1}{2}(A + B)} \qquad \frac{c - a}{c + a} = \frac{\tan\frac{1}{2}(C - A)}{\tan\frac{1}{2}(C + A)} \qquad \frac{b - c}{b + c} = \frac{\tan\frac{1}{2}(B - C)}{\tan\frac{1}{2}(B + C)}$$

Solve $\triangle ABC$.

9. $\angle B = 68°, b = 14, c = 18$ **10.** $\angle A = 52°, \angle B = 33°, a + b = 50$

Area of a Triangle   The following formulas can be used to find the area of a   **4.5**
triangle when two sides and the included angle are known:

$$K = \frac{1}{2}ab\sin C \qquad K = \frac{1}{2}bc\sin A \qquad K = \frac{1}{2}ac\sin B$$

These formulas are used when two angles and the included side are known:

$$K = \frac{c^2\sin A\sin B}{2\sin C} \qquad K = \frac{a^2\sin B\sin C}{2\sin A} \qquad K = \frac{b^2\sin A\sin C}{2\sin B}$$

Find the area of $\triangle ABC$ to the nearest ten square units.

11. $\angle A = 48°, b = 24, c = 18$ **12.** $\angle A = 53°, \angle B = 100°, c = 20$

Heron's Formula   When three sides of a triangle are given, Heron's   **4.6**
formula can be used to find the area of the triangle and the altitude.

$$K = \sqrt{s(s - a)(s - b)(s - c)}, \text{ where } s = \frac{a + b + c}{2}$$

13. Find the area of $\triangle ABC$ to the nearest unit, if $a = 24, b = 28$, and $c = 35$.

14. Find the length of the altitude to side BC in $\triangle ABC$, to the nearest unit, if
$a = 30, b = 48$, and $c = 60$.

Vectors in the Plane   A vector has direction and magnitude. Vectors can   **4.7**
be added or subtracted using either parallelogram addition or tail-to-head addition. To
find the sum of two vectors, use the law of cosines.

15. Find the magnitudes of the x- and y-components of **v,** if **v** has a direction
angle of $135°$ and $\|\mathbf{v}\| = 20$.

16. The air speed of an airplane is 400 mi/h, and its heading is $150°$. A wind is
blowing from the south at 20 mi/h. Find the plane's ground speed and course.

Solve $\triangle ABC$. Give lengths to two significant digits and angle measures to the nearest degree.

1. $\angle A = 38°, b = 19, c = 23$

2. $a = 12, b = 17, c = 21$

3. $\angle A = 37°, \angle B = 28°, c = 18$

4. $\angle B = 62°, a = 14, c = 23$

5. $\angle A = 49°, b = 23, c = 19$

6. $a = 15, b = 19, c = 22$

7. $\angle A = 51°, \angle B = 63°, c = 26$

8. $\angle A = 103°, a = 49, b = 35$

9. $\angle B = 37°, a = 44, b = 38$

10. $\angle C = 10°, a = 15, c = 40$

11. $a = 12, b = 16, c = 19$

12. $a = 21, b = 17, c = 15$

Determine the area of each triangle to the nearest ten square units.

13. $\angle B = 82°, a = 12, c = 15$

14. $\angle A = 73°, b = 42, c = 67$

15. $\angle A = 46°, \angle B = 37°, c = 43$

16. $\angle A = 51°, \angle C = 29°, b = 32$

Use Heron's formula to find the area of each triangle to the nearest ten square units.

17. $a = 12, b = 14, c = 19$

18. $a = 15, b = 16, c = 11$

19. $a = 14, b = 16, c = 21$

20. $a = 24, b = 27, c = 31$

Find the altitude to side BC in $\triangle ABC$, to the nearest unit.

21. $a = 32, b = 50, c = 64$

22. $a = 18, b = 21, c = 16$

23. Find the magnitude of the x- and y-components of **u,** if **u** has a direction angle of $45°$ and $\|\mathbf{u}\| = 40$.

24. Find the magnitude and direction of **u** and **v,** if $\|\mathbf{u}\| = 10$, $\|\mathbf{v}\| = 15$, and if the angle between **u** and **v** measures $110°$. Express the direction of **w** as the measure of the angle between **v** and **w**.

25. An airplane has an air speed of 600 mi/h and a heading of $70°$. A wind is blowing from the south at 40 mi/h. Find, to two significant digits, the plane's ground speed and course.

Challenge

Prove that for any $\triangle ABC$, $a = b \cos C + c \cos B$.

Select the best choice for each question.

1. In the triangle shown, $\angle A = 30°$, $\angle B = 65°$, and $a = 5$. Find b.

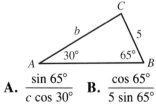

   A. $\dfrac{\sin 65°}{c \cos 30°}$ B. $\dfrac{\cos 65°}{5 \sin 65°}$

   C. $\dfrac{5 \sin 65°}{\sin 30°}$ D. $\dfrac{\sin 30°}{5 \sin 65°}$

   E. $\dfrac{\sin 30°}{\cos 65°}$

2. The expression $|2x - 1| \le 9$ is equivalent to
   A. $4 \le x \le 5$ B. $x \le 5$
   C. $-4 \le x \le 10$ D. $-4 \le x \le 5$
   E. $-8 \le x \le 10$

3. $8x^2 - 3x = 1$ has
   A. one real number solution
   B. two real number solutions
   C. one real number solution and one complex number solution
   D. two complex number solutions
   E. one complex number solution

4. A painter can paint a house in 3 hr. The painter's assistant can complete the task in 6 hr. How long would it take to paint the house if the painter and the assistant worked together?
   A. $4\frac{1}{2}$ hr   B. 5 hr   C. 2 hr
   D. 3 hr   E. 1 hr

5. If $3p - a = 6p - d$, then p equals
   A. $\dfrac{d - a}{3}$ B. $\dfrac{a - d}{3}$ C. $3a + d$
   D. $d - \dfrac{1}{3}a$ E. $\dfrac{d + a}{3}$

6. In $\triangle ABC$, if $a = 5$, $c = 9$, and $\cos B = \dfrac{1}{15}$, find b.
   A. 15   B. 10   C. 9
   D. 20   E. No solution

7. Find the area of an equilateral triangle if the measure of one side is 6.
   A. 15   B. $9\sqrt{3}$ C. 18
   D. $36\sqrt{3}$ E. $6\sqrt{3}$

8. The area of parallelogram $ABCD$ is
   A. $\dfrac{2\sqrt{2}}{15}$
   B. $\dfrac{15\sqrt{2}}{4}$
   C. $\dfrac{4\sqrt{2}}{15}$ D. $\dfrac{15\sqrt{3}}{2}$ E. $\dfrac{15\sqrt{2}}{2}$

9. Find the area of a square whose side measures $3\sqrt{7}$.
   A. 63   B. $6\sqrt{7}$ C. $9\sqrt{7}$
   D. 42   E. 23

10. Which of the following is not a function?
    I. $y = 3x$
    II. $x = -6$
    III. $x = 5y$
    A. I only   B. II only   C. III only
    D. I and II only   E. I and III only

11. $\cos 45° \sin 30° - \tan 45° =$
    A. $\dfrac{\sqrt{2} - 2}{2}$ B. $\dfrac{-\sqrt{2} - 2}{2}$
    C. $\dfrac{\sqrt{2} - 4}{4}$ D. $\dfrac{\sqrt{6} - 4}{4}$
    E. $\dfrac{-\sqrt{6} - 4}{4}$

1. Use the law of cosines to solve $\triangle ABC$ given $a = 11$, $b = 14$, and $\angle C = 34°$. Round the length of side c to two significant digits and the lengths of angle measures to the nearest degree.

2. Determine which real number(s) must be excluded from the domain of
$$f(x) = \frac{2}{x - 3}$$

3. Express $132°18'25''$ in decimal degrees to the nearest hundredth of a degree.

4. Prove $\dfrac{\dfrac{1}{\sin \theta} + \dfrac{1}{\cos \theta}}{\dfrac{1}{\sin \theta} - \dfrac{1}{\cos \theta}} = \dfrac{\cos \theta + \sin \theta}{\cos \theta - \sin \theta}$.

5. Find $\cos(-48°)$ if $\cos 48° = 0.6691$.

6. Find the area of the triangle with sides 5m, 7m, and 3m to the nearest square meter.

7. Given right triangle ABC with $\angle C = 90°$, $a = 11$ cm, and $c = 21$ cm, find b to the nearest centimeter.

A belt on a lawnmower runs a pulley of radius 3 in. at 100 rpm. Find

8. the angular velocity of the pulley in radians per second

9. the linear velocity of the belt in inches per second

10. Graph $y = 2 \sec \left(x + \frac{\pi}{8}\right)$ over a two-period interval.

11. Check graphically to see whether or not $2 \cos^2 x = 1 + \cos 2x$ is an identity.

12. Sketch the graph of $y = 3 \cot 2x$ over the interval $-\pi \le x \le \pi$.

13. Find the magnitude and direction of the sum $\mathbf{w}$ of two vectors $\mathbf{u}$ and $\mathbf{v}$ of magnitudes 6 and 1, respectively, if the angle between them measures $62°$. Round the magnitude of the nearest whole number and the direction to the nearest degree.

14. State whether or not the relation $\{(1, 6), (2, 3), (1, 7)\}$ is a function.

15. Convert $(\sin \theta + \cos \theta + 1)(\sin \theta + \cos \theta - 1)$ to $2 \sin \theta \cos \theta$.

16. Using the law of cosines, find the length of a to two significant digits given $\angle A = 37°$, $b = 10$, and $c = 14$.

17. If $0° \le \theta \le 360°$, find the value(s) of θ that satisfy $\cos \theta = -1$ and $\tan \theta = 0$.

18. A cat playtoy hangs at rest from a spring. It is then pulled down 12 inches and released. The toy bounces back and forth through its rest position making one full cycle every 4 s. Using the sine function as a model, write an equation for the function and then sketch its graph.

19. Verify the following identity for the given angle measure:
$$\sin^2 \frac{\pi}{6} \cos \frac{\pi}{6} \sec \frac{\pi}{6} = 1 - \cos^2 \frac{\pi}{6}$$

20. Compute to two significant digits the unknown sides for $\triangle ABC$ if $\angle C = 42°$, $\angle B = 87°$, and $c = 21$ in.

21. Determine whether the function $y = x^2 - 5$ is symmetric with respect to the x-axis, to the y-axis, to the origin, to more than one of these, or to none of these.

22. Graph $y = \cos x - \sin 2x$ over the interval. $-2\pi \le x \le 2\pi$.

23. Given $\angle D = 58°$, $\angle F = 77°$, and $d = 35$ in., compute the area of $\triangle DEF$ to the nearest 10 square inches.

24. A hip roof is in the form of four congruent isosceles triangles. If each triangle has a 30-ft base and a height of 12 ft, find the measure of each base angle.

25. Let θ be an angle in standard position. In which quadrant(s) can θ lie when $\cos \theta > 0$ and $\sin \theta < 0$?

26. Determine the amplitude and period for the function $y = -2 \sin (-2x)$. Then sketch its graph over the interval $-\pi \le x \le \pi$.

27. Find the exact values of the other five trigonometric functions for an angle in standard position in quadrant II whose cosine is $-\frac{4}{5}$.

28. Determine the amplitude, period, phase shift, and vertical shift for the function $y = -2 \cos (2x + 2\pi) - 3$.

29. The shadow of a tower is 200 ft when the sun's elevation is $75.8°$. Find the height of the tower to three significant digits.

30. Find θ, where $0° \le \theta \le 360°$, to the nearest tenth of one degree given that $\cos \theta = -0.5253$ and $\sin \theta$ is positive.

31. Use the law of tangents to find the unknown angles of $\triangle ABC$ to the nearest degree if $\angle B = 65°$, $a = 32$, and $c = 17$.

32. Prove $\dfrac{\cot^2 \alpha}{\sec^2 \alpha} = \dfrac{1 - \sin^2 \alpha}{\tan^2 \alpha}$.

33. Find x if the distance between $(2, 2)$ and $(x, 10)$ is 10.

34. Determine the number of solutions to $\triangle ABC$ if $\angle A = 36°$, $a = 48$, and $b = 54$. If one or more solutions exist, solve the triangle or triangles.

Determine whether each function is even, odd, or neither even nor odd.

35. $g(x) = 4x^3 + 2x^2$ **36.** $f(x) = 0.5x^2 - 3$ **37.** $h(x) = -2x^3 + x$

38. A sailboat on Lake Michigan is 3.7 miles due north of buoy A. Buoy B is 9.3 miles due west of buoy A. How far apart are the sailboat and buoy B?

39. Name three angles (two positive and one negative) that are coterminal with an angle of $-57°$.

If $\angle A = 65°$ and $\angle B = 32°$, compute

40. $\tan \frac{1}{2}(A - B)$ **41.** $\tan \frac{1}{2}(A + B)$

42. Write the expression $\dfrac{1}{1 - \sin \alpha} + \dfrac{1}{1 + \sin \alpha}$ in terms of $\cos \alpha$.

Use the fact that sine and cosine are periodic functions and cofunctions to determine the following values.

43. If $\cos 67° = 0.3907$, find $\sin 23°$.

44. If $\sin 192° = -0.2079$, find $\sin 912°$.

45. If $\sin 33° = 0.5446$, find $\sin (-327°)$.

46. If $\cos 160° = -0.9397$, find $\cos (-200°)$.

47. Give the measure of the reference angle θ' for $\theta = 344°$.

48. Use the definitions of the trigonometric functions to prove $\sec^2 \alpha - \tan^2 \alpha = 1$.

49. What is the area, in acres, of a triangular lot that is 296 ft on one side and 242 ft on an adjacent side, if the angle between these two sides measures 37.5°? (1 acre $= 43,560$ ft^2)

50. Prove $\dfrac{1 - \sin \beta}{\cos \beta} = \dfrac{\cos \beta}{1 + \sin \beta}$.

51. Find the measure of the reference angle θ' for $\theta = 1227°$

52. A ladder rests against a building at a point that is 39 ft from the ground. If the ladder makes a 50° angle with the ground, what is the length of the ladder?

5 | Trigonometric Identities

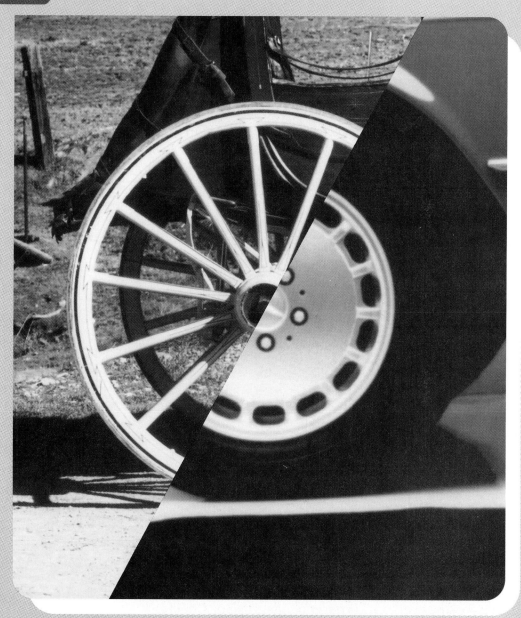

For hundreds of years, 4-wheel vehicles were maneuvered with 2-wheel steering. In the 1980s, Japanese engineers created a technological revolution with the introduction of 4-wheel steering.

Cosine: Sum and Difference Identities

Objective: To develop and use formulas for the cosine of a sum or difference of two angle measures

Before introducing the sum and difference identities, the trigonometric functions of angles whose measures are multiples of 30°, 45°, or 60° are reviewed.

Preview

Recall that the lengths of the sides of a 45°-45°-90° triangle are in the ratio $1:1:\sqrt{2}$, where $\sqrt{2}$ corresponds to the hypotenuse. The lengths of the sides of a 30°-60°-90° triangle are in the ratio $1:\sqrt{3}:2$, where 2 corresponds to the hypotenuse, and $\sqrt{3}$ to the side opposite the 60° angle. You can use these facts and reference angles to find the values of the trigonometric functions of special angles.

EXAMPLE Find the exact values of the six trigonometric functions of 120°.

The reference angle is 60°. Let $r = 2$.
Then $x = -1$ and $y = \sqrt{3}$.

$$\sin 120° = \frac{\sqrt{3}}{2} \qquad \csc 120° = \frac{2}{\sqrt{3}} = \frac{2\sqrt{3}}{3}$$

$$\cos 120° = \frac{-1}{2} = -\frac{1}{2} \qquad \sec 120° = \frac{2}{-1} = -2$$

$$\tan 120° = \frac{\sqrt{3}}{-1} = -\sqrt{3} \qquad \cot 120° = \frac{-1}{\sqrt{3}} = -\frac{\sqrt{3}}{3}$$

Find the exact values of the six trigonometric functions of each angle.

1. 225°   2. 330°   3. 240°   4. −60°   5. $\dfrac{3\pi}{4}$ 6. $-\dfrac{7\pi}{6}$

In previous chapters, the identities that were presented involved only one angle (usually called θ). Now, identities that involve two angles, such as α and β, will be introduced. The first is the *difference identity for cosine*.

Difference Identity for Cosine

$$\cos (\alpha - \beta) = \cos \alpha \cos \beta + \sin \alpha \sin \beta$$

To prove this identity, place angles α and β in standard position, as shown in the figure at the right. The terminal sides of angles α and β intersect the unit circle O at points D and F, respectively, and the measure of central angle DOF is $\alpha - \beta$. Note that the coordinates of points D and F are

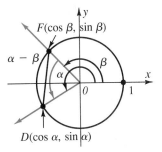

$$D(\cos \alpha, \sin \alpha) \qquad F(\cos \beta, \sin \beta)$$

The figure at the right shows $\angle AOC$, with measure $\alpha - \beta$, in standard position. Note that the coordinates of points A and C are

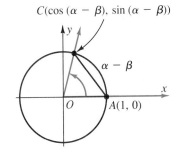

$$A(1, 0) \qquad C(\cos (\alpha - \beta), \sin (\alpha - \beta))$$

Since $\triangle AOC$ and $\triangle DOF$ are congruent (by SAS), $\overline{AC}$ and $\overline{DF}$ are congruent and $AC = DF$. Use the distance formula to express AC and DF in terms of the coordinates of points A, C, D, and F.

First, find AC.

$$
\begin{aligned}
AC &= \sqrt{[\cos (\alpha - \beta) - 1]^2 + [\sin (\alpha - \beta) - 0]^2} \qquad \textit{Distance formula} \\
&= \sqrt{\cos^2 (\alpha - \beta) - 2 \cos (\alpha - \beta) + 1 + \sin^2 (\alpha - \beta)} \\
&= \sqrt{\cos^2 (\alpha - \beta) + \sin^2 (\alpha - \beta) + 1 - 2 \cos (\alpha - \beta)} \\
&= \sqrt{1 + 1 - 2 \cos (\alpha - \beta)} \qquad \textit{sin}^2\,\theta + \textit{cos}^2\,\theta = 1 \\
&= \sqrt{2 - 2 \cos (\alpha - \beta)}
\end{aligned}
$$

Then, find DF.

$$
\begin{aligned}
DF &= \sqrt{(\cos \alpha - \cos \beta)^2 + (\sin \alpha - \sin \beta)^2} \qquad \textit{Distance formula} \\
&= \sqrt{\cos^2 \alpha - 2 \cos \alpha \cos \beta + \cos^2 \beta + \sin^2 \alpha - 2 \sin \alpha \sin \beta + \sin^2 \beta} \\
&= \sqrt{\cos^2 \alpha + \sin^2 \alpha + \cos^2 \beta + \sin^2 \beta - 2 \cos \alpha \cos \beta - 2 \sin \alpha \sin \beta} \\
&= \sqrt{1 + 1 - 2 \cos \alpha\, 2 \cos \beta - 2 \sin \alpha \sin \beta} \qquad \textit{sin}^2\,\theta + \textit{cos}^2\,\theta = 1 \\
&= \sqrt{2 - 2 \cos \alpha \cos \beta - 2 \sin \alpha \sin \beta}
\end{aligned}
$$

Since $AC = DF$, the proof is completed as follows:

$$
\begin{aligned}
\sqrt{2 - 2 \cos (\alpha - \beta)} &= \sqrt{2 - 2 \cos \alpha \cos \beta - 2 \sin \alpha \sin \beta} \\
2 - 2 \cos (\alpha - \beta) &= 2 - 2 \cos \alpha \cos \beta - 2 \sin \alpha \sin \beta \\
-2 \cos (\alpha - \beta) &= -2 \cos \alpha \cos \beta - 2 \sin \alpha \sin \beta \\
\cos (\alpha - \beta) &= \cos \alpha \cos \beta + \sin \alpha \sin \beta
\end{aligned}
$$

Note that $\cos (\alpha - \beta) = \cos [-(\alpha - \beta)] = \cos (\beta - \alpha)$, so the identity holds for any values of α and β.

EXAMPLE 1 **Find the exact value of cos 15°.**

Since $15° = 45° - 30°$, let $\alpha = 45°$ and $\beta = 30°$.

$$\cos 15° = \cos (45° - 30°)$$
$$= \cos 45° \cos 30° + \sin 45° \sin 30° \quad \textit{Difference identity for cosine}$$
$$= \frac{\sqrt{2}}{2}\left(\frac{\sqrt{3}}{2}\right) + \frac{\sqrt{2}}{2}\left(\frac{1}{2}\right) = \frac{\sqrt{6}}{4} + \frac{\sqrt{2}}{4} = \frac{\sqrt{6} + \sqrt{2}}{4}$$

The *sum identity for cosine* follows from the difference identity for cosine.

Sum Identity for Cosine

$$\cos (\alpha + \beta) = \cos \alpha \cos \beta - \sin \alpha \sin \beta$$

To prove the sum identity for cosine, write β as $-(-\beta)$. Then

$$\cos (\alpha + \beta) = \cos [\alpha - (-\beta)]$$
$$= \cos \alpha \cos (-\beta) + \sin \alpha \sin (-\beta)$$
$$= \cos \alpha \cos \beta - \sin \alpha \sin \beta \qquad \begin{array}{l} \textit{cos } (-\beta) = \textit{cos } \beta; \\ \textit{sin } (-\beta) = -\textit{sin } \beta \end{array}$$

EXAMPLE 2 **Find the exact value of $\cos \dfrac{7\pi}{12}$.**

Since $\dfrac{7\pi}{12} = \dfrac{4\pi}{12} + \dfrac{3\pi}{12} = \dfrac{\pi}{3} + \dfrac{\pi}{4}$, let $\alpha = \dfrac{\pi}{3}$ and $\beta = \dfrac{\pi}{4}$.

$$\cos \frac{7\pi}{12} = \cos \left(\frac{\pi}{3} + \frac{\pi}{4}\right)$$
$$= \cos \frac{\pi}{3} \cos \frac{\pi}{4} - \sin \frac{\pi}{3} \sin \frac{\pi}{4} \quad \textit{Sum identity for cosine}$$
$$= \frac{1}{2}\left(\frac{\sqrt{2}}{2}\right) - \frac{\sqrt{3}}{2}\left(\frac{\sqrt{2}}{2}\right) = \frac{\sqrt{2}}{4} - \frac{\sqrt{6}}{4} = \frac{\sqrt{2} - \sqrt{6}}{4}$$

It is important to note that $\cos (\alpha - \beta)$ is *not* equal to $\cos \alpha - \cos \beta$.
To illustrate, $\cos (45° - 30°) = \dfrac{\sqrt{6} + \sqrt{2}}{4}$ (from Example 1), but

$\cos 45° - \cos 30° = \dfrac{\sqrt{2}}{2} - \dfrac{\sqrt{3}}{2} = \dfrac{\sqrt{2} - \sqrt{3}}{2}$. Similarly, $\cos (\alpha + \beta)$ is *not*
equal to $\cos \alpha + \cos \beta$.

EXAMPLE 3 **Find the exact values of cos $(\alpha + \beta)$ and cos $(\alpha - \beta)$ if sin $\alpha = \dfrac{3}{5}$, tan $\beta = -\dfrac{5}{12}$, $0 < \alpha < \dfrac{\pi}{2}$, and $\dfrac{\pi}{2} < \beta < \pi$.**

Find cos α, sin β, and cos β before using the identities for cosine.

To find cos α, note that sin $\alpha = \dfrac{3}{5}$
and α is in quadrant I. Use $y = 3$ and
$r = 5$, and calculate x.

$$x = \pm \sqrt{5^2 - 3^2} = \pm \sqrt{16} = \pm 4$$

Since α is in quadrant I, $x = 4$.

Thus, cos $\alpha = \dfrac{4}{5}$.

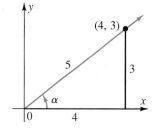

To find sin β and cos β, note that

tan $\beta = -\dfrac{5}{12}$ and β is in quadrant II.

Use $x = -12$ and $y = 5$, and calculate r.

$$r = \sqrt{(-12)^2 + 5^2} = 13$$

Thus, sin $\beta = \dfrac{5}{13}$ and cos $\beta = -\dfrac{12}{13}$.

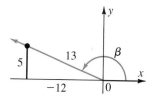

Then, cos $(\alpha + \beta) = \dfrac{4}{5}\left(-\dfrac{12}{13}\right) - \dfrac{3}{5}\left(\dfrac{5}{13}\right)$

$$= -\dfrac{48}{65} - \dfrac{15}{65} = -\dfrac{63}{65}$$

Also, cos $(\alpha - \beta) = \dfrac{4}{5}\left(-\dfrac{12}{13}\right) + \dfrac{3}{5}\left(\dfrac{5}{13}\right)$

$$= -\dfrac{48}{65} + \dfrac{15}{65} = -\dfrac{33}{65}$$

The above sum and difference identities can be used to prove other identities.

EXAMPLE 4 **Prove: sin $\theta = $ cos $\left(\dfrac{\pi}{2} - \theta\right)$**

$$
\begin{array}{c|c}
\sin \theta & \cos\left(\dfrac{\pi}{2} - \theta\right) \\
& \cos \dfrac{\pi}{2} \cos \theta + \sin \dfrac{\pi}{2} \sin \theta \\
& 0 (\cos \theta) + 1(\sin \theta) \\
& = \sin \theta
\end{array}
$$

Therefore, sin $\theta = $ cos $\left(\dfrac{\pi}{2} - \theta\right)$.

CLASS EXERCISES

Simplify each expression by writing it in terms of the cosine of one angle.

1. $\cos 80° \cos 20° + \sin 80° \sin 20°$

2. $\cos 20° \cos 45° - \sin 20° \sin 45°$

3. $\cos 30° \cos 35° - \sin 30° \sin 35°$

4. $\cos 85° \cos 30° + \sin 85° \sin 30°$

5. $\cos \dfrac{7\pi}{11} \cos \dfrac{2\pi}{11} - \sin \dfrac{7\pi}{11} \sin \dfrac{2\pi}{11}$

6. $\cos \dfrac{7\pi}{6} \cos \dfrac{5\pi}{6} + \sin \dfrac{7\pi}{6} \sin \dfrac{5\pi}{6}$

Tell whether each of the following statements is true or false.

7. $\cos 57° = \cos (40° + 17°)$

8. $\cos (-20°) = \cos 30° - \cos 50°$

9. $\cos 65° = \cos 30° + \cos 35°$

10. $\cos 70° = \cos 80° - \cos 10°$

PRACTICE EXERCISES

Use the sum or the difference identity for cosine to find the exact value of each trigonometric function.

1. $\cos 75°$

2. $\cos 105°$

3. $\cos 165°$

4. $\cos 195°$

5. $\cos 345°$

6. $\cos 285°$

7. $\cos \left(\dfrac{5\pi}{3} + \dfrac{\pi}{4} \right)$

8. $\cos \left(\dfrac{7\pi}{6} + \dfrac{\pi}{4} \right)$

Find the exact value of $\cos (\alpha + \beta)$, given $0 < \alpha < \dfrac{\pi}{2}$, $0 < \beta < \dfrac{\pi}{2}$.

9. $\cos \alpha = \dfrac{4}{5}, \cos \beta = \dfrac{12}{13}$

10. $\sin \alpha = \dfrac{3}{5}, \sin \beta = \dfrac{5}{13}$

Find the exact value of $\cos (\alpha + \beta)$, given $\dfrac{\pi}{2} < \alpha < \pi$, $\dfrac{\pi}{2} < \beta < \pi$.

11. $\sin \alpha = \dfrac{7}{25}, \tan \beta = -\dfrac{3}{4}$

12. $\tan \alpha = -\dfrac{15}{8}, \sin \beta = \dfrac{5}{13}$

Find the exact value of $\cos (\alpha - \beta)$, given $\pi < \alpha < \dfrac{3\pi}{2}$, $\dfrac{\pi}{2} < \beta < \pi$.

13. $\cos \alpha = -\dfrac{7}{25}, \tan \beta = -\dfrac{3}{4}$

14. $\cos \alpha = -\dfrac{5}{13}, \sin \beta = \dfrac{15}{17}$

Find the exact value of $\cos (\alpha + \beta)$, given $\dfrac{3\pi}{2} < \alpha < 2\pi$, $\pi < \beta < \dfrac{3\pi}{2}$.

15. $\cos \alpha = \dfrac{15}{17}, \tan \beta = \dfrac{4}{3}$

16. $\cos \alpha = \dfrac{7}{25}, \sin \beta = -\dfrac{5}{13}$

Use the sum or the difference identity for cosine to prove each identity.

17. $\cos (360° + \alpha) = \cos \alpha$

18. $\cos (180° + \alpha) = -\cos \alpha$

19. $\cos (270° - \beta) = -\sin \beta$

20. $\cos (180° - \beta) = -\cos \beta$

21. $\cos\left(\dfrac{\pi}{3} - \beta\right) = \dfrac{\cos\beta + \sqrt{3}\sin\beta}{2}$

22. $\cos\left(\beta - \dfrac{\pi}{6}\right) = \dfrac{\sqrt{3}\cos\beta + \sin\beta}{2}$

23. $\cos\left(\dfrac{3\pi}{4} + \alpha\right) = -\dfrac{\sqrt{2}}{2}(\cos\alpha + \sin\alpha)$

24. $\cos 2\alpha = \cos^2\alpha - \sin^2\alpha$

Use the sum or the difference identity for cosine to find the exact value of each trigonometric function.

25. $\cos(-285°)$ **26.** $\cos(-105°)$ **27.** $\cos\left(-\dfrac{17\pi}{12}\right)$ **28.** $\cos\left(-\dfrac{5\pi}{12}\right)$

Prove each identity.

29. $\cos(\alpha - \beta) - \cos(\alpha + \beta) = 2\sin\alpha\sin\beta$

30. $\cos(\alpha + \beta) + \cos(\alpha - \beta) = 2\cos\alpha\cos\beta$

Applications

31. Surveying Part of a surveyor's diagram is shown at the right. Find the exact value of the cosine of $\angle AOD$.

32. Aviation A jet is flying at an altitude of 12 mi, and its radar unit senses objects on the ground up to 13 mi from the plane in one direction and up to 20 mi from the plane in the opposite direction, as shown in the figure. Find the exact value of $\cos(\alpha + \beta)$.

EXTRA

You can use the law of cosines, $b^2 = a^2 + c^2 - 2ac\cos(\alpha + \beta)$, to prove the sum identity for cosine for acute angles.

$$b^2 = a^2 + c^2 - 2ac\cos(\alpha + \beta)$$

$$\cos(\alpha + \beta) = \dfrac{a^2 + c^2 - b^2}{2ac}$$

$$\cos(\alpha + \beta) = \dfrac{(h^2 + y^2) + (h^2 + x^2) - (x + y)^2}{2ac}$$

$$\cos(\alpha + \beta) = \dfrac{2h^2 - 2xy}{2ac} = \dfrac{h^2 - xy}{ac} = \dfrac{h}{c}\left(\dfrac{h}{a}\right) - \dfrac{x}{c}\left(\dfrac{y}{a}\right) = \cos\alpha\cos\beta - \sin\alpha\sin\beta$$

Use the diagram at the right to derive the difference identity for cosine in a similar way. Start with this statement:

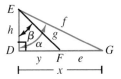

$$e^2 = f^2 + g^2 - 2fg\cos(\alpha - \beta)$$

Sine: Sum and Difference Identities

Objective: To develop and use formulas for the sine of a sum or difference of two angle measures

The cosine function, the cotangent function, and the cosecant function are *cofunctions* of the sine function, the tangent function, and the secant function, respectively.

Preview

Any trigonometric function of θ, where $0° < \theta < 90°$, is equal to the cofunction of its complementary angle.

EXAMPLES   **a.** $\sin 50° = \cos (90° - 50°)$ **b.** $\cot \dfrac{\pi}{6} = \tan \left(\dfrac{\pi}{2} - \dfrac{\pi}{6} \right)$

$\qquad\qquad\qquad\qquad\qquad = \cos 40° \qquad\qquad\qquad\qquad = \tan \dfrac{\pi}{3}$

Express each function of an angle in terms of the cofunction of its complementary angle.

1. $\sec 22.2°$ **2.** $\tan 15.5°$ **3.** $\csc 17°50'$ **4.** $\sin 18°12'$ **5.** $\cos \dfrac{\pi}{3}$ **6.** $\cot \dfrac{\pi}{4}$

It can be shown that the sine and the cosine are cofunctions for all angles. That is,

$$\cos (90° - \theta) = \sin \theta \qquad \text{and} \qquad \cos \theta = \sin (90° - \theta)$$

for *all* values of θ. The difference identity for cosine is used to prove these identities.

$$\begin{aligned}
\cos (90° - \theta) &= \cos 90° \cos \theta + \sin 90° \sin \theta \\
&= 0 \, (\cos \theta) + 1(\sin \theta) \\
&= 0 + \sin \theta \\
&= \sin \theta
\end{aligned}$$

$$\begin{aligned}
\cos \theta &= \cos [90° - (90° - \theta)] \qquad \textit{Substitute } 90° - (90° - \theta) \textit{ for } \theta. \\
&= \cos 90° \cos (90° - \theta) + \sin 90° \sin (90° - \theta) \\
&= 0 \, [\cos (90° - \theta)] + 1[\sin (90° - \theta)] \\
&= 0 + \sin (90° - \theta) \\
&= \sin (90° - \theta)
\end{aligned}$$

Sum Identity for Sine

$$\sin(\alpha + \beta) = \sin \alpha \cos \beta + \cos \alpha \sin \beta$$

Cofunctions and the difference identity for cosine are used in the proof.

$$
\begin{aligned}
\sin(\alpha + \beta) &= \cos[90° - (\alpha + \beta)] \\
&= \cos[(90° - \alpha) - \beta] \\
&= \cos(90° - \alpha)\cos \beta + \sin(90° - \alpha)\sin \beta \\
&= \sin \alpha \cos \beta + \cos \alpha \sin \beta
\end{aligned}
$$

sin θ = cos (90° − θ)

EXAMPLE 1   **Find the exact value of sin 165°.**

Since $165° = 120° + 45°$, let $\alpha = 120°$ and $\beta = 45°$.

$$
\begin{aligned}
\sin 165° &= \sin(120° + 45°) \\
&= \sin 120° \cos 45° + \cos 120° \sin 45° \qquad \textit{Sum identity for sine} \\
&= \frac{\sqrt{3}}{2}\left(\frac{\sqrt{2}}{2}\right) + \left(-\frac{1}{2}\right)\left(\frac{\sqrt{2}}{2}\right) = \frac{\sqrt{6}}{4} - \frac{\sqrt{2}}{4} = \frac{\sqrt{6} - \sqrt{2}}{4}
\end{aligned}
$$

Another important identity is the *difference identity for sine*.

Difference Identity for Sine

$$\sin(\alpha - \beta) = \sin \alpha \cos \beta - \cos \alpha \sin \beta$$

The proof follows from the sum identity for sine.

$$
\begin{aligned}
\sin(\alpha - \beta) &= \sin[\alpha + (-\beta)] \\
&= \sin \alpha \cos(-\beta) + \cos \alpha \sin(-\beta) \\
&= \sin \alpha \cos \beta - \cos \alpha \sin \beta \quad \textit{cos}(-\beta) = \textit{cos }\beta; \textit{ sin}(-\beta) = -\textit{sin }\beta
\end{aligned}
$$

EXAMPLE 2   **Find the exact value of $\sin \dfrac{\pi}{12}$.**

Since $\dfrac{\pi}{12} = \dfrac{4\pi}{12} - \dfrac{3\pi}{12} = \dfrac{\pi}{3} - \dfrac{\pi}{4}$, let $\alpha = \dfrac{\pi}{3}$ and $\beta = \dfrac{\pi}{4}$.

$$
\begin{aligned}
\sin \frac{\pi}{12} &= \sin\left(\frac{\pi}{3} - \frac{\pi}{4}\right) \\
&= \sin \frac{\pi}{3} \cos \frac{\pi}{4} - \cos \frac{\pi}{3} \sin \frac{\pi}{4} \qquad \textit{Difference identity for sine} \\
&= \frac{\sqrt{3}}{2}\left(\frac{\sqrt{2}}{2}\right) - \frac{1}{2}\left(\frac{\sqrt{2}}{2}\right) = \frac{\sqrt{6}}{4} - \frac{\sqrt{2}}{4} = \frac{\sqrt{6} - \sqrt{2}}{4}
\end{aligned}
$$

Use the values of α and β from Examples 1 and 2 to verify the fact that

$$\sin(\alpha + \beta) \neq \sin\alpha + \sin\beta \qquad \text{and} \qquad \sin(\alpha - \beta) \neq \sin\alpha - \sin\beta$$

EXAMPLE 3 **Find the exact values of sin (α + β) and sin (α − β) if sin α = $\dfrac{\sqrt{3}}{2}$, tan β = $-\dfrac{15}{8}$, $\dfrac{\pi}{2} < \alpha < \pi$, and $\dfrac{3\pi}{2} < \beta < 2\pi$.**

Find cos α, sin β, and cos β before using the identities for sine.

To find cos α, note that sin α = $\dfrac{\sqrt{3}}{2}$
and α is in quadrant II. Use y = $\sqrt{3}$ and r = 2, and calculate x.

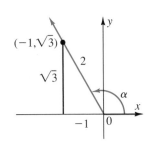

$$x = \pm\sqrt{2^2 - (\sqrt{3})^2} = \pm 1$$

Since α is in quadrant II, x = −1.

Thus, cos α = $-\dfrac{1}{2}$.

To find sin β and cos β, note that
tan β = $-\dfrac{15}{8}$ and β is in quadrant IV.
Use y = −15 and x = 8, and find r.

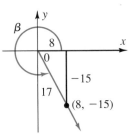

$$r = \sqrt{8^2 + (-15)^2} = 17$$

Thus, sin β = $-\dfrac{15}{17}$ and cos β = $\dfrac{8}{17}$.

Then, sin (α + β) = $\dfrac{\sqrt{3}}{2}\left(\dfrac{8}{17}\right) + \left(-\dfrac{1}{2}\right)\left(-\dfrac{15}{17}\right)$ *Sum identity for sine*

$$= \dfrac{8\sqrt{3}}{34} + \dfrac{15}{34}$$

$$= \dfrac{8\sqrt{3} + 15}{34}$$

sin (α − β) = $\dfrac{\sqrt{3}}{2}\left(\dfrac{8}{17}\right) - \left(-\dfrac{1}{2}\right)\left(-\dfrac{15}{17}\right)$ *Difference identity for sine*

$$= \dfrac{8\sqrt{3}}{34} - \dfrac{15}{34}$$

$$= \dfrac{8\sqrt{3} - 15}{34}$$

The sum and difference identities for sine can be used to prove other identities.

EXAMPLE 4   **Prove: $\sin(\pi + \theta) = -\sin\theta$**

$$\begin{array}{c|c} \sin(\pi + \theta) & -\sin\theta \\ \sin\pi\cos\theta + \cos\pi\sin\theta & \\ 0(\cos\theta) + (-1)(\sin\theta) & \\ & -\sin\theta = \end{array}$$ *Work with the left side.*

Thus, $\sin(\pi + \theta) = -\sin\theta$.

CLASS EXERCISES

Simplify each expression by writing it in terms of the sine of one angle.

1. $\sin 30° \cos 40° + \cos 30° \sin 40°$

2. $\sin 25° \cos 20° - \cos 25° \sin 20°$

3. $\sin 65° \cos 25° - \cos 65° \sin 25°$

4. $\sin 50° \cos 35° + \cos 50° \sin 35°$

5. $\sin 100° \cos 12° + \cos 100° \sin 12°$

6. $\sin 105° \cos 85° - \cos 105° \sin 85°$

7. $\sin\dfrac{6\pi}{7}\cos\dfrac{2\pi}{7} + \cos\dfrac{6\pi}{7}\sin\dfrac{2\pi}{7}$

8. $\sin\dfrac{7\pi}{9}\cos\dfrac{4\pi}{9} - \cos\dfrac{7\pi}{9}\sin\dfrac{4\pi}{9}$

9. $\sin\dfrac{2\pi}{3}\cos\dfrac{\pi}{3} - \cos\dfrac{2\pi}{3}\sin\dfrac{\pi}{3}$

10. $\sin\dfrac{\pi}{3}\cos\dfrac{4\pi}{3} + \cos\dfrac{\pi}{3}\sin\dfrac{4\pi}{3}$

11. $\sin\dfrac{\pi}{3}\cos\dfrac{\pi}{6} + \cos\dfrac{\pi}{3}\sin\dfrac{\pi}{6}$

12. $\sin\dfrac{\pi}{3}\cos\dfrac{\pi}{6} - \cos\dfrac{\pi}{3}\sin\dfrac{\pi}{6}$

13. $\sin\dfrac{5\pi}{4}\cos\dfrac{\pi}{2} - \cos\dfrac{5\pi}{4}\sin\dfrac{\pi}{2}$

14. $\sin\dfrac{5\pi}{4}\cos\dfrac{\pi}{2} + \cos\dfrac{5\pi}{4}\sin\dfrac{\pi}{2}$

Tell whether each of the following statements is true or false.

15. $\sin 75° = \sin 50° \cos 25° - \cos 50° \sin 25°$

16. $\sin 70° = \sin 80° \cos 10° - \cos 80° \sin 10°$

17. $\sin 40° = \sin 50° - \sin 10°$

18. $\sin 35° = \sin 20° + \sin 15°$

19. $\sin 50° = \sin 70° \cos 20° - \cos 70° \sin 20°$

20. $\sin 105° = \sin 90° \cos 15° - \cos 90° \sin 15°$

21. $\sin 120° = \sin 70° \cos 50° + \sin 70° \cos 50°$

22. $\sin 25° = \sin 10° \cos 15° + \sin 15° \cos 10°$

PRACTICE EXERCISES

Use the sum or the difference identity for sine to find the exact value of each trigonometric function.

1. $\sin 15°$

2. $\sin 75°$

3. $\sin 195°$

4. $\sin 255°$

5. $\sin 375°$

6. $\sin 345°$

7. $\sin\left(\dfrac{\pi}{3} + \dfrac{\pi}{4}\right)$

8. $\sin\left(\dfrac{\pi}{4} + \dfrac{4\pi}{3}\right)$

Find the exact value of $\sin(\alpha + \beta)$.

9. $\sin\alpha = \dfrac{12}{13}$, $\sin\beta = \dfrac{7}{25}$, $0 < \alpha < \dfrac{\pi}{2}$, $0 < \beta < \dfrac{\pi}{2}$

10. $\cos\alpha = -\dfrac{3}{5}$, $\cos\beta = \dfrac{8}{17}$, $\dfrac{\pi}{2} < \alpha < \pi$, $0 < \beta < \dfrac{\pi}{2}$

11. $\cos\alpha = \dfrac{15}{17}$, $\tan\beta = \dfrac{4}{3}$, $\dfrac{3\pi}{2} < \alpha < 2\pi$, $\pi < \beta < \dfrac{3\pi}{2}$

12. $\cos\alpha = -\dfrac{7}{25}$, $\tan\beta = -\dfrac{3}{4}$, $\pi < \alpha < \dfrac{3\pi}{2}$, $\dfrac{\pi}{2} < \beta < \pi$

Find the exact value of $\sin(\alpha - \beta)$.

13. $\cos\alpha = \dfrac{3}{5}$, $\tan\beta = \dfrac{12}{5}$, $0 < \alpha < \dfrac{\pi}{2}$, $0 < \beta < \dfrac{\pi}{2}$

14. $\cos\alpha = \dfrac{4}{5}$, $\tan\beta = \dfrac{5}{12}$, $0 < \alpha < \dfrac{\pi}{2}$, $0 < \beta < \dfrac{\pi}{2}$

15. $\cos\alpha = \dfrac{7}{25}$, $\sin\beta = \dfrac{5}{13}$, $\dfrac{3\pi}{2} < \alpha < 2\pi$, $0 < \beta < \dfrac{\pi}{2}$

16. $\cos\alpha = -\dfrac{5}{13}$, $\sin\beta = \dfrac{15}{17}$, $\pi < \alpha < \dfrac{3\pi}{2}$, $\dfrac{\pi}{2} < \beta < \pi$

Use the sum or the difference identity for sine to prove each identity.

17. $\sin(180° - \alpha) = \sin\alpha$

18. $\sin(\alpha - 180°) = -\sin\alpha$

19. $\sin(210° + \alpha) = \dfrac{-\cos\alpha - \sqrt{3}\sin\alpha}{2}$

20. $\sin(270° + \beta) = -\cos\beta$

21. $\sin(45° + \alpha) = \dfrac{\sqrt{2}}{2}(\cos\alpha + \sin\alpha)$

22. $\sin\left(\dfrac{\pi}{2} + \beta\right) = \cos\beta$

23. $\sin\left(\dfrac{\pi}{3} - \beta\right) = \dfrac{1}{2}(\sqrt{3}\cos\beta - \sin\beta)$

24. $\sin(2\beta) = 2\sin\beta\cos\beta$

25. $\sin\left(\beta - \dfrac{7\pi}{6}\right) = -\dfrac{1}{2}(\sqrt{3}\sin\beta - \cos\beta)$

26. $\sin(-\beta) = -\sin\beta$

Find the exact value of each trigonometric function using the sine identities.

27. $\sin(-105°)$ **28.** $\sin(-285°)$ **29.** $\sin\left(-\dfrac{5\pi}{12}\right)$ **30.** $\sin\left(-\dfrac{17\pi}{12}\right)$

Prove each identity.

31. $\sin(\alpha + \beta) + \sin(\alpha - \beta) = 2 \sin \alpha \cos \beta$

32. $\sin(\alpha + \beta) - \sin(\alpha - \beta) = 2 \cos \alpha \sin \beta$

Applications

Mechanics   When a weight is suspended from a spring that has been stretched down and the weight is pushed farther down and released, the equation of motion can be written in the form $y = A \sin(\omega\tau + \phi)$, where A and ω are constants and ϕ is a radian measure.

Given each equation, use the sum identity of sines to rewrite it in the form $y = A \sin \omega t \cos \phi + B \cos \omega t \sin \phi$. Then simplify if possible.

33. $y = 5 \sin\left(4t + \dfrac{2\pi}{3}\right)$ **34.** $y = 2 \sin\left(3t + \dfrac{5\pi}{6}\right)$

35. $y = 3 \sin\left(2t + \dfrac{\pi}{2}\right)$ **36.** $y = 6 \sin\left(3t + \dfrac{4\pi}{3}\right)$

EXTRA

Use the portion of the unit-circle diagram at the right to derive the sum identity for sine for acute angles.

1. The measure of $\angle FBE$ is β. Why?

2. Complete the derivation of the sum identity for sine.

$$\begin{aligned}
\sin(\alpha + \beta) &= BC \\
&= BF + FC \\
&= BE(\cos \ \underline{?}\ ) + FC \\
&= \sin \ \underline{?}\ \cos \ \underline{?}\ + EG \\
&= \sin \ \underline{?}\ \cos \ \underline{?}\ + OE(\sin \ \underline{?}\ ) \\
&= \sin \ \underline{?}\ \cos \ \underline{?}\ + \cos \ \underline{?}\ \sin \ \underline{?}
\end{aligned}$$

3. Use the diagram to derive the sum identity for cosine for acute angles.

Tangent: Sum and Difference Identities

5.3

Objective: To develop and use formulas for the tangent of a sum or difference of two angle measures

To use tangent identities, you may have to rationalize denominators.

Preview

Recall that $a^2 - b^2$ is equal to the product of the *conjugates* $a + b$ and $a - b$. Conjugates can be used to rationalize denominators.

EXAMPLE $\dfrac{4}{1 + \sqrt{2}} = \left(\dfrac{4}{1 + \sqrt{2}}\right)\left(\dfrac{1 - \sqrt{2}}{1 - \sqrt{2}}\right) = \dfrac{4 - 4\sqrt{2}}{1 - 2} = -4 + 4\sqrt{2}$

Simplify by rationalizing the denominator.

1. $\dfrac{2}{1 - \sqrt{3}}$ 2. $\dfrac{\sqrt{2}}{1 + \sqrt{2}}$ 3. $\dfrac{1 - \sqrt{2}}{3 + \sqrt{2}}$ 4. $\dfrac{\sqrt{2} - \sqrt{6}}{\sqrt{2} + \sqrt{6}}$ 5. $\dfrac{\sqrt{2} + 3}{1 - 2\sqrt{2}}$

The sum identities for sine and cosine lead to the *sum identity for tangent*.

Sum Identity for Tangent

$$\tan (\alpha + \beta) = \dfrac{\tan \alpha + \tan \beta}{1 - \tan \alpha \tan \beta}, \quad \tan \alpha \tan \beta \neq 1$$

Therefore, $\tan (\alpha + \beta) = \dfrac{\sin (\alpha + \beta)}{\cos (\alpha + \beta)}$ $\tan \theta = \dfrac{\sin \theta}{\cos \theta}$

$$= \dfrac{\sin \alpha \cos \beta + \cos \alpha \sin \beta}{\cos \alpha \cos \beta - \sin \alpha \sin \beta}$$

$$= \dfrac{\dfrac{\sin \alpha \cos \beta}{\cos \alpha \cos \beta} + \dfrac{\cos \alpha \sin \beta}{\cos \alpha \cos \beta}}{\dfrac{\cos \alpha \cos \beta}{\cos \alpha \cos \beta} - \dfrac{\sin \alpha \sin \beta}{\cos \alpha \cos \beta}}$$ *Divide each term by* $\cos \alpha \cos \beta$.

$$= \dfrac{\dfrac{\sin \alpha}{\cos \alpha} + \dfrac{\sin \beta}{\cos \beta}}{1 - \left(\dfrac{\sin \alpha}{\cos \alpha}\right)\left(\dfrac{\sin \beta}{\cos \beta}\right)} = \dfrac{\tan \alpha + \tan \beta}{1 - \tan \alpha \tan \beta}$$

EXAMPLE 1 **Find the exact value of tan 345°.**

Since $345° = 300° + 45°$, let $\alpha = 300°$ and $\beta = 45°$.

$\tan 345° = \tan (300° + 45°)$

$$= \frac{\tan 300° + \tan 45°}{1 - \tan 300° \tan 45°} \qquad \text{\textit{Sum identity for tangent}}$$

$$= \frac{-\sqrt{3} + 1}{1 - (-\sqrt{3})(1)}$$

$$= \frac{-\sqrt{3} + 1}{1 + \sqrt{3}}$$

$$= \left(\frac{1 - \sqrt{3}}{1 + \sqrt{3}}\right)\left(\frac{1 - \sqrt{3}}{1 - \sqrt{3}}\right) \qquad \text{\textit{Simplify using the conjugate.}}$$

$$= \frac{1 - 2\sqrt{3} + 3}{1 - 3} = \frac{4 - 2\sqrt{3}}{-2} = -2 + \sqrt{3}$$

The *difference identity for tangent* can be derived from the sum identity.

> ### Difference Identity for Tangent
>
> $$\tan (\alpha - \beta) = \frac{\tan \alpha - \tan \beta}{1 + \tan \alpha \tan \beta}, \quad \tan \alpha \tan \beta \neq -1$$

The proof makes use of the sum identity for tangent and the identity $\tan (-\beta) = -\tan \beta$. The proof of $\tan (-\beta) = -\tan \beta$ is given first.

$$\tan (-\beta) = \frac{\sin (-\beta)}{\cos (-\beta)} = \frac{-\sin \beta}{\cos \beta} = -\tan \beta$$

Then, $\tan (\alpha - \beta) = \tan [\alpha + (-\beta)]$

$$= \frac{\tan \alpha + \tan (-\beta)}{1 - \tan \alpha \tan (-\beta)} \qquad \text{\textit{Sum identity for tangent}}$$

$$= \frac{\tan \alpha - \tan \beta}{1 + \tan \alpha \tan \beta} \qquad \text{\textit{tan} }(-\beta) = -\text{\textit{tan} }\beta$$

EXAMPLE 2 **Find the exact value of $\tan \dfrac{5\pi}{12}$.**

Since $\dfrac{5\pi}{12} = \dfrac{8\pi}{12} - \dfrac{3\pi}{12} = \dfrac{2\pi}{3} - \dfrac{\pi}{4}$, let $\alpha = \dfrac{2\pi}{3}$ and $\beta = \dfrac{\pi}{4}$.

$$\tan \frac{5\pi}{12} = \tan \left(\frac{2\pi}{3} - \frac{\pi}{4}\right)$$

$$\tan \frac{5\pi}{12} = \frac{\tan \dfrac{2\pi}{3} - \tan \dfrac{\pi}{4}}{1 + \tan \dfrac{2\pi}{3} \tan \dfrac{\pi}{4}} \qquad \textit{Difference identity for tangent}$$

$$= \frac{-\sqrt{3} - 1}{1 + (-\sqrt{3})(1)} = \frac{-\sqrt{3} - 1}{1 - \sqrt{3}}$$

$$= \left(\frac{-\sqrt{3} - 1}{1 - \sqrt{3}}\right)\left(\frac{1 + \sqrt{3}}{1 + \sqrt{3}}\right) = \frac{-4 - 2\sqrt{3}}{-2} = 2 + \sqrt{3}$$

Note that $\tan(\alpha + \beta) \neq \tan\alpha + \tan\beta$ and $\tan(\alpha - \beta) \neq \tan\alpha - \tan\beta$.

EXAMPLE 3 **Find the exact value of $\tan(\alpha + \beta)$ if $\sin\alpha = \dfrac{4}{5}$, $\cos\beta = -\dfrac{5}{13}, 0 < \alpha < \dfrac{\pi}{2}$, and $\dfrac{\pi}{2} < \beta < \pi$.**

To find $\tan\alpha$, note that $\sin\alpha = \dfrac{4}{5}$ and α is in quadrant I. Use $y = 4$ and $r = 5$, and find x.

$$x = \pm\sqrt{5^2 - 4^2} = \pm 3$$

Since α is in quadrant I, $x = 3$ and $\tan\alpha = \dfrac{4}{3}$.

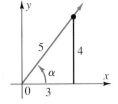

To find $\tan\beta$, note that $\cos\beta = -\dfrac{5}{13}$ and β is in quadrant II. Use $x = -5$ and $r = 13$, and find y.

$$y = \pm\sqrt{13^2 - (-5)^2} = \pm 12$$

Since β is in quadrant II, $y = 12$ and $\tan\beta = -\dfrac{12}{5}$.

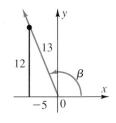

$$\tan(\alpha + \beta) = \frac{\tan\alpha + \tan\beta}{1 - \tan\alpha \tan\beta}$$

$$= \frac{\dfrac{4}{3} + \left(-\dfrac{12}{5}\right)}{1 - \left(\dfrac{4}{3}\right)\left(-\dfrac{12}{5}\right)} \qquad \textit{To simplify, multiply by } \dfrac{15}{15}.$$

$$= \frac{20 - 36}{15 + 48} = -\frac{16}{63}$$

The sum and difference identities can be used to prove other identities.

EXAMPLE 4 **Prove: $\tan(\pi + \theta) = \tan\theta$**

$$
\begin{array}{c|c}
\tan(\pi + \theta) & \tan\theta \\[6pt]
\dfrac{\tan\pi + \tan\theta}{1 - \tan\pi\tan\theta} & \\[10pt]
\dfrac{0 + \tan\theta}{1 - (0)(\tan\theta)} & \\[10pt]
\dfrac{\tan\theta}{1} & \\[6pt]
\tan\theta \;\; = & \quad\text{Thus, } \tan(\pi + \theta) = \tan\theta.
\end{array}
$$

CLASS EXERCISES

Write each expression in terms of the tangent of one angle.

1. $\dfrac{\tan 40° + \tan 20°}{1 - \tan 40° \tan 20°}$

2. $\dfrac{\tan 70° - \tan 10°}{1 + \tan 70° \tan 10°}$

3. $\dfrac{\tan \dfrac{5\pi}{9} - \tan \dfrac{4\pi}{9}}{1 + \tan \dfrac{5\pi}{9} \tan \dfrac{4\pi}{9}}$

4. $\dfrac{\tan \dfrac{\pi}{8} + \tan \dfrac{\pi}{8}}{1 - \tan \dfrac{\pi}{8} \tan \dfrac{\pi}{8}}$

5. $\dfrac{\tan \dfrac{\pi}{4} - \tan \dfrac{\pi}{4}}{1 + \tan \dfrac{\pi}{4} \tan \dfrac{\pi}{4}}$

6. $\dfrac{\tan \dfrac{3\pi}{4} + \tan \dfrac{\pi}{4}}{1 - \tan \dfrac{3\pi}{4} \tan \dfrac{\pi}{4}}$

Tell whether each of the following statements is true or false.

7. $\tan 100° = \dfrac{\tan 90° + \tan 10°}{1 + \tan 90° \tan 10°}$

8. $\tan 65° = \dfrac{\tan 40° + \tan 25°}{1 - \tan 40° \tan 25°}$

9. $\tan 45° = \dfrac{\tan 40° + \tan 5°}{1 - \tan 40° \tan 5°}$

10. $\tan 80° = \dfrac{\tan 95° - \tan 15°}{1 - \tan 95° \tan 15°}$

11. $\tan 75° = \dfrac{\tan 80° - \tan 5°}{1 + \tan 70° \tan 5°}$

12. $\tan 120° = \dfrac{\tan 60° + \tan 60°}{1 - \tan 60° \tan 60°}$

13. $\tan 25° = \dfrac{\tan 45° - \tan 20°}{1 + \tan 20° \tan 45°}$

14. $\tan 50° = \dfrac{\tan 65° - \tan 15°}{1 + \tan 15° \tan 65°}$

PRACTICE EXERCISES

Use the sum or the difference identity for tangent to find the exact value.

1. $\tan 255°$ **2.** $\tan 15°$ **3.** $\tan 105°$ **4.** $\tan 285°$

5. $\tan 165°$ **6.** $\tan 195°$ **7.** $\tan\left(\dfrac{5\pi}{3} + \dfrac{\pi}{4}\right)$ **8.** $\tan\left(\dfrac{4\pi}{3} + \dfrac{3\pi}{4}\right)$

Find the exact value of tan (α + β).

9. $\cos \alpha = \dfrac{3}{5}$, $\sin \beta = -\dfrac{5}{13}$, $\dfrac{3\pi}{2} < \alpha < 2\pi$, $\pi < \beta < \dfrac{3\pi}{2}$

10. $\cos \alpha = \dfrac{7}{25}$, $\sin \beta = \dfrac{5}{13}$, $\dfrac{3\pi}{2} < \alpha < 2\pi$, $0 < \beta < \dfrac{\pi}{2}$

11. $\tan \alpha = \dfrac{8}{15}$, $\cos \beta = -\dfrac{4}{5}$, $0 < \alpha < \dfrac{\pi}{2}$, $\dfrac{\pi}{2} < \beta < \pi$

12. $\cos \alpha = -\dfrac{5}{13}$, $\tan \beta = \dfrac{15}{8}$, $\dfrac{\pi}{2} < \alpha < \pi$, $\pi < \beta < \dfrac{3\pi}{2}$

Find the exact value of tan (α − β).

13. $\cos \alpha = \dfrac{7}{25}$, $\cos \beta = -\dfrac{4}{5}$, $0 < \alpha < \dfrac{\pi}{2}$, $\dfrac{\pi}{2} < \beta < \pi$

14. $\cos \alpha = \dfrac{3}{5}$, $\cos \beta = \dfrac{5}{13}$, $0 < \alpha < \dfrac{\pi}{2}$, $\dfrac{3\pi}{2} < \beta < 2\pi$

15. $\tan \alpha = -\dfrac{4}{3}$, $\sin \beta = \dfrac{24}{25}$, $\dfrac{\pi}{2} < \alpha < \pi$, $\dfrac{\pi}{2} < \beta < \pi$

16. $\tan \alpha = \dfrac{12}{5}$, $\sin \beta = \dfrac{3}{5}$, $\pi < \alpha < \dfrac{3\pi}{2}$, $\dfrac{\pi}{2} < \beta < \pi$

Use the sum or the difference identity for tangent to prove each identity.

17. $\tan (45° - \alpha) = \dfrac{1 - \tan \alpha}{1 + \tan \alpha}$

18. $\tan (\alpha + 45°) = \dfrac{1 + \tan \alpha}{1 - \tan \alpha}$

19. $\tan (180° + \alpha) = \tan \alpha$

20. $\tan (180° - \alpha) = -\tan \alpha$

21. $\tan (360° + \alpha) = \tan \alpha$

22. $\tan (360° - \alpha) = -\tan \alpha$

23. $\tan \left(\dfrac{3\pi}{4} + \alpha \right) = \dfrac{\tan \alpha - 1}{\tan \alpha + 1}$

24. $\tan \left(\dfrac{3\pi}{4} - \alpha \right) = \dfrac{\tan \alpha + 1}{\tan \alpha - 1}$

25. $\tan \left(\alpha - \dfrac{5\pi}{4} \right) = \dfrac{\tan \alpha - 1}{\tan \alpha + 1}$

26. $\tan \left(\alpha + \dfrac{7\pi}{4} \right) = \dfrac{\tan \alpha - 1}{\tan \alpha + 1}$

Find the exact value of each trigonometric function using the tangent identities.

27. $\tan (-15°)$

28. $\tan (-75°)$

29. $\tan \left(-\dfrac{5\pi}{12} \right)$

30. $\tan \left(-\dfrac{\pi}{12} \right)$

Prove each identity.

31. $\tan (\beta + 45°) + \tan (\beta - 45°) = 2 \tan 2\beta$

32. $\tan (\beta + 45°) + \tan (45° - \beta) = 2 \left(\dfrac{1 + \tan^2 \beta}{1 - \tan^2 \beta} \right)$

Applications

Coordinate Geometry Recall from algebra that the slope m of a line is

$$m = \frac{\text{change in the } y\text{-coordinates}}{\text{change in the } x\text{-coordinates}}$$

In the figure at the right, $m = \frac{y_1 - 0}{x_1 - 0} = \frac{y_1}{x_1}$.

But by definition of tangent, $\tan \theta = \frac{y_1}{x_1}$.

Therefore, the slope of line t is equal to $\tan \theta$.

33. Let θ be the acute angle formed by the intersection of lines n and p, which form angles α and β, respectively, with the x-axis. Let m_1 be the slope of n and let m_2 be the slope of p. Show that $\tan \theta = \frac{m_2 - m_1}{1 + m_1 m_2}$.

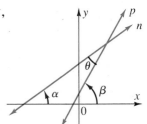

34. Use the formula in Exercise 33 to find $\tan \theta$ if $m_1 = \frac{1}{2}$ and $m_2 = \frac{2}{3}$.

TEST YOURSELF

Use a sum or difference identity to find the exact value of each.

1. $\cos 75°$	2. $\cos 255°$	5.1
3. $\sin 15°$	4. $\sin 105°$	5.2
5. $\tan 375°$	6. $\tan 75°$	5.3

7. Find $\cos (\alpha + \beta)$ if $\sin \alpha = \frac{3}{5}$, $\cos \beta = \frac{3}{5}$, $0 < \alpha < \frac{\pi}{2}$, and $\frac{3\pi}{2} < \beta < 2\pi$.   5.1

8. Find $\sin (\alpha - \beta)$ if $\cos \alpha = -\frac{8}{17}$, $\sin \beta = -\frac{7}{25}$, $\frac{\pi}{2} < \alpha < \pi$, and $\pi < \beta < \frac{3\pi}{2}$.   5.2

9. Find $\tan (\alpha - \beta)$ if $\sin \alpha = -\frac{4}{5}$, $\sin \beta = -\frac{7}{25}$, $\pi < \alpha < \frac{3\pi}{2}$, and $\frac{3\pi}{2} < \beta < 2\pi$.   5.3

Prove each identity.

10. $\cos \left(\frac{3\pi}{2} + \alpha \right) = \sin \alpha$ 11. $\cos \left(\frac{\pi}{2} + \alpha \right) = -\sin \alpha$ 5.1

12. $\tan (\alpha - 180°) = \tan \alpha$ 13. $\tan (360° - \beta) = -\tan \beta$ 5.3

Double-Angle Identities

Objective: To develop and use double-angle identities

Some cases of the sum and difference identities are used so often that they are given special names. The first group consists of the *double-angle identities*.

Preview

A navigator on a ship uses a method known as *doubling the angle on the bow* to determine his position relative to a fixed point, such as a lighthouse. He begins at a point *A*, where the acute angle between the path of his ship and the lighthouse is α. He then measures the distance the ship moves to a point *B*, where the angle between the path of the ship and the lighthouse is 2α. The distance from *A* to *B* is equal to the distance from *B* to the lighthouse at *L*.

1. What is the measure of ∠*BLA*?

2. Why is *AB* equal to *BL*?

3. If α were 60°, and the distance from *A* to *B* were 3 mi, what would be the distance from *A* to *L*?

The identities for sin (α + β), cos (α + β), and tan (α + β) can be used to derive identities for sin 2α, cos 2α, and tan 2α.

$$\sin (\alpha + \beta) = \sin \alpha \cos \beta + \cos \alpha \sin \beta$$
$$\sin (\alpha + \alpha) = \sin \alpha \cos \alpha + \cos \alpha \sin \alpha \qquad \textit{Let } \beta = \alpha.$$
$$\sin 2\alpha = 2 \sin \alpha \cos \alpha$$

$$\cos (\alpha + \beta) = \cos \alpha \cos \beta - \sin \alpha \sin \beta$$
$$\cos (\alpha + \alpha) = \cos \alpha \cos \alpha - \sin \alpha \sin \alpha \qquad \textit{Let } \beta = \alpha.$$
$$\cos 2\alpha = \cos^2 \alpha - \sin^2 \alpha$$

$$\tan (\alpha + \beta) = \frac{\tan \alpha + \tan \beta}{1 - \tan \alpha \tan \beta}$$
$$\tan (\alpha + \alpha) = \frac{\tan \alpha + \tan \alpha}{1 - \tan \alpha \tan \alpha} \qquad \textit{Let } \beta = \alpha.$$
$$\tan 2\alpha = \frac{2 \tan \alpha}{1 - \tan^2 \alpha}$$

Two alternate forms of $\cos 2\alpha$ can be derived using $\cos^2 \alpha + \sin^2 \alpha = 1$.

Alternate Form 1 (cosine form)

$$\begin{aligned} \cos 2\alpha &= \cos^2 \alpha - \sin^2 \alpha \\ &= \cos^2 \alpha - (1 - \cos^2 \alpha) \\ &= 2 \cos^2 \alpha - 1 \end{aligned}$$

Alternate Form 2 (sine form)

$$\begin{aligned} \cos 2\alpha &= \cos^2 \alpha - \sin^2 \alpha \\ &= (1 - \sin^2 \alpha) - \sin^2 \alpha \\ &= 1 - 2 \sin^2 \alpha \end{aligned}$$

Double-Angle Identities

$$\sin 2\alpha = 2 \sin \alpha \cos \alpha \qquad\qquad \cos 2\alpha = \cos^2 \alpha - \sin^2 \alpha$$

$$\tan 2\alpha = \frac{2 \tan \alpha}{1 - \tan^2 \alpha}, \tan \alpha \neq \pm 1 \qquad \cos 2\alpha = 2 \cos^2 \alpha - 1$$

$$\cos 2\alpha = 1 - 2 \sin^2 \alpha$$

EXAMPLE 1 If $\sin \alpha = -\dfrac{4}{5}$ and $270° < \alpha < 360°$, find the exact values of each:

a. $\sin 2\alpha$ b. $\cos 2\alpha$ c. $\tan 2\alpha$

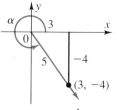

Sin $\alpha = -\frac{4}{5}$ and α is in quadrant IV.

Use $y = -4$ and $r = 5$, and find the value of x.

$$x = \pm \sqrt{(5)^2 - (-4)^2} = \pm \sqrt{9} = \pm 3$$

Since α is the quadrant IV, $x = 3$. Thus, $\cos \alpha = \frac{3}{5}$ and $\tan \alpha = -\frac{4}{3}$.

a. $\sin 2\alpha = 2 \left(-\frac{4}{5} \right) \left(\frac{3}{5} \right) = -\frac{24}{25}$ $\sin 2\alpha = 2 \sin \alpha \cos \alpha$

b. $\cos 2\alpha = 1 - 2 \left(-\frac{4}{5} \right)^2 = -\frac{7}{25}$ $\cos 2\alpha = 1 - 2 \sin^2 \alpha$

c. $\tan 2\alpha = \dfrac{2 \left(-\frac{4}{3} \right)}{1 - \left(-\frac{4}{3} \right)^2} = \dfrac{24}{7}$ $\tan 2\alpha = \dfrac{2 \tan \alpha}{1 - \tan^2 \alpha}$

Formulas for multiples of angle measures follow from the double-angle identities.

EXAMPLE 2 **Derive an identity for $\cos 3\theta$ in terms of $\cos \theta$.**

$$\begin{aligned} \cos 3\theta &= \cos (2\theta + \theta) \\ &= \cos 2\theta \cos \theta - \sin 2\theta \sin \theta \\ &= (2 \cos^2 \theta - 1) \cos \theta - (2 \sin \theta \cos \theta) \sin \theta \\ &= 2 \cos^3 \theta - \cos \theta - 2 \sin^2 \theta \cos \theta \\ &= 2 \cos^3 \theta - \cos \theta - 2 (1 - \cos^2 \theta) \cos \theta \\ &= 2 \cos^3 \theta - \cos \theta - 2 \cos \theta + 2 \cos^3 \theta \\ &= 4 \cos^3 \theta - 3 \cos \theta \end{aligned}$$

The double-angle identities can be used to prove other identities.

EXAMPLE 3 **Prove:** $\sin 2\alpha = \dfrac{2 \tan \alpha}{1 + \tan^2 \alpha}$

$$\sin 2\alpha \ \bigg| \ \dfrac{2 \tan \alpha}{1 + \tan^2 \alpha}$$

$$\dfrac{2 \tan \alpha}{\sec^2 \alpha} \qquad 1 + \tan^2 \alpha = \sec^2 \alpha$$

$$\dfrac{\dfrac{2 \sin \alpha}{\cos \alpha}}{\dfrac{1}{\cos^2 \alpha}} \qquad \tan \alpha = \dfrac{\sin \alpha}{\cos \alpha}; \ \sec^2 \alpha = \dfrac{1}{\cos^2 \alpha}$$

$$2 \sin \alpha \cos \alpha \qquad \textit{Multiply numerator and denominator by } \cos^2 \alpha.$$

$$= \sin 2\alpha \qquad \textit{Thus, } \sin 2\alpha = \dfrac{2 \tan \alpha}{1 + \tan^2 \alpha}.$$

CLASS EXERCISES

Tell whether each of the following statements is true or false.

1. $\cos 2(20°) = 2 \cos^2 40° - 1$

2. $\cos 2(25°) = 1 - 2 \sin^2 25°$

3. $\sin (-50°) = 2 \sin (-25°) \cos (-25°)$

4. $\sin 2(42°) = 2 \sin 84° \cos 84°$

5. $\cos 70° = \cos^2 35° - \sin^2 35°$

6. $\cos 40° = 2 \cos^2 40° - 1$

7. $\tan 2(45°) = \dfrac{2 \tan 45°}{1 - \tan^2 45°}$

8. $\tan (-70°) = \dfrac{2 \tan (-35°)}{1 - \tan^2 (-35°)}$

Write each expression in terms of a trigonometric function of one angle.

9. $2 \sin 35° \cos 35°$

10. $1 - 2 \sin^2 40°$

11. $\dfrac{2 \tan 22.5°}{1 - \tan^2 22.5°}$

12. $2 \sin 67.5° \cos 67.5°$

PRACTICE EXERCISES

Use the double-angle identities to find the exact value of each trigonometric function.

1. If $\cos \alpha = \frac{3}{5}$ and $0° < \alpha < 90°$, find $\cos 2\alpha$.

2. If $\sin \alpha = \frac{3}{5}$ and $0° < \alpha < 90°$, find $\cos 2\alpha$.

3. If $\tan \alpha = \frac{4}{3}$ and $0° < \alpha < 90°$, find $\tan 2\alpha$.

4. If $\tan \alpha = \frac{3}{4}$ and $0° < \alpha < 90°$, find $\tan 2\alpha$.

5. If $\sin \alpha = \frac{4}{5}$ and $90° < \alpha < 180°$, find $\sin 2\alpha$.

6. If $\cos \alpha = \frac{4}{5}$ and $270° < \alpha < 360°$, find $\sin 2\alpha$.

7. If $\sin \alpha = \frac{5}{13}$ and $90° < \alpha < 180°$, find $\sin 2\alpha$.

8. If $\cos \alpha = \frac{12}{13}$ and $270° < \alpha < 360°$, find $\sin 2\alpha$.

9. If $\sin \theta = \frac{3}{5}$ and $90° < \theta < 180°$, find $\cos 2\theta$.

10. If $\sin \theta = -\frac{12}{13}$ and $180° < \theta < 270°$, find $\cos 2\theta$.

11. If $\cos \theta = \frac{5}{13}$ and $270° < \theta < 360°$, find $\tan 2\theta$.

12. If $\cos \theta = \frac{4}{5}$ and $0° < \theta < 90°$, find $\tan 2\theta$.

13. If $\tan \theta = -\frac{24}{7}$ and $90° < \theta < 180°$, find $\sin 2\theta$.

14. If $\tan \theta = -\frac{12}{5}$ and $90° < \theta < 180°$, find $\sin 2\theta$.

15. If $\tan \theta = -\frac{4}{3}$ and $90° < \theta < 180°$, find $\cos 2\theta$.

16. If $\tan \theta = \frac{7}{24}$ and $180° < \theta < 270°$, find $\sin 2\theta$.

17. Derive an identity for $\sin 3\theta$ in terms of $\sin \theta$.

18. Derive an identity for $\tan 3\theta$ in terms of $\tan \theta$.

Prove each identity.

19. $\cos 2\beta = \dfrac{1 - \tan^2 \beta}{\sec^2 \beta}$

20. $\tan \beta = \dfrac{2 \sin \beta \cos \beta}{1 + \cos 2\beta}$

21. $\cot \beta = \dfrac{1 + \cos 2\beta}{2 \sin \beta \cos \beta}$

22. $\sec 2\theta = \dfrac{1}{1 - 2 \sin^2 \theta}$

23. $\dfrac{1}{\cos \theta} = \dfrac{\sin 2\theta \cos \theta - \cos 2\theta \sin \theta}{\sin \theta \cos \theta}$

24. $\cot \alpha - \tan \alpha = \dfrac{2 \cos 2\alpha}{\sin 2\alpha}$

25. $\sin^2 \beta = \dfrac{1}{2}(1 - \cos 2\beta)$

26. $\cos^2 \alpha = \dfrac{1}{2}(\cos 2\alpha + 1)$

27. $\sec 2\beta = \dfrac{\cot \beta + \tan \beta}{\cot \beta - \tan \beta}$

28. $(\sin \theta + \cos \theta)^2 = \sin 2\theta + 1$

29. $(\cos \theta - \sin \theta)^2 = 1 - \sin 2\theta$

30. $\cos 2\alpha = \cos^4 \alpha - \sin^4 \alpha$

31. $\sin 4\alpha = 4 \sin \alpha \cos \alpha \cos 2\alpha$

32. $\cos 4\beta = 1 - 8 \sin^2 \beta \cos^2 \beta$

33. **Indirect Measurement** A pole casts a shadow of 25 ft at one time and a shadow of 10 ft at a later time when the angle of elevation is twice as large. Find the height *h* of the pole, to the nearest foot.

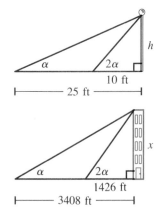

34. **Indirect Measurement** The World Trade Center in New York City casts a 3408-ft shadow at one time and a 1426-ft shadow at a later time when the angle of elevation is twice as large. Find the height *x* of the World Trade Center, to the nearest foot.

Trigonometry in Tidal Motion

The gravitational effect of the moon and the sun on a large body of water causes tides. The water level is lowest at low tide and highest at high tide, and tides repeat every 12.4 hours. The *height of the tide* at a given time is the difference between the height of the water at that time and the height at low tide. The relationship between the height of the tide and the time elapsed since the last high tide can sometimes be approximated using the cosine function

$$y = \frac{a}{2} + \frac{a}{2}\cos\left(\frac{2\pi x}{12.4}\right)$$

where *y* is the height of the tide in meters, *a* is the number of meters by which high tide exceeds low tide, and *x* is the number of hours elapsed since the most recent high tide.

1. High tide at Tampa, Florida exceeds low tide by about 1 meter. Complete the second row of the tide table below by substituting 1 for *a* in the equation above. Round each entry to the nearest 0.1 meter.

x = time elapsed since last high tide (in hours)	0	1	2	3	4	5	6.2	7	8	9	10	11	12.4
y = height of the tide (in meters)	1												

2. Use the information in the table in Exercise 1 to graph the ordered pairs (x, y). Join the points in a smooth curve to complete a graph showing the approximate heights of the tide at Tampa during a 12.4-hour period. Compare this graph with the graph of $y = \cos x$.

5.5 Half-Angle Identities

Objective: To develop and use half-angle identities

A sonic boom can be described in terms of a function of a half-angle.

Preview

Supersonic speed is described using a Mach number, named after Ernst Mach, an Austrian physicist who contributed to the study of sound. A plane traveling at the speed of sound (about 742.82 mph at 32°F at sea level) is traveling at Mach 1. When the speed of a plane is greater than that of sound (greater than Mach 1), a listener in the area will hear a sonic boom. The Mach number M is defined as

$$M = \frac{\text{speed of plane}}{\text{speed of sound}}$$

A sonic boom is created by sound waves that form a cone with vertex angle θ. Mathematicians have shown that if

$M > 1$, then $\sin\left(\dfrac{\theta}{2}\right) = \dfrac{1}{M}$.

1. On July 28, 1976 Capt. Eldon W Joersz (USAF) flew a Lockheed SR-71A at 2193.16 mph. Calculate his Mach number to the nearest tenth.

2. If $\theta = \dfrac{\pi}{4}$, find the Mach number to the nearest tenth.

3. If $\theta = \dfrac{\pi}{8}$, find the Mach number to the nearest tenth.

4. If θ is halved, what is the effect on the Mach number?

The half-angle identities for sine and cosine can be derived by setting θ equal to $\dfrac{\alpha}{2}$ in the double-angle identities for cosine.

Since $\sin \dfrac{\alpha}{2}$ does not have two values, the sign is either positive *or* negative, depending upon the quadrant in which $\dfrac{\alpha}{2}$ lies.

$$\cos 2\left(\frac{\alpha}{2}\right) = 1 - 2\sin^2\left(\frac{\alpha}{2}\right)$$

$$\cos \alpha = 1 - 2\sin^2\left(\frac{\alpha}{2}\right)$$

$$2\sin^2\left(\frac{\alpha}{2}\right) = 1 - \cos \alpha$$

$$\sin^2\left(\frac{\alpha}{2}\right) = \frac{1 - \cos \alpha}{2}$$

$$\sin \frac{\alpha}{2} = \pm \sqrt{\frac{1 - \cos \alpha}{2}}$$

For $\cos \frac{\pi}{2}$, the sign is also determined by the quadrant in which $\frac{\alpha}{2}$ lies.

$$\cos 2 \left(\frac{\alpha}{2}\right) = 2 \cos^2 \left(\frac{\alpha}{2}\right) - 1$$

$$\cos \alpha = 2 \cos^2 \left(\frac{\alpha}{2}\right) - 1$$

$$2 \cos^2 \left(\frac{\alpha}{2}\right) = 1 + \cos \alpha$$

$$\cos^2 \left(\frac{\alpha}{2}\right) = \frac{1 + \cos \alpha}{2}$$

$$\cos \frac{\alpha}{2} = \pm \sqrt{\frac{1 + \cos \alpha}{2}}$$

The half-angle identity for tangent is derived using the identity $\tan \theta = \frac{\sin \theta}{\cos \theta}$.

$$\tan \frac{\alpha}{2} = \frac{\sin \dfrac{\alpha}{2}}{\cos \dfrac{\alpha}{2}}$$

$$= \frac{\pm \sqrt{\dfrac{1 - \cos \alpha}{2}}}{\pm \sqrt{\dfrac{1 + \cos \alpha}{2}}} = \pm \sqrt{\frac{\dfrac{1 - \cos \alpha}{2}}{\dfrac{1 + \cos \alpha}{2}}}$$

$$= \pm \sqrt{\frac{1 - \cos \alpha}{1 + \cos \alpha}}$$

There are two alternate forms for $\tan \frac{\alpha}{2}$.

$$\tan \frac{\alpha}{2} = \frac{1 - \cos \alpha}{\sin \alpha}, \ \sin \alpha \neq 0 \qquad \text{and} \qquad \tan \frac{\alpha}{2} = \frac{\sin \alpha}{1 + \cos \alpha}, \ \cos \alpha \neq -1$$

These forms are often preferred because the expressions on the right are not preceded by $\pm$ symbols; that is, each identity gives the correct sign directly. Exercises 23 and 24 ask for the derivations of these identities.

Half-Angle Identities

$$\sin \frac{\alpha}{2} = \pm \sqrt{\frac{1 - \cos \alpha}{2}} \qquad \tan \frac{\alpha}{2} = \pm \sqrt{\frac{1 - \cos \alpha}{1 + \cos \alpha}}, \ \cos \alpha \neq -1$$

$$\cos \frac{\alpha}{2} = \pm \sqrt{\frac{1 + \cos \alpha}{2}} \qquad \tan \frac{\alpha}{2} = \frac{1 - \cos \alpha}{\sin \alpha}, \ \sin \alpha \neq 0$$

$$\tan \frac{\alpha}{2} = \frac{\sin \alpha}{1 + \cos \alpha}, \ \cos \alpha \neq -1$$

EXAMPLE 1 **Find the exact value of each trigonometric function:**

　　a. $\sin \dfrac{7\pi}{12}$　　**b.** $\cos 165°$

a. Let $\dfrac{\alpha}{2} = \dfrac{7\pi}{12}$. Then angle $\dfrac{\alpha}{2}$ lies in quadrant II, where the sine is positive.

Also, $\alpha = 2\left(\dfrac{7\pi}{12}\right) = \dfrac{7\pi}{6}$.

$$\sin \dfrac{7\pi}{12} = \sqrt{\dfrac{1 - \cos \dfrac{7\pi}{6}}{2}} \qquad \sin \dfrac{\alpha}{2} = \sqrt{\dfrac{1 - \cos \alpha}{2}}$$

$$= \sqrt{\dfrac{1 - \left(-\dfrac{\sqrt{3}}{2}\right)}{2}} = \sqrt{\dfrac{2 + \sqrt{3}}{4}} = \dfrac{\sqrt{2 + \sqrt{3}}}{2}$$

b. Let $\dfrac{\alpha}{2} = 165°$. Then angle $\dfrac{\alpha}{2}$ lies in quadrant II, where the cosine is negative. Also, $\alpha = 2(165°) = 330°$.

$$\cos 165° = -\sqrt{\dfrac{1 + \cos 330°}{2}} \qquad \cos \dfrac{\alpha}{2} = -\sqrt{\dfrac{1 + \cos \alpha}{2}}$$

$$= -\sqrt{\dfrac{1 + \dfrac{\sqrt{3}}{2}}{2}} = -\sqrt{\dfrac{2 + \sqrt{3}}{4}} = -\dfrac{\sqrt{2 + \sqrt{3}}}{2}$$

You can use the half-angle identities to find the value of a trigonometric function of a half-angle if you are given the value of one trigonometric function of the angle and the quadrant in which the terminal side lies.

EXAMPLE 2 **Find the exact value of** $\tan \dfrac{\alpha}{2}$ **if** $\tan \alpha = -\dfrac{4}{3}$ **and** α **is in quadrant IV.**

$\text{Tan } \alpha = -\dfrac{4}{3}$ and α is in quadrant IV. Use $y = -4$ and $x = 3$, and find r.

$$r = \sqrt{3^2 + (-4)^2} = \sqrt{25} = 5$$

Thus, $\sin \alpha = -\dfrac{4}{5}$ and $\cos \alpha = \dfrac{3}{5}$.

$$\tan \dfrac{\alpha}{2} = \dfrac{-\dfrac{4}{5}}{1 + \dfrac{3}{5}} = \dfrac{-\dfrac{4}{5}}{\dfrac{8}{5}} = -\dfrac{4}{8} = -\dfrac{1}{2} \qquad \tan \dfrac{\alpha}{2} = \dfrac{\sin \alpha}{1 + \cos \alpha}$$

Show that the other forms for $\tan \dfrac{\alpha}{2}$ yield the same result.

Care must be taken to determine the correct quadrant for a half-angle.

EXAMPLE 3 **Find the exact value of $\sin \dfrac{\theta}{2}$ if $\cos \theta = -\dfrac{3}{5}$, θ lies in quadrant III, and $0° < \theta < 360°$.**

Since θ lies in quadrant III and $0° < \theta < 360°$, it must be true that $180° < \theta < 270°$. Thus, $90° < \dfrac{\theta}{2} < 135°$ and $\dfrac{\theta}{2}$ lies in quadrant II, where the sine is positive.

$$\sin \frac{\theta}{2} = \sqrt{\frac{1 - \left(-\dfrac{3}{5}\right)}{2}} = \sqrt{\frac{5+3}{10}} = \sqrt{\frac{4}{5}} = \frac{2\sqrt{5}}{5} \qquad \sin \frac{\theta}{2} = \sqrt{\frac{1 - \cos \theta}{2}}$$

Half-angle identities can be used to prove other identities.

EXAMPLE 4 **Prove: $-\cos \theta = 2 \sin^2 \left(\dfrac{\theta}{2}\right) - 1$**

$$-\cos \theta \ \Bigg|\ 2 \sin^2 \left(\frac{\theta}{2}\right) - 1$$

$$\Bigg|\ 2 \left(\pm \sqrt{\frac{1 - \cos \theta}{2}}\right)^2 - 1$$

$$\Bigg|\ 2 \left(\frac{1 - \cos \theta}{2}\right) - 1$$

$$\Bigg|\ 1 - \cos - 1$$

$$= -\cos \theta \qquad \text{Thus, } -\cos \theta = 2 \sin^2 \left(\frac{\theta}{2}\right) - 1.$$

CLASS EXERCISES

Tell whether each statement is true or false.

1. $\cos 10° = \sqrt{\dfrac{1 + \sin 5°}{2}}$

2. $\sin 20° = \sqrt{\dfrac{1 - \cos 40°}{2}}$

3. $\tan 200° = \sqrt{\dfrac{\sin 100°}{1 + \cos 100°}}$

4. $\tan 10° = \sqrt{\dfrac{1 - \cos 20°}{\sin 20°}}$

5. $\tan 15° = \sqrt{\dfrac{1 - \cos 30°}{1 + \cos 30°}}$

6. $\tan 100° = -\sqrt{\dfrac{\sin 50°}{1 + \cos 50°}}$

7. $\sin 60° = \sqrt{\dfrac{1 + \cos 120°}{2}}$

8. $\cos 40° = -\sqrt{\dfrac{1 + \cos 40°}{2}}$

Match each expression on the left with an equivalent one on the right.

9. $\cos 40°$

A. $\sqrt{\dfrac{1 - \cos 40°}{2}}$

10. $\sin 20°$

B. $\sqrt{\dfrac{1 - \cos 40°}{1 + \cos 40°}}$

11. $\tan 20°$

C. $\sqrt{\dfrac{1 + \cos 80°}{2}}$

PRACTICE EXERCISES

Use the half-angle identities to find the exact value of each function.

1. $\tan 105°$ **2.** $\cos 105°$ **3.** $\sin 22.5°$ **4.** $\tan 22.5°$

5. $\cos 67.5°$ **6.** $\sin 67.5°$ **7.** $\sin \dfrac{5\pi}{8}$ **8.** $\cos \dfrac{5\pi}{8}$

9. $\cos \dfrac{7\pi}{8}$ **10.** $\tan \dfrac{7\pi}{8}$ **11.** $\tan \dfrac{\pi}{12}$ **12.** $\sin \dfrac{\pi}{12}$

Find the exact value of each trigonometric function. Assume $0° < \theta < 360°$.

13. $\cos \dfrac{\theta}{2}$ if $\cos \theta = \dfrac{4}{5}$ and θ lies in quadrant I

14. $\sin \dfrac{\theta}{2}$ if $\cos \theta = \dfrac{3}{5}$ and θ lies in quadrant I

15. $\cos \dfrac{\theta}{2}$ if $\cos \theta = \dfrac{1}{2}$ and θ lies in quadrant I

16. $\sin \dfrac{\theta}{2}$ if $\cos \theta = \dfrac{\sqrt{2}}{2}$ and θ lies in quadrant I

17. $\tan \dfrac{\theta}{2}$ if $\tan \theta = 2$ and θ lies in quadrant III

18. $\tan \dfrac{\theta}{2}$ if $\tan \theta = -2$ and θ lies in quadrant II

19. $\tan \dfrac{\theta}{2}$ if $\sin \theta = -\dfrac{24}{25}$ and θ lies in quadrant IV

20. $\tan \dfrac{\theta}{2}$ if $\sin \theta = -\dfrac{8}{17}$ and θ lies in quadrant IV

21. $\sin \theta$ if $\cos 2\theta = \dfrac{5}{13}$ and θ lies in quadrant I

22. $\cos \theta$ if $\cos 2\theta = -\dfrac{12}{13}$ and θ lies in quadrant III

Prove each identity.

23. $\tan \dfrac{\alpha}{2} = \dfrac{\sin \alpha}{1 + \cos \alpha}$

24. $\tan \dfrac{\alpha}{2} = \dfrac{1 - \cos \alpha}{\sin \alpha}$

25. $\cos \alpha = 2 \cos^2 \left(\dfrac{\alpha}{2} \right) - 1$

26. $\sin \beta = 2 \sin \dfrac{\beta}{2} \cos \dfrac{\beta}{2}$

27. $\csc \beta = \dfrac{1 + \tan^2 \left(\dfrac{\beta}{2} \right)}{2 \tan \left(\dfrac{\beta}{2} \right)}$

28. $\tan^2 \left(\dfrac{\alpha}{2} \right) = \dfrac{(1 - \cos \alpha)^2}{\sin^2 \alpha}$

29. $\cos \alpha = \dfrac{\cot \dfrac{\alpha}{2} - \tan \dfrac{\alpha}{2}}{\tan \dfrac{\alpha}{2} + \cot \dfrac{\alpha}{2}}$

30. $\sec \beta = \dfrac{1 + \tan^2 \left(\dfrac{\beta}{2} \right)}{1 - \tan^2 \left(\dfrac{\beta}{2} \right)}$

31. $\cos 2\theta = 1 - 8 \sin^2 \left(\dfrac{\theta}{2} \right) \cos^2 \left(\dfrac{\theta}{2} \right)$

32. $\tan^2 \left(\dfrac{\theta}{2} \right) = 1 - \dfrac{2 \cos \theta}{1 + \cos \theta}$

Applications

33. Geometry The area A of an isosceles triangle can be expressed as $A = \left(s \sin \dfrac{\alpha}{2} \right) \left(s \cos \dfrac{\alpha}{2} \right)$, where s is the length of each of the two congruent sides, and α is the angle formed by those sides. Show that this formula is equivalent to $A = \dfrac{1}{2} s^2 \sin \alpha$.

34. Geometry Use one of the formulas in Exercise 33 to find the area of an isosceles triangle, to the nearest square centimeter, if $s = 10$ cm and $\alpha = 48°$.

TRIGONOMETRY IN GEOGRAPHY

The earth is nearly spherical, with a radius of about 3958 mi. A *great circle* is formed by a plane passing through the center O of a sphere and cutting the surface. The *equator* is a great circle on the earth, and *parallels* are smaller circles formed by planes intersecting the earth parallel to the equator.

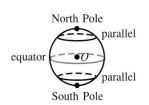

Meridians are great circles that intersect the geographic North and South Poles. In the figure at the right, $\angle XOY$ determines a unique great circle containing shorter $\overset{\frown}{XY}$.

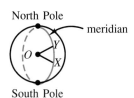

If the measure of $\angle XOY$ is 10°, what is the length of $\overset{\frown}{XY}$, to the nearest mile?

Product/Sum Identities

Objective: To develop and use product/sum identities

By adding and subtracting sum and difference identities for sine and cosine, a new set of identities can be derived.

Preview

Planes usually fly along great-circle routes, since this is the shortest distance between two points on a sphere. A great circle is the circle formed by the intersection of a sphere and a plane through its center. Latitude and longitude can be used to find the great-circle distance between two cities, A and B, in the Northern Hemisphere using the formula

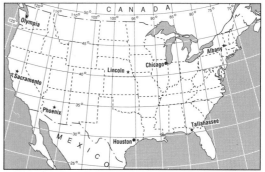

$$\cos d = \cos \theta \cos a \cos b + \sin a \sin b$$

where

θ = the positive longitudinal difference between cities A and B
a = latitude of city A
b = latitude of city B
d = degree measure of the great-circle arc between A and B ($0° < d \leq 180°$).

Find the degree measure, to the nearest tenth of a degree, of the great-circle arc between the given cities. Assume that the earth is spherical.

1. New York City (latitude 40.8°; longitude 74°) and Los Angeles (latitude 34.1°; longitude 118.3°)

2. Boston (latitude 42.4°; longitude 71.1°) and Miami (latitude 25.8°; longitude 80.2°)

3. Use the results in Exercises 1 and 2 to find the great-circle distance, to the nearest 10 mi, between each pair of cities. Assume that a 1° arc of a great circle has a length of approximately 69.1 mi.

At times it is necessary to express a trigonometric sum as a product or a product as a sum.

$$2 \cos \alpha \cos \beta = \cos (\alpha - \beta) + \cos (\alpha + \beta)$$
$$2 \sin \alpha \sin \beta = \cos (\alpha - \beta) - \cos (\alpha + \beta)$$
$$2 \sin \alpha \cos \beta = \sin (\alpha + \beta) + \sin (\alpha - \beta)$$
$$2 \cos \alpha \sin \beta = \sin (\alpha + \beta) - \sin (\alpha - \beta)$$

Each of the product identities is found by adding or subtracting the sum and difference identities for sine or cosine.

$$\begin{aligned} \cos \alpha \cos \beta + \sin \alpha \sin \beta &= \cos (\alpha - \beta) \quad &\textit{Add } \cos (\alpha + \beta) \textit{ to } \cos (\alpha - \beta). \\ \cos \alpha \cos \beta - \sin \alpha \sin \beta &= \cos (\alpha + \beta) \\ \hline 2 \cos \alpha \cos \beta \quad\quad\quad\quad &= \cos (\alpha - \beta) + \cos (\alpha + \beta) \end{aligned}$$

$$\begin{aligned} \cos \alpha \cos \beta + \sin \alpha \sin \beta &= \cos (\alpha - \beta) \quad &\textit{Subtract } \cos (\alpha + \beta) \textit{ from} \\ \cos \alpha \cos \beta - \sin \alpha \sin \beta &= \cos (\alpha + \beta) \quad &\cos (\alpha - \beta). \\ \hline 2 \sin \alpha \sin \beta &= \cos (\alpha - \beta) - \cos (\alpha + \beta) \end{aligned}$$

The other two identities can be verified in a similar manner.

EXAMPLE 1 **Express 2 cos 40° sin 15° as a difference.**

$$\begin{aligned} 2 \cos 40° \sin 15° &= \sin (40° + 15°) - \sin (40° - 15°) \quad &2 \cos \alpha \sin \beta \\ &= \sin 55° - \sin 25° \quad &= \sin (\alpha + \beta) - \sin (\alpha - \beta) \end{aligned}$$

The product/sum identities can also be used to express sums or differences as products.

EXAMPLE 2 **Express cos 27° + cos 61° as a product.**

Let $\alpha + \beta = 61°$ and $\alpha - \beta = 27°$.

$$\begin{aligned} \alpha + \beta &= 61° \quad &\textit{Add the two equations and solve for } \alpha. \\ \alpha - \beta &= 27° \\ \hline 2\alpha &= 88° \\ \alpha &= 44° \end{aligned}$$

$$44° + \beta = 61°, \beta = 17° \quad\quad \alpha + \beta = 61° \quad\quad \textit{Solve for } \beta.$$

Then use the identity $\cos (\alpha - \beta) + \cos (\alpha + \beta) = 2 \cos \alpha \cos \beta$.

$$\cos 27° + \cos 61° = 2 \cos 44° \cos 17°$$

Verify the fact that if $\alpha + \beta = 27°$ and $\alpha - \beta = 61°$ in Example 2, you obtain the same result.

There are alternate forms for the product/sum identities that can often be more easily applied. To derive these alternate identities, let $x = \alpha + \beta$ and $y = \alpha - \beta$. Add these equations to solve for α and subtract to solve for β.

$$\begin{array}{ll} x = \alpha + \beta & x = \alpha + \beta \\ \underline{y = \alpha - \beta} & \underline{y = \alpha - \beta} \\ x + y = 2\alpha & x - y = 2\beta \\ \dfrac{x + y}{2} = \alpha & \dfrac{x - y}{2} = \beta \end{array}$$

Substitute these values into the product/sum identities to obtain the alternate forms.

$$\cos x + \cos y = 2 \cos \left(\frac{x + y}{2}\right) \cos \left(\frac{x - y}{2}\right) \qquad \sin x + \sin y = 2 \sin \left(\frac{x + y}{2}\right) \cos \left(\frac{x - y}{2}\right)$$

$$\cos x - \cos y = -2 \sin \left(\frac{x + y}{2}\right) \sin \left(\frac{x - y}{2}\right) \qquad \sin x - \sin y = 2 \cos \left(\frac{x + y}{2}\right) \sin \left(\frac{x - y}{2}\right)$$

EXAMPLE 3   **Express sin 70° + sin 24° as a product.**

$$\sin x + \sin y = 2 \sin \left(\frac{x + y}{2}\right) \cos \left(\frac{x - y}{2}\right)$$

$$\sin 70° + \sin 24° = 2 \sin \left(\frac{70° + 24°}{2}\right) \cos \left(\frac{70° - 24°}{2}\right) \quad x = 70° \text{ and } y = 24°$$

$$= 2 \sin 47° \cos 23°$$

The product/sum identities can be used to prove other identities.

EXAMPLE 4   **Prove: $\cot \theta = \dfrac{\sin 4\theta + \sin 6\theta}{\cos 4\theta - \cos 6\theta}$**

$$\cot \theta \left|\begin{array}{l} \dfrac{\sin 4\theta + \sin 6\theta}{\cos 4\theta - \cos 6\theta} \\[2ex] \dfrac{2 \sin \dfrac{4\theta + 6\theta}{2} \cos \dfrac{4\theta - 6\theta}{2}}{-2 \sin \dfrac{4\theta + 6\theta}{2} \sin \dfrac{4\theta - 6\theta}{2}} \\[3ex] \dfrac{2 \sin 5\theta \cos (-\theta)}{-2 \sin 5\theta \sin (-\theta)} \\[2ex] \dfrac{\cos (-\theta)}{-\sin (-\theta)} \\[2ex] \dfrac{\cos \theta}{\sin \theta} \end{array}\right.$$

$$\sin x + \sin y = 2 \sin \left(\frac{x + y}{2}\right) \cos \left(\frac{x - y}{2}\right)$$

$$\cos x - \cos y = -2 \sin \left(\frac{x + y}{2}\right) \sin \left(\frac{x - y}{2}\right)$$

$$\cos (-\theta) = \cos \theta; \sin (-\theta) = -\sin \theta$$

$$= \cot \theta$$

Thus, $\cot \theta = \dfrac{\sin 4\theta + \sin 6\theta}{\cos 4\theta - \cos 6\theta}$

CLASS EXERCISES

Match each expression on the left with an equivalent expression on the right.

1. $2 \sin 20° \sin 5°$

2. $2 \sin 15° \sin 10°$

3. $2 \cos 10° \cos 40°$

4. $2 \sin 100° \cos 50°$

A. $\cos 50° + \cos (-30°)$

B. $\sin 150° + \sin 50°$

C. $\cos 15° - \cos 25°$

D. $\cos 5° - \cos 25°$

Express each product as a sum or difference.

5. $2 \sin 50° \cos 20°$

6. $2 \sin 25° \sin 5°$

7. $2 \cos 40° \cos 10°$

8. $2 \cos 30° \sin 10°$

Express each sum or difference as a product.

9. $\cos 70° + \cos 10°$

10. $\sin 70° - \sin 10°$

11. $\sin 4x + \sin 2x$

12. $\sin 10x - \sin 6x$

13. $\cos 4x - \cos 8x$

14. $\cos 12x + \cos 4x$

PRACTICE EXERCISES

Express each product as a sum or difference.

1. $2 \sin 42° \sin 18°$

2. $2 \cos 26° \cos 16°$

3. $2 \cos 5x \sin 2x$

4. $2 \sin 6x \sin x$

5. $2 \sin \dfrac{\pi}{12} \cos \dfrac{\pi}{6}$

6. $2 \cos \dfrac{\pi}{8} \sin \dfrac{\pi}{4}$

7. $\cos \dfrac{3\pi}{8} \cos \dfrac{5\pi}{8}$

8. $\sin \dfrac{\pi}{12} \sin \dfrac{\pi}{4}$

Express each sum or difference as a product.

9. $\cos 38° + \cos 106°$

10. $\sin 46° + \sin 98°$

11. $\sin 12\theta + \sin 3\theta$

12. $\sin 15\theta - \sin 4\theta$

13. $\cos \dfrac{6\pi}{7} - \cos \dfrac{2\pi}{7}$

14. $\cos \dfrac{7\pi}{9} + \cos \dfrac{2\pi}{9}$

Prove each identity.

15. $\cot \beta = \dfrac{\cos 3\beta + \cos \beta}{\sin 3\beta - \sin \beta}$

16. $\cot \alpha = \dfrac{\cos 5\alpha + \cos 3\alpha}{\sin 5\alpha - \sin 3\alpha}$

17. $\tan 5\beta = \dfrac{\cos 8\beta - \cos 2\beta}{\sin 2\beta - \sin 8\beta}$

18. $\tan 2\alpha = \dfrac{\cos \alpha - \cos 3\alpha}{\sin 3\alpha - \sin \alpha}$

19. $\dfrac{1}{\tan \alpha} = \dfrac{\cos 4\alpha + \cos 2\alpha}{\sin 4\alpha - \sin 2\alpha}$

20. $\dfrac{\tan 6\beta}{\tan \beta} = \dfrac{\sin 7\beta + \sin 5\beta}{\sin 7\beta - \sin 5\beta}$

21. $-\tan \theta = \dfrac{\cos 3\theta - \cos \theta}{\sin 3\theta + \sin \theta}$

22. $\dfrac{\csc \theta}{\sec \theta} = \dfrac{\sin 5\theta + \sin 3\theta}{\cos 3\theta - \cos 5\theta}$

23. $-\dfrac{\csc 5\theta}{\sec 5\theta} = \dfrac{\sin 8\theta - \sin 2\theta}{\cos 8\theta - \cos 2\theta}$

24. $\dfrac{\csc 3\theta}{\sec 3\theta} = \dfrac{\cos 4\theta + \cos 2\theta}{\sin 4\theta + \sin 2\theta}$

25. $\cos^2 y - \sin^2 x = \cos (x + y) \cos (x - y)$

26. $\sin x \cos x + \sin y \cos y = \sin (x + y) \cos (x - y)$

Applications

Computer The user is asked to supply two angle values. The program then evaluates both sides of $2 \sin \alpha \cos \beta = \sin (\alpha + \beta) + \sin (\alpha - \beta)$. The results are then compared.

27. If the displayed values are nearly the same, differing only in their tenth decimal places, what conclusion(s) might you make? Explain.

```
10  HOME: INPUT"ENTER TWO
    ANGLE MEASURES": A,B
20  R1 = A*0.017453292 :
    R2 = B*0.017453292
30  L = 2*SIN(R1)*COS(R2)
40  R = SIN(R1 + R2) + SIN(R1 - R2)
50  PRINT: PRINT"2 SIN A COS B",
    "=", "SIN"(A + B) + "SIN"(A - B)
60  PRINT L,R
70  END
```

TEST YOURSELF

1. Use double-angle identities to find the exact value of $\cos 2\theta$ if $\sin \theta = \frac{5}{13}$ and $90° < \theta < 180°$.   5.4

2. Use double-angle identities to find the exact value of $\tan 2\theta$ if $\tan \theta = \frac{5}{4}$ and $180° < \theta < 270°$.

3. Use half-angle identities to find the exact value of $\sin 105°$.   5.5

4. Use half-angle identities to find the exact value of $\tan \frac{5\pi}{8}$.

5. Find the exact value of $\tan \frac{\alpha}{2}$ if $\cos \alpha = -\frac{2}{3}$ and α lies in quadrant III $0° < \alpha < 360°$.

6. Find the exact value of $\cos \frac{\alpha}{2}$ if $\cos \alpha = -\frac{12}{13}$ and α lies in quadrant II $0° < \alpha < 360°$.

7. Prove: $\csc 2\alpha + \cot 2\alpha = \cot \alpha$

8. Prove: $\sec 2\theta = \dfrac{\csc^2 \theta}{\csc^2 \theta - 2}$

9. Express as a product:
$\sin 6x + \sin 4x$

10. Express as a product:
$\cos 25° - \cos 15°$ 5.6

5.6 Product/Sum Identities **241**

APPLICATION: Aviation

Did you know that in many professions or occupations there are certain "rules of thumb" that are approximately true and are passed down from one generation to the next generation? Some "rules of thumb" are based on legend, some on customs and folklore, and still others on mathematical principles. Airplane pilots, for example, use experience, good judgment, common sense, and some "rules of thumb." The "rule of thumb" considered here will have its basis in trigonometry.

An airline pilot, unlike a motorist in an automobile, cannot safely make sharp or sudden changes in the direction of his or her aircraft. A rate of turn must be selected that is both safe and effective. For many types of aircraft, a rate of turn called the *standard-rate balanced turn* is used. This rate is defined to be exactly 3° per second. Hence, a pilot using a standard-rate balanced turn requires 120 s to turn the aircraft through a complete revolution of 360°.

When turning the aircraft, a pilot needs to choose an *angle of bank*, that is, the tilt of the plane's wings from the horizontal. The angle is determined by both the plane's speed and its rate of turn.

Mathematicians and pilots have developed a formula that can be used to calculate the angle of bank for a standard-rate balanced turn.

$$\tan \theta = 2.3855 \times 10^{-3} v$$ *Formula for the angle of bank, where v is the speed of the plane in miles per hour*

EXAMPLE   Suppose you are piloting an aircraft through a standard-rate balanced turn of 3° traveling 135 mi/h. What angle of bank do you need?

$\tan \theta = 2.3855 \times 10^{-3} v$
$\tan \theta = 2.3855 \times 10^{-3} \times 135$ *Calculation-ready form*
$\tan \theta = 0.3320$
$\quad \theta = 18°$

The angle of bank is approximately 18°.

	Speed	Angle of bank (degrees)
The table shows the	60	8
result of using the formula	80	11
to find the angle of bank	100	13.4
at different speeds for a	120	16
standard-rate balanced turn.	140	18.5
	160	21
	180	23
	200	25.5

At times, a pilot may need a quick and reasonably accurate way of determining the angle of bank. A "rule of thumb" used to find the angle of bank for standard-rate balanced turns at various speeds is

$$\text{angle of bank} = \frac{\text{speed in miles per hour}}{10} + 5$$

In the example, the "rule of thumb" would have given an angle of bank of 18.5°.

EXERCISES

Using the formula $\tan \theta = 2.3855 \times 10^{-3} \, v$, calculate the angle of bank for a standard-rate balanced turn at various speeds.

1. 90 mi/h   **2.** 150 mi/h   **3.** 250 mi/h   **4.** 350 mi/h

Using the "rule of thumb," estimte the bank angle for a standard-rate balanced turn at the following speeds.

5. 110 mi/h   **6.** 145 mi/h   **7.** 300 mi/h   **8.** 400 mi/h

9. If you know that the angle of bank is 22° for a standard-rate balanced turn, use $\tan \theta = 2.3855 \times 10^{-3} \, v$ to find the plane's speed in miles per hour.

10. The angle of bank of an airplane is 31°. Approximate the plane's speed in miles per hour using the "rule of thumb" formula.

11. Find the amount of error that results from using the "rule of thumb" rather than the formula to find the measure of the angle of bank for a standard-rate balanced turn at 90 mi/h. Express this error as a percent of the angle measure calculated using the formula.

12. Find the percent of error made by using the "rule of thumb" rather than the formula for the following velocities (all expressed in miles per hour): 60, 120, 150, 250, 350, 600, 700.

13. Based on your results in Exercises 11 and 12, do you think that the "rule of thumb" is most accurate for high, intermediate, or low speeds?

Vocabulary

difference identity	sum identity	other identities
for cosine (208)	for cosine (210)	double-angle (227)
for sine (215)	for sine (215)	half-angle (232)
for tangent (221)	for tangent (220)	sum/product (238)

Cosine: Sum and Difference Identities   5.1

$$\cos(\alpha + \beta) = \cos\alpha\cos\beta - \sin\alpha\sin\beta \qquad \cos(\alpha - \beta) = \cos\alpha\cos\beta + \sin\alpha\sin\beta$$

1. Find the exact value of $\cos 255°$.   **2.** Find the exact value of $\cos\frac{\pi}{12}$.

3. Find the exact value of $\cos(\alpha - \beta)$ if $\tan\alpha = \frac{4}{3}$, $\sin\beta = -\frac{1}{2}$, $\pi < \alpha < \frac{3\pi}{2}$, and $\frac{3\pi}{2}, < \beta < 2\pi$.

4. Prove: $\cos(5\pi + \theta) = -\cos\theta$

Sine: Sum and Difference Identities   5.2

$$\sin(\alpha + \beta) = \sin\alpha\cos\beta + \cos\alpha\sin\beta \qquad \sin(\alpha - \beta) = \sin\alpha\cos\beta - \cos\alpha\sin\beta$$

5. Find the exact value of $\sin 105°$.   **6.** Find the exact value of $\sin\frac{5\pi}{12}$.

7. Find the exact value of $\sin(\alpha + \beta)$ if $\cos\alpha = -\frac{3}{5}$, $\cos\beta = \frac{7}{25}$, $\pi < \alpha < \frac{3\pi}{2}$, and $0 < \beta < \frac{\pi}{2}$.

8. Prove: $\sin(270° + \alpha) = -\cos\alpha$

Tangent: Sum and Difference Identities   5.3

$$\tan(\alpha + \beta) = \frac{\tan\alpha + \tan\beta}{1 - \tan\alpha\tan\beta}, \quad \tan\alpha\tan\beta \neq 1$$

$$\tan(\alpha - \beta) = \frac{\tan\beta - \tan\beta}{1 + \tan\alpha\tan\beta}, \quad \tan\alpha\tan\beta \neq -1$$

9. Find the exact value of $\tan 75°$.   **10.** Find the exact value of $\tan\frac{7\pi}{12}$.

11. Find the exact value of $\tan(\alpha - \beta)$ if $\cos\alpha = -\frac{1}{2}$, $\cos\beta = -\frac{1}{2}$, $\frac{\pi}{2} < \alpha < \pi$, and $\pi < \beta < \frac{3\pi}{2}$.

12. Prove: $\tan(3\pi - \alpha) = -\tan\alpha$

Double-Angle Identities 5.4

$$\sin 2\alpha = 2 \sin \alpha \cos \alpha \qquad \cos 2\alpha = \cos^2 \alpha - \sin^2 \alpha$$
$$\tan 2\alpha = \frac{2 \tan \alpha}{1 - \tan^2 \alpha}, \ \tan \alpha \neq \pm 1 \qquad \cos 2\alpha = 2 \cos^2 \alpha - 1$$
$$\cos 2\alpha = 1 - 2 \sin^2 \alpha$$

If $\sin \alpha = -\dfrac{3}{5}$ and α lies in quadrant IV, find the exact value of each expression.

13. $\sin 2\alpha$ **14.** $\cos 2\alpha$ **15.** $\tan 2\alpha$

16. Prove: $\tan \theta = \csc 2\theta - \cot 2\theta$

Half-Angle Identities 5.5

$$\sin \frac{\alpha}{2} = \pm \sqrt{\frac{1 - \cos \alpha}{2}} \qquad \tan \frac{\alpha}{2} = \pm \sqrt{\frac{1 - \cos \alpha}{1 + \cos \alpha}}, \ \cos \alpha \neq -1$$

$$\cos \frac{\alpha}{2} = \pm \sqrt{\frac{1 + \cos \alpha}{2}} \qquad \tan \frac{\alpha}{2} = \frac{\sin \alpha}{1 + \cos \alpha}, \ \cos \alpha \neq -1$$

$$\tan \frac{\alpha}{2} = \frac{1 - \cos \alpha}{\sin \alpha}, \ \sin \alpha \neq 0$$

If $\sin \theta = -\dfrac{12}{13}$ and θ lies in quadrant III, find the exact value of each expression. Assume that $0° < \theta < 360°$.

17. $\sin \dfrac{\theta}{2}$ **18.** $\cos \dfrac{\theta}{2}$ **19.** $\tan \dfrac{\theta}{2}$

20. Prove: $\cot \dfrac{\alpha}{2} = \dfrac{1}{\csc \alpha - \cot \alpha}$

Product/Sum Identities 5.6

$$2 \cos \alpha \cos \beta = \cos (\alpha - \beta) + \cos (\alpha + \beta)$$
$$2 \sin \alpha \sin \beta = \cos (\alpha - \beta) - \cos (\alpha + \beta)$$
$$2 \sin \alpha \cos \beta = \sin (\alpha + \beta) + \sin (\alpha - \beta)$$
$$2 \cos \alpha \sin \beta = \sin (\alpha + \beta) - \sin (\alpha - \beta)$$

There are also alternate forms for the product/sum identities.

21. Express $2 \sin 9x \sin 4x$ as a sum. **22.** Express $2 \cos 50° \sin 34°$ as a sum.

23. Express $\sin 73° + \sin 41°$ as a product.

24. Express $\cos (-2x) - \cos (-4x)$ as a product.

25. Prove: $\dfrac{\sin 2x + \sin 6x}{\sin 10x - \sin 2x} = \dfrac{\cos (-2x)}{\cos 6x}$ **26.** Prove: $\tan 3\alpha = \dfrac{\cos \alpha - \cos 5\alpha}{\sin 5\alpha - \sin \alpha}$

Use sum and difference identities to find the value of each trigonometric function.

1. $\cos 75°$ **2.** $\sin 285°$ **3.** $\tan 375°$

4. Find the exact value of $\cos(\alpha + \beta)$ if $\sin \alpha = \frac{5}{13}$, $\tan \beta = \frac{7}{24}$, $0 < \alpha < \frac{\pi}{2}$, and $\pi < \beta < \frac{3\pi}{2}$.

5. Find the exact value of $\sin(\alpha - \beta)$ if $\cos \alpha = -\frac{1}{2}$, $\sin \beta = \frac{3}{5}$, $\frac{\pi}{2} < \alpha < \pi$, and $\frac{\pi}{2} < \beta < \pi$.

6. Find the exact value of $\tan(\alpha + \beta)$ if $\sin \alpha = -\frac{4}{5}$, $\sin \beta = \frac{\sqrt{2}}{2}$, $\pi < \alpha < \frac{3\pi}{2}$, and $\frac{\pi}{2} < \beta < \pi$.

7. If $\cos \theta = \frac{7}{25}$ and $0° < \theta < 90°$, find $\sin 2\theta$.

8. If $\tan \theta = -\frac{12}{5}$ and $270° < \theta < 360°$, find $\cos 2\theta$.

9. If $\cos \theta = -0.5$ and $90° < \theta < 180°$, find $\tan 2\theta$.

If $\cos \alpha = -\dfrac{3}{5}$ and α lies in quadrant III, find the exact value of each expression. Assume that $0° < \alpha < 360°$.

10. $\sin \dfrac{\alpha}{2}$ **11.** $\tan \dfrac{\alpha}{2}$

12. Express $2 \cos 7x \sin x$ as a sum.   **13.** Express $\cos 76° - \cos 14°$ as a product.

14. Prove: $\sin 2\alpha = \dfrac{2}{\sec \alpha \csc \alpha}$

15. Prove: $\cos 2\alpha = \cos^4 \alpha - \sin^4 \alpha$

16. Prove: $\cot 2\beta = \dfrac{\cos \beta + \cos 3\beta}{\sin 3\beta + \sin \beta}$

Challenge

Prove: $\left(\sin \dfrac{\alpha}{2} + \cos \dfrac{\alpha}{2}\right)^2 = 1 + \sin \alpha$

Select the best choice for each question.

1. $\sin 15° =$

A. $\dfrac{\sqrt{3} + \sqrt{2}}{4}$ **B.** $\dfrac{\sqrt{6} - \sqrt{2}}{4}$

C. $\dfrac{\sqrt{6} - \sqrt{2}}{2}$ **D.** $\dfrac{\sqrt{3} - \sqrt{2}}{4}$

E. $\dfrac{\sqrt{2} + \sqrt{6}}{4}$

2. $\cos(\alpha + \beta)$ equals
A. $\cos \alpha + \cos \beta$
B. $\sin \alpha \sin \beta + \cos \alpha \cos \beta$
C. $\cos \alpha \cos \beta - \sin \alpha \sin \beta$
D. $\sin \alpha \cos \beta - \sin \beta \cos \alpha$
E. $\cos \alpha \sin \alpha + \cos \beta \sin \beta$

3. If $f(x) = x^2 + x + 1$,
then $f(a + b) =$
A. $a^2 + a + b^2 + b + 2$
B. $a^2 + b^2 + a + b + 1$
C. $2a^2 + 2ab + 2b^2 + 1$
D. $a^2 + 2ab + b^2 + a + b + 1$
E. $a^2 + 2ab + b^2 + 1$

4. If $\sin \theta = -\frac{3}{5}$, where $\pi < \theta < \frac{3\pi}{2}$,
then $\tan 2\theta =$

A. $\frac{3}{8}$ **B.** 3 **C.** $\frac{24}{7}$ **D.** $\frac{8}{3}$ **E.** $-\frac{24}{7}$

5. The equation of a circle with radius 3 and center $(2, -3)$ is
A. $(x - 2)^2 + (y + 3)^2 = 9$
B. $(x + 2)^2 + (y - 3)^2 = 3$
C. $(x - 3)^2 + (y + 2)^2 = 9$
D. $x^2 + y^2 = 9$
E. $(x - 2)^2 + (y - 3)^2 = 3$

6. $\cos(7\pi - x)$ is equivalent to
A. $\cos 7\pi$ **B.** $\sin(7\pi - x)$
C. $-\sin x$ **D.** $-\cos x$
E. $\sin 7\pi$

7. If $A = \frac{1}{2}h(b_1 + b_2)$, then $b_2 =$

A. $\dfrac{A - 2hb_1}{2h}$ **B.** $\dfrac{A - hb_1}{h}$

C. $A - hb_1$ **D.** $2A - 2b_1$

E. $\dfrac{2A - hb_1}{h}$

8. Solve for x.
$$\dfrac{6}{x^2 - 6x + 8} - \dfrac{x}{x - 2} = \dfrac{9}{x - 4}$$
A. $x = -8, x = 3$
B. $x = 6, x = 4$
C. $x = 8, x = -4$
D. $x = -6, x = 5$
E. $x = 9$

9. Evaluate: $\begin{vmatrix} 3 & -1 & 4 \\ 0 & 2 & 2 \\ 3 & 1 & -2 \end{vmatrix}$

A. 24 **B.** -48 **C.** -14
D. -42 **E.** -12

10. The area of a square is 25 ft^2.
What is its perimeter?
A. 10 **B.** 20 **C.** 40 **D.** 25
E. not enough information

11. For $0 \le x \le \pi$, the following represents the graph of

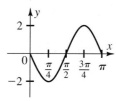

A. $y = 2 \cos x$ **B.** $y = \sin x$
C. $y = -2 \sin 2x$
D. $y = -2 \sin x$
E. $y = -2 \cos 2x$

Simplify.

Examples $\dfrac{\sqrt{3} + \sqrt{6}}{\sqrt{8}} = \dfrac{\sqrt{3} + \sqrt{6}}{\sqrt{8}} \cdot \dfrac{\sqrt{2}}{\sqrt{2}} = \dfrac{\sqrt{6} + \sqrt{12}}{\sqrt{16}} = \dfrac{\sqrt{6} + 2\sqrt{3}}{4}$

$\sqrt[3]{\dfrac{1}{3x}} = \sqrt[3]{\dfrac{1}{3x} \cdot \dfrac{3^2 x^2}{3^2 x^2}} = \dfrac{\sqrt[3]{9x^2}}{\sqrt{27x^3}} = \dfrac{\sqrt[3]{9x^2}}{3x}$

1. $\dfrac{2 + \sqrt{3}}{\sqrt{3}}$ **2.** $\dfrac{5}{3\sqrt{2x}}$ **3.** $\dfrac{2 - \sqrt{7}}{\sqrt{8}}$ **4.** $\sqrt[3]{\dfrac{x}{y}}$ **5.** $\sqrt[5]{\dfrac{3}{16}}$ **6.** $\sqrt[4]{\dfrac{2x}{y^3}}$

Solve each radical equation for x.

Example $\sqrt[3]{2x + 3} = -3$

$2x + 3 = -27$ *Cube each side.*

$x = -15$

7. $\sqrt{x - 3} = 5$ **8.** $-\sqrt{x + 2} = -7$ **9.** $\sqrt[3]{x - 2} = 5$

10. $\sqrt{2x + 4} = 8$ **11.** $\sqrt{x^2 + 9} = x + 1$ **12.** $\sqrt[3]{7x + 1} = 4$

Solve each equation using the quadratic formula $x = \dfrac{-b \pm \sqrt{b^2 - 4ac}}{2a}$.

Example $5x^2 - 38x + 56 = 0$ $a = 5, b = -38, c = 56$

$x = \dfrac{-(-38) \pm \sqrt{(-38)^2 - (4)(5)(56)}}{(2)(5)}$

$= \dfrac{38 \pm \sqrt{1444 - 1120}}{10} = \dfrac{38 \pm \sqrt{324}}{10} = \dfrac{38 \pm 18}{10} = 5.6 \text{ or } 2$

13. $x^2 + 3x = -2$ **14.** $2x^2 - 7x - 15 = 0$ **15.** $2x^2 + 4x - 6 = 0$

16. $6x^2 + x - 1 = 0$ **17.** $x^2 = 3x$ **18.** $x^2 - 6x + 9 = 0$

Find the inverse of each function.

Example $y = 2x + 6$

$x = 2y + 6$ *Interchange x and y.*

$x - 6 = 2y$ *Solve for y.*

$\dfrac{1}{2}x - 3 = y$

19. $y = 3x - 5$ **20.** $y = 4x + 3$ **21.** $y = \dfrac{1}{2}x - \dfrac{1}{2}$ **22.** $y = \dfrac{1}{3}x + 9$

248   Chapter 5 Trigonometric Identities

6 Inverse Trigonometric Functions

Light is electromagnetic radiation that may be perceived by the normal human eye. Trigonometry plays an important role in understanding its path. Lightning is a visible flash of light that results from a large-scale natural electric discharge in the atmosphere.

6.1 Inverse Relations and Functions

Objectives To graph functions and relations and their inverses
To evaluate expressions involving the inverse sine and
inverse cosine

Recall that a relation is a set of ordered pairs, and that a function is a relation in which each element in the domain is paired with a unique element in the range.

Preview

A public television station recorded the number of people who donated to the station following fund-raising appeals. It found that the number of people, *y*, who gave per week could be approximated by the equation

$$y = 18 + 5x$$

where *x* is the number of advertising spots for the appeal during that week.

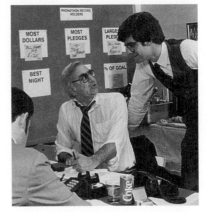

1. Graph the equation *y* = 18 + 5*x*.

2. Interchange *x* and *y* in the equation *y* = 18 + 5*x* and solve the new equation for *y*. What do *x* and *y* represent in the new equation? How do the domain and range of the first relation compare to those of the second?

3. Graph the equation you obtained in Exercise 2 and the line *y* = *x* on the same coordinate plane as you graphed *y* = 18 + 5*x*. Fold the paper along the line *y* = *x*. What assumptions can be made about the graph of *y* = 18 + 5*x* and the graph of the equation you wrote for Exercise 2?

Consider the relation

$$\{(1, 2), (2, 2), (3, 4), (4, 6), (5, 8)\}$$

The domain is $\{1, 2, 3, 4, 5\}$ and the range is $\{2, 4, 6, 8\}$. The inverse of this relation is formed by interchanging the first and second elements in each pair of the relation. The result is

$$\{(2, 1), (2, 2), (4, 3), (6, 4), (8, 5)\} \qquad \textit{Inverse of relation above}$$

The domain of the inverse, $\{2, 4, 6, 8\}$, is the range of the original relation, and the range of the inverse, $\{1, 2, 3, 4, 5\}$, is the domain of the original relation. The original relation is a function, but the inverse relation is not, since the first element, 2, is paired with both 1 and 2.

When a relation is defined by an equation in two variables, its inverse can often be formed by first interchanging the variables.

EXAMPLE 1 **Find the inverse of the relation $y = \pm\sqrt{x}$, and state whether or not the inverse is a function.**

$$y = \pm\sqrt{x}$$
$$x = \pm\sqrt{y} \qquad \textit{Interchange x and y.}$$
$$x^2 = y \qquad \textit{Square both sides and solve for y.}$$
$$y = x^2 \qquad \textit{The inverse of } y = \pm\sqrt{x}$$

In this case the original relation is not a function, since there are two values of y for each nonzero value of x. However, the inverse, $y = x^2$, is a function since for each x there is only one y.

When the inverse of a function is also a function, it is called the **inverse function** of the original function. If $f(x)$ is a function, then its inverse function is denoted by $f^{-1}(x)$. Keep in mind that the symbol -1 does *not* indicate the reciprocal, $\dfrac{1}{f(x)}$, when it is used in this way.

EXAMPLE 2 **Find the inverse of the function $f(x) = 2x + 4$.**

$$y = 2x + 4 \qquad \textit{Write f (x) = 2x + 4 as y = 2x + 4.}$$
$$x = 2y + 4 \qquad \textit{Interchange x and y.}$$
$$x - 4 = 2y \qquad \textit{Solve for y.}$$
$$y = \frac{x}{2} - 2$$

The inverse of $f(x) = 2x + 4$ is $f^{-1}(x) = \dfrac{x}{2} - 2$. It is also a function, since there is only one value of y for each value of x.

The graphs of the functions in Example 2 are shown at the right. If the coordinate plane were folded along the line $y = x$, the graphs of $f(x) = 2x + 4$ and $f^{-1}(x) = \dfrac{x}{2} - 2$ would coincide. Therefore, $f(x)$ and $f^{-1}(x)$ are symmetric with respect to the line $y = x$. This symmetry about $y = x$ exists for any pair of inverse functions or relations.

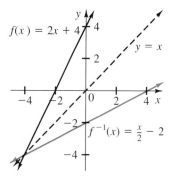

The trigonometric functions also have inverses. To find the inverse of $y = \sin x$, you may interchange the variables to obtain $x = \sin y$, but you will not be able to solve this equation algebraically for y in terms of x. Therefore, the inverse of $y = \sin x$ is written as follows:

$$y = \sin^{-1} x \qquad \text{\textit{Read: y is a number whose sine is x.}}$$

(Remember that $\sin^{-1} x$ is *not* the reciprocal of $\sin x$. The reciprocal of $\sin x$ is written as either $(\sin x)^{-1}$ or $\dfrac{1}{\sin x}$.) The inverse of each of the other trigonometric functions can be formed in the same way. For example, the inverse of $y = \cos x$ is

$$y = \cos^{-1} x \qquad \text{\textit{Read: y is a number whose cosine is x.}}$$

The inverse of the sine function can be graphed by interchanging the x- and y-coordinates of points on the graph of the sine function.

EXAMPLE 3 **Graph $y = \sin x$ and $y = \sin^{-1} x$ on the same coordinate plane.**

Sketch the graph of $y = \sin x$. Then interchange the coordinates of points on the graph to find the corresponding points on the graph of $y = \sin^{-1} x$. The coordinates of some of the points are shown below.

$y = \sin x$	$y = \sin^{-1} x$
$(-\pi, 0)$	$(0, -\pi)$
$\left(-\dfrac{\pi}{2}, -1\right)$	$\left(-1, -\dfrac{\pi}{2}\right)$
$(0, 0)$	$(0, 0)$
$\left(\dfrac{\pi}{2}, 1\right)$	$\left(1, \dfrac{\pi}{2}\right)$
$(\pi, 0)$	$(0, \pi)$
$\left(\dfrac{3\pi}{2}, -1\right)$	$\left(-1, \dfrac{3\pi}{2}\right)$
$(2\pi, 0)$	$(0, 2\pi)$

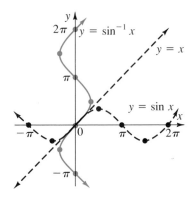

The vertical line test shows that $y = \sin^{-1} x$ is not a function. For example, there are an infinite number of values of $y = \sin^{-1} x$ for $x = 0$:

$$y = \sin^{-1} 0 = \cdots, -2\pi, -\pi, 0, \pi, 2\pi, \cdots$$

or

$$y = \sin^{-1} 0 = n\pi, \quad \text{where } n \text{ is an integer}$$

The inverse of $y = \cos x$ can be graphed in a similar manner (see Exercise 33). The relation $y = \cos^{-1} x$ is not a function, since there are an infinite number of values of y for any given value of x, where $-1 \le x \le 1$.

An alternate way of writing $\sin^{-1} x$ and $\cos^{-1} x$ is as arcsin x and arccos x. This alternate form is used in Example 4.

EXAMPLE 4   **Evaluate arccos $\dfrac{\sqrt{2}}{2}$ in degrees.**

Let $y = \arccos \dfrac{\sqrt{2}}{2}$. Then $\cos y = \dfrac{\sqrt{2}}{2}$.

Therefore, $y = \cdots,\ -315°,\ -45°,\ 45°,\ 315°,\ \cdots$

or

$$y = 45° + n\,(360°) \quad \text{or} \quad y = -45° + n\,(360°),$$

where n is an integer.

CLASS EXERCISES

Find the inverse of each relation. Is the relation a function? Is the inverse a function?

1. $\{(2, 4), (3, 6), (4, 8), (5, 10)\}$ **2.** $\{(1, 3), (1, 4), (1, 5), (1, 6)\}$

3. $\{(-2, 3), (-1, 3), (0, 3), (1, 3), (2, 3)\}$ **4.** $\{(5, 3), (5, 4), (4, 4), (3, 2)\}$

Find the inverse of each function.

5. $y = 4x$ **6.** $y = x + 3$ **7.** $y = -x$ **8.** $y = x - 1$

PRACTICE EXERCISES

Graph each relation and its inverse.

1. $y = x^3$ **2.** $y = 2x$

Find the inverse of each function.

3. $y = 10x$ **4.** $y = 6x$ **5.** $y = x - 8$ **6.** $y = x + 13$

7. $y = -7x$ **8.** $y = -9x$ **9.** $y = 2x + 1$ **10.** $y = 4x - 3$

11. $y = 5x - 4$ **12.** $y = 2x + 5$ **13.** $y = \frac{2}{3}x + 6$ **14.** $y = \frac{2}{3}x - 5$

Evaluate each expression in degrees.

15. $\sin^{-1} 1$ **16.** $\cos^{-1} 0$ **17.** arccos (-1) **18.** arcsin (-1)

Evaluate each expression in radians.

19. $\cos^{-1} 0$ **20.** $\sin^{-1} (-1)$ **21.** $\sin^{-1} 1$ **22.** $\cos^{-1} (-1)$

23. arccos (-0.5) **24.** arcsin $\left(-\dfrac{\sqrt{3}}{2}\right)$ **25.** arcsin (-0.5) **26.** arccos $\dfrac{\sqrt{2}}{2}$

Find the inverse of each function. Graph each function and its inverse.

27. $y = 4x$

28. $y = \frac{1}{2}x$

29. $y = x - 1$

30. $y = x + 3$

31. $y = 2x - 4$

32. $y = 3x + 5$

33. Graph $y = \cos x$ and its inverse on the same coordinate plane.

34. Given $f(x) = 3x - 24$, find $f^{-1}(x)$. Then find both $f(f^{-1}(x))$ and $f^{-1}(f(x))$. What is the result? *Hint*: To find $f(f^{-1}(x))$, use the expression you found for $f^{-1}(x)$ as a replacement for x in $3x - 24$.

35. Given $f(x) = x^3 + 1$, find $f^{-1}(x)$, $f(f^{-1}(x))$ and $f^{-1}(f(x))$. What is the result?

36. Based on your results in Exercises 34 and 35, write a rule for testing to see if two functions are inverses of each other.

Applications

37. Science The function that expresses Celsius temperature in terms of Fahrenheit temperature is given by $C(x) = \frac{5}{9}(x - 32)$. Find the inverse function that expresses Fahrenheit temperature in terms of Celsius.

38. Geometry The formula for expressing the circumference of a circle in terms of its radius is $C(r) = 2\pi r$. Find the inverse function.

EXTRA

The *horizontal-line test* provides a quick way to tell whether or not the inverse of a function is also a function. Specifically, the inverse of a function is also a function if and only if no horizontal line can be found that intersects the graph of the given function in more than one point. Thus, the function defined by the equation $y = x^3 + 3x^2$ has an inverse that is not a function. However, if the domain of $y = x^3 + 3x^2$ is restricted to the interval $-2 \leq x \leq 0$, the restricted function's inverse is also a function.

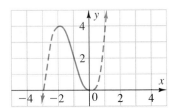

1. Describe at least two other ways to restrict the domain of $y = x^3 + 3x^2$ so that the inverse of the restricted function is also a function.

2. Sketch the graph of the function $y = \sin x$ for $-2\pi \leq x \leq 2\pi$. Use the horizontal-line test to determine whether or not the inverse of $y = \sin x$ is a function over that interval.

6.2 The Inverse Sine and Cosine Functions

Objectives: To define the inverse Sine and Cosine functions
To evaluate expressions involving inverse Sines and Cosines

The inverses of the sine and cosine functions are not functions, but the domains of the sine and cosine can be restricted so that their inverses *are* functions.

Preview

The exact values of the sine and cosine functions can be given for some angle measures. For example, $\cos 45° = \dfrac{\sqrt{2}}{2}$ and $\sin\left(-\dfrac{\pi}{3}\right) = -\dfrac{\sqrt{3}}{2}$.

Give the exact value of each sine or cosine function.

1. $\cos 0°$ **2.** $\sin 30°$ **3.** $\cos 60°$ **4.** $\sin 135°$ **5.** $\cos(-45°)$ **6.** $\sin 2\pi$

If the domain of the sine function is restricted to values from $-\dfrac{\pi}{2}$ through $\dfrac{\pi}{2}$, the inverse of this restricted function *is* a function. The domain can be restricted in other ways, but the interval $-\dfrac{\pi}{2} \le x \le \dfrac{\pi}{2}$ is chosen because it gives a range that is the same as that of the unrestricted function $y = \sin x$. That is, the range of the restricted sine function is $-1 \le y \le 1$. This function is often written $y = \text{Sin } x$, where the capital S distinguishes it from $y = \sin x$.

The inverse of $y = \text{Sin } x$ is the **inverse Sine** function, $y = \text{Sin}^{-1} x$. The capital S distinguishes this function from the relation $y = \sin^{-1} x$. The inverse Sine function is also commonly denoted by $y = \text{Arcsin } x$.

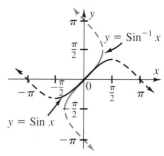

$$y = \text{Sin}^{-1} x \quad \text{if and only if} \quad \sin y = x \quad \text{and} \quad -\dfrac{\pi}{2} \le y \le \dfrac{\pi}{2}.$$

Thus, the domain of $y = \text{Sin}^{-1} x$ is $-1 \le x \le 1$ and the range is $-\dfrac{\pi}{2} \le y \le \dfrac{\pi}{2}$.

6.2 The Inverse Sine and Cosine Functions   **255**

The values in the range of $\text{Sin}^{-1} x$ are called the **principal values** of $\sin^{-1} x$. Similarly, if the domain of the cosine function is restricted to values of x from 0 through π, then the inverse of the restricted function is also a function. The restricted cosine function is denoted $y = \text{Cos } x$, and its inverse, the **inverse Cosine** function is denoted $y = \text{Cos}^{-1} x$ or $y = \text{Arccos } x$.

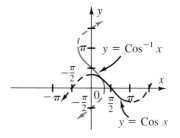

> $y = \text{Cos}^{-1} x$ if and only if $\cos y = x$ and $0 \le y \le \pi$.

Thus, the domain of $y = \text{Cos}^{-1} x$ is $-1 \le x \le 1$ and the range is $0 \le y \le \pi$. The values in the range of $\text{Cos}^{-1} x$ are called the *principal values* of $\cos^{-1} x$.

EXAMPLE 1 **Evaluate in radians:** **a.** $\text{Arccos } \dfrac{\sqrt{3}}{2}$ **b.** $\text{Sin}^{-1}\left(-\dfrac{\sqrt{2}}{2}\right)$

a. Let $y = \text{Arccos } \dfrac{\sqrt{3}}{2}$. Then $\cos y = \dfrac{\sqrt{3}}{2}$.

Since $\cos \dfrac{\pi}{6} = \dfrac{\sqrt{3}}{2}$ and $0 \le \dfrac{\pi}{6} \le \pi$, $y = \text{Arccos } \dfrac{\sqrt{3}}{2} = \dfrac{\pi}{6}$.

b. Let $y = \text{Sin}^{-1}\left(-\dfrac{\sqrt{2}}{2}\right)$. Then $\sin y = -\dfrac{\sqrt{2}}{2}$.

Since $\sin\left(-\dfrac{\pi}{4}\right) = -\dfrac{\sqrt{2}}{2}$ and $-\dfrac{\pi}{2} \le -\dfrac{\pi}{4} \le \dfrac{\pi}{2}$,

$y = \text{Sin}^{-1}\left(-\dfrac{\sqrt{2}}{2}\right) = -\dfrac{\pi}{4}$.

A scientific calculator can be used to find the values of inverse trigonometric functions. On some calculators, Sin^{-1} and Cos^{-1} keys are provided. On others, the *inv* key and the *sin* and *cos* keys are used. It is important to note that a calculator displays only the principal values, which are given in decimal form. To illustrate, in Example 1b a calculator would display the answer as -45 in degree mode or as -0.785398163 in radian mode.

EXAMPLE 2 **Find the approximate value of $\text{Sin}^{-1} 0.6823$ in degrees.**

Be sure the calculator is in degree mode.

$\text{Sin}^{-1} 0.6823 = 43°$ *Enter 0.6823 and use the Sin^{-1} or the* inv *and* sin *keys.*

EXAMPLE 3 **Evaluate:** $\sin\left(\text{Cos}^{-1}\dfrac{\sqrt{3}}{2}\right)$

Let $y = \text{Cos}^{-1}\dfrac{\sqrt{3}}{2}$.

Then $\cos y = \dfrac{\sqrt{3}}{2}$ and $y = \text{Cos}^{-1}\dfrac{\sqrt{3}}{2} = \dfrac{\pi}{6}.$ $0 \le \dfrac{\pi}{6} \le \pi$

So, $\sin\left(\text{Cos}^{-1}\dfrac{\sqrt{3}}{2}\right) = \sin\dfrac{\pi}{6} = \dfrac{1}{2}.$

From the definitions of the inverse Sine and inverse Cosine functions, it is clear that $\sin(\text{Arcsin } x) = x$ and $\cos(\text{Arccos } x) = x$. However, it is not always true that $\text{Arcsin}(\sin x) = x$ or that $\text{Arccos}(\cos x) = x$. The next example illustrates this.

EXAMPLE 4 **Evaluate:** $\text{Arcsin}\left(\sin\dfrac{7\pi}{4}\right)$

$\text{Arcsin}\left(\sin\dfrac{7\pi}{4}\right) = \text{Arcsin}\left(-\dfrac{\sqrt{2}}{2}\right)$ $\sin\dfrac{7\pi}{4} = -\dfrac{\sqrt{2}}{2}$

Let $y = \text{Arcsin}\left(-\dfrac{\sqrt{2}}{2}\right)$. Then $\sin y = -\dfrac{\sqrt{2}}{2}$. The principal value is $-\dfrac{\pi}{4}.$

CLASS EXERCISES

Thinking Critically

1. Explain the difference between $y = \text{Cos}^{-1} x$ and $y = \cos^{-1} x$.

2. For what values of x is $\text{Sin}^{-1}(\sin x)$ equal to x?

Find each value.

3. $\text{Arcsin}\dfrac{1}{2}$ **4.** $\text{Arccos}\dfrac{1}{2}$ **5.** $\text{Arccos}\dfrac{\sqrt{2}}{2}$ **6.** $\text{Arcsin}\dfrac{\sqrt{3}}{2}$

7. $\text{Sin}^{-1} 0$ **8.** $\text{Cos}^{-1} 0$ **9.** $\text{Sin}^{-1} 1$ **10.** $\text{Cos}^{-1} 1$

PRACTICE EXERCISES   Use appropriate technology.

Evaluate each expression in radians.

1. $\text{Arcsin}\left(-\dfrac{\sqrt{3}}{2}\right)$ **2.** $\text{Arccos}\left(-\dfrac{\sqrt{3}}{2}\right)$ **3.** $\text{Arccos}\dfrac{1}{2}$ **4.** $\text{Arcsin}\left(-\dfrac{1}{2}\right)$

5. $\text{Sin}^{-1}\left(-\dfrac{\sqrt{3}}{2}\right)$ **6.** $\text{Cos}^{-1}\dfrac{\sqrt{2}}{2}$ **7.** $\text{Sin}^{-1}(-1)$ **8.** $\text{Cos}^{-1}(-1)$

Find the approximate value of each expression to the nearest degree.

9. Arcsin 0.8123   **10.** Arccos 0.4927   **11.** Arccos 0.9427   **12.** Arcsin 0.72

13. $\text{Sin}^{-1} 0.6142$ **14.** $\text{Cos}^{-1} 0.3129$ **15.** $\text{Cos}^{-1}(-0.7123)$ **16.** $\text{Sin}^{-1}(-0.85)$

Find the exact value of each expression.

17. $\sin (\text{Sin}^{-1} 0.4)$

18. $\cos (\text{Cos}^{-1} 0.8667)$

19. $\sin \left(\text{Cos}^{-1} \dfrac{\sqrt{2}}{2} \right)$

20. $\sin \left(\text{Cos}^{-1} \left(-\dfrac{1}{2} \right) \right)$

21. $\cos \left(\text{Arcsin} \dfrac{\sqrt{3}}{2} \right)$

22. $\cos \left(\text{Arcsin} \left(-\dfrac{\sqrt{3}}{2} \right) \right)$

23. $\text{Arcsin} \left(\sin \dfrac{2\pi}{3} \right)$

24. $\text{Arccos} \left(\cos \left(-\dfrac{4\pi}{3} \right) \right)$

25. $\text{Arcsin} \left(\sin \dfrac{\pi}{6} \right)$

26. $\text{Arccos} \left(\cos \left(-\dfrac{\pi}{3} \right) \right)$

27. $\text{Arccos} \left(\sin \dfrac{3\pi}{4} \right)$

28. $\text{Arcsin} \left(\cos \dfrac{\pi}{4} \right)$

Approximate each number to four decimal places.

29. $\sin (\text{Arccos } 0.4127)$

30. $\cos (\text{Arcsin } 0.7459)$

31. $\cos (\text{Arcsin } 0.6148)$

32. $\sin (\text{Arccos } 0.1292)$

33. $\sin [\text{Cos}^{-1} (-0.8234)]$

34. $\cos [\text{Sin}^{-1} (-0.6927)]$

35. $\cos [\text{Sin}^{-1} (-0.5123)]$

36. $\sin [\text{Cos}^{-1} (-0.8139)]$

Graph each equation.

37. $y = \text{Sin}^{-1} x + 2$

38. $y = \text{Sin}^{-1} x - 1$

39. $y = 2 \text{ Arccos } x$

40. $y = 3 \text{ Arccos } x$

Find the exact value of each expression.

41. $\sin \left(\text{Arccos} \dfrac{\sqrt{3}}{2} + \pi \right)$

42. $\sin \left(\text{Arcsin} \dfrac{1}{2} - \pi \right)$

43. $\cos \left(\text{Arccos} \dfrac{\sqrt{2}}{2} + \dfrac{\pi}{2} \right)$

44. $\cos \left(\text{Arcsin} \left(-\dfrac{\sqrt{3}}{2} \right) - \dfrac{\pi}{2} \right)$

45. If $-1 \le x \le 0$, is it true that Arcsin $x <$ Arccos x? Explain.

46. For what values of x is it true that Arcsin $x =$ Arccos x?

47. For what values of x is it true that Arcsin $x \le$ Arccos x?

48. **Mechanics** The distance x that a slider moves when it is attached to a rod mechanism is given by $x = s - s \cos (\text{Arcsin } 0.4)$, where s is the length of the rod. Find x when s is 10 cm.

49. **Engineering** The ends of two vertical supports of a bridge under construction are connected by a cross-member that is 10 ft long. One end of the cross-member is attached at point A, 8 ft from the ground, and the other end is attached to point B, 2 ft from the ground. Express the measure of α, the acute angle that the cross-member makes with the right vertical support in terms of the inverse Cosine function.

BIOGRAPHY: IRMGARD FLUGGE-LOTZ (1903–1974)

Irmgard Lotz was born in Hameln, Germany. Since her mother's family was in the construction business and her father was a mathematician, it was natural for Irmgard to develop an interest in technical subjects. In 1923, she began the study of fluid dynamics and applied mathematics at the Technische Hochschule of Hanover. After receiving her undergraduate degree in 1927, she continued her studies at Hanover and received a doctorate in engineering in 1929. Later, she was employed at the Aerodynamische Versuchsanstalt (AVA) at Göttingen as a junior research engineer, where she worked with Ludwig Prandtl, one of the founders of the science of aerodynamics.

In 1948, as Irmgard Flugge-Lotz, she and her husband, Wilhelm Flugge-Lotz, joined the faculty at Stanford University in California. She was later appointed Stanford's first female full professor of engineering mechanics, aeronautics, and astronautics. She was also appointed a fellow of the American Institute of Aeronautics and Astronautics (AIAA), becoming just the second woman so honored. The Society of Woman Engineers honored her with its Achievement Award that same year.

Other Inverse Trigonometric Functions

Objectives: To define the inverse Tangent, Cotangent, Secant, and Cosecant functions

To evaluate expressions involving these functions

You have seen that the domains of the sine and cosine functions can be restricted so that their inverses are also functions. It is also possible to restrict the domains of the other four trigonometric functions in such a way that their inverses are functions.

Preview

The exact values of the secant, cosecant, tangent, and cotangent functions can be given for some angle measures. For example, $\tan(-60°) = -\sqrt{3}$.

Give the exact value of each function.

1. $\cot 30°$ 2. $\tan(-45°)$ 3. $\sec 120°$ 4. $\csc(-30°)$ 5. $\cot 60°$

6. $\tan \dfrac{\pi}{6}$ 7. $\tan \dfrac{\pi}{4}$ 8. $\cot \dfrac{\pi}{4}$ 9. $\sec \dfrac{3\pi}{4}$ 10. $\csc\left(-\dfrac{\pi}{3}\right)$

The inverses of the tangent, cotangent, secant, and cosecant functions are *not* functions.

Function	$y = \tan x$	$y = \cot x$	$y = \sec x$	$y = \csc x$
Inverse Relation	$y = \tan^{-1} x$	$y = \cot^{-1} x$	$y = \sec^{-1} x$	$y = \csc^{-1} x$

When the domains of $y = \tan x$ and $y = \cot x$ are appropriately restricted, their inverses, the **inverse Tangent** and **inverse Cotangent** are functions. The values of $y = \text{Tan}^{-1} x$ and $y = \text{Cot}^{-1} x$ are called the *principal values*. For $y = \text{Tan}^{-1} x$, the principal values are between $-\dfrac{\pi}{2}$ and $\dfrac{\pi}{2}$. For $y = \text{Cot}^{-1} x$, the principal values are between 0 and π. These domains are chosen so that $y = \text{Tan } x$ and $y = \text{Cot } x$ are continuous and have the same ranges as the unrestricted tangent and cotangent functions, respectively.

$y = \text{Tan}^{-1} x$ if and only if $\tan y = x$ and $-\dfrac{\pi}{2} < y < \dfrac{\pi}{2}$

$y = \text{Cot}^{-1} x$ if and only if $\cot y = x$ and $0 < y < \pi$

The graphs of these functions are shown at the top of the next page.

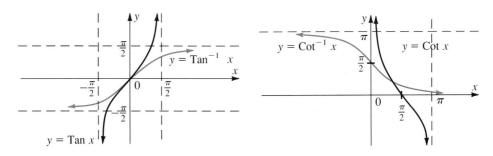

The **inverse Secant** and **inverse Cosecant** functions are the inverses of the restricted secant and cosecant functions, $y = \text{Sec } x$ and $y = \text{Csc } x$. The domains of these functions are restricted in the same way as the Sine and Cosine, respectively, except where the secant and cosecant are not defined.

$$y = \text{Sec}^{-1} x \quad \text{if and only if} \quad \sec y = x \quad \text{and} \quad 0 \le y \le \pi, y \ne \frac{\pi}{2}$$

$$y = \text{Csc}^{-1} x \quad \text{if and only if} \quad \csc y = x \quad \text{and} \quad -\frac{\pi}{2} \le y \le \frac{\pi}{2}, y \ne 0$$

The graphs of these functions are shown below.

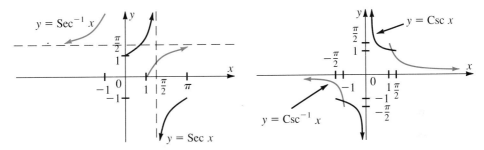

Here are the domains and ranges of the trigonometric functions and their inverses.

Function	$y = \text{Sin } x$	$y = \text{Sin}^{-1} x$	$y = \text{Cos } x$	$y = \text{Cos}^{-1} x$
Domain	$-\frac{\pi}{2} \le x \le \frac{\pi}{2}$	$-1 \le x \le 1$	$0 \le x \le \pi$	$-1 \le x \le 1$
Range	$-1 \le y \le 1$	$-\frac{\pi}{2} \le y \le \frac{\pi}{2}$	$-1 \le y \le 1$	$0 \le y \le \pi$
Function	$y = \text{Tan } x$	$y = \text{Tan}^{-1} x$	$y = \text{Cot } x$	$y = \text{Cot}^{-1} x$
Domain	$-\frac{\pi}{2} < x < \frac{\pi}{2}$	all real numbers	$0 < x < \pi$	all real numbers
Range	all real numbers	$-\frac{\pi}{2} < y < \frac{\pi}{2}$	all real numbers	$0 < y < \pi$
Function	$y = \text{Sec } x$	$y = \text{Sec}^{-1} x$	$y = \text{Csc } x$	$y = \text{Csc}^{-1} x$
Domain	$0 \le x \le \pi, x \ne \frac{\pi}{2}$	$|x| \ge 1$	$-\frac{\pi}{2} \le x \le \frac{\pi}{2}, x \ne 0$	$|x| \ge 1$
Range	$|y| \ge 1$	$0 \le y \le \pi, y \ne \frac{\pi}{2}$	$|y| \ge 1$	$-\frac{\pi}{2} \le y \le \frac{\pi}{2}, y \ne 0$

The inverse Tangent, inverse Cotangent, inverse Secant, and inverse Cosecant may also be written as Arctangent, Arccotangent, Arcsecant, and Arccosecant. This notation is shown in Example 1.

EXAMPLE 1 **Evaluate:** Arctan (-1)

Let $y =$ Arctan (-1). Then tan $y = -1$.
Since $\tan\left(-\dfrac{\pi}{4}\right) = -1$ and $-\dfrac{\pi}{2} < -\dfrac{\pi}{4} < \dfrac{\pi}{2}$,
$y =$ Arctan $(-1) = -\dfrac{\pi}{4}$.

When there is no Tan^{-1} key on a calculator, the *inv* key and the *tan* key are used to find approximate values of the inverse Tangent.

EXAMPLE 2 **Find the approximate value of Tan^{-1} 2.246 in degrees.**

Tan^{-1} 2.246 = 66°   *Use the degree mode. Enter 2.246 and use the* Tan^{-1} *key or the* inv *and* tan *keys.*

It is usually necessary to use the reciprocal key $\left(\dfrac{1}{x} \text{ or } x^{-1}\right)$ to find approximate values of the inverse Cotangent, inverse Secant, and inverse Cosecant.

EXAMPLE 3 **Find the approximate value to the nearest degree:**
  a. Sec^{-1} 1.843   **b.** Cot^{-1} (-1.862)

  a. Since $\sec x = \dfrac{1}{\cos x}$ and the domain of $y =$ Sec x is the same as
  that of $y =$ Cos x $\left(\text{if } x \neq \dfrac{\pi}{2}\right)$, use the $\dfrac{1}{x}$ key.

  Sec^{-1} 1.843 = 57°   *Enter 1.843 and use the* $\dfrac{1}{x}$ *and the* Cos^{-1}
  keys, or the $\dfrac{1}{x}$, *inv, and* cos *keys.*

  b. Since $\cot x = \dfrac{1}{\tan x}$ (when both functions are defined), an angle
  with cotangent -1.862 has a tangent of $\dfrac{1}{-1.862}$. However, the
  domain of $y =$ Cot x is $0 < x < \pi$, while the domain of
  $y =$ Tan x is $-\dfrac{\pi}{2} < x < \dfrac{\pi}{2}$. When the $\dfrac{1}{x}$, *inv*, and *tan* keys are
  used to find Cot^{-1} (-1.862), the calculator gives the principal
  value of the angle with tangent $\dfrac{1}{-1.862}$. To find the principal
  value of the angle with cotangent -1.862, you must add 180°.

$$\text{Cot}^{-1}(-1.862) = \text{Tan}^{-1}\left(\frac{1}{-1.862}\right) + 180°$$
$$= -28° + 180°$$
$$= 152°$$

The following identities follow from the definitions:

$$\tan(\text{Tan}^{-1} x) = x \qquad \cot(\text{Cot}^{-1} x) = x$$
$$\sec(\text{Sec}^{-1} x) = x \qquad \csc(\text{Csc}^{-1} x) = x$$

Identities involving inverse trigonometric functions can be proved using the definitions of the trigonometric functions and other known identities.

EXAMPLE 4 **Prove: $\csc(\text{Arcsin } v) = \dfrac{1}{v}$**

$$
\begin{array}{c|c}
\csc(\text{Arcsin } v) & \dfrac{1}{v} \\[2mm]
\csc\left(\text{Arccsc}\,\dfrac{1}{v}\right) & \qquad \sin x = \dfrac{1}{\csc x} \\[2mm]
\dfrac{1}{v} \;=\; & \csc(\text{Arccsc } x) = x
\end{array}
$$

CLASS EXERCISES

Find the exact value of each expression in radians.

1. Arctan 1   **2.** $\text{Tan}^{-1}\sqrt{3}$ **3.** $\text{Cot}^{-1} 1$ **4.** Arccot $\sqrt{3}$

5. Arcsec 1   **6.** Arccsc 1   **7.** $\text{Csc}^{-1}\sqrt{2}$ **8.** $\text{Sec}^{-1} 2$

PRACTICE EXERCISES   Use appropriate technology.

Find the exact value of each expression in radians.

1. $\text{Sec}^{-1}\sqrt{2}$ **2.** $\text{Csc}^{-1}(-2)$ **3.** Arctan$(-\sqrt{3})$ **4.** Arccot(-1)

Find the approximate value of each expression to the nearest degree.

5. Arctan 2.253   **6.** Arccot 3.341   **7.** $\text{Sec}^{-1}(-1.242)$ **8.** Csc^{-1} 1.294

Find the exact value of each expression.

9. $\tan(\text{Tan}^{-1} 0.6)$ **10.** $\cot(\text{Cot}^{-1} 0.3)$ **11.** $\sec(\text{Sec}^{-1} 5)$

12. $\csc(\text{Csc}^{-1} 6)$ **13.** $\tan\left[\text{Arcsin}\left(-\dfrac{1}{2}\right)\right]$ **14.** $\sin(\text{Arctan }\sqrt{3})$

15. $\cos[\text{Arctan}(-1)]$ **16.** $\sin[\text{Arccot}(-1)]$ **17.** $\sin(\text{Cot}^{-1}\sqrt{2})$

18. $\cos[\text{Cot}^{-1}(-\sqrt{2})]$ **19.** $\sec(\text{Csc}^{-1} 4)$ **20.** $\sin\left(\text{Tan}^{-1}\dfrac{12}{5}\right)$

Prove.

21. $\text{Sec}^{-1} 2 + \text{Tan}^{-1} \dfrac{\sqrt{3}}{3} = \dfrac{\pi}{2}$

22. $\text{Cot}^{-1} \dfrac{\sqrt{3}}{3} - \text{Csc}^{-1} 2 = \dfrac{\pi}{6}$

23. $\cos (\text{Arcsec } w) = \dfrac{1}{w}$

24. $\cot (\text{Arctan } z) = \dfrac{1}{z}$

Find the exact value of each expression.

25. $\sin (\text{Tan}^{-1} 1 + \text{Csc}^{-1} 2)$

26. $\tan (\text{Tan}^{-1} 2 + \text{Cot}^{-1} 1)$

Sketch each graph.

27. $y = 1 + \text{Arctan } x$

28. $y = \text{Arcsec } x - 2$

Applications

29. Physics   The maximum angle by which a lever rotates around its pivot when it is attached to an elliptical cam is $\text{Arctan } \dfrac{1}{7}$. Find the measure of the angle to the nearest tenth of one degree.

30. Construction   A steel ball is suspended from a primary cable attached to a crane 21 ft above the ground. A second cable is used to draw the ball upward. At point A, 10 ft above the ground, the ball is released. (See the figure.) Express α in terms of the Arctangent function. Then find the value of α.

TEST YOURSELF

Evaluate each expression in radians.

1. $\arcsin \dfrac{1}{2}$

2. $\arccos \left(-\dfrac{\sqrt{2}}{2} \right)$

3. $\arccos \dfrac{\sqrt{3}}{2}$

4. $\arcsin 0$ **6.1**

Find the approximate value of each expression to the nearest degree.

5. $\text{Sin}^{-1} 0.2463$

6. $\text{Csc}^{-1} (-1.6124)$

7. $\text{Arcsin } 0.8136$ **6.2**

Approximate each number to four decimal places.

8. $\sin (\text{Cos}^{-1} 0.3262)$

9. $\cos (\text{Cos}^{-1} 0.7125)$

10. $\sin (\text{Sin}^{-1} 0.1344)$

Find the exact value of each expression.

11. $\sec \left(\text{Arccsc } \dfrac{2\sqrt{3}}{3} \right)$

12. $\tan [\text{Cot}^{-1} (-1)]$

13. $\csc \left(\text{Tan}^{-1} \dfrac{3}{4} \right)$ **6.3**

6.4

Solving Trigonometric Equations: Using Special Angles

Objectives: To define conditional equations

To find exact solutions to trigonometric equations

Trigonometric equations that are true for all values of a variable for which the equation is defined are called *trigonometric identities*. These identities can be useful in solving certain other trigonometric equations.

Preview

In order to review some of the important trigonometric identities, match each expression on the left with an equivalent expression on the right.

1. $\sin 2x$

2. $1 + \tan^2 x$

3. $1 - \cos^2 x$

4. $\tan x \cos x$

5. $\csc^2 x - 1$

6. $\cos 2x$

7. $\cos x$

a. $\sin x$

b. $\cot^2 x$

c. $2 \sin x \cos x$

d. $\sec^2 x$

e. $\sin x \cot x$

f. $\sin^2 x$

g. $\cos^2 x - \sin^2 x$

An equation that is true for at least one value of the variable, but not all values, is called a *conditional equation*. The equation $\cos x = 0$ is a conditional equation, since it is true for some values of x, such as $-\dfrac{3\pi}{2}, -\dfrac{\pi}{2}, \dfrac{\pi}{2},$ and $\dfrac{3\pi}{2}$, but not for others. The rules of algebra can often be used to solve conditional trigonometric equations.

EXAMPLE 1 **Solve $2 \sin x - 1 = 0$, where $0 \le x < 2\pi$.**

$$2 \sin x - 1 = 0$$
$$2 \sin x = 1$$
$$\sin x = \frac{1}{2}$$

Over the interval $0 \le x < 2\pi$, the only values of x for which $\sin x = \dfrac{1}{2}$ is true are $x = \dfrac{\pi}{6}$ and $x = \dfrac{5\pi}{6}$.

Check to show that $\dfrac{\pi}{6}$ and $\dfrac{5\pi}{6}$ are solutions.

$$2 \sin \dfrac{\pi}{6} - 1 \overset{?}{=} 0 \qquad\qquad 2 \sin \dfrac{5\pi}{6} - 1 \overset{?}{=} 0$$

$$2\left(\dfrac{1}{2}\right) - 1 \overset{?}{=} 0 \qquad\qquad 2\left(\dfrac{1}{2}\right) - 1 \overset{?}{=} 0$$

$$0 = 0 \checkmark \qquad\qquad\qquad 0 = 0 \checkmark$$

When an equation contains two or more trigonometric functions, it is sometimes helpful to express the terms using one function only.

EXAMPLE 2 **Solve $2 \cos^2 x + \sin x + 1 = 0$, where $0 \le x < 360°$.**

$$2 \cos^2 x + \sin x + 1 = 0$$
$$2(1 - \sin^2 x) + \sin x + 1 = 0 \qquad \textit{Express $\cos^2 x$ in terms of}$$
$$2 \sin^2 x - \sin x - 3 = 0 \qquad\quad \textit{$\sin x$: $\cos^2 x = 1 - \sin^2 x$.}$$
$$(2 \sin x - 3)(\sin x + 1) = 0 \qquad \textit{Factor.}$$
$$2 \sin x - 3 = 0 \quad \text{or} \quad \sin x + 1 = 0$$
$$\sin x = \dfrac{3}{2} \quad \text{or} \qquad \sin x = -1$$

Since $\sin x$ can be no greater than 1, $\sin x = \dfrac{3}{2}$ has no solution.

If $\sin x = -1$, then $x = 270°$. Check in the original equation to show that $270°$ is a solution.

A double-angle identity is used to solve the equation in the next example.

EXAMPLE 3 **Solve $\sin 2x = \sqrt{3} \cos x$, where $0 \le x < 2\pi$.**

$$\sin 2x = \sqrt{3} \cos x$$
$$2 \sin x \cos x = \sqrt{3} \cos x \qquad\qquad \textit{$\sin 2x = 2 \sin x \cos x$}$$
$$2 \sin x \cos x - \sqrt{3} \cos x = 0 \qquad\qquad \textit{Write the equation with all}$$
$$\textit{nonzero terms on one side.}$$
$$\cos x (2 \sin x - \sqrt{3}) = 0 \qquad\qquad \textit{Factor.}$$
$$\cos x = 0 \quad \text{or} \quad 2 \sin x - \sqrt{3} = 0$$
$$\cos x = 0 \quad \text{or} \qquad\qquad \sin x = \dfrac{\sqrt{3}}{2}$$
$$x = \dfrac{\pi}{2} \text{ or } x = \dfrac{3\pi}{2} \quad \text{or} \qquad x = \dfrac{\pi}{3} \text{ or } x = \dfrac{2\pi}{3}$$

Check in the original equation to show that $\dfrac{\pi}{3}, \dfrac{\pi}{2}, \dfrac{2\pi}{3}$, and $\dfrac{3\pi}{2}$ are solutions.

CLASS EXERCISES

Tell whether each equation is an identity or a conditional equation.

1. $\sin x = 0.5$ **2.** $\sec x \cos x = 1$ **3.** $\tan x = 15$

4. $\sin x \cos x = 1$ **5.** $1 + \tan x = \cot x$ **6.** $\sin x = \tan x \cos x$

Find the exact solutions to each equation for the interval $0 \le x < 2\pi$.

7. $\sin x = -\dfrac{1}{2}$ **8.** $\cos x = \dfrac{\sqrt{3}}{2}$ **9.** $\sec x = 2$

PRACTICE EXERCISES

Find the exact solutions to each equation for the interval $0 \le x < 2\pi$.

1. $54 \sin x = 27$ **2.** $36 \cos x = 18$ **3.** $10 \cos x = 5\sqrt{2}$

4. $18 \sin x = 9\sqrt{2}$ **5.** $4 \csc x = -8$ **6.** $3 \sec x = -6$

Find the exact solutions to each equation for the interval $0° \le x < 360°$.

7. $2 + 2 \cos x = 0$ **8.** $2 \sin x - 2 = 0$ **9.** $1 + \sqrt{3} \tan x = 0$

10. $\sqrt{3} + \cot x = 0$ **11.** $2 + \sec x = 0$ **12.** $2 + \sqrt{2} \csc x = 0$

Find the exact solutions to each equation for the interval $0 \le x < 2\pi$.

13. $7 \cos x + 12 = 6 \cos x + 13$ **14.** $15 \sin x + 19 = 14 \sin x + 18$

15. $4 \tan x - 5 = 5 \tan x - 4$ **16.** $5 \cot x + 12 = 6 \cot x + 13$

17. $4 \sin^2 x = 3$ **18.** $2 \cos^2 x = 1$

19. $\cot^2 x = 1$ **20.** $\csc^2 x = 2$

21. $2 \sin^2 x + 3 \sin x + 1 = 0$ **22.** $2 \cos^2 x + 3 \cos x + 1 = 0$

23. $\tan^2 x = \tan x$ **24.** $\cot^2 x - \cot x = 0$

25. $\tan x = \cot x$ **26.** $\cos x = \sec x$

27. $2 \cos^2 x + \sin x = 2$ **28.** $4 \sin^2 x - 8 \cos x + 1 = 0$

29. $\sec^2 x - 2 \tan x = 0$ **30.** $\cos 2x + 3 \cos x = -2$

31. $\cos x = \sin 2x$ **32.** $2 \sin x - \sin 2x = 0$

33. $\cos 2x = 2 \sin^2 x$ **34.** $\sin^2 x + 3 \cos x = 3$

35. $2 \sin^3 x - \sin^2 x - \sin x = 0$ **36.** $2 \cos^3 x + \cos^2 x - \cos x = 0$

37. $2 \sin 2x = -\tan 2x$ **38.** $\cos^2 2x = \frac{1}{2} \cos 2x$

39. $\dfrac{\sec x}{1 + \sec x} = \dfrac{\sec^2 x}{2 + \sec x}$ **40.** $\dfrac{\sin x}{\sin x - 1} = \dfrac{\sin^2 x}{\sin x - 3}$

41. Physics Two friction wheels rotate in contact with each other. In the figure, points A and B are on unit circles. The elevation of A is given by the expression $6 + \sin\theta$, and the elevation of B is given by $6 + \sin 2\theta$, where θ is the angle of rotation of the larger wheel from the initial position shown. Find the values of θ, where $0 \leq \theta < 2\pi$, for which the elevations of the two points are equal.

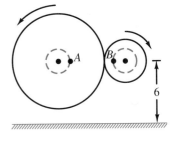

42. Physics When the weight shown is pulled down a distance of 1 in. from its rest position and released, it oscillates a distance of 2 in. between its lowest and highest position. If $t = 0$ at the lowest position, then

$$y = 1 - \cos\left(\frac{\pi t}{2}\right)$$

where y is the distance of the weight from its lowest position and t is the number of seconds elapsed. If $0 \leq t < 4$, find t when $y = 1.5$.

CHALLENGE

The path of a ball kicked from rest can be described by the function

$$f(x) = (\tan\theta)x - \frac{16x^2}{v^2\cos^2\theta}$$

where θ is the angle at which the ball is kicked;
v is the initial velocity in ft/s; and
x and $f(x)$ are the ball's horizontal and vertical locations along its path.

Suppose the ball is kicked with an initial velocity of 8 ft/s.

1. How far does the ball travel if the angle is 30°? 45°? 60°?

2. What value of θ yields the greatest total horizontal distance?

3. Write an expression to represent the ball's horizontal distance (d).

4. The vertex of the parabola is the point of maximum height. Write an expression for the maximum height of the ball.

5. Research field goals in football. Determine an initial velocity and angle that would achieve a successful field goal.

6.5 Trigonometric Equations: Approximate Solutions

Objective: To find approximate values of solutions to trigonometric equations

 A scientific calculator can be used to approximate the values of solutions to trigonometric equations.

Preview

EXAMPLE Find, to the nearest tenth of a degree, all values of x such that $0° \leq x < 360°$ and sin x = 0.4826.

sin x = 0.4826
  x = 28.9°
  and
  x = 180° − 28.9° = 151.1°

Use the Sin^{-1} key or the inv *and* sin *keys to obtain the principal value. Since sin x is positive, there is also a value of x between 90° and 180°.*

Find, to the nearest degree, all values of x such that $0° \leq x < 360°$ and the given statement is true.

1. cos x = 0.8823   **2.** cos x = −0.3623   **3.** tan x = 2.544

4. sin x = −0.3410   **5.** sin x = 0.9672   **6.** tan x = −0.2547

7. sec x = 2.398   **8.** csc x = −1.253   **9.** cot x = −3.265

The equations considered in the preceding lesson had exact solutions. However, solutions to most trigonometric equations must be represented as approximations.

EXAMPLE 1 Find, to the nearest tenth of a degree, the solutions to $8 \cos x + 3 = 5 \cos x + 5$, where $0° \leq x < 360°$.

$$8 \cos x + 3 = 5 \cos x + 5$$
$$3 \cos x = 2$$
$$\cos x = \frac{2}{3}$$

  x = 48.2°
  and
  x = 360° − 48.2°
   = 311.8°

Use the Cos^{-1} key or the inv *and* cos *keys to find the principal value. Since cos x is positive, there is also a value of x between 270° and 360°.*

Check to show that 48.2° and 311.8° are approximate solutions.

EXAMPLE 2   **Find, to the nearest tenth of a degree, the solutions to**
$12 \sin^2 x + 7 \sin x + 1 = 0$, where $0° \le x < 360°$.

$$12 \sin^2 x + 7 \sin x + 1 = 0$$
$$(4 \sin x + 1)(3 \sin x + 1) = 0$$

$$4 \sin x + 1 = 0 \qquad \text{or} \qquad 3 \sin x + 1 = 0$$

$$\sin x = -\frac{1}{4} \quad \text{or} \qquad \sin x = -\frac{1}{3}$$

$$x = -14.5° \qquad\qquad x = -19.5° \quad \textit{Principal values}$$

These values are not solutions over the interval $0° \le x < 360°$. However,

$x = 180° - (-14.5°) = 194.5°$ *Since $\sin x = -\frac{1}{4}$ or $\sin x = -\frac{1}{3}$,*
$x = 180° - (-19.5°) = 199.5°$ *x is between 180° and 360°.*
$x = 360° + (-19.5°) = 340.5°$
$x = 360° + (-14.5°) = 345.5°$

Check that $194.5°$, $199.5°$, $340.5°$, and $345.5°$ are approximate solutions.

If an equation cannot be solved by factoring, the quadratic formula can sometimes be used. The solutions to $ax^2 + bx + c = 0$ are given by

$$x = \frac{-b \pm \sqrt{b^2 - 4ac}}{2a}, a \ne 0$$

EXAMPLE 3   **Find, to the nearest tenth of a degree, the solutions to**
$3 \tan^2 x + 5 \tan x - 1 = 0$, where $0° \le x < 360°$.

$$3 \tan^2 x + 5 \tan x - 1 = 0$$

$$\tan x = \frac{-5 \pm \sqrt{5^2 - 4(3)(-1)}}{2(3)} \qquad \textit{Use the quadratic formula.}$$

$$\tan x = \frac{-5 + \sqrt{37}}{6} \quad \text{or} \quad \tan x = \frac{-5 - \sqrt{37}}{6}$$

$$\tan x = 0.1805 \quad \text{or} \quad \tan x = -1.847$$

$$x = 10.2° \qquad \text{or} \qquad x = -61.6° \quad \textit{Principal values}$$

The principal value $10.2°$ is between $0°$ and $90°$. There is a value of x between $180°$ and $270°$ where $\tan x$ is also positive.

$$x = 180° + 10.2° = 190.2°$$

The principal value $-61.6°$ is not a solution, since it is not between $0°$ and $360°$, but there are two values of x, one between $90°$ and $180°$, the other between $270°$ and $360°$, where $\tan x$ is also negative.

$$x = 180° + (-61.6°) = 118.4°$$
$$x = 360° + (-61.6°) = 298.4°$$

Check that $10.2°$, $118.4°$, $190.2°$, and $298.4°$ are approximate solutions.

A graphing calculator or a computer can be used to determine whether solutions to a trigonometric equation are approximately correct. To illustrate, the solutions to the equation in Example 3 can be checked by graphing $y = 3 \tan^2 x + 5 \tan x - 1$ over the interval $0° \le x < 360°$. The points where the graph intersects the x-axis are the points where $3 \tan^2 x + 5 \tan x - 1 = 0$. Therefore, the x-coordinates of those points are the solutions to the equation.

x scl: $\frac{\pi}{2}$ y scl:1

CLASS EXERCISES

Find, to the nearest degree, the solutions to each equation, where $0° \le x < 360°$.

1. $7 \sin x + 8 = \sin x + 10$

2. $6 \cos x - 5 = 9 \cos x - 4$

3. $8 \tan x + 15 = 3 \tan x + 11$

4. $7 \cot x + 2 = \cot x + 8$

5. $3 \sec x - 7 = \sec x + 2$

6. $13 \csc x + 25 = 5 \csc x + 7$

7. $8 \cos 2x + 3 = 2 \cos 2x - 1$

8. $10 \sin 2x + 5 = 5 \sin 2x + 8$

PRACTICE EXERCISES   Use appropriate technology.

Find, to the nearest degree, the solutions to each equation, where $0° \le x < 360°$. Verify the solutions by graphing.

1. $8 \sin x - 1 = 2 \sin x + 3$

2. $18 \cos x - 58 = 31 \cos x - 54$

3. $16 \cos x + 29 = 7 \cos x + 22$

4. $73 \sin x - 64 = 29 \sin x - 51$

5. $25 \cos^2 x - 1 = 0$

6. $36 \sin^2 x - 1 = 0$

7. $6 \sin^2 x - 5 \sin x - 1 = 0$

8. $12 \cos^2 x - 8 \cos x + 1 = 0$

9. $12 \cos^2 x + 7 \cos x + 1 = 0$

10. $8 \sin^2 x - 2 \sin x - 1 = 0$

11. $8 \sin^2 x + 5 \sin x - 1 = 0$

12. $9 \sin^2 x + 7 \sin x - 2 = 0$

13. $12 \cos x - 11 \cos x + 2 = 0$

14. $7 \sin^2 x - 4 \sin x - 1 = 0$

15. $\sec^2 x - 12 \sec x + 2 = 0$

16. $\csc^2 x - 15 \csc x + 1 = 0$

17. $\csc^2 x + 15 \csc x + 8 = 0$

18. $\sec^2 x + 18 \sec x + 9 = 0$

19. $5 \sin 2x + 2 = 8 \sin 2x + 1$

20. $8 \cos 2x + 7 = \cos 2x + 4$

21. $10 \cos 2x - 19 = 6 \cos 2x - 18$

22. $22 \sin 2x - 5 = 10 \sin 2x + 3$

23. $8 \tan 2x - 9 = 7 \tan 2x + 13$

24. $12 \cot 2x - 7 = 2 \cot 2x - 10$

25. $35 \sin^2 x - 20 \sin x + 1 = 0$

26. $24 \cos^2 x + 11 \cos x + 1 = 0$

Find, to the nearest hundredth of a radian, the solutions to each equation, where $0 \le x < 2\pi$.

27. $\sec^2 x + 7 \sec x + 10 = 0$

28. $\csc^2 x + 10 \csc x + 16 = 0$

29. $\csc^2 x - 9 \csc x - 10 = 0$

30. $\sec^2 x - 14 \sec x - 15 = 0$

31. $8 \tan^2 11x - 17 \tan 11x + 3 = 0$

32. $12 \cot^2 13x - 19 \cot 13x + 1 = 0$

33. $5 \cos^2 (3x + 1) + 2 \cos (3x + 1) - 1 = 0$

34. $7 \sin^2 (2x + 3) + \sin (2x + 3) - 2 = 0$

35. $\tan^2 (7x - 5) + 6 \tan (7x - 5) + 10 = 0$

36. $\cot^2 (5x + 8) - 7 \cot (5x + 8) - 12 = 0$

Applications

37. Physics A pressure wave is transmitted down a long tube, and the pressure p is given by the equation

$$p = 2 \cos [2\pi(x - 0.5t)]$$

where x is measured in meters and t in seconds. A pressure sensor is located 1 m from the opening of the tube. If the wave originates when $x = 0$ and $t = 0$, find the first value of t for which the sensor reads 0.5.

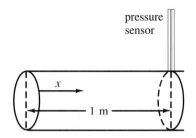

38. Physics Point A rotates on a shaft, and point B is on a slider that moves horizontally. If x is the distance moved by the horizontal slider when the wheel rotates through an angle θ, it can be shown that

$$x = 1.54 (1 - \cos \theta) + 2 \sin^2 \theta$$

Find θ, $0° \le \theta \le 360°$, if $x = 2$.

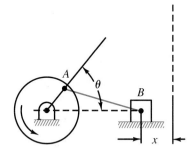

CHALLENGE

Choose any acute angle θ. Use a calculator to find $\tan \theta$, square the result, add 1, and then multiply by $\cos^2 \theta$. Explain the result. Make up another calculator sequence that always gives the same result for any value of acute angle θ.

Rotation of Axes

Objectives: To derive and use formulas for rotations of axes
To graph equations using rotated axes

The graphs of $x^2 - y^2 = 2$ and $y = \dfrac{1}{x}$ can be made to coincide with one another by rotating the plane of one of the graphs about the origin.

$x^2 - y^2 = 2$

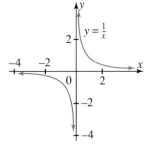

$y = \dfrac{1}{x}$

Preview

1. Copy the graph of $x^2 - y^2 = 2$ on a piece of lightweight paper, and place the copy over the graph of $y = \dfrac{1}{x}$ shown above so that the origins of the coordinate planes coincide. Through what angle must you rotate the graph of $x^2 - y^2 = 2$ in order to make it coincide with the graph of $y = \dfrac{1}{x}$?

2. Draw the graphs of the equations $\dfrac{x^2}{9} + \dfrac{y^2}{4} = 1$ and $\dfrac{x^2}{4} + \dfrac{y^2}{9} = 1$. Through what angle must you rotate the first graph in order to make it coincide with the second graph?

When the xy-coordinate axes are rotated about the origin through an angle θ, this motion is called a **rotation of axes** and θ is called the **angle of rotation.** The new axes are labeled the x'- and y'-axes, and each point P on the plane has two pairs of coordinates, (x, y) and (x', y'). The sum identities for sine and cosine can be used to show how the xy- and $x'y'$-coordinates of a point P are related.

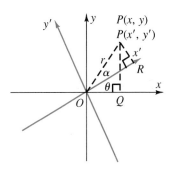

$$\cos(\theta + \alpha) = \frac{x}{r}$$

$$\cos(\theta + \alpha) = \frac{OQ}{OP}$$

$$x = r\cos(\theta + \alpha)$$
$$x = r(\cos\theta\cos\alpha - \sin\theta\sin\alpha) \qquad \textit{Sum identity for cosine}$$
$$x = r\cos\theta\cos\alpha - r\sin\theta\sin\alpha$$
$$x = r\left(\frac{x'}{r}\right)\cos\theta - r\left(\frac{y'}{r}\right)\sin\theta \qquad \cos\alpha = \frac{OR}{OP} = \frac{x'}{r};\ \sin\alpha = \frac{PR}{OP} = \frac{y'}{r}$$
$$x = x'\cos\theta - y'\sin\theta$$

Similarly

$$\sin(\theta + \alpha) = \frac{y}{r}$$

$$\sin(\theta + \alpha) = \frac{PQ}{OP}$$

$$y = r\sin(\theta + \alpha)$$
$$y = r(\sin\theta\cos\alpha + \cos\theta\sin\alpha) \qquad \textit{Sum identity for sine}$$
$$y = r\sin\theta\cos\alpha + r\cos\theta\sin\alpha$$
$$y = r\left(\frac{x'}{r}\right)\sin\theta + r\left(\frac{y'}{r}\right)\cos\theta \qquad \cos\alpha = \frac{OR}{OP} = \frac{x'}{r};\ \sin\alpha = \frac{PR}{OP} = \frac{y'}{r}$$
$$y = x'\sin\theta + y'\cos\theta$$

This system of equations

$$\begin{cases} x = x'\cos\theta - y'\sin\theta \\ y = x'\sin\theta + y'\cos\theta \end{cases}$$

can be solved to find x' and y' in terms of x, y, and functions of θ

$$x' = x\cos\theta + y\sin\theta$$
$$y' = -x\sin\theta + y\cos\theta$$

You will be asked to derive the above two formulas in Exercise 28.

The second pair of equations above are used to find the new coordinates of a point after the xy-plane has been rotated a given number of degrees.

EXAMPLE 1   **If the coordinate axes are rotated 30° about the origin, find the new coordinates of the point for which the old coordinates are (4, 8).**

$$x' = 4\cos 30° + 8\sin 30° \qquad x' = x\cos\theta + y\sin\theta$$
$$= 4\left(\frac{\sqrt{3}}{2}\right) + 8\left(\frac{1}{2}\right)$$
$$= 2\sqrt{3} + 4$$

$$y' = -4\sin 30° + 8\cos 30° \qquad y' = -x\sin\theta + y\cos\theta$$
$$= -4\left(\frac{1}{2}\right) + 8\left(\frac{\sqrt{3}}{2}\right)$$
$$= -2 + 4\sqrt{3}$$

The new coordinates are $(2\sqrt{3} + 4, -2 + 4\sqrt{3})$.

If an equation in x and y is given, along with an angle of rotation, an equation in terms of x' and y' can be written.

EXAMPLE 2   **A curve with equation $xy = 2$ is rotated about the origin through an angle of 45°. Find the equation of the curve in the new coordinate system.**

$$xy = 2$$

$$(x' \cos 45° - y' \sin 45°)(x' \sin 45° + y' \cos 45°) = 2$$

$$\left[x'\left(\frac{\sqrt{2}}{2}\right) - y'\left(\frac{\sqrt{2}}{2}\right) \right]\left[x'\left(\frac{\sqrt{2}}{2}\right) + y'\left(\frac{\sqrt{2}}{2}\right) \right] = 2$$

$$\frac{(x')^2}{2} - \frac{(y')^2}{2} = 2$$

$$\frac{(x')^2}{4} - \frac{(y')^2}{4} = 1$$

$x = x' \cos\theta - y' \sin\theta;$
$y = x' \sin\theta + y' \cos\theta$

The choice of $\theta = 45°$ in Example 2 transformed the equation into one that can be readily recognized as that of a hyperbola. It is not usually obvious what angle of rotation should be chosen to transform an equation into one with a more familiar form. However, it is frequently helpful to eliminate an xy-term. It can be shown that an equation of the form $Ax^2 + Bxy + Cy^2 + Dx + Ey + F = 0$, where $B \neq 0$, can be transformed into an equation without an xy-term if the axes are rotated through an angle such that

$$\text{Cot } 2\theta = \frac{A - C}{B} \qquad \text{or} \qquad \theta = \frac{1}{2}\text{Cot}^{-1}\frac{A - C}{B}$$

EXAMPLE 3   **Determine the angle of rotation needed to transform the equation $x^2 + xy + y^2 = 3$ into an equation without an xy-term. Find the new equation and draw its graph, showing both sets of axes.**

In the equation $x^2 + xy + y^2 = 3$, $A = 1$, $B = 1$, and $C = 1$.

Therefore, $\quad \theta = \dfrac{1}{2}\text{Cot}^{-1}\dfrac{1 - 1}{1} \qquad$ *Substitute in $\theta = \frac{1}{2}\text{Arccot}\frac{A - C}{B}$.*

$$= \frac{1}{2}\text{Cot}^{-1} 0$$

$$= \frac{1}{2}(90°) = 45°$$

Calculate the values of x and y in terms of x' and y'.

$$x = x' \cos 45° - y' \sin 45° \qquad x = x' \cos\theta - y' \sin\theta$$

$$= x'\left(\frac{\sqrt{2}}{2}\right) - y'\left(\frac{\sqrt{2}}{2}\right)$$

$$y = x' \sin 45° + y' \cos 45° \qquad y = x' \sin\theta + y' \cos\theta$$

$$= x'\left(\frac{\sqrt{2}}{2}\right) + y'\left(\frac{\sqrt{2}}{2}\right)$$

Then substitute these expressions for x and y in the original equation. After the left side of the equation is simplified, the following equation is obtained.

$$\frac{(x')^2}{2} + \frac{(y')^2}{6} = 1$$

The new equation is clearly that of an ellipse. It is easier to recognize and to sketch the new equation than the original equation. Rotations preserve all distances and angles, so they also preserve the shapes of geometric figures. If you wish to study the geometric properties of the graph with the equation $x^2 + xy + y^2 = 3$, you may do so by studying the graph of $\frac{x^2}{2} + \frac{y^2}{6} = 1$, which is easier to sketch.

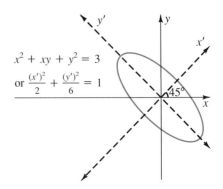

$x^2 + xy + y^2 = 3$

or $\frac{(x')^2}{2} + \frac{(y')^2}{6} = 1$

CLASS EXERCISES

Describe the shape of the graph of each equation.

1. $\dfrac{x^2}{9} - \dfrac{y^2}{25} = 4$ **2.** $\dfrac{x^2}{4} + \dfrac{y^2}{9} = 1$ **3.** $\dfrac{y^2}{16} - \dfrac{x^2}{9} = 1$ **4.** $x^2 - y = 3$

Find the angle of rotation needed to transform each equation into one without an xy-term.

5. $3x^2 + 5xy + 3y^2 = 19$ **6.** $5x^2 - xy + 4y^2 = 8$

7. $10x^2 - 12xy - 2y^2 + 3y = 23$ **8.** $x^2 - xy + 3y = 9$

PRACTICE EXERCISES

Find the new coordinates of each point for the given angle of rotation.

1. $(5, -3)$, $45°$ **2.** $(6, 9)$, $30°$ **3.** $(8, -5)$, $60°$ **4.** $(3, 7)$, $45°$

5. $(-6, -7)$, $30°$ **6.** $(-7, -10)$, $60°$ **7.** $(-3, 1)$, $135°$ **8.** $(11, -4)$, $120°$

The xy-axes are rotated through the given angle θ. Find an equation of the curve in the new coordinate system.

9. $x^2 + 4xy + y^2 = 4$; $\theta = 45°$ **10.** $2x^2 - 3xy + 2y^2 = 8$; $\theta = 45°$

11. $2x^2 + 3\sqrt{3}xy - y^2 = 35$; $\theta = 30°$ **12.** $x^2 - 2\sqrt{3}xy - y^2 = 32$; $\theta = 60°$

13. $5x^2 + 6xy + 5y^2 = 8$; $\theta = 45°$ **14.** $3x^2 + 4xy + 3y^2 = 12$; $\theta = 45°$

Find the angle of rotation needed to transform the given equation into one without an *xy*-term. Find the new equation and draw its graph, showing both sets of axes.

15. $5x^2 + xy + 5y^2 = 4$

16. $xy = 10$

17. $xy = -6$

18. $x^2 + 5xy + y^2 = 10$

19. $5x^2 - 6xy + 5y^2 = 30$

20. $x^2 - \sqrt{3}xy = 5$

21. $\sqrt{3}xy - y^2 = 9$

22. $19xy = -73$

23. $6x^2 - xy + 7y^2 = 4$

24. $x^2 + 2\sqrt{3}xy - y^2 = 8$

25. $x^2 + 3xy - 5y^2 = 8$

26. $14x^2 - 9xy - 2y^2 = 18$

27. If only even powers of x and y appear in an equation, and the axes are rotated 180°, what can be said about the new equation?

28. Derive the formulas $x' = x \cos \theta + y \sin \theta$ and $y' = -x \sin \theta + y \cos \theta$ by solving the system of equations $\begin{cases} x = x' \cos \theta - y' \sin \theta \\ y = x' \sin \theta + y' \cos \theta. \end{cases}$

Applications

29. Engineering The equation of the shape of an elliptical cam is $13x^2 + 6\sqrt{3}xy + 7y^2 = 32$. Determine the angle of rotation needed to eliminate the *xy*-term from the equation. Then sketch the shape of the cam, showing both the *xy*- and the *x′y′*-axes.

30. Geometry The graph of the equation $x \cos \alpha + y \sin \alpha = d$ is a line. Show that the perpendicular distance from the origin to this line is $|d|$ by rotating the axes through the angle α. *Hint*: Prove $x' = d$.

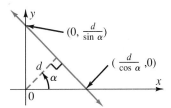

TEST YOURSELF

Find the exact solutions to each equation in radians, for $0 \le x < 2\pi$.

1. $7 \tan x = 5 \tan x + 2$ **2.** $\sin^2 x + \sin x = 0$ **3.** $\cos^2 x + \cos x = 2$ 6.4

Approximate the solutions to each equation in degrees, for $0° \le x < 360°$.

4. $7 \sin x + 2 = 3 \sin x + 5$ **5.** $5 \tan^2 x - 4 \tan x - 1 = 0$ 6.5

6. Find the new coordinates of the point $(4, -7)$ when the axes are rotated through a 30° angle.   6.6

7. Determine the angle of rotation needed to transform $5x^2 + \sqrt{3}xy + 4y^2 + 3y = 9$ into an equation without an *xy*-term.

APPLICATION:
Rainbow

Did you know that the natural phenomenon known as a *rainbow* can only be observed when the sun is low in the sky and behind you and the rain clouds are ahead of you? Scientists and philosophers have been theorizing about rainbows since the time of Aristotle, about 2500 years ago.

Rainbows are caused by the reflection and refraction of the sun's rays striking drops of water. When an incident light ray passes at an angle from one medium (such as air) into another medium (such as water), part of the ray is *reflected* away from the boundary at an angle equal to the angle of incidence. The remainder of the ray enters the new medium, bending away from its original path. This bending, called *refraction*, arises because light rays travel more slowly in water than in a vacuum. For a ray of pure red light

$$\frac{\text{velocity in a vacuum}}{\text{velocity in water}} = 1.332$$

Thus, red light travels about 1.3 times as fast in a vacuum as in water. The ratio

$$\frac{\text{velocity in a vacuum}}{\text{velocity in medium}}$$

is called the *index of refraction* of the medium. The velocity of light in air and in a vacuum are virtually the same, so the index of refraction of air is one.

The path of the refracted ray depends upon both the angle of incidence and the indices of refraction of the two media involved. Willebrord Snell experimentally formulated this relationship as *Snell's Law* in about 1621.

$$n_1 \sin \theta_i = n_2 \sin \theta_r$$

where n_1 and n_2 are the indices of refraction of the first medium and the second medium, respectively, θ_i is the angle of incidence, and θ_r is the angle of refraction.

A raindrop acts like a tiny prism, reflecting back part of the sun's light and refracting the remainder. An incident ray enters the drop, rebounding (in part) inside the drop, and emerges from the drop, usually in an entirely different direction from which it began.

EXAMPLE The index of refraction of water for red light is 1.332. Use Snell's Law to find θ_{r_2}, the angle that refracted ray r_2 makes as it emerges from the raindrop shown below. The angle of incidence of ray i_1 is 30°.

$$n_1 \sin \theta_i = n_2 \sin \theta_r$$
$$1 \sin 30° = 1.332 \sin \theta_{r_1}$$

$$\frac{\sin 30°}{1.332} = \sin \theta_{r_1}$$
$$0.3754 = \sin \theta_{r_1}$$
$$\sin^{-1} 0.3754 = 22.05°$$

Thus, $\theta_{r_1} = 22.05°$.

It can be shown that θ_{r_1} equals θ_{i_2}
Therefore, $\theta_{i_2} = 22.05°$ also.

Use Snell's Law again to find θ_{r_2}

$$1.332 \sin \theta_{i_2} = 1 \sin \theta_{r_2}$$
$$1.332 \sin 22.05° = \sin \theta_{r_2}$$
$$0.5 = \sin \theta_{r_2}$$
$$\theta_{r_2} = \sin^{-1} 0.5 = 30°$$

Refracted ray r_2 emerges from the raindrop at an angle of 30°.

The index of refraction of water for violet light is 1.344. So, if the refracted ray in the example had been a ray of violet light rather than red, the angle of refraction θ_{r_1} would have been a bit smaller. This would have caused r_2 to emerge from the raindrop in a slightly different direction from that of the red light ray. By sending red and violet rays into separate directions, the raindrop *disperses* the two colors. In a similar fashion, white light (a mixture of *all* colors) is dispersed by a raindrop into a full spectrum of colors. This dispersion of white light by raindrops, magnified many many times, results in the perception of a rainbow.

EXERCISES

1. An incident ray of violet light strikes a raindrop at an angle of 35°. Find the angle of refraction θ_{r_1}.

2. In the example, θ_{i_1} and θ_{r_2} are both 30°. Is this a coincidence? Explain.

3. Suppose that $\theta_{r_2} = 21°$. Find the angle of incidence θ_{i_1} if the color of the light is red.

4. Use your knowledge of geometry or trigonometry to explain why $\theta_{r_1} = \theta_{i_2}$.

Vocabulary

angle of rotation (273)
inverse Cosecant (261)
inverse Cosine (256)
inverse Cotangent (260)
inverse function (251)

inverse Secant (261)
inverse Sine (255)
inverse Tangent (260)
principal value (255)
rotation of axes (273)

Inverse Relations and Functions   The inverse of $y = \sin x$ is
$y = \sin^{-1} x$, and the inverse of $y = \cos x$ is $y = \cos^{-1} x$.   6.1

1. Find the inverse of the function $y = x - \frac{3}{4}$.

Evaluate each expression in radians.

2. arccos 0.5

3. arcsin 1

The Inverse Sine and Cosine Functions   The appropriately restricted sine   6.2
and cosine functions, $y = \text{Sin } x$ and $y = \text{Cos } x$, have inverses that are also
functions. The inverse functions are denoted $y = \text{Sin}^{-1} x$ (or $y = \text{Arcsin } x$) and
$y = \text{Cos}^{-1} x$ (or $y = \text{Arccos } x$).

$y = \text{Sin}^{-1} x$ domain: $-1 \le x \le 1$ range: $-\frac{\pi}{2} \le y \le \frac{\pi}{2}$

$y = \text{Cos}^{-1} x$ domain: $-1 \le x \le 1$ range: $0 \le y \le \pi$

4. Find the exact value of $\text{Cos}^{-1}\left(-\frac{\sqrt{2}}{2}\right)$ in radians.

5. Find the exact value of cos (Arcsin 0.5).

6. Find the approximate value of $\text{Sin}^{-1} 0.4765$ to the nearest degree.

Other Inverse Trigonometric Functions   If the domains of the tangent,   6.3
cotangent, secant, and cosecant functions are appropriately restricted, their
inverses are also functions.

$y = \text{Tan}^{-1} x$ domain: all real numbers   range: $-\frac{\pi}{2} < y < \frac{\pi}{2}$

$y = \text{Cot}^{-1} x$ domain: all real numbers   range: $0 < y < \pi$

$y = \text{Sec}^{-1} x$ domain: $|x| \ge 1$ range: $0 \le y \le \pi, y \ne \frac{\pi}{2}$

$y = \text{Csc}^{-1} x$ domain: $|x| \ge 1$ range: $-\frac{\pi}{2} \le y \le \frac{\pi}{2}, y \ne 0$

7. Find the exact value of $\text{Tan}^{-1}(-1)$ in radians.

8. Find the approximate value of $\text{Arccsc}\,(-4.321)$ to the nearest degree.

9. Find the exact value of $\tan\,[\text{Arcsec}\,(-2)]$.

6.4
Solving Trigonometric Equations: Using Special Angles   Many
trigonometric equations can be solved using algebraic properties and known identities.

Find the exact solutions to each equation for the interval $0 \le x < 2\pi$.

10. $12 \sec x + 15 = 10 \sec x + 19$ **11.** $2 \sin^2 x + \sin x - 1 = 0$

12. $2 \csc^2 x + 3 \csc x - 2 = 0$ **13.** $\sin 2x + \cos x = 2 \cos x$

6.5
Trigonometric Equations: Approximate Solutions   When the solutions
to a trigonometric equation are not exact, a calculator can be used.

Find, to the nearest degree, the solutions to each equation for the interval $0° \le x < 360°$.

14. $8 \cos x + 9 = 2 \cos x + 11$ **15.** $5 \cos^2 x - \cos x - 1 = 0$

6.6
Rotations of Axes   If the x- and y-axes are rotated through an angle θ, a new
$x'y'$-coordinate system is obtained. The coordinates (x, y) and (x', y') of a point are related by the following two systems of equations:

$$x = x' \cos \theta - y' \sin \theta \qquad x' = x \cos \theta + y \sin \theta$$
$$y = x' \sin \theta + y' \cos \theta \qquad y' = -x \sin \theta + y \cos \theta$$

An equation of the form $Ax^2 + Bxy + Cy^2 + Dx + Ey + F = 0$, where $B \ne 0$, can be transformed into an equation without an xy-term if the axes are rotated through an angle $\theta = \dfrac{1}{2} \text{Cot}^{-1} \dfrac{A - C}{B}$.

Find the new coordinates of each point for the given angle of rotation.

16. $(22, -17), 45°$ **17.** $(15, 19), 120°$

The xy-axes are rotated through the given angle θ. Find an equation of the curve in the new coordinate system.

18. $5x^2 + 3xy - 6y^2 = 11; \theta = 90°$ **19.** $4x^2 + 5xy - 2y^2 = 9; \theta = 90°$

Find the angle of rotation needed to transform the given equation into one without an xy-term. Find the new equation and draw its graph, showing both the xy- and $x'y'$-axes.

20. $x^2 + xy + y^2 = 6$ **21.** $xy = 8$

1. Find the inverse of the function $y = 10x - 7$.

2. Evaluate $\arcsin\left(-\dfrac{\sqrt{2}}{2}\right)$ in degrees.

3. Evaluate $\arccos\dfrac{\sqrt{3}}{2}$ in radians.

4. Find the exact value of $\operatorname{Sin}^{-1} 1$ in radians.

5. Find the exact value of $\operatorname{Tan}^{-1}(-1)$ in degrees.

6. Find the exact value of $\csc[\operatorname{Arccsc}(-2)]$.

7. Find the exact value of $\sin\left[\operatorname{Arccos}\left(-\dfrac{1}{2}\right)\right]$.

Find the approximate value of each expression to the nearest degree.

8. Arcsin 0.6193

9. $\operatorname{Tan}^{-1}(-4.562)$

Find the exact solutions to each equation in radians, for the interval $0 \leq x < 2\pi$.

10. $13 \tan x - 9 = 11 \tan x - 11$

11. $2 \cos^2 x - \cos x - 1 = 0$

Find, to the nearest degree, the solutions to each equation for the interval $0° \leq x < 360°$.

12. $7 \sin x - 9 = 2 \sin x - 7$

13. $3 \tan^2 x - 4 \tan x - 7 = 0$

14. If the x- and y-axes are rotated about the origin through an angle of 120°, what are the coordinates of the point $(-5, 7)$ in the new coordinate system?

15. Determine the angle of rotation that will transform $x^2 + 5xy + y^2 = 7$ into an equation without an xy-term. Find the new equation and draw its graph, showing both the xy- and $x'y'$-axes.

CHALLENGE

If $-1 \leq x \leq 1$, which of the following expressions is equal to $\sin(2 \operatorname{Arcsin} x)$?

a. $2\sqrt{1 - x^2}$ **b.** $2x(1 - x^2)$ **c.** $2x\sqrt{1 - x^2}$ **d.** $2x$

In each item you are to compare a quantity in Column 1 with a quantity in Column 2. Write the letter of the correct answer from these choices:

A. The quantity in Column 1 is greater than the quantity in Column 2.
B. The quantity in Column 2 is greater than the quantity in Column 1.
C. The quantity in Column 1 is equal to the quantity in Column 2.
D. The relationship cannot be determined from the information given.

Notes: Information centered over both columns refers to one or both of the quantities being compared. A symbol that appears in both columns has the same meaning in each column. All variables represent real numbers. Most figures are not drawn to scale.

	Column 1	Column 2
1.	$x^2 + 2x - 5$	6

$$\sin^2 x - 1 = 0,\ 0° \le x < 360°$$

	Column 1	Column 2
2.	x	45°

$$y = \text{Arcsin}\left(-\tfrac{1}{2}\right)$$

3.	y	$-\dfrac{\pi}{6}$

$$y = \sin\left(\text{Arcsin}\ \tfrac{2}{5}\right)$$

4.	y	$\dfrac{1}{5}$

$$5x - 2y = 8$$

5.	The slope of the line	3
6.	$\dfrac{9x^2 - 15x}{3x}$	$3x - 5$
7.	$\dfrac{x}{y}$	xy

$$x - y = 7$$
$$2x + y = 5$$

8.	y	x

$O \perp Q$
$N \perp P$

	Column 1	Column 2
9.	y	$\dfrac{x}{2}$

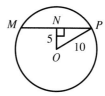

10.	$MN + NP$	$20\sqrt{3}$
11.	The area of a circle with radius 2	The area of a circle with radius $\dfrac{x}{2}$

$$y = 4 \sin 2x$$

12.	The maximum value of the function	The amplitude of the function

$$y = \text{Sin}^{-1}x$$

13.	$\cos y$	$\sqrt{1 - x^2}$

1. Express $\cos 52° + \cos 88°$ as a product.

Check graphically to see whether or not each equation is an identity.

2. $\cos x \sin 2x - \sin x \cos 2x = \sin x$ **3.** $\tan^2 \alpha + \cot^2 \alpha = \dfrac{\tan \alpha}{\cot \alpha} - \dfrac{\cot \alpha}{\tan \alpha}$

Find the exact solutions to each equation in radians, for the interval $0 \le x \le 2\pi$.

4. $\sqrt{2} \cos x - 1 = 0$ **5.** $\sin^2 x - 5 \sin x = -4$

Prove each identity.

6. $\sin 2\alpha \cot \alpha = 1 + \cos 2\alpha$ **7.** $\sin \dfrac{\theta}{2} \cos \dfrac{\theta}{2} = \dfrac{1}{2} (\sin \theta)$

8. Find the inverse of the function $y = x + 2$. Graph the function and its inverse.

9. Draw an arrow representing the vector from $A(3, 5)$ to $B(8, 17)$ and find its magnitude.

10. Evaluate $\sin (\alpha + \beta)$ if $\cos \alpha = \dfrac{4}{5}$ and $\tan \beta = \dfrac{5}{12}, 0 < \alpha < \dfrac{\pi}{2}, 0 < \beta < \dfrac{\pi}{2}$.

11. Find the exact value of $\sin \left(\text{Arccos} \left(- \dfrac{\sqrt{3}}{2} \right) \right)$.

12. Sketch the graph of the function $y = \cos \left(2x + \dfrac{\pi}{2} \right)$.

Approximate the solutions to each equation, where $0° \le x < 360°$.

13. $2 \sin 3x - 1 = 0$ **14.** $2 \csc^2 x + 10 \csc x + 11 = 0$

15. Find the area, in acres, of a triangular plot of land with sides 475 ft, 628 ft, and 710 ft. (*Note*: 1 acre $= 43,560$ ft^2)

16. Express $\dfrac{\tan 85° + \tan 52°}{1 - \tan 85° \tan 52°}$ in terms of a single tangent.

17. $-\dfrac{7\pi}{12}$ is an angle in standard position. Find the measure of its reference angle θ'.

18. Express $2 \cos \dfrac{\pi}{6} \sin \dfrac{\pi}{12}$ as a sum or difference.

Benoit B. Mandelbrot developed the concept of fractals, which are curves that have sections that are small-scale replicas of the whole. A computer can use complex numbers to generate pictures of fractals.

Polar Coordinates

7.1

Objectives: To define polar coordinates
To convert rectangular coordinates to polar coordinates and polar coordinates to rectangular coordinates

Points on the Cartesian, or rectangular, coordinate plane are represented by ordered pairs of real numbers (x, y).

Preview

The distance d between the origin $(0, 0)$ and any point with *rectangular coordinates* (x, y) is given by the distance formula

$$d = \sqrt{(x - 0)^2 + (y - 0)^2} = \sqrt{x^2 + y^2}$$

Find the distance from the origin to each point.

1. $(5, -9)$ **2.** $(0, 4)$ **3.** $(-1, -1)$ **4.** $(-4, 0)$

5. $(-\sqrt{7}, 1)$ **6.** $(-1, \sqrt{11})$ **7.** $(-\sqrt{2}, \sqrt{3})$ **8.** $(3\sqrt{5}, 2\sqrt{7})$

There is another way to locate points in the plane. In the **polar coordinate system,** the reference system consists of a point O, called the **pole** (or origin), and a ray called the **polar axis,** with O its initial point.

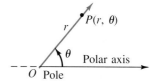

The **polar coordinates** of a point P in the polar coordinate system are an ordered pair (r, θ), where $|r|$ is the distance from the pole to P and θ is the measure of the angle from the polar axis to ray OP. The pole, O, has polar coordinates $(0, \theta)$, where θ is arbitrary. Angle θ can be measured in degrees or in radians.

EXAMPLE 1   **Graph point Q with polar coordinates $(2, 30°)$.**

Sketch a 30° angle in standard position. Use the polar axis as its initial side.

Then locate on the ray the point that is 2 units from the pole.

Both r and θ can be either positive or negative. If r is *negative*, the point (r, θ) is |r| units from the pole on the ray opposite the terminal side of the angle with measure θ. If θ is *negative*, the angle is measured in a clockwise direction from the polar axis. If θ is in standard position, then a point P with coordinates (r, θ) is located as follows, where ray OT makes an angle θ with the polar axis.

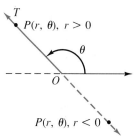

If $r > 0$, P lies on ray OT at a distance r from O.
If $r < 0$, P lies on the ray opposite ray OT at a distance |r| from O.
If $r = 0$, P is the pole O.

In general, $(-r, \theta) = (r, \theta \pm 180°)$, if θ is in degrees, or $(-r, \theta) = (r, \theta \pm \pi)$, if θ is in radians.

EXAMPLE 2   **Graph the point D with polar coordinates $(-3, -60°)$.**

Draw $\theta = -60°$ in standard position. Since r is negative, locate the point |-3| units from the pole on the ray *opposite* the terminal side of the angle. Note that D can also be represented by $(3, -60° + 180°) = (3, 120°)$ or by $(3, -60° - 180°) = (3, -240°)$.

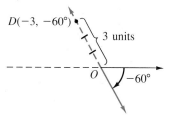

In the rectangular coordinate system, each point is represented by a unique ordered pair. In the polar system, points may be named with an infinite number of pairs of polar coordinates. For example, the polar coordinates $(3, 105°)$, $(-3, -75°)$, $(-3, 285°)$, and $(3, -255°)$ all represent point P.

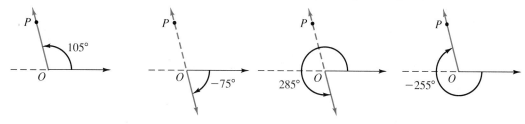

Generally, any point (r, θ) can be represented by $(r, \theta + n \cdot 360°)$ or $(r, \theta + 2\pi n)$, where n is an integer.

EXAMPLE 3   **If $-2\pi \le \theta \le 2\pi$, name four pairs of polar coordinates for point R.**

Point R is represented by

$$\left(5, \frac{3\pi}{4}\right), \left(-5, -\frac{\pi}{4}\right), \left(-5, \frac{7\pi}{4}\right), \text{ or } \left(5, -\frac{5\pi}{4}\right).$$

If a rectangular coordinate system is superimposed upon a polar coordinate system and the units are assumed to be the same, then for any point P, a correspondence can be established between its rectangular coordinates (x, y) and its polar coordinates (r, θ).

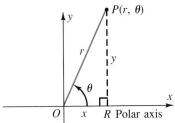

Notice that the two origins coincide. Also, the positive x-axis of the rectangular system coincides with the polar axis of the polar system. In triangle OPR, $\cos \theta = \dfrac{x}{r}$ and $\sin \theta = \dfrac{y}{r}$. Therefore, $x = r \cos \theta$ and $y = r \sin \theta$.

> The rectangular coordinates (x, y) of a point P that has polar coordinates (r, θ) can be found using the following conversion formulas:
> $$x = r \cos \theta \qquad y = r \sin \theta$$

EXAMPLE 4   **Find the rectangular coordinates of the following points:**

    a. $A(4, -50°)$ **b.** $B\left(-5, \dfrac{2\pi}{3}\right)$

a. $x = r \cos \theta$ $y = r \sin \theta$

       $= 4 \cos (-50°)$ $= 4 \sin (-50°)$

       $= 2.57$ *To the nearest hundredth* $= -3.06$

The rectangular coordinates of A are $(2.57, -3.06)$.

b. $x = r \cos \theta$ $y = r \sin \theta$

       $= -5 \cos \dfrac{2\pi}{3} = \dfrac{5}{2}$ $= -5 \sin \dfrac{2\pi}{3} = -\dfrac{5\sqrt{3}}{2}$

The rectangular coordinates of B are $\left(\dfrac{5}{2}, -\dfrac{5\sqrt{3}}{2}\right)$.

When a point is identified with rectangular coordinates (x, y), a pair of polar coordinates (r, θ) for the point can be found using the Pythagorean theorem and the Arctangent function. The principal values of the inverse tangent relation are between $-\dfrac{\pi}{2}$ and $\dfrac{\pi}{2}$, so if the point is in the first or fourth quadrant, $\theta = \text{Arctan } \dfrac{y}{x}$. If the point is in the second or third quadrant, add π radians (or $180°$) to Arctan $\dfrac{y}{x}$.

For $x < 0$, $\theta = \left(\text{Arctan } \dfrac{y}{x}\right) + \pi$

A pair of polar coordinates (r, θ) of a point named by the rectangular coordinates (x, y) can be found using the following formulas:

$$r = \sqrt{x^2 + y^2}$$

$\theta = \text{Arctan }\dfrac{y}{x}$, if $x > 0$

$\theta = \left(\text{Arctan }\dfrac{y}{x}\right) + \pi$, or $\left(\text{Arctan }\dfrac{y}{x}\right) + 180°$, if $x < 0$

$\theta = \dfrac{\pi}{2}$ or $\theta = 90°$, if $x = 0$ and $y > 0$

$\theta = \dfrac{3\pi}{2}$ or $\theta = 270°$, if $x = 0$ and $y < 0$

$\theta = $ any real number, if $x = 0$ and $y = 0$

EXAMPLE 5   **Find polar coordinates of point C with rectangular coordinates $(-2, 2)$.**

$r = \sqrt{(-2)^2 + 2^2}$ $r = \sqrt{x^2 + y^2}$
$\quad = \sqrt{8} = 2\sqrt{2}$

Since $x < 0$, $\theta = \text{Arctan }\left(\dfrac{2}{-2}\right) + \pi$

$\qquad\qquad = \text{Arctan }(-1) + \pi$

$\qquad\qquad = -\dfrac{\pi}{4} + \pi = \dfrac{3\pi}{4}$

Polar coordinates of C are $\left(2\sqrt{2}, \dfrac{3\pi}{4}\right)$, or $(2\sqrt{2}, 135°)$.

As you know, a point represented by a unique pair of rectangular coordinates (x, y) can be represented by an infinite number of pairs of polar coordinates. When the equation $r = \sqrt{x^2 + y^2}$ is used, a *nonnegative* value of r is always obtained. It is possible to represent the same point using $r = -\sqrt{x^2 + y^2}$ and an appropriate value of θ. To illustrate, show that point C in Example 5 could also be represented by the polar coordinates $(-2\sqrt{2}, -45°)$.

You can change equations from rectangular to polar form and the reverse.

EXAMPLE 6   **a. Express $y = 5$ in polar-coordinate form.**
b. Express $r = 3 \cos \theta$ in rectangular form.

a. $\qquad y = 5$
$\qquad r \sin \theta = 5$ *Replace y with r sin θ.*
$\qquad\qquad r = \dfrac{5}{\sin \theta}$

b.
$$r = 3 \cos \theta$$
$$r^2 = 3 r \cos \theta \qquad \textit{Multiply both sides by r.}$$
$$x^2 + y^2 = 3 x \qquad \textit{Replace } r^2 \textit{ with } x^2 + y^2 \textit{ and r cos } \theta \textit{ with x.}$$

CLASS EXERCISES

Estimate polar coordinates for each point T.

1.   **2.**   **3.**   **4.**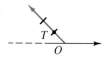

Find the rectangular coordinates of each point.

5. $(3, 90°)$ **6.** $(5, 270°)$ **7.** $(-3, 180°)$ **8.** $(-1, 0°)$

Find polar coordinates for each point. Give θ to the nearest degree.

9. $(5, -5)$ **10.** $(-6, 6)$ **11.** $(3, 4)$ **12.** $(-4, 3)$

PRACTICE EXERCISES

Plot each point with the given polar coordinates.

1. $(3, 45°)$ **2.** $(2, 60°)$ **3.** $(-2, 270°)$ **4.** $(-1, 180°)$

5. $(4, -420°)$ **6.** $(1, -435°)$ **7.** $(0, 300°)$ **8.** $\left(-4, -\dfrac{5\pi}{6}\right)$

9. $\left(-3, -\dfrac{2\pi}{3}\right)$ **10.** $\left(0, \dfrac{7\pi}{4}\right)$ **11.** $\left(3, \dfrac{5\pi}{2}\right)$ **12.** $\left(3, \dfrac{\pi}{2}\right)$

If $-360° \le \theta \le 360°$, give four pairs of polar coordinates for each point P.

13.   **14.**   **15.**   **16.**

Find rectangular coordinates of each point with the given polar coordinates.

17. $(0, 90°)$ **18.** $(0, 180°)$ **19.** $(2, 130°)$

20. $(3, 80°)$ **21.** $(3, -70°)$ **22.** $(2, -115°)$

Find polar coordinates for each point with the given rectangular coordinates.

23. $(0, -1)$ **24.** $(0, 2)$ **25.** $(2, -2)$

26. $(-2, -2)$ **27.** $(-3, \sqrt{3})$ **28.** $(3, -\sqrt{3})$

Write each equation in polar-coordinate form.

29. $x = 4$ **30.** $x^2 + y^2 = 49$ **31.** $x^2 + y^2 = 36$ **32.** $5x - 2y = 1$

Write each equation in rectangular-coordinate form.

33. $r = 10 \sin \theta$ **34.** $r = -6 \cos \theta$ **35.** $r = 6 \csc \theta$ **36.** $r = 3 \sec \theta$

Write each equation in polar-coordinate form.

37. $4x^2 + 4y^2 = -3y$ **38.** $3x^2 + 3y^2 = -7x$

39. $y^2 + (x - 6)^2 = 36$ **40.** $x^2 + (y - 5)^2 = 25$

Write each equation in rectangular-coordinate form.

41. $r = \dfrac{1}{2 + \sin \theta}$ **42.** $r = \dfrac{1}{3 - \cos \theta}$

43. $r = 6 \tan \theta \sec \theta$ **44.** $r = 8 \cot \theta \csc \theta$

Applications

45. Physics The radiation pattern of an antenna is given by the equation $r = 50(1 + \cos \theta)$. Write the equation in rectangular-coordinate form.

46. Physics A satellite has an orbit with the equation $r = \dfrac{200{,}000}{43 - 2 \cos \theta}$. Write the equation in rectangular-coordinate form.

TRIGONOMETRY AND PASCAL'S TRIANGLE

Pascal's triangle, a special triangular array of numbers, appears in algebra, in set theory, and in probability. When tan $(n\theta)$ is expressed in terms of powers of tan θ, Pascals's triangle appears again.

	Coefficients
$\tan \theta \;\; = \dfrac{\mathbf{1} \tan^1 \theta}{\mathbf{1}}$	1 1
$\tan 2\theta = \dfrac{\mathbf{2} \tan^1 \theta}{\mathbf{1} - \mathbf{1} \tan^2 \theta}$	1 2 1
$\tan 3\theta = \dfrac{\mathbf{3} \tan^1 \theta - \mathbf{1} \tan^3 \theta}{\mathbf{1} - \mathbf{3} \tan^2 \theta}$	1 3 3 1
$\tan 4\theta = \dfrac{\mathbf{4} \tan^1 \theta - \mathbf{4} \tan^3 \theta}{\mathbf{1} - \mathbf{6} \tan^2 \theta + \mathbf{1} \tan^4 \theta}$	1 4 6 4 1

Note the following patterns: Odd powers of tan θ appear in the numerator. Even powers of tan θ appear in the denominator. The signs alternate in numerator and denominator. Pascal's triangle is formed by alternately selecting coefficients from the denominator and numerator. For $n = 5$, express tan $(n\theta)$ and the coefficients of Pascal's triangle.

7.2 Graphs of Polar Equations

Objective: To graph polar equations

Graphs of polar equations depict many everyday occurrences, such as the manner in which microphones pick up sound.

Preview

Microphones are designed to pick up sound in a variety of patterns. *Uni*directional microphones pick up sound from one direction, whereas *bi*directional microphones pick up sound from two directions.

The polar-coordinate diagrams below show the pickup patterns of a unidirectional microphone and a bidirectional microphone. In each case, the microphone is placed at the pole. The two parts, or *leaves*, of the pickup pattern in the second diagram are arranged as far away from one another as possible in order to avoid interference.

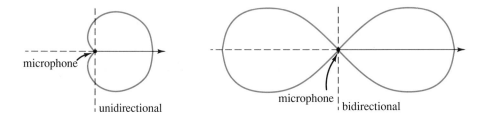

microphone

unidirectional

microphone bidirectional

Draw a pickup pattern for each type of microphone. Assume the element of the microphone is placed at the pole to avoid interference.

1. A *tri*directional microphone

2. A *quadri*directional microphone

3. An *omni*directional microphone (from all directions)

A *polar equation* involves polar coordinates, and a *polar graph* is a graph of the set of all points (r, θ) that satisfy a given polar equation.

EXAMPLE 1 **Graph $\theta = 60°$.**

In the equation $\theta = 60°$, the variable r can take on any real value. Thus, the graph is a straight line through the origin that forms an angle of $60°$ with the polar axis. Note that on this line the points for which $r > 0$ lie in quadrant I, while the points for which $r < 0$ are in quadrant III.

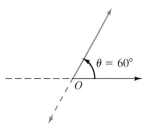

Another familiar graph is shown in Example 2.

EXAMPLE 2 **Graph $r = 4$. Then express the equation $r = 4$ using rectangular coordinates.**

In the equation $r = 4$, the variable θ can have any real value. Thus, the graph is a circle with center $(0,0)$ and radius 4.

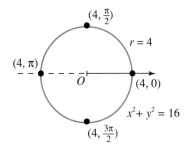

$r = \sqrt{x^2 + y^2}$

$4 = \sqrt{x^2 + y^2}$ *Substitute 4 for r.*

$16 = x^2 + y^2$ *Square each side.*

$x^2 + y^2 = 16$ is the equation of a circle with center $(0, 0)$ and radius 4.

It is helpful to set up a table of values before graphing a polar equation. Let θ range from $0°$ to $360°$ and find corresponding values for r.

EXAMPLE 3 **Graph $r = 4 \cos \theta$. Then express the equation in rectangular form.**

First construct a table of values, $0° \le \theta \le 360°$. Round $4 \cos \theta$ to the nearest tenth. Then graph the points and connect them with a smooth curve.

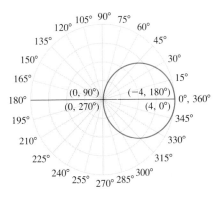

θ	$r = 4 \cos \theta$	θ	$r = 4 \cos \theta$
0°	4	210°	-3.5
30°	3.5	225°	-2.8
45°	2.8	240°	-2
60°	2	270°	0
90°	0	300°	2
120°	-2	315°	2.8
135°	-2.8	330°	3.5
150°	-3.5	360°	4
180°	-4		

Notice that a pattern develops in the table of values for $r = 4 \cos \theta$. The point $(4, 0°)$ is the same as the point $(-4, 180°)$, the point $(3.5, 30°)$ is the same as the point $(-3.5, 210°)$, and so on. The graph appears to be a circle with center at $(2, 0)$. To check, express the equation in rectangular form.

$$r = 4 \cos \theta$$
$$r^2 = 4r \cos \theta \qquad \textit{Multiply each side by r.}$$
$$x^2 + y^2 = 4x \qquad \textit{Substitute } x^2 + y^2 \textit{ for } r^2 \textit{ and x for r cos } \theta.$$
$$x^2 - 4x + y^2 = 0$$
$$x^2 - 4x + 4 + y^2 = 4 \qquad \textit{Complete the square on x.}$$
$$(x - 2)^2 + y^2 = 4$$

This is the equation of a circle with center $(2, 0)$ and radius 2.

In general, the graph of any polar equation of the form $r = 2a \cos \theta$ is a circle with radius a and center $(a, 0)$. Two other types of curves will be studied in this lesson. The first is the **limaçon.** Limaçons are curves represented by polar equations of the form $r = a \pm b \cos \theta$ or $r = a \pm b \sin \theta$.

If $a < b$, the graph has an extra loop.	If $a = b$, the graph is heart-shaped and is called a **cardioid.**	If $a > b$, the graph has no extra loop.

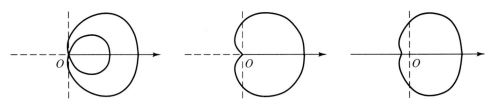

EXAMPLE 4 **Graph $r = 1 + 4 \cos \theta$.**

$$1 + 4 \cos \theta = 0 \qquad \textit{Let r = 0 in 1 + 4 cos } \theta = r.$$
$$\theta = 104.48° \quad \text{or} \quad \theta = 255.52°$$

Then graph the points. Since $1 < 4$ (that is, $a < b$), the graph has an extra loop.

θ	r	θ	r
0°	5	210°	−2.5
30°	4.5	225°	−1.8
45°	3.8	240°	−1
60°	3	256°	0
90°	1	270°	1
104°	0	300°	3
120°	−1	315°	3.8
135°	−1.8	330°	4.5
150°	−2.5	360°	5
180°	−3		

![icon] If you have access to a graphing utility with polar graphing capability, you can graph such an equation simply by entering it in its original form. If your graphing utility is not designed for polar graphs, however, you may still be able to graph the equation on a rectangular screen. By substituting the given value of r into the rectangular conversion formulas you can obtain expressions for x and y.

EXAMPLE 5 **Graph $r = 4 + 4 \cos \theta$.**

Since $x = r \cos \theta$ and $y = r \sin \theta$, substitution yields the equations $x = (4 + 4 \cos \theta) \cos \theta$ and $y = (4 + 4 \cos \theta) \sin \theta$. If you access the parametric mode of your graphing utility, you can enter both equations and graph the result.

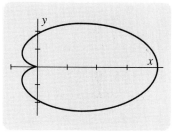

x scl: 2 y scl: 2

Another special type of curve is the **rose.** Rose curves are represented by polar equations of the form $r = a \cos n\theta$ or $r = a \sin n\theta$, where n is a positive integer. If n is odd, the rose has n leaves. If n is even, the rose has $2n$ leaves.

EXAMPLE 6 **Graph $r = 4 \cos 8\theta$.**

Since n is 8, an even number, the rose has $2n$ or 16 leaves.

You may achieve surprising results with your graphing utility by experimenting with different values of a and n in one or both of the parametric equations.

x scl: 2 y scl: 2

CLASS EXERCISES

Match each equation with a description on the right.

1. $r = 3 + 3 \sin \theta$
2. $r = 2$
3. $r = 3 \cos 2\theta$
4. $r = 3 + 2 \sin \theta$
5. $r = 2 - 4 \cos \theta$
6. $r = -2 \sin 2\theta$
7. $r = 3(1 + 2 \sin \theta)$
8. $r = 6(1 + \cos \theta)$

a. Limaçon with an extra loop
b. Limaçon with no extra loop
c. Cardioid
d. Rose
e. Circle

Graph each polar equation.

1. $\theta = 135°$

2. $\theta = -\dfrac{2\pi}{3}$

Graph each polar equation. Then express the equation in rectangular form.

3. $r = 5$ **4.** $r = 3$ **5.** $r = -2$ **6.** $r = -4$

Graph each polar equation. Then express the equation in rectangular form.

7. $r = 6 \cos \theta$ **8.** $r = 8 \cos \theta$ **9.** $r = -8 \cos \theta$

10. $r = -6 \cos \theta$ **11.** $r = 8 \sin \theta$ **12.** $r = 4 \sin \theta$

Graph each equation.

13. $r = 4 + 3 \sin \theta$ (limaçon) **14.** $r = 5 + 2 \sin \theta$ (limaçon)

15. $r = 6 - 4 \cos \theta$ (limaçon) **16.** $r = 7 - 3 \cos \theta$ (limaçon)

17. $r = 4 + 4 \sin \theta$ (cardioid) **18.** $r = 3 + 3 \cos \theta$ (cardioid)

19. $r = 3 - 3 \cos \theta$ (cardioid) **20.** $r = 2 - 2 \sin \theta$ (cardioid)

21. $r = 1 - \cos \theta$ (cardioid) **22.** $r = 1 + \sin \theta$ (cardioid)

23. $r = 6 \sin 2\theta$ (four-leaved rose) **24.** $r = \cos 2\theta$ (four-leaved rose)

25. $r = 4 \cos 3\theta$ (three-leaved rose) **26.** $r = \sin 5\theta$ (five-leaved rose)

27. $r = 5 \cos 4\theta$ (eight-leaved rose) **28.** $r = 2 \sin 4\theta$ (eight-leaved rose)

Another special curve is called the *lemniscate*. Graph each polar equation to find the shape of a lemniscate.

29. $r^2 = \cos 2\theta$

30. $r^2 = \sin 2\theta$

The *Spiral of Archimedes* is also a special curve. Graph each polar equation to find the shape of an Archimedean spiral.

31. $r = \theta$

32. $r = -3\theta$

Applications

33. Physics The orbit of a satellite has the equation $r = \dfrac{2}{2 + \cos \theta}$. Sketch the graph of its orbit. Is the path of the orbit a circle, a parabola, an ellipse, or a hyperbola?

34. Physics   The path of a comet has the equation $r = \dfrac{2}{1 - \sin \theta}$. Sketch the graph of its path. Is the path a circle, a parabola, an ellipse, or a hyperbola?

35. Coordinate Geometry   Find the points of intersection of the graphs of $r = \cos \theta$ and $r = \sin \theta$.

EXTRA

Rotations of Polar Graphs   The equation for a *lemniscate* is of the form $r^2 = a^2 \cos 2\theta$ or $r^2 = a^2 \sin 2\theta$. A lemniscate always has two leaves, each of length a.

The graph below is a lemniscate with the equation $r^2 = 25 \cos 2\theta$. A table of values is shown. Note that the values of r in the table are found by taking the square root of each side of the equation.

$$r^2 = 25 \cos 2\theta$$
$$r = \pm 5\sqrt{\cos 2\theta}$$

θ	$r = \pm 5\sqrt{\cos 2\theta}$
0°	±5
15°	±4.7
30°	±3.5
45°	0
45° < θ < 135°	undefined
135°	0
150°	±3.5
165°	±4.7
180°	±5

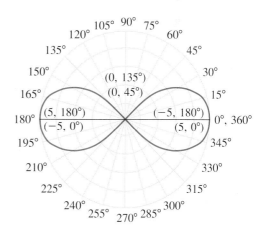

1. Construct a table of values of the equation $r^2 = 25 \cos 2(\theta - 45°)$. Then graph the equation.

2. Through how many degrees would the graph of $r^2 = 25 \cos 2\theta$ (shown above) have to be rotated in a counterclockwise direction in order to obtain the graph you drew for Exercise 1?

3. Construct a table of values for $r^2 = 25 \sin 2\theta$. Then graph the equation. How does this graph compare with that of $r^2 = 25 \cos 2(\theta - 45°)$?

4. To explain your results in Exercise 3, show that $r^2 = 25 \sin 2\theta$ is equivalent to $r^2 = 25 \cos 2(\theta - 45°)$. *Hint*: Start with $r^2 = 25 \sin 2\theta$ and use the fact that $\sin 2\theta = \cos(90° - 2\theta)$.

Sums and Differences of Complex Numbers

Objectives: To introduce the complex numbers

To define addition and subtraction of complex numbers

The solutions to many equations are members of the set of real numbers, which includes the integers, the rational numbers, and the irrational numbers. However, a new set of numbers, called the *complex numbers*, must be introduced before certain types of equations can be solved.

Preview

Determine which of the equations at the right do *not* have a solution that is a member of the given set of numbers.

1. The set of integers

2. The set of rational numbers

3. The set of real numbers

a. $2x - 4 = 3$ b. $x^2 = 4$
c. $3x^2 = 4$ d. $x^2 + 4 = 2$
e. $3x + 4 = 2$ f. $x^2 + 4 = 3$
g. $\dfrac{x^3}{2} + 4 = 0$ h. $x^2 + \dfrac{3}{4} = 1$

The equations $x^2 + 1 = 0$ and $x^2 = -1$ are equivalent. Since the square root of a negative number is not a real number, the equation $x^2 = -1$ has no solution in the real-number system. In order to solve an equation of this type, the **imaginary number** i is defined by the equation

$$i = \sqrt{-1}$$

It follows that $i^2 = -1$. Therefore, the solutions of $x^2 = -1$ are $x = i$ and $x = -i$. Using imaginary numbers, the square root of a negative number can be expressed as the product of a real number and the number i. For example, $\sqrt{-4} = \sqrt{4}\sqrt{-1} = 2\sqrt{-1} = 2i$. For any *positive* real number b,

$$\sqrt{-b} = \sqrt{b}\sqrt{-1} = \sqrt{b}\,i = i\sqrt{b}$$

It is customary to write $i\sqrt{b}$ rather than $\sqrt{b}i$ to avoid confusion with $\sqrt{bi}$.

EXAMPLE 1 **Simplify:** **a.** $\sqrt{-9}$ **b.** $\sqrt{-24}$

a. $\sqrt{-9} = i\sqrt{9}$
$\phantom{\sqrt{-9}} = i(3)$
$\phantom{\sqrt{-9}} = 3i$

b. $\sqrt{-24} = i\sqrt{24}$
$\phantom{\sqrt{-24}} = i\sqrt{4}\sqrt{6}$
$\phantom{\sqrt{-24}} = i(2)\sqrt{6}$
$\phantom{\sqrt{-24}} = 2i\sqrt{6}$

A **complex number** is a number that can be written in the form $a + bi$, where a and b are real numbers and $i = \sqrt{-1}$. The *real* part of a complex number $a + bi$ is a, and the *imaginary* part is bi. A complex number $a + bi$ is *imaginary* if $b \neq 0$, *pure imaginary* if $a = 0$ and $b \neq 0$, and *real* if $b = 0$.

Complex Numbers

Imaginary Numbers	Real Numbers
$2 + 3i$ $6 - i$ $1 - \sqrt{-8}$ **Pure Imaginary Numbers** $-2i\sqrt{5}$	19 -5 $9.\overline{3}$ **Irrational Numbers** $0.212212221...$ 0 $\pi\sqrt{7}$ $\frac{13}{3}$ 10.2

The powers of i follow a cyclic pattern. A few of these powers are listed below.

$$i^1 = i$$
$$i^2 = -1$$
$$i^3 = i^2 \cdot i = -i$$
$$i^4 = i^2 \cdot i^2 = (-1)(-1) = 1$$

$$i^5 = i^4 \cdot i = 1i = i$$
$$i^6 = i^4 \cdot i^2 = 1(-1) = -1$$
$$i^7 = i^4 \cdot i^3 = 1(-i) = -i$$
$$i^8 = i^4 \cdot i^4 = 1(1) = 1$$

Notice that the numbers above repeat in cycles of four. Any higher integral powers of i can be factored into powers of i^4 and a single factor of i, i^2, or i^3. Since $i^4 = 1$, every integral power of i is equal to i, -1, $-i$, or 1.

EXAMPLE 2   **Simplify:**   **a.** i^{18} **b.** i^{223}

     a. $i^{18} = i^{16} \cdot i^2 = (i^4)^4 i^2 = (1)^4(-1) = 1(-1) = -1$

     b. $i^{223} = i^{220} \cdot i^3 = (i^4)^{55} \cdot i^3 = (1)^{55}(-i) = (1)(-i) = -i$

Two complex numbers are equal if and only if their real parts are equal and their imaginary parts are equal. That is,

$$a + bi = c + di \quad \text{if and only if} \quad a = c \text{ and } b = d$$

Complex-number arithmetic has many properties that are similar to real-number arithmetic. For example, the definition of addition for complex numbers uses the commutative, associative, and distributive properties.

> **Addition of Complex Numbers**   For all real numbers a, b, c, and d
> $$(a + bi) + (c + di) = (a + c) + (b + d)i$$

Notice that when two complex numbers are added, the real parts are grouped and then added, as are the imaginary parts.

EXAMPLE 3 **Find the sum and simplify:**

    a. $(3 + 2i) + (5 + 7i)$ **b.** $(-4 + 7i) + (4 - 7i)$

    a. $(3 + 2i) + (5 + 7i) = (3 + 5) + (2 + 7)i = 8 + 9i$

    b. $(-4 + 7i) + (4 - 7i) = (-4 + 4) + (7 - 7)i = 0 + 0i = 0$

As suggested in part *b* of Example 3, every complex number has a unique *additive inverse*. To subtract complex number $c + di$ from $a + bi$, add the additive inverse of $c + di$, which is $-(c + di)$ or $(-c - di)$.

Subtraction of Complex Numbers   For all real numbers a, b, c, and d

$$(a + bi) - (c + di) = (a + bi) + (-c - di) = (a - c) + (b - d)i$$

EXAMPLE 4

$$\begin{aligned}(11 + 5i) - (-7 - 6i) &= (11 + 5i) + (7 + 6i) \\ &= (11 + 7) + (5 + 6)i \\ &= 18 + 11i\end{aligned}$$

CLASS EXERCISES

Thinking Critically

1. Is the set of complex numbers closed under addition? That is, is the sum of two complex numbers always a complex number?

2. Is the set of imaginary numbers closed under addition?

Simplify.

3. $\sqrt{-81}$ 4. $\sqrt{-100}$ 5. $\sqrt{-144}$ 6. $\sqrt{-121}$

7. $(2 + 3i) + (4 + 6i)$ 8. $(1 + 7i) + (2 + i)$

9. $(7 + 4i) - (5 + 3i)$ 10. $(9 + 2i) - (6 + i)$

In which set(s) of numbers listed at the right does each of the following numbers belong?

11. $-3i$	12. $19.131131113\ldots$	Complex
13. 17	14. $7 + 3i$	Imaginary
15. $6 - 2i$	16. $6i$	Pure Imaginary
17. i^2	18. i^4	Real
19. $\sqrt{7}$	20. 7.125	Rational
		Irrational

PRACTICE EXERCISES

Simplify.

1. $\sqrt{-25}$ **2.** $\sqrt{-9}$ **3.** $\sqrt{-\frac{1}{49}}$ **4.** $\sqrt{-\frac{36}{25}}$

5. $\sqrt{-28}$ **6.** $\sqrt{-20}$ **7.** $\sqrt{-150}$ **8.** $\sqrt{-180}$

9. $\sqrt{-9} + \sqrt{-9}$ **10.** $\sqrt{-7} + \sqrt{-7}$ **11.** $\sqrt{-4} - \sqrt{-8}$ **12.** $\sqrt{-9} - \sqrt{-18}$

13. $\sqrt{-16} + \sqrt{-49}$ **14.** $\sqrt{-4} - \sqrt{-125}$ **15.** i^{25} **16.** i^{37}

17. i^{104} **18.** i^{84} **19.** i^{246} **20.** i^{263}

Find each sum or difference and simplify.

21. $(6 + 9i) + (8 + 5i)$ **22.** $(7 + 10i) + (4 + 6i)$

23. $(-5 - 3i) + (2 - 7i)$ **24.** $(9 - 11i) + (-14 - 2i)$

25. $(7 + 15i) - (14 - i)$ **26.** $(9 + 16i) - (12 - 4i)$

27. $(-4 - 7i) - (10 - 5i)$ **28.** $(-14 - 6i) + (14 - 5i)$

29. $(-12 - \sqrt{-15}) + (-5 + \sqrt{-15})$ **30.** $(-8 - \sqrt{-9}) - (6 - \sqrt{-9})$

31. $(12 + \sqrt{-5}) - (12 - \sqrt{-6})$ **32.** $(15 + \sqrt{-49}) - (9 - \sqrt{-1})$

33. $(2 - 3\sqrt{-72}) + (-4 + 2\sqrt{-18})$ **34.** $(-5 + \sqrt{-12}) - (7 + 2\sqrt{-27})$

Let n be a positive integer and compute the following.

35. i^{4n} **36.** i^{4n+1} **37.** i^{4n+3} **38.** i^{4n+4}

Solve each equation for x.

39. $(3 - 4i) + x = (-7 + i)$ **40.** $(-5 + 3i) + x = (-9 - 4i)$

41. $(-4 + 6i) + x = (-1 - 3i)$ **42.** $(7 - 9i) + x = (-3 + 5i)$

43. $(-5 - 4i) + x = 2i$ **44.** $(7 - 2i) + x = 4i$

Find the real values of x and y that make each statement true.

45. $5x + yi = -10 + 7i$ **46.** $6x + yi = -24 + 4i$

47. $4x + 3yi = 7i$ **48.** $6x + 4yi = 15i$

49. Determine whether $x = 3 - 2i$ is a solution of the equation
$7 - 5i - x = 4 + 3i$.

50. Determine whether $(-2 + 5i, 9 + i)$ is a solution of the equation
$3x + 2y = 17i$.

Simplify.

51. $(9 + i^9) + (6 - 3i^3)$

52. $(-2 + i^7) + (3 + 2i^2)$

53. $(7 - i^8) - (5 - 3i^7)$

54. $(6 - i^6) - (-2 + 2i^{22})$

55. $(3 + i^{118}) - (-2 - i^{68})$

56. $(-1 - 5i^{14}) - (-3 + 17i^{17})$

Applications

57. Algebra The solutions of a quadratic equation are $(5 + 6i)$ and $(5 - 6i)$. Find the sum of the two solutions.

58. Algebra Using $x + yi$, $w + zi$, and $u + vi$, show that the associative property holds for the addition of complex numbers.

Electricity In an electrical circuit, current can be obstructed by resistors, capacitors, and inductances. Resistance R (in ohms) from resistors is represented by a real number. Reactance X (in ohms), from capacitors X_C and inductances X_L, is represented by a pure imaginary number. Impedance Z, which is the total effective resistance, is the sum of the resistances and reactances.

$$Z = R + Xi \qquad \text{where } X = X_L - X_C$$

The magnitude of the impedance is $|Z|$ (in ohms) where $|Z| = \sqrt{R^2 + X^2}$.

59. A circuit has a resistance of 8 ohms and a reactance of 6 ohms. Find the magnitude of the impedance.

60. In a circuit, the resistance is 8.055 ohms and the reactance from capacitors is 3.701 ohms and from inductances 9.099 ohms. Find the magnitude of the impedance.

BIOGRAPHY: Abraham De Moivre

Abraham De Moivre was born in France in 1667, studied in Paris, and moved to London in 1688. He lived there until his death in 1754. As his writings attest, he was interested in a wide variety of mathematical topics, including trigonometry. His best friends included Isaac Newton and other notable mathematicians of that time. De Moivre was elected a fellow of the prestigious Royal Society, a member of the Berlin Academy of Sciences, and a foreign associate of the Paris Academy of Sciences.

Research the life of Abraham De Moivre and investigate some areas of mathematics, other than trigonometry, in which he made contributions.

Products and Quotients of Complex Numbers

Objective: To find the product and the quotient of two complex numbers

Multiplying complex numbers is similar to multiplying binomials.

Preview

The *FOIL* method can be used to multiply two binomials. FOIL means to add the products of the *First*, the *Outer*, the *Inner*, and the *Last* terms.

EXAMPLE   **Multiply:** $(x + 2)(3x - 7)$

$$
\begin{array}{cccc}
& F & O & I & L \\
(x + 2)(3x - 7) = & x(3x) + x(-7) + 2(3x) + 2(-7) \\
& = 3x^2 - 7x + 6x - 14 \\
& = 3x^2 - x - 14
\end{array}
$$

Multiply.

1. $(x - 9)(x + 4)$ **2.** $(2x - 3)(3x - 5)$ **3.** $(x + 2y)(3x + 5y)$

4. $(0.5a - 1)(0.2a + 4)$ **5.** $(n + \sqrt{2})(n - \sqrt{2})$ **6.** $(2x - \sqrt{3})(3x - \sqrt{3})$

To multiply two complex numbers in the form $a + bi$, use the FOIL method and the fact that $i^2 = -1$.

EXAMPLE 1   **Find the product and simplify:**
  a. $(a + bi)(c + di)$ **b.** $(3 + 2i)(5 - 4i)$
  c. $\sqrt{-4}\,(\sqrt{-16})$ **d.** $(5 - 3i)(5 + 3i)$

a. $(a + bi)(c + di) = a\,(c) + a(di) + bi(c) + bi(di)$
$$= ac + adi + bci + bdi^2$$
$$= ac + bd(-1) + (ad + bc)i = (ac - bd) + (ad + bc)i$$

b. $(3 + 2i)(5 - 4i) = 3(5) + 3(-4i) + 2i(5) + 2i(-4i)$
$$= 15 + (-12i) + 10i + (-8i^2)$$
$$= 15 - 2i - 8(-1) = 23 - 2i$$

c. $\sqrt{-4}\,(\sqrt{-16}) = i\sqrt{4}\,(i\sqrt{16}) = 2i(4i) = 8i^2 = 8(-1) = -8$

d. $(5 - 3i)(5 + 3i) = 5(5) + 5(3i) + [-3i(5)] + [-3i(3i)]$
$$= 25 + 15i + (-15i) + (-9i^2)$$
$$= 25 - 9(-1) = 25 + 9 = 34$$

In part (c) of Example 1, note that $\sqrt{-4} \cdot \sqrt{-16} \neq \sqrt{(-4)(-16)}$. If $x \geq 0$ and $y \geq 0$, then $\sqrt{x} \cdot \sqrt{y} = \sqrt{xy}$. However, $\sqrt{-x} \cdot \sqrt{-y} \neq \sqrt{(-x)(-y)}$. Therefore, square roots of negative numbers should be expressed in terms of i before they are multiplied.

In part d of Example 1, the complex numbers $5 - 3i$ and $5 + 3i$ are called *conjugates*. In general, complex numbers of the form $a + bi$ and $a - bi$ are called **conjugates** of each other. They differ only in the sign of the imaginary part. The product of a complex number and its conjugate is a real number, since

$$(a + bi)(a - bi) = a^2 - abi + abi - b^2 i^2$$
$$= a^2 + 0 - b^2(-1)$$
$$= a^2 + b^2$$

Conjugates can be used to simplify the reciprocal of a complex number.

EXAMPLE 2 **Find the reciprocal of $3 - i$ and simplify.**

$$\frac{1}{3 - i} = \left(\frac{1}{3 - i}\right)\left(\frac{3 + i}{3 + i}\right)$$

Multiply the numerator and denominator by the conjugate of the denominator.
$(a + bi)(a - bi) = a^2 + b^2$

$$= \frac{3 + i}{3^2 + 1^2}$$

$$= \frac{3 + i}{9 + 1}$$

$$= \frac{3 + i}{10}, \text{ or } \frac{3}{10} + \frac{1}{10}i$$

The quotient of two complex numbers can be simplified using the conjugate of the denominator, as illustrated in Example 3.

EXAMPLE 3 **Find the quotient and simplify:** **a.** $\dfrac{a + bi}{c + di}$ **b.** $\dfrac{4 + 5i}{3 + 2i}$

a. $\dfrac{a + bi}{c + di} = \left(\dfrac{a + bi}{c + di}\right)\left(\dfrac{c - di}{c - di}\right)$

$$= \frac{ac - adi + bci - bdi^2}{c^2 + d^2}$$

$$= \frac{ac - adi + bci - (bd)(-1)}{c^2 + d^2}$$

$$= \frac{ac - adi + bci + bd}{c^2 + d^2}$$

$$= \frac{(ac + bd) + (bc - ad)i}{c^2 + d^2}, \text{ or } \frac{ac + bd}{c^2 + d^2} + \frac{bc - ad}{c^2 + d^2}i$$

b. $\dfrac{4 + 5i}{3 + 2i} = \left(\dfrac{4 + 5i}{3 + 2i}\right)\left(\dfrac{3 - 2i}{3 - 2i}\right)$

$\qquad = \dfrac{12 - 8i + 15i - 10i^2}{3^2 + 2^2}$

$\qquad = \dfrac{12 + 7i - 10(-1)}{9 + 4}$

$\qquad = \dfrac{22 + 7i}{13}, \text{ or } \dfrac{22}{13} + \dfrac{7}{13}i$

Many of the properties of real-number arithmetic hold for complex numbers. For example, the addition and multiplication of complex numbers are commutative, associative, and distributive.

CLASS EXERCISES

Thinking Critically

1. Is the set of complex numbers closed under multiplication?

2. Is the set of imaginary numbers closed under multiplication?

Multiply.

3. $i(2 - i)$ **4.** $i(3 + i)$ **5.** $(2 + i)(3 + 2i)$

6. $(4 + i)(1 - 2i)$ **7.** $\sqrt{-100} \cdot \sqrt{-64}$ **8.** $\sqrt{-36} \cdot \sqrt{-100}$

State the conjugate of each complex number and find the product of each number and its conjugate.

9. $6 + i$ **10.** $2 - 4i$ **11.** $15i$ **12.** 12 **13.** $-i$

PRACTICE EXERCISES

Multiply.

1. $i(4 - 2i)$ **2.** $i(-3 + 6i)$

3. $(2 + i)(3 - 2i)$ **4.** $(-4 - 2i)(-2 + 3i)$

5. $(-7 - 4i)(-3 - 3i)$ **6.** $(-6 - 2i)(-2 - 5i)$

7. $\sqrt{-6} \cdot \sqrt{-600}$ **8.** $\sqrt{-10} \cdot \sqrt{-640}$

9. $\sqrt{-3} \cdot \sqrt{-8}$ **10.** $\sqrt{-2} \cdot \sqrt{-16}$

11. $(4 + 5i)^2$ **12.** $(6 - 5i)^2$

13. $(4 - 3i)^2$ **14.** $(3 - 5i)^2$

15. $(2 + 3i)(2 - 3i)$ **16.** $(4 - 5i)(4 + 5i)$

17. $(\sqrt{5} + i)(\sqrt{5} - 4i)$ **18.** $(\sqrt{3} - 5i)(\sqrt{3} + i)$

Find the reciprocal and simplify.

19. $5 + i$ **20.** $6 - i$ **21.** $3 - 4i$ **22.** $-5 + 2i$ **23.** $-6 - 3i$

24. $-5 - 3i$ **25.** $-4i$ **26.** $-5i$ **27.** $3i$ **28.** $4i$

Find each quotient and simplify.

29. $\dfrac{1 + i}{1 - i}$ **30.** $\dfrac{1 - i}{1 + i}$ **31.** $\dfrac{3 + 4i}{3 - 4i}$ **32.** $\dfrac{5 + 7i}{5 - 7i}$

33. $\dfrac{2 + 6i}{-3 - i}$ **34.** $\dfrac{3 + 7i}{-4 - i}$ **35.** $\dfrac{-4 - 5i}{-2 + 9i}$ **36.** $\dfrac{-6 - 5i}{-3 + 7i}$

Solve for x.

37. $(1 - 5i)x = i$ **38.** $(2 + 3i)x = i$ **39.** $(2 + i)x + 5 - i = 3i$

40. $ix + 3 - 5i = 4$ **41.** $ix + 2 = 3i$ **42.** $(1 - i)x - 1 = i$

Solve each equation. Recall that the quadratic formula is $x = \dfrac{-b \pm \sqrt{b^2 - 4ac}}{2a}$.

43. $x^2 - 2x + 5 = 0$ **44.** $x^2 - 2x + 19 = 0$ **45.** $4x^2 - 4x + 7 = 0$

46. $2x^2 - x + 3 = 0$ **47.** $x^2 + 26 = 0$ **48.** $2x^2 + 5 = 0$

Simplify.

49. $(3 + 2i)^3$ **50.** $(3 - 2i)^4$

51. $(4 + \sqrt{-49}) \div (-\sqrt{-1})^{13}$ **52.** $(7 - \sqrt{-7}) \div (\sqrt{-1})^{11}$

53. Find x and y such that $(2 + xi)(3 - 6i) = 30 + yi$.

54. Find x and y such that $\dfrac{x + 2i}{1 - 3i} = -\dfrac{1}{5} + yi$.

Prove each statement. *Hint*: Let $a + bi$ and $c + di$ represent the two numbers.

55. The sum of the conjugates of two complex numbers equals the conjugate of their sum.

56. The product of the conjugates of two complex numbers equals the conjugate of their product.

Applications

Physics   The voltage in an electric circuit can be represented by the formula $V = IZ$, where I is the current (in amps) and Z is the impedance (in ohms).

57. If the current in a circuit is $(3 + 5i)$ amps and the impedance is $(6 - 2i)$ ohms, find the voltage.

58. If the voltage in a circuit is $(20 + 2i)$ volts, and the current is $(2 + i)$ amps, find the impedance.

59. The total voltage of one circuit is $(10 + 5i)$ volts. The total voltage of a second circuit is $(6 - 2i)$ volts. Find the ratio of the total voltage of the first circuit to that of the second circuit. Simplify your answer.

60. Computer This program prints the product of two complex numbers of the form $A + Bi$ and $C + Di$. The user supplies the values of A, B, C, and D. Use the program to check your answers for Exercises 1–18.

```
10 HOME
20 INPUT "ENTER THE VALUES OF
   A, B, C, AND D"; A, B, C, D
30 F = A*C
40 M = A*D + C*B
50 L = B*D*-1
60 PRINT "THE PRODUCT IS" F + L" + "M"I"
70 END
```

TEST YOURSELF

Find the rectangular coordinates of each point with the given polar coordinates.

1. $(-3, 150°)$

2. $(2, -40°)$

7.1

Find polar coordinates for each point with the given rectangular coordinates.

3. $(\sqrt{3}, -1)$

4. $(2, 0)$

Write each equation in polar-coordinate form.

5. $y = 5$

6. $x^2 = 5y$

Write each equation in rectangular-coordinate form.

7. $r = 5 \sec \theta$

8. $r = 3$

Graph each polar equation.

9. $r = 3 + 4 \cos \theta$

10. $r = -2 \sin 2\theta$

7.2

Simplify.

11. $\sqrt{-40}$

12. $\sqrt{-75}$

13. i^{19}

14. i^{36}

7.3

Find each sum or difference and simplify.

15. $(10 + 4i) + (12 - 3i)$

16. $(-7 + 2i) - (-8 + 5i)$

Find each product or quotient and simplify.

17. $(-9 + i)(2 - 3i)$

18. $\dfrac{2 + 4i}{1 - 3i}$

7.4

7.5 Complex Numbers in Polar Form

Objective: To express complex numbers in polar form

It is useful to be able to plot complex numbers on a coordinate plane.

Preview

The *Argand diagram* on which complex numbers are graphed (pictured at the right) was named after Jean-Robert Argand (1768–1822), a little-known Swiss bookkeeper. He wrote a short book on the geometric representation of complex numbers and received credit for the idea, even though De Moivre, Euler, and Gauss had previously thought of complex numbers as points on the plane.

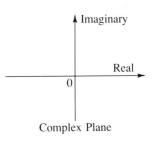

1. Why do you think Argand chose the horizontal axis as the real axis, and the vertical axis as the imaginary axis?

2. Can you think of any other men or women who made significant contributions to science, mathematics, or society outside their field of study?

When a complex number $a + bi$ is represented by a point in the plane, the real part is associated with the horizontal axis, and the imaginary part is associated with the vertical axis. Thus, the complex number $a + bi$ is represented by the ordered pair (a, b) and plotted in the usual manner.

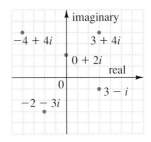

Each point in the **complex plane** determines a unique directed line segment, or vector, that begins at the origin and ends at that point. The vector OP represents $a + bi$, or (a, b), and has both length (magnitude) and direction. The length of a vector $\mathbf{z}$ is denoted by $\|\mathbf{z}\|$. By the Pythagorean theorem, $\|\mathbf{z}\| = \sqrt{a^2 + b^2}$. Because it measures distance from the origin, $\|\mathbf{z}\|$ is equivalent to the **absolute value** of $a + bi$.

For any complex number $a + bi$, $|a + bi| = \sqrt{a^2 + b^2}$

The vector $\overrightarrow{OP}$ forms an angle θ with the real axis, and the measure of θ is called the **argument** of the complex number. Since the principal value of Arctan θ is between $-\dfrac{\pi}{2}$ and $\dfrac{\pi}{2}$, the value of θ is found as follows:

$$\theta = \text{Arctan } \frac{b}{a} \quad \text{if } a > 0$$

$$\theta = \text{Arctan } \frac{b}{a} + \pi \quad (\text{or Arctan } \frac{b}{a} + 180°) \quad \text{if } a < 0$$

$$\theta = \frac{\pi}{2} \quad (\text{or } 90°) \quad \text{if } a = 0 \text{ and } b > 0$$

$$\theta = \frac{3\pi}{2} \quad (\text{or } 270°) \quad \text{if } a = 0 \text{ and } b < 0$$

$$\theta = \text{any real number} \quad \text{if } a = 0 \text{ and } b = 0$$

EXAMPLE 1 Find $\|z\|$ and θ for each vector z. Then graph the vector.
  a. $z = 3 + 4i$ **b.** $z = -4$

a. $\|z\| = |3 + 4i|$

$\quad = \sqrt{3^2 + 4^2} \quad |a + bi| = \sqrt{a^2 + b^2}$

$\quad = 5$

$\theta = \text{Arctan } \frac{4}{3} \qquad a > 0$

$\quad = 53°$

b. $\|z\| = |-4 + 0i|$

$\quad = \sqrt{(-4)^2 + 0^2}$

$\quad = 4$

$\theta = \text{Arctan } \frac{0}{-4} + 180° \qquad a < 0$

$\quad = 180°$

The complex number $a + bi$ is said to be in **rectangular form.** A useful form, called the **polar form,** or *trigonometric form,* of a complex number, can be derived using vector $\overrightarrow{OP}$. Recall that vector $\overrightarrow{OP}$ has length $r = \sqrt{a^2 + b^2}$ and it forms an angle θ with the real axis. In the triangle shown at the right, $\cos \theta = \dfrac{a}{r}$, $\sin \theta = \dfrac{b}{r}$, and $\tan \theta = \dfrac{b}{a}$. Therefore, $a = r \cos \theta$ and $b = r \sin \theta$. Substituting these values for a and b yields

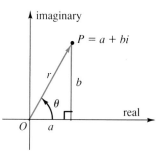

$$a + bi = r \cos \theta + (r \sin \theta)i = r(\cos \theta + i \sin \theta)$$

The **polar form** of the complex number $a + bi$ is

$$r(\cos \theta + i \sin \theta) \quad \text{where } r = \sqrt{a^2 + b^2} \text{ and } \tan \theta = \frac{b}{a}, a \neq 0$$

The quantity r is called the **modulus** and θ is called the *argument*, as mentioned earlier, $r(\cos \theta + i \sin \theta)$ is sometimes written as **r cis θ.**

$$r(\cos \theta + i \sin \theta)$$
$$r(c \quad \quad i\, s \quad \theta)$$

EXAMPLE 2 **Express each complex number in polar form:**

        a. $-1 + i\sqrt{3}$ **b.** $-5 - 3i$ **c.** $-5i$

a. $r = \sqrt{(-1)^2 + (\sqrt{3})^2}$ $a = -1, b = \sqrt{3}$ $\theta = \left(\text{Arctan} \dfrac{\sqrt{3}}{-1} \right) + \pi$

    $= 2$ $= \dfrac{2\pi}{3}$, or $120°$

    Thus, $-1 + i\sqrt{3} = 2\,(\cos 120° + i \sin 120°)$.

b. $r = \sqrt{(-5)^2 + (-3)^2}$ $a = -5, b = -3$ $\theta = \left(\text{Arctan} \dfrac{-3}{-5} \right) + \pi$

    $= \sqrt{34}$ $= 211°$

    Thus, $-5 - 3i = \sqrt{34}\,(\cos 211° + i \sin 211°)$.

c. Since $a = 0$ and $b = -5$, the number is on the imaginary axis.

    $r = \sqrt{0^2 + (-5)^2} = 5$

    $\theta = \dfrac{3\pi}{2}$, or $270°$ $a = 0, b < 0$

    Thus, $-5i = 5\left(\cos \dfrac{3\pi}{2} + i \sin \dfrac{3\pi}{2} \right)$.

To change a complex number from polar to rectangular form, find a and b.

EXAMPLE 3 **Express each complex number in rectangular form:**

    a. $4(\cos 225° + i \sin 225°)$ **b.** $6\left(\cos \dfrac{2\pi}{3} + i \sin \dfrac{2\pi}{3} \right)$

    a. $a = 4 \cos 225°$ $r = 4, \theta = 225°$ $b = 4 \sin 225°$

            $= 4\left(-\dfrac{\sqrt{2}}{2} \right) = -2\sqrt{2}$ $= 4\left(-\dfrac{\sqrt{2}}{2} \right) = -2\sqrt{2}$

    Thus, $4\,(\cos 225° + i \sin 225°) = -2\sqrt{2} - 2i\,\sqrt{2}$.

b. $a = 6 \cos \dfrac{2\pi}{3}$ $\qquad r = 6, \theta = \dfrac{2\pi}{3}$ $\qquad b = 6 \sin \dfrac{2\pi}{3}$

$\qquad\qquad = 6\left(-\dfrac{1}{2}\right) = -3$ $\qquad\qquad\qquad\qquad = 6\left(\dfrac{\sqrt{3}}{2}\right) = 3\sqrt{3}$

$\qquad$ Thus, $6\left(\cos \dfrac{2\pi}{3} + i \sin \dfrac{2\pi}{3}\right) = -3 + 3i\sqrt{3}.$

 Some calculators have a rectangular-to-polar and polar-to-rectangular conversion feature. Consult your owner's manual for instructions.

CLASS EXERCISES

Express each complex number as an ordered pair and then graph it.

1. $-3 - 4i$ **2.** $-4 + i$ **3.** $3 - 3i$ **4.** $4i$ **5.** -2

Express each complex number in polar form.

6. $-2 + 2i$ **7.** $-3 + 3i\sqrt{3}$ **8.** $-5 - 5i$ **9.** 5 **10.** $5i$

PRACTICE EXERCISES $\qquad$ Use appropriate technology.

Calculate $\|z\|$ and θ. Then graph each complex number.

1. $z = -1 - i$ **2.** $z = -1 + i$ **3.** $z = -2 + 2i$ **4.** $z = 2 - 2i$

5. $z = \sqrt{3} - i\sqrt{3}$ **6.** $z = -\sqrt{5} + i\sqrt{5}$ **7.** $z = 4i$ **8.** $z = -2i$

9. $z = -5$ **10.** $z = 4$ **11.** $z = -5 + 5i\sqrt{5}$ **12.** $z = \sqrt{2} - 2i\sqrt{2}$

Express each complex number in polar form.

13. $1 - i$ **14.** $1 + i$ **15.** $\sqrt{3} + i$ **16.** $1 - i\sqrt{3}$

17. 4 **18.** $-6i$ **19.** $2\sqrt{3} - 2i$ **20.** $2 - 2i\sqrt{3}$

Express each complex number in rectangular form.

21. $3 (\cos 45° + i \sin 45°)$ $\qquad\qquad$ **22.** $2 (\cos 30° + i \sin 30°)$

23. $\cos (-60°) + i \sin (-60°)$ $\qquad\qquad$ **24.** $\cos (-240°) + i \sin (-240°)$

25. $5 (\cos 30° + i \sin 30°)$ $\qquad\qquad$ **26.** $4 (\cos 60° + i \sin 60°)$

27. $2 [\cos (-45°) + i \sin (-45°)]$ $\qquad\qquad$ **28.** $3 [\cos (-30°) + i \sin (-30°)]$

29. $\cos 0 + i \sin 0$ $\qquad\qquad$ **30.** $3 (\cos \pi + i \sin \pi)$

31. $3 \operatorname{cis} \left(\dfrac{2\pi}{3}\right)$ **32.** $5 \operatorname{cis} \left(\dfrac{3\pi}{4}\right)$ **33.** $4 \operatorname{cis} \left(-\dfrac{3\pi}{4}\right)$ **34.** $2 \operatorname{cis} \left(-\dfrac{5\pi}{6}\right)$

The conjugate of a given complex number z is sometimes indicated as z̄. Given $z = a + bi$ and $\bar{z} = a - bi$, prove each of the following statements.

35. $\|z\| = \sqrt{z \cdot \bar{z}}$ **36.** $\|z\| = \|\bar{z}\|$ **37.** $2\|z\| = \|2z\|$

38. Let $z_1 = a + bi$ and $z_2 = c + di$. Prove that $\|z_1 + z_2\| \leq \|z_1\| + \|z_2\|$.

39. The inequality in Exercise 38 is called the *triangle inequality*. Relate z_1 and z_2 to two sides of a triangle and explain why this title is appropriate. Show that $\|z_2 - z_1\|$ is the distance from z_2 to z_1 in the complex number plane.

Applications

Geometry   Complex numbers can be added on the complex number plane using the parallelogram law for vector addition. If $a + bi$ and $c + di$ are represented as vectors drawn from the origin to points (a, b) and (c, d), their sum is represented by the diagonal of the parallelogram.

Find each sum graphically. Check algebraically.

40. $4 + i, -4 + 5i$

41. $3 + 2i, -2 + 4i$

42. $6 - i, 3 - 2i$

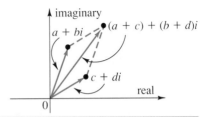

HISTORICAL NOTE

This table gives some of the high points in the history of complex numbers.

Century	Mathematicians	Contributions
1st	Heron	First wrote the square root of a negative number
16th	Jerome Cardan Rafael Bombelli	Used square roots of negative numbers to solve certain equations
17th	René Descartes Abraham De Moivre	Coined the terms *real* and *imaginary* Introduced the important relation $(\cos \theta + i \sin \theta)^n = \cos n\theta + i \sin n\theta$
18th	Leonhard Euler	Denoted $\sqrt{-1}$ by i
19th	Karl Friedrich Gauss Augustin Louis Cauchy	Introduced the term *complex number* Introduced the terms *conjugate* and *modulus*
	William Rowan Hamilton	Defined complex numbers as pairs of real numbers

Find out about other high points in the history of complex numbers.

Multiplying and Dividing Complex Numbers in Polar Form

Objective: To multiply and divide complex numbers in polar form

It is sometimes necessary to find an angle coterminal with a given angle when the product or quotient of complex numbers is computed.

Preview

Coterminal angles are angles in standard position whose terminal sides coincide. Their measures differ by an integral multiple of 360°. Thus, angles coterminal with angle θ have measures equal to $n \cdot 360° + \theta$, where n is an integer.

Find an angle between 0° and 360° that is coterminal with each angle.

1. $-75°$ **2.** $395°$ **3.** $700°$ **4.** $-550°$ **5.** $1025°$ **6.** $-900°$

The addition or subtraction of two complex numbers expressed in rectangular form is a straightforward procedure. However, the multiplication or division of two complex numbers can require many steps. If the complex numbers are expressed in polar form, multiplication and division may be less complicated.

Let z_1 and z_2 be complex numbers, where $z_1 = r_1(\cos \theta_1 + i \sin \theta_1)$ and $z_2 = r_2(\cos \theta_2 + i \sin \theta_2)$. Then

$$z_1 z_2 = r_1(\cos \theta_1 + i \sin \theta_1) \cdot r_2(\cos \theta_2 + i \sin \theta_2)$$
$$= r_1 r_2(\cos \theta_1 + i \sin \theta_1)(\cos \theta_2 + i \sin \theta_2)$$
$$= r_1 r_2(\cos \theta_1 \cos \theta_2 + i \cos \theta_1 \sin \theta_2 + i \sin \theta_1 \cos \theta_2 + i^2 \sin \theta_1 \sin \theta_2)$$
$$= r_1 r_2(\cos \theta_1 \cos \theta_2 + i \cos \theta_1 \sin \theta_2 + i \sin \theta_1 \cos \theta_2 - \sin \theta_1 \sin \theta_2)$$
$$= r_1 r_2[(\cos \theta_1 \cos \theta_2 - \sin \theta_1 \sin \theta_2) + i(\cos \theta_1 \sin \theta_2 + \sin \theta_1 \cos \theta_2)]$$
$$= r_1 r_2[\cos (\theta_1 + \theta_2) + i \sin (\theta_1 + \theta_2)] \qquad \textit{Use the sum identities}$$
$$\textit{for sine and cosine.}$$

The product of two complex numbers $z_1 = r_1(\cos \theta_1 + i \sin \theta_1)$ and $z_2 = r_2(\cos \theta_2 + i \sin \theta_2)$ is given by the formula

$$z_1 z_2 = r_1 r_2[\cos (\theta_1 + \theta_2) + i \sin (\theta_1 + \theta_2)]$$

Notice that the modulus of the product of two complex numbers is the product of the two moduli. The argument of the product is the sum of the two arguments.

EXAMPLE 1 **Find the product and simplify:**
$$3(\cos 27° + i \sin 27°) \cdot 5(\cos 82° + i \sin 82°)$$

$3(\cos 27° + i \sin 27°) \cdot 5(\cos 82° + i \sin 82°)$
$$= 3 \cdot 5[\cos (27° + 82°) + i \sin (27° + 82°)]$$
$$= 15(\cos 109° + i \sin 109°)$$

The procedure for the division of two complex numbers in polar form is very similar to the procedure for multiplication. Let z_1 and z_2 be complex numbers, where $z_1 = r_1(\cos \theta_1 + i \sin \theta_1)$ and $z_2 = r_2(\cos \theta_2 + i \sin \theta_2)$. Then

$$\frac{z_1}{z_2} = \frac{r_1(\cos \theta_1 + i \sin \theta_1)}{r_2(\cos \theta_2 + i \sin \theta_2)}$$

$$= \frac{r_1(\cos \theta_1 + i \sin \theta_1)}{r_2(\cos \theta_2 + i \sin \theta_2)} \cdot \left(\frac{\cos \theta_2 - i \sin \theta_2}{\cos \theta_2 - i \sin \theta_2} \right)$$

$$= \frac{r_1}{r_2} \left[\frac{\cos \theta_1 \cos \theta_2 - i \cos \theta_1 \sin \theta_2 + i \sin \theta_1 \cos \theta_2 - i^2 \sin \theta_1 \sin \theta_2}{\cos^2 \theta_2 - i^2 \sin^2 \theta_2} \right]$$

$$= \frac{r_1}{r_2} \left[\frac{(\cos \theta_1 \cos \theta_2 + \sin \theta_1 \sin \theta_2) + i(\sin \theta_1 \cos \theta_2 - \cos \theta_1 \sin \theta_2)}{\cos^2 \theta_2 + \sin^2 \theta_2} \right]$$

$$= \frac{r_1}{r_2} [\cos (\theta_1 - \theta_2) + i \sin (\theta_1 - \theta_2)] \qquad cos^2\ \theta_2 + sin^2\ \theta_2 = 1;$$

$$cos\ \theta_1\ cos\ \theta_2 + sin\ \theta_1\ sin\ \theta_2 = cos\ (\theta_1 - \theta_2);$$
$$sin\ \theta_1\ cos\ \theta_2 - cos\ \theta_1\ sin\ \theta_2 = sin\ (\theta_1 - \theta_2)$$

> The quotient of two complex numbers $z_1 = r_1(\cos \theta_1 + i \sin \theta_1)$ and $z_2 = r_2(\cos \theta_2 + i \sin \theta_2)$ is given by the formula
>
> $$\frac{z_1}{z_2} = \frac{r_1}{r_2} [\cos (\theta_1 - \theta_2) + i \sin (\theta_1 - \theta_2)], z_2 \neq 0$$

Notice that the modulus of the quotient of two complex numbers is the quotient of their moduli, and the argument of the quotient is the difference of their arguments.

EXAMPLE 2 **Find the quotient and simplify:** $\dfrac{9(\cos 54° + i \sin 54°)}{18(\cos 77° + i \sin 77°)}$

$\dfrac{9(\cos 54° + i \sin 54°)}{18(\cos 77° + i \sin 77°)}$
$$= \frac{9}{18} [\cos (54° - 77°) + i \sin (54° - 77°)]$$

$$= \frac{1}{2} [\cos (-23°) + i \sin (-23°)]$$

$$= \frac{1}{2} (\cos 337° + i \sin 337°) \qquad \textit{A 337° angle is coterminal with a } -23° \textit{ angle.}$$

You may be required to express a complex number in polar form before using the formulas discussed in this lesson.

EXAMPLE 3   **Calculate the quotient $\dfrac{3 + 5i}{2 + 4i}$ using the formulas in this lesson.**

First express $3 + 5i$ and $2 + 4i$ in polar form.

For $3 + 5i$: $r = \sqrt{3^2 + 5^2} = \sqrt{34}$ For $2 + 4i$: $r = \sqrt{2^2 + 4^2} = 2\sqrt{5}$
$\qquad\qquad\quad \theta = \text{Arctan } \frac{5}{3} = 59°$ $\qquad\qquad\qquad\qquad\qquad \theta = \text{Arctan } \frac{4}{2} = 63°$

$3 + 5i = \sqrt{34}\,(\cos 59° + i \sin 59°)$ $\qquad\quad 2 + 4i = 2\sqrt{5}\,(\cos 63° + i \sin 63°)$

Therefore, $\dfrac{3 + 5i}{2 + 4i} = \dfrac{\sqrt{34}\,(\cos 59° + i \sin 59°)}{2\sqrt{5}\,(\cos 63° + i \sin 63°)}$

$\qquad\qquad\qquad = \dfrac{\sqrt{170}}{10}\,[\cos (59° - 63°) + i \sin (59° - 63°)]$

$\qquad\qquad\qquad = \dfrac{\sqrt{170}}{10}\,[\cos (-4°) + i \sin (-4°)] = \dfrac{\sqrt{170}}{10}\,[\cos 356° + i \sin 356°)$

CLASS EXERCISES

Find r and θ for each product or quotient.

1. $2(\cos 4° + i \sin 4°) \cdot 3(\cos 10° + i \sin 10°)$

2. $6(\cos 30° + i \sin 30°) \cdot 5(\cos 4° + i \sin 4°)$

3. $3(\cos 15° + i \sin 15°) \cdot 6(\cos 2° + i \sin 2°)$

4. $2(\cos 150° + i \sin 150°) \cdot 6(\cos 2° + i \sin 2°)$

5. $16(\cos 80° + i \sin 80°) \div 2(\cos 100° + i \sin 100°)$

6. $30(\cos 70° + i \sin 70°) \div 5(\cos 10° + i \sin 10°)$

7. $\dfrac{14(\cos 60° + i \sin 60°)}{2(\cos 50° + i \sin 50°)}$ $\qquad\qquad$ 8. $\dfrac{21(\cos 90° + i \sin 90°)}{3(\cos 110° + i \sin 110°)}$

PRACTICE EXERCISES   Use appropriate technology.

Find each product or quotient.

1. $2(\cos 70° + i \sin 70°) \cdot 3(\cos 60° + i \sin 60°)$

2. $4(\cos 80° + i \sin 80°) \cdot 3(\cos 50° + i \sin 50°)$

3. $4\left(\cos \dfrac{2\pi}{3} + i \sin \dfrac{2\pi}{3}\right) \cdot 6\left(\cos \dfrac{\pi}{4} + i \sin \dfrac{\pi}{4}\right)$

4. $2\left(\cos \dfrac{5\pi}{6} + i \sin \dfrac{5\pi}{6}\right) \cdot 4\left(\cos \dfrac{\pi}{3} + i \sin \dfrac{\pi}{3}\right)$

5. $7(\cos 63° + i \sin 63°) \cdot 12(\cos 89° + i \sin 89°)$

6. $6(\cos 48° + i \sin 48°) \cdot 15(\cos 94° + i \sin 94°)$

7. $10(\cos 176° + i \sin 176°) \cdot 5(\cos 216° + i \sin 216°)$

8. $9(\cos 214° + i \sin 214°) \cdot 11(\cos 215° + i \sin 215°)$

9. $7\left(\cos \dfrac{5\pi}{2} + i \sin \dfrac{5\pi}{2}\right) \cdot 8\left(\cos \dfrac{3\pi}{4} + i \sin \dfrac{3\pi}{4}\right)$

10. $6\left(\cos \dfrac{7\pi}{3} + i \sin \dfrac{7\pi}{3}\right) \cdot 9\left(\cos \dfrac{5\pi}{6} + i \sin \dfrac{5\pi}{6}\right)$

11. $8(\cos 30° + i \sin 30°) \div 2(\cos 10° + i \sin 10°)$

12. $12(\cos 40° + i \sin 40°) \div 4(\cos 15° + i \sin 15°)$

13. $20(\cos 86° + i \sin 86°) \div 5(\cos 61° + i \sin 61°)$

14. $36(\cos 104° + i \sin 104°) \div 9(\cos 79° + i \sin 79°)$

15. $7\left(\cos \dfrac{2\pi}{3} + i \sin \dfrac{2\pi}{3}\right) \div 2\left(\cos \dfrac{\pi}{6} + i \sin \dfrac{\pi}{6}\right)$

16. $8\left(\cos \dfrac{3\pi}{4} + i \sin \dfrac{3\pi}{4}\right) \div 5\left(\cos \dfrac{\pi}{2} + i \sin \dfrac{\pi}{2}\right)$

Calculate each product or quotient in polar form. Check by repeating the calculations using the rectangular form of the numbers.

17. $8(\cos 20° + i \sin 20°) \cdot 2(\cos 25° + i \sin 25°)$

18. $7(\cos 60° + i \sin 60°) \cdot 4(\cos 135° + i \sin 135°)$

19. $\dfrac{10(\cos 330° + i \sin 330°)}{7(\cos 180° + i \sin 180°)}$
        20. $\dfrac{9(\cos 240° + i \sin 240°)}{5(\cos 120° + i \sin 120°)}$

21. $(3 + 4i)(\sqrt{3} - i)$
        22. $(4\sqrt{3} - 4i)(3 + 3i)$

23. $(3 + i)(2i)$
        24. $(6 + 2i)(-i)\ 2\sqrt{10}$

25. $\dfrac{3 + i}{2i}$
   26. $\dfrac{6 + 2i}{-i}$
   27. $\dfrac{3 + 4i}{\sqrt{3} - i}$
   28. $\dfrac{4\sqrt{3} - 4i}{3 + 3i}$

29. Use repeated multiplication to show that $\operatorname{cis} \dfrac{\pi}{3}$ is a solution of $z^6 = 1$.

30. Use repeated multiplication to show that $\operatorname{cis} 45°$ is a solution of $z^8 = 1$.

The polar form of $z = a + bi$ is $z = r \operatorname{cis} \theta$. The conjugate of z, denoted $\bar{z}$, is $a - bi$. Prove each of the following statements.

31. $\bar{z} = r \operatorname{cis}(-\theta)$
      32. $\dfrac{1}{z} = \dfrac{1}{r} \operatorname{cis}(-\theta)$
      33. $r^2 = z \cdot \bar{z}$

Technology A circle in a complex plane is defined by $\|\mathbf{z}\| = r$, where r is the radius of the circle. Since $\|\mathbf{z}\| = \sqrt{a^2 + b^2}$, such a circle can be graphed on a set of coordinate axes by substituting $\sqrt{a^2 + b^2}$ for $\|\mathbf{z}\|$ and solving for y.

For example, the equation $\|\mathbf{z}\| = 5$ can be written as $\sqrt{x^2 + y^2} = 5$, or $y = \pm \sqrt{25 - x^2}$. You can graph this equation on a graphing utility by entering the positive and negative parts as separate functions and graphing them together. The trace feature of your utility will verify that $(3, -4)$ is a point on the graph; thus $3 - 4i$ is a solution of the equation.

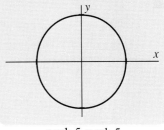

x scl: 5 y scl: 5

34. Use the trace feature of your graphing utility to find four other points on the circle $\|\mathbf{z}\| = 5$. Name the points in complex form.

35. Write the equation $\|\mathbf{z}\| = 8$ in polar form and graph it on your utility. Find the polar and rectangular coordinates of four points on the graph.

36. Write the following equations in the form $y = \pm \sqrt{}$ and graph them on your utility.

    a. $\|\mathbf{z} - 2\| = 5$ **b.** $\|\mathbf{z} - (3 + i)\| = 2$

CHALLENGE

Given: A, B, C, and D are complex numbers.

$A = r_1(\cos \alpha + i \sin \alpha)$

$B = r_2(\cos \beta + i \sin \beta)$

$C = 1(\cos 0° + i \sin 0°) = 1$

Also, triangle AOC is similar to triangle DOB.

Show that $D = A \cdot B$.

Hint: Let the modulus of D be r_3. Find a relationship between r_1, r_2, and r_3.

De Moivre's Theorem

7.7

Objective: To use De Moivre's theorem to evaluate powers of complex numbers

De Moivre's theorem will enable you to find powers of a complex number expressed in polar form and to derive certain identities using complex numbers in polar form.

Preview

The angle-sum and double-angle identities can be used to derive identities for multiples of an angle measure.

EXAMPLE **Derive an identity for sin 3θ in terms of sin θ and cos θ.**

$$\begin{aligned} \sin 3\theta &= \sin (2\theta + \theta) \\ &= \sin 2\theta \cos \theta + \cos 2\theta \sin \theta \\ &= (2 \sin \theta \cos \theta)\cos \theta + (\cos^2 \theta - \sin^2 \theta) \sin \theta \\ &= 2 \sin \theta \cos^2 \theta + \cos^2 \theta \sin \theta - \sin^3 \theta \\ &= 3 \cos^2 \theta \sin \theta - \sin^3 \theta \end{aligned}$$

Derive an identity for each expression in terms of sin θ and cos θ.

1. cos 3θ **2.** sin 4θ **3.** cos 4θ

To square a complex number expressed in polar form, multiply the number by itself using the product rule from the last lesson.

$$\begin{aligned} [r(\cos \theta + i \sin \theta)]^2 &= [r(\cos \theta + i \sin \theta)][r(\cos \theta + i \sin \theta)] \\ &= r \cdot r[\cos (\theta + \theta) + i \sin (\theta + \theta)] \\ &= r^2(\cos 2\theta + i \sin 2\theta) \end{aligned}$$

Notice that 2 is both the exponent of r and the coefficient of θ.

To cube a complex number expressed in polar form, use the number as a factor three times.

$$\begin{aligned} [r(\cos \theta + i \sin \theta)]^3 &= [r(\cos \theta + i \sin \theta)][r(\cos \theta + i \sin \theta)][r(\cos \theta + i \sin \theta)] \\ &= [r(\cos \theta + i \sin \theta)][r^2(\cos 2\theta + i \sin 2\theta)] \\ &= r^3[\cos (\theta + 2\theta) + i \sin (\theta + 2\theta)] \\ &= r^3(\cos 3\theta + i \sin 3\theta) \end{aligned}$$

Similarly, $[r(\cos \theta + i \sin \theta)]^4 = r^4(\cos 4\theta + i \sin 4\theta)$, and so forth. *De Moivre's theorem* provides a general formula for finding any power of a complex number in polar form.

> **De Moivre's Theorem** If z is a complex number in polar form, and n is a positive integer, then
>
> $$z^n = [r(\cos \theta + i \sin \theta)]^n = r^n(\cos n\theta + i \sin n\theta)$$

EXAMPLE 1 **Evaluate $[2(\cos 72° + i \sin 72°)]^5$. Then express the answer in rectangular form.**

$$\begin{aligned}[2(\cos 72° + i \sin 72°)]^5 &= 2^5[\cos 5(72°) + i \sin 5(72°)]^5 \quad \textit{Apply De Moivre's}\\ &= 32(\cos 360° + i \sin 360°) \qquad\qquad \textit{theorem.}\\ &= 32[1 + i(0)] = 32\end{aligned}$$

In order to apply De Moivre's theorem, complex numbers must be in polar form.

EXAMPLE 2 **Find $(1 + i\sqrt{3})^4$. Express the answer in rectangular form.**

First write $1 + i\sqrt{3}$ in polar form.

$$r = \sqrt{1^2 + (\sqrt{3})^2} = 2 \qquad\qquad \theta = \text{Arctan} \frac{\sqrt{3}}{1} = 60°$$

Therefore, $\quad 1 + i\sqrt{3} = 2(\cos 60° + i \sin 60°)$

$$\begin{aligned}\text{Then,} \quad (1 + i\sqrt{3})^4 &= [2(\cos 60° + i \sin 60°)]^4\\ &= 2^4[\cos 4(60°) + i \sin 4(60°)] \quad \textit{Apply De Moivre's theorem.}\\ &= 16(\cos 240° + i \sin 240°)\\ &= 16\left[-\frac{1}{2} + \left(-\frac{\sqrt{3}}{2}\right)i\right] = -8 - 8i\sqrt{3}\end{aligned}$$

De Moivre's theorem also applies when n is a negative integer, if $a + bi \neq 0$.

EXAMPLE 3 **Find $(-1 + i)^{-4}$. Express the answer in rectangular form.**

First write $-1 + i$ in polar form. $r = \sqrt{(-1)^2 + 1^2} = \sqrt{2}$ and

$$\theta = \text{Arctan} \frac{1}{-1} + \pi = \frac{3\pi}{4}, \text{ so } -1 + i = \sqrt{2}\left(\cos \frac{3\pi}{4} + i \sin \frac{3\pi}{4}\right).$$

$$\begin{aligned}\text{Then,} \quad (-1 + i)^{-4} &= (\sqrt{2})^{-4}\left[\cos(-4)\left(\frac{3\pi}{4}\right) + i \sin(-4)\left(\frac{3\pi}{4}\right)\right]\\ &= \frac{1}{4}[\cos(-3\pi) + i \sin(-3\pi)] = \frac{1}{4}[-1 + i(0)] = -\frac{1}{4}\end{aligned}$$

The result in Example 3 can be checked as follows:

$$(-1 + i)^{-4} = \frac{1}{(-1 + i)^4}$$

$$= \frac{1}{(-1 + i)^2(-1 + i)^2}$$

$$= \frac{1}{(1 - 2i + i^2)(1 - 2i + i^2)}$$

$$= \frac{1}{(-2i)(-2i)} = \frac{1}{4i^2} = \frac{1}{-4} = -\frac{1}{4}$$

Trigonometric identities that involve the representation of cos $n\theta$ and sin $n\theta$ in terms of cos θ and sin θ, where n is a positive integer, can be derived using De Moivre's theorem.

EXAMPLE 4 **Derive the identities for cos 2θ and sin 2θ in terms of cos θ and sin θ.**

$$(\cos \theta + i \sin \theta)^2 = \cos 2\theta + i \sin 2\theta \quad \textit{De Moivre's theorem}$$
$$\cos^2\theta + 2i \cos \theta \sin \theta + i^2\sin^2\theta = \cos 2\theta + i \sin 2\theta$$
$$(\cos^2\theta - \sin^2\theta) + 2i \cos \theta \sin \theta = \cos 2\theta + i \sin 2\theta$$

Now use the last equation above and the definition of equality for two complex numbers.

$$\cos^2\theta - \sin^2\theta = \cos 2\theta \quad \textit{The real parts are equal.}$$
$$2 \cos \theta \sin \theta = \sin 2\theta \quad \textit{The imaginary parts are equal.}$$

CLASS EXERCISES

Match each expression on the left with an equivalent expression on the right.

1. $[2(\cos 10° + i \sin 10°)]^6$ **a.** $81(\cos 60° + i \sin 60°)$

2. $[4(\cos 10° + i \sin 10°)]^3$ **b.** $9(\cos 108° + i \sin 108°)$

3. $[3(\cos 15° + i \sin 15°)]^4$ **c.** $64(\cos 60° + i \sin 60°)$

4. $[9(\cos 15° + i \sin 15°)]^2$ **d.** $\cos 108° + i \sin 108°$

5. $(\cos 12° + i \sin 12°)^9$ **e.** $64(\cos 30° + i \sin 30°)$

6. $[3(\cos 54° + i \sin 54°)]^2$ **f.** $81(\cos 30° + i \sin 30°)$

PRACTICE EXERCISES

Apply De Moivre's theorem to evaluate each power. Express your answers in rectangular form.

1. $[3(\cos 40° + i \sin 40°)]^3$ **2.** $[4(\cos 30° + i \sin 30°)]^5$

3. $[5(\cos 15° + i \sin 15°)]^3$ **4.** $[5(\cos 45° + i \sin 45°)]^3$

5. $\{2[\cos{(-20°)} + i\sin{(-20°)}]\}^3$ **6.** $\{2[\cos{(-10°)} + i\sin{(-10°)}]\}^3$

7. $(\cos{30°} + i\sin{30°})^3$ **8.** $(\cos{45°} + i\sin{45°})^6$

9. $[6(\cos{200°} + i\sin{200°})]^3$ **10.** $[3(\cos{240°} + i\sin{240°})]^3$

11. $(2 - 2i)^4$ **12.** $(1 - i\sqrt{3})^5$ **13.** $(\sqrt{3} + i)^3$ **14.** $(-1 + i)^4$

15. $(-1 + i)^6$ **16.** $(2\sqrt{2} - 2i\sqrt{2})^4$ **17.** $(3 - 3i)^3$ **18.** $(-\sqrt{3} - i)^3$

19. i^{-5} **20.** $(-i)^{-4}$ **21.** $(2\operatorname{cis}60°)^{-3}$ **22.** $(4\operatorname{cis}30°)^{-4}$

Use De Moivre's theorem to derive an identity for each expression in terms of $\sin\theta$ and $\cos\theta$.

23. $\cos 3\theta$ **24.** $\sin 3\theta$

25. $\sin 4\theta$ **26.** $\cos 4\theta$

Apply De Moivre's theorem and then simplify. Express answers in rectangular form.

27. $\dfrac{(1 + i)^3}{1 + i\sqrt{3}}$ **28.** $\dfrac{(2 - 2i)^4}{-1 - i\sqrt{3}}$

Applications

29. Algebra Show that $\cos 45° + i\sin 45°$ is a solution of $x^4 + x^2 = -1 + i$.

30. Algebra Show that $\cos\dfrac{\pi}{6} + i\sin\dfrac{\pi}{6}$ is a solution of $x^{12} + x^6 + x^3 - i = 0$.

EXTRA

To solve some equations involving cosines, use the fact that $\cos a = \cos b$ if and only if $a = b + k(360°)$ or $a = -b + k(360°)$, where k is an integer.

EXAMPLE Solve: $\cos{(660° - 3x)} = \cos 2x$

$660° - 3x = 2x + k(360°)$ or $660° - 3x = -2x + k(360°)$

$\dfrac{660° - k(360°)}{5} = x$ or $660° - k(360°) = x$

Substitute 1 for k to obtain two solutions.

$x = \dfrac{660° - 1(360°)}{5} = \dfrac{300°}{5} = 60°$ $x = 660° - 1(360°) = 300°$

1. Use $k = 2$ and $k = 3$ to find additional solutions of $\cos{(660° - 3x)} = \cos 2x$.

2. Find at least four solutions of the equation $\cos{(150° - 2x)} = \cos 4x$, using the method outlined above.

Roots of Complex Numbers

Objective: To find the n distinct roots of a complex number

To find any possible real nth roots of a number a, you can solve an equation of the form $x^n - a = 0$, or $x^n = a$.

Preview

The real solutions of a quadratic equation such as $x^2 - 4 = 0$ can be found by factoring or by taking square roots.

$$x^2 - 4 = 0$$
$(x + 2)(x - 2) = 0$ *Factor.*
$x + 2 = 0$ or $x - 2 = 0$
$x = -2$ or $x = 2$

$$x^2 - 4 = 0$$
$$x^2 = 4$$
$x = \pm 2$ *Take the square roots.*
$x = -2$ or $x = 2$

Find the real solution(s) of each equation.

1. $x^2 - 169 = 0$ **2.** $x^2 - 12 = 0$ **3.** $x^3 - 8 = 0$

4. $x^3 + 27 = 0$ **5.** $x^4 - 256 = 0$ **6.** $x^5 + 1 = 0$

Consider the cubic (third degree) equation $x^3 = 64$. If the cube root of each side is taken, the real root $x = 4$ is obtained. There are, however, two other roots of 64 that are complex numbers. The *complex roots theorem*, which can be shown to follow from De Moivre's theorem, is used to find all of the nth roots of a complex number.

Complex Roots Theorem   If n is a positive integer, then $a + bi = r(\cos \theta + i \sin \theta)$ has n distinct roots, which are given by

$$\sqrt[n]{r}\left[\cos \frac{1}{n}(\theta + k \cdot 360°) + i \sin \frac{1}{n}(\theta + k \cdot 360°)\right] \quad \text{or}$$

$$\sqrt[n]{r}\left[\cos \frac{1}{n}(\theta + 2k\pi) + i \sin \frac{1}{n}(\theta + 2k\pi)\right] \quad \text{where}$$

$$k = 0, 1, 2, \ldots, n - 1$$

EXAMPLE 1 **Find the three complex roots of 64. That is, find all of the complex roots of $x^3 = 64$. Express the roots in rectangular form.**

$64 = 64 + 0i = 64(\cos 0° + i \sin 0°)$ *Express 64 in polar form.*

$x^3 = 64(\cos 0° + i \sin 0°)$

$x = \sqrt[3]{64}\left[\cos \frac{1}{3}(0° + k \cdot 360°) + i \sin \frac{1}{3}(0° + k \cdot 360°)\right]$ *$\sqrt[3]{64}$ is the real cube root of 64.*

   $= 4[\cos (0° + k \cdot 120°) + i \sin (0° + k \cdot 120°)]$

For $k = 0$: $x = 4[\cos (0° + 0 \cdot 120°) + i \sin (0° + 0 \cdot 120°)]$
                  $= 4(\cos 0° + i \sin 0°) = 4(1 + 0) = 4$

For $k = 1$: $x = 4[\cos (0° + 1 \cdot 120°) + i \sin (0° + 1 \cdot 120°)]$
                  $= 4(\cos 120° + i \sin 120°) = 4\left(-\frac{1}{2} + i\frac{\sqrt{3}}{2}\right) = -2 + 2i\sqrt{3}$

For $k = 2$: $x = 4[\cos (0° + 2 \cdot 120°) + i \sin (0° + 2 \cdot 120°)]$
                  $= 4(\cos 240° + i \sin 240°) = 4\left[\left(-\frac{1}{2} + i\left(-\frac{\sqrt{3}}{2}\right)\right)\right] = -2 - 2i\sqrt{3}$

Thus, the three cube roots of 64 are 4, $-2 + 2i\sqrt{3}$, and $-2 - 2i\sqrt{3}$.
Check to show that each number, raised to the third power, equals 64.

The complex roots theorem can be used to solve equations of higher degree.

EXAMPLE 2 **Solve: $x^4 + 81 = 0$**

$x^4 + 81 = 0$ is equivalent to $x^4 = -81$. Express -81, or $-81 + 0i$, in polar form.

$$r = \sqrt{(-81)^2 + 0^2} = 81 \qquad \theta = \text{Arctan}\,\frac{0}{-81} + \pi = \pi$$

Thus, $-81 = 81(\cos \pi + i \sin \pi)$

and $x = \sqrt[4]{81}\left[\cos \frac{1}{4}(\pi + 2k\pi) + i \sin \frac{1}{4}(\pi + 2k\pi)\right]$

        $= 3\left[\cos\left(\frac{\pi}{4} + \frac{2k\pi}{4}\right) + i \sin\left(\frac{\pi}{4} + \frac{2k\pi}{4}\right)\right]$

For $k = 0$: $x = 3\left(\cos \frac{\pi}{4} + i \sin \frac{\pi}{4}\right) = 3\left(\frac{\sqrt{2}}{2} + i\frac{\sqrt{2}}{2}\right) = \frac{3\sqrt{2}}{2} + \frac{3i\sqrt{2}}{2}$

For $k = 1$: $x = 3\left(\cos \frac{3\pi}{4} + i \sin \frac{3\pi}{4}\right) = 3\left(-\frac{\sqrt{2}}{2} + i\frac{\sqrt{2}}{2}\right) = -\frac{3\sqrt{2}}{2} + \frac{3i\sqrt{2}}{2}$

For $k = 2$: $x = 3\left(\cos \frac{5\pi}{4} + i \sin \frac{5\pi}{4}\right) = 3\left(-\frac{\sqrt{2}}{2} - i\frac{\sqrt{2}}{2}\right) = -\frac{3\sqrt{2}}{2} - \frac{3i\sqrt{2}}{2}$

For $k = 3$: $x = 3\left(\cos \frac{7\pi}{4} + i \sin \frac{7\pi}{4}\right) = 3\left(\frac{\sqrt{2}}{2} - i\frac{\sqrt{2}}{2}\right) = \frac{3\sqrt{2}}{2} - \frac{3i\sqrt{2}}{2}$

To check, raise each of the four roots to the fourth power.

Notice that the roots of $x^4 + 81 = 0$ are cyclic in nature. In fact, when graphed on the complex plane, the nth roots of a complex number are equally spaced around a circle. The figure at the right shows the four fourth roots of -81.

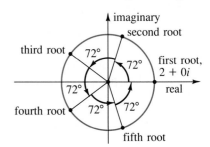

If one nth root of a complex number is known, then all the nth roots can be graphed on the complex plane as follows:

Step 1 With its center at the origin, draw a circle with radius $\sqrt[n]{r}$.
Step 2 Graph the known root.
Step 3 To locate the other roots, divide the circle into n arcs of equal length.

The measure of the angle between the consecutive roots is $\dfrac{360°}{n}$, or $\dfrac{2\pi}{n}$.

EXAMPLE 3   **Graph the five fifth roots of 32.**

Step 1 Since $\sqrt[5]{32} = 2$, a circle with radius 2 should be drawn.
Step 2 Graph the first root, $2 + 0i$.
Step 3 The five roots are equally spaced $\dfrac{360°}{5} = 72°$ apart.

The complex roots theorem can also be used to find the roots of a complex number, $a + bi$, where $b \neq 0$.

EXAMPLE 4   **Find the five fifth roots of $-1 - i$. Express your answers in polar form.**

First express $-1 - i$ in polar form.

$$r = \sqrt{(-1)^2 + (-1)^2} = \sqrt{2} \qquad \theta = \text{Arctan}\,\frac{-1}{-1} + 180° = 225°$$

Thus, $-1 - i = \sqrt{2}\,(\cos 225° + i \sin 225°)$

The five fifth roots are given by

$$\sqrt[10]{2}\left[\cos\left(\frac{225°}{5} + \frac{k \cdot 360°}{5}\right) + i \sin\left(\frac{225°}{5} + \frac{k \cdot 360°}{5}\right) \right] \quad \text{or}$$
$$\sqrt[10]{2}\,[\cos\,(45° + k \cdot 72°) + i \sin\,(45° + k \cdot 72°)]$$

For $k = 0$: $x = \sqrt[10]{2}\,(\cos 45° + i \sin 45°)$
For $k = 1$: $x = \sqrt[10]{2}\,(\cos 117° + i \sin 117°)$
For $k = 2$: $x = \sqrt[10]{2}\,(\cos 189° + i \sin 189°)$ *The five fifth*
For $k = 3$: $x = \sqrt[10]{2}\,(\cos 261° + i \sin 261°)$ *roots of $-1 - i$*
For $k = 4$: $x = \sqrt[10]{2}\,(\cos 333° + i \sin 333°)$

CLASS EXERCISES

Given the number n of nth roots of a complex number, give the measure of the angle between any two consecutive roots on the complex plane.

1. $n = 3$ **2.** $n = 4$ **3.** $n = 5$ **4.** $n = 6$ **5.** $n = 8$ **6.** $n = 9$

7. Show that the nth *roots of unity* (nth roots of 1) are of the form

$$\cos \frac{k \cdot 360°}{n} + i \sin \frac{k \cdot 360°}{n}.$$

8. Find the cube roots of unity.

PRACTICE EXERCISES   Use appropriate technology.

Find the indicated roots. Express the roots in rectangular form.

1. the square roots of $9i$ **2.** the square roots of $16i$

3. the cube roots of 27   **4.** the cube roots of -27

5. the cube roots of $-64i$ **6.** the cube roots of $-8i$

7. the fourth roots of $-1 - i\sqrt{3}$ **8.** the cube roots of $1 - i$

9. the fourth roots of 16   **10.** the cube roots of $-27i$

11. the square roots of 1   **12.** the fourth roots of 1

13. the sixth roots of 1   **14.** the eighth roots of 1

Graph the roots of each equation on the complex plane.

15. $x^4 = 1$ **16.** $x^{10} = 1$ **17.** $x^3 = 27$

18. $x^6 = 64$ **19.** $x^5 = -243$ **20.** $x^3 = -125$

Solve each equation.

21. $x^3 - 27 = 0$ **22.** $x^3 + 64 = 0$

23. $x^5 + 1 = 0$ **24.** $x^4 - 1 = 0$

25. $x^3 + 8 = 0$ **26.** $x^3 + 27 = 0$

27. $16x^4 = i$ **28.** $x^4 = -i$

29. $x^3 - (32 + 32i\sqrt{3}) = 0$ **30.** $x^4 = -1 - i\sqrt{3}$

31. $x^5 - 32 = 0$ **32.** $x^5 + 32 = 0$

33. $x^6 = -1$ **34.** $x^6 = 1$

35. $x^3 + \left(\dfrac{1}{2} - \dfrac{i\sqrt{3}}{2} \right) = 0$ **36.** $x^3 - (1 + i\sqrt{3}) = 0$

Find the indicated roots.

37. $(1 + i)^{\frac{4}{3}}$

38. $\left(\dfrac{27\sqrt{2}}{2} + \dfrac{27i\sqrt{2}}{2}\right)^{\frac{2}{3}}$

39. $(4\sqrt{2} - 4i\sqrt{2})^{\frac{2}{3}}$

40. $(-8 + 8i\sqrt{3})^{\frac{3}{4}}$

Applications

Computer This program calculates N nth roots of a complex number R cis W. The values of N, R, and W are supplied by the user.

41. Use the program to find the 3rd roots of 27 cis 120°.

42. Use the program to find the 6th roots of 64 cis 120°.

43. Use the program to find the fifth roots of unity.

```
10   HOME : INPUT "ENTER THE NUMBER OF
     ROOTS ";N
20   INPUT "ENTER THE VALUE OF R ";R:
     INPUT "ENTER THE VALUE OF W ";W
30   PRINT : PRINT "THE "N" ROOTS OF
     "R" (CIS "W") ARE: "
40   FOR K = 0 TO N - 1
50   PRINT : PRINT  R ^ (1 / N);"(COS ";
     (1/N) * (W + K * 360);" + i SIN ";
     (1/N) * (W + K * 360);")"
60   NEXT K
70   END
```

TEST YOURSELF

Express each complex number in polar form.

1. $3 - 3i$

2. $5 + 5i$

7.5

Express each complex number in rectangular form.

3. $4(\cos 135° + i \sin 135°)$

4. $2(\cos 60° + i \sin 60°)$

Find each product or quotient.

5. $4(\cos 194° + i \sin 194°) \cdot 10(\cos 160° + i \sin 160°)$

6. $9(\cos 107° + i \sin 107°) \div 3(\cos 62° + i \sin 62°)$

7.6

Evaluate each power. Express your answers in rectangular form.

7. $[4(\cos 10° + i \sin 10°)]^{12}$

8. $(6 + 6i)^4$

7.7

Find the indicated roots. Express the roots in rectangular form.

9. the cube roots of -27

10. the fourth roots of 16

7.8

TECHNOLOGY:
Chaos Theory

Early models of population change assumed a constant growth rate, unlimited food supply, and no predators. Since these ideal conditions exist only in the minds of statisticians, another model was developed to reflect real conditions.

This newer model uses the equation $P_{k+1} = rP_k(1 - P_k)$ and the *iterative process*, which means that the equation is applied to itself repeatedly. P represents the mean population in a given year k, and r is the growth rate. If 0 represents extinction and 1 represents the maximum population, then the equation will generate the following table, showing the *equilibrium* (stabilized) state of the population, given an initial population of 0.4.

Rate of Growth	Equilibrium Population
1.5	.33
2.0	.50
2.5	.60
3.0	.66

To see how the iterative process works on a calculator, choose a rate of growth and define Y_1 or F_1 using the population equation above. For example, a rate of 1.2 would yield the function $Y_1 = 1.2x (1 - x)$. Store an initial value of 0.4 in X, and then store the value of Y_1 or F_1 in X. Every time you press ENTER or EXE the calculator iterates the function, moving gradually closer to the equilibrium population figure.

Oddly, if $r > 3$, a new process called *bifurcation* begins, in which equilibrium numbers alternate. As r increases, bifurcations compound, finally becoming continuous and producing the appearance of chaos. Bifurcation theory is the study of the point at which an equation seems to pass from order to chaos through bifurcation; it is central to the study of turbulence, that moment when heated water starts to boil or a quiet stream turns to churning rapids.

EXERCISES

1. What is the equilibrium state of an initial population of 0.2 if the rate of growth is 2.6? 2.7? 2.8?

2. Find the equilibrium state(s) of a population of 0.4 with a rate of 3.5.

Vocabulary

absolute value of a
  complex number (309)
argument (309)
cardioid (294)
cis θ (310)
complex number (299)
complex plane (308)
complex roots theorem (322)
conjugate (304)
De Moivre's theorem (319)
imaginary number (298)

limaçon (294)
modulus (310)
polar axis (286)
polar coordinates (286)
polar coordinate system (286)
polar form of a complex number (309)
pole (286)
pure imaginary number (299)
rectangular form of a complex number (309)
rose (295)

Polar Coordinates A point on the plane has rectangular coordinates of
the form (x, y) and polar coordinates of the form (r, θ). To change
from polar to rectangular coordinates, use the formulas $x = r \cos \theta$ and
$y = r \sin \theta$. To change from rectangular to polar coordinates, use
$r = \sqrt{x^2 + y^2}$, $\theta = \text{Arctan} \frac{y}{x}$ (if $x > 0$), and $\theta = \text{Arctan} \frac{y}{x} + 180°$ (if $x < 0$).

   7.1

1. Graph the point $(-4, 45°)$ on the polar coordinate system.

2. Find the rectangular coordinates of the point $(2, 45°)$.

3. Find the polar coordinates of the point $(1, \sqrt{3})$.

Graphs of Polar Equations Some curves have special names.

   7.2

Circle	$r = 2a \cos \theta$
Limaçon (with extra loop)	$r = a \pm b \cos \theta$ or $r = a \pm b \sin \theta, a < b$
Limaçon (without extra loop)	$r = a \pm b \cos \theta$ or $r = a \pm b \sin \theta, a > b$
Cardioid	$r = a \pm a \cos \theta$ or $r = a \pm a \sin \theta$
Rose	$r = a \cos n\theta$ or $r = a \sin n\theta$, n is a positive integer

Graph each equation.

4. $r = 1 + 2 \sin \theta$ 5. $r = -6 \sin 2\theta$ 6. $r = 2 + 2 \sin \theta$

Sums and Differences of Complex Numbers The definitions of
addition and subtraction of two complex numbers $a + bi$ and $c + di$, are:

   7.3

$$(a + bi) + (c + di) = (a + c) + (b + d)i$$
$$(a + bi) - (c + di) = (a + bi) + (-c - di) = (a - c) + (b - d)i$$

Simplify.

7. $\sqrt{-20}$ **8.** $(-4 + 3i) + (9 - 11i)$ **9.** $(-2 - i) - (6 - 8i)$

Products and Quotients of Complex Numbers   To multiply two complex numbers, use the FOIL method and the fact that $i^2 = -1$. The two complex numbers $a + bi$ and $a - bi$ are conjugates, and the quotient of two complex numbers can be simplified by multiplying the dividend and the divisor by the conjugate of the divisor.

7.4

10. Multiply and simplify: $(-1 - 3i)(4 + 5i)$

11. Divide and simplify: $(5 + 6i) \div (-3 - 2i)$

Complex Numbers in Polar Form   The polar form of a complex number $a + bi$ is $r(\cos \theta + i \sin \theta)$, where $r = \sqrt{a^2 + b^2}$ and $\tan \theta = \dfrac{b}{a}$, $a \neq 0$. The rectangular form of a complex number can be found using the formulas $a = r \cos \theta$ and $b = r \sin \theta$.

7.5

12. Express in polar form: $3 + 3i$

13. Express in rectangular form: $6(\cos 150° + i \sin 150°)$

Multiplying and Dividing Complex Numbers in Polar Form   The product and the quotient of two complex numbers $z_1 = r_1(\cos \theta_1 + i \sin \theta_1)$ and $z_2 = r_2(\cos \theta_2 + i \sin \theta_2)$ are given by the following formulas:

7.6

$$z_1 z_2 = r_1 r_2 [\cos (\theta_1 + \theta_2) + i \sin (\theta_1 + \theta_2)]$$

$$\frac{z_1}{z_2} = \frac{r_1}{r_2} [\cos (\theta_1 - \theta_2) + i \sin (\theta_1 - \theta_2)]$$

14. Multiply: $12(\cos 30° + i \sin 30°) \cdot 6(\cos 150° + i \sin 150°)$

15. Divide: $12(\cos 100° + i \sin 100°) \div 6(\cos 40° + i \sin 40°)$

De Moivre's Theorem   If z is a complex number in polar form, and n is a positive integer, then $z^n = r^n(\cos n\theta + i \sin n\theta)$.

7.7

Evaluate each power. Express your answers in rectangular form.

16. $[3(\cos 15° + i \sin 15°)]^2$ **17.** $[5(\cos 10° + i \sin 10°)]^6$

Roots of Complex Numbers   The complex roots theorem states that if n is a positive integer, then $a + bi = r(\cos \theta + i \sin \theta)$ has n distinct roots, given by $\sqrt[n]{r} \left[\cos \dfrac{1}{n} (\theta + k \cdot 360°) + i \sin \dfrac{1}{n} (\theta + k \cdot 360°) \right]$, where $k = 0, 1, 2, \ldots, n - 1$.

18. Find the cube roots of -1.   **19.** Find the fourth roots of 625.

Find the rectangular coordinates of each point.

1. $(2, -135°)$

2. $(-3, 120°)$

Find the polar coordinates of each point.

3. $(-3, -3)$

4. $(-1, \sqrt{3})$

5. Graph the polar equation $r = 2 - 2 \sin \theta$.

6. Simplify: i^{60}

7. Find the sum: $(6 + 3i) + (11 - 4i)$

8. Find the difference: $(2 - 5i) - (-4 - i)$

9. Multiply: $(9 - 6i)(-1 + i)$

10. Divide: $(2 + 3i) \div (5 - i)$

Express each complex number in polar form.

11. $-\sqrt{3} + i$

12. $4 + 4i$

13. Express $2(\cos 225° + i \sin 225°)$ in rectangular form.

Multiply.

14. $8(\cos 96° + i \sin 96°) \cdot 4(\cos 129° + i \sin 129°)$

15. $3(\cos 264° + i \sin 264°) \cdot 12(\cos 126° + i \sin 126°)$

Divide.

16. $16(\cos 199° + i \sin 199°) \div 2(\cos 49° + i \sin 49°)$

17. $20(\cos 214° + i \sin 214°) \div 4(\cos 274° + i \sin 274°)$

Calculate each power. Express your answers in rectangular form.

18. $[4(\cos 20° + i \sin 20°)]^3$

19. $[2(\cos 60° + i \sin 60°)]^4$

20. Find the two square roots of $25i$.

Challenge

Find the points of intersection of the graphs of $r = 6 \cos 2\theta$ and $r = 6 \sin \theta$.

Select the best choice for each question.

1. Solve for x: $2x^2 - 4x + 3 = 0$
   A. $4 \pm 2i\sqrt{2}$ B. $-4 \pm i\sqrt{2}$
   C. $\dfrac{2 \pm i\sqrt{2}}{2}$ D. $\dfrac{4 \pm i\sqrt{7}}{4}$
   E. $\dfrac{2 \pm i\sqrt{7}}{4}$

2. $(-3 + 2i)(5 - 5i) =$
   A. $5 + 5i$ B. $-5(1 - 5i)$
   C. $5 - 5i$ D. $-5(1 + 5i)$
   E. $-5(1 + i)$

3. Which of the following is (are) true
   for the function f defined by
   $$f(x) = \begin{cases} x^2 + 2 & \text{if } x > 0 \\ -5 & \text{if } x = 0 \\ x + 1 & \text{if } x < 0 \end{cases}$$
   I. $f(-1) < f(0)$
   II. x can be any real number.
   III. $f(4) = 16$
   A. I only   B. II only   C. III only
   D. I and II only   E. II and III only

4. $4 - 4i$ is equivalent to
   A. $2\sqrt{2}\left[\cos\left(-\dfrac{\pi}{4}\right) + i \sin\left(-\dfrac{\pi}{4}\right)\right]$
   B. $\sqrt{2}\left[\cos\left(-\dfrac{\pi}{4}\right) + i \sin\left(-\dfrac{\pi}{4}\right)\right]$
   C. $\sqrt{6}\left[\cos\left(-\dfrac{\pi}{3}\right) + i \sin\left(-\dfrac{\pi}{3}\right)\right]$
   D. $4\sqrt{2}\left[\cos\left(-\dfrac{\pi}{4}\right) + i \sin\left(-\dfrac{\pi}{4}\right)\right]$
   E. $4\sqrt{2}\left[\cos\left(-\dfrac{\pi}{6}\right) + i \sin\left(-\dfrac{\pi}{6}\right)\right]$

5. $i^{36} \cdot i^{25} =$
   A. 1   B. $-i$ C. i D. -1 E. 0

6. Evaluate $(-\sqrt{3} + i)^6$.
   A. $3 - 2i$ B. -64 C. $12 - 12i$
   D. 36   E. 54

7. $\dfrac{2 + i}{5 - 2i} =$
   A. $\dfrac{8}{29} + \dfrac{9}{29}i$ B. $8 + 9i$
   C. $-\dfrac{8}{29} + 9i$ D. $-\dfrac{8}{29} - \dfrac{1}{29}i$
   E. $8 + \dfrac{9}{29}i$

8. Find x.
   A. 4   B. 2
   C. 5   D. 3
   E. 6
   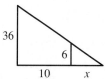

9. Which of the following represents
   the equation of the line which
   contains the point $(5, -2)$ and is
   parallel to the line $-3x + y = 2$?
   A. $y = 3x + 2$ B. $y = -3x - 2$
   C. $y = -3x - 2$ D. $y = 3x - 1$
   E. $y = 3x - 17$

10. Write the complex number
    $5\left(\cos\dfrac{\pi}{3} + i \sin\dfrac{\pi}{3}\right)$ in the
    form $a + bi$.
    A. $5 + 5i$ B. $\dfrac{5\sqrt{3}}{2} + \dfrac{5}{2}i$
    C. $-5 - 5i$ D. $\dfrac{5}{2} + \dfrac{5\sqrt{3}}{2}i$
    E. $-\dfrac{5\sqrt{3}}{2} - \dfrac{5}{2}i$

11. Three-fourths of a number is
    10 less than twice the number.
    Find the number.
    A. 5   B. 12   C. 4   D. 8   E. 24

12. The total surface area S of a
    cylinder is $2\pi r^2 + 2\pi rh$. Find
    the radius r if $S = 30\pi$ and $h = 2$.
    A. 5   B. -5 C. 3   D. -3 E. 15

Simplify.

Examples $-2^4 = -1(2^4) = -1(2)(2)(2)(2) = -16$
$(-2)^4 = (-2)(-2)(-2)(-2) = 16$

1. -3^5

2. $(-4)^3$

3. $-\left(\frac{1}{2}\right)^4$

4. $-\left(\frac{1}{4}\right)^2$

5. $\left(-\frac{1}{4}\right)^2$

6. 0.2^3

Simplify.

Example $(5x^2y^4)^3(x^2y)^2 = (5)^3(x^2)^3(y^4)^3(x^2)^2(y)^2 = 125x^6y^{12}x^4y^2 = 125x^{10}y^{14}$

7. $\dfrac{6x^5}{2x^2}$

8. $\dfrac{9x^9y^6}{3x^2y^{-1}}$

9. $\left(\dfrac{x}{y^{-2}}\right)^3$

10. $(-2xy)(3x^2y)$

11. $(3a^2b^3)^{-3}(27a^8b^{10})$

12. $(x^{-4})\left(\dfrac{3x^{10}y}{y^{-2}}\right)$

Evaluate each expression.

Example $5(2)^4 - 10 \div 2 = 5(16) - 10 \div 2 = 80 - 5 = 75$

13. $(10 \div 2)(3 + 4)$

14. $10 \div [2(3 + 4)]$

15. $3 - 2(5)$

16. $(3 - 2)(5)$

17. $(2 + 3)^2 \div 5 + 1$

18. $(2 + 3)^2 \div (5 + 1)$

Evaluate each expression for $x = -2$ and $y = 3$.

Example $x^2 - 4y$
$(-2)^2 - 4(3) = 4 - 12 = -8$

19. $x^2 - y$

20. $xy - 2y^2$

21. $\dfrac{x}{y} + \dfrac{y}{2}$

22. $-3x - 2y$

23. $5x^2 + 2y$

24. $-3x^2 - 2y^2$

Express each number in scientific notation.

Example $0.000000247 = 2.47 \times 10^{-7}$

25. 625000

26. 0.0000387

27. 63800000

28. 0.000375

29. 37.2×10^6

30. 0.625×10^4

8 Exponential and Logarithmic Functions

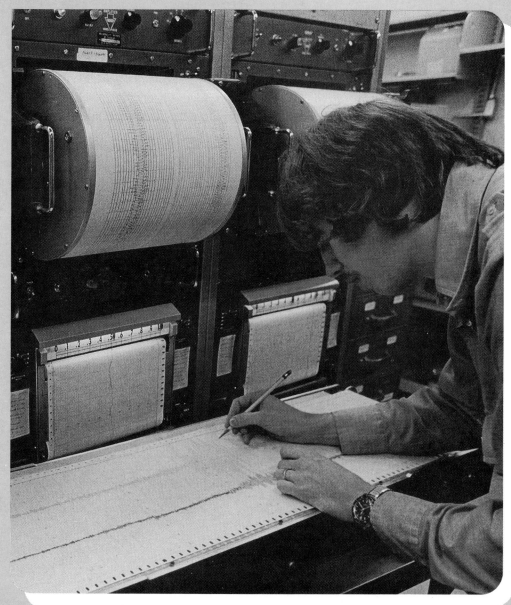

Seismology is the study of earthquakes and the mechanical properties of the earth. A seismogram is a record made by a seismograph. Its period and amplitude reflect the intensity, direction, and duration of any movement of the ground.

8.1 Real Exponents

Objectives: To evaluate expressions containing real number exponents
To solve exponential equations

Before introducing real number exponents, it will be helpful to review the properties of rational number exponents.

Preview

If a and b are positive real numbers, and m and n are rational numbers, then the following properties hold true:

$$a^m a^n = a^{m+n} \qquad \frac{a^m}{a^n} = a^{m-n} \qquad (a^m)^n = a^{mn} \qquad (ab)^m = a^m b^m$$

$$\left(\frac{a}{b}\right)^m = \frac{a^m}{b^m} \qquad a^{-m} = \frac{1}{a^m} \qquad a^{\frac{1}{n}} = \sqrt[n]{a} \qquad a^{\frac{m}{n}} = \sqrt[n]{a^m} = (\sqrt[n]{a})^m$$

EXAMPLE **Simplify:** **a.** $(2x)^{-3}$ **b.** $(27x^6)^{\frac{2}{3}}$ **c.** $(81x)^{\frac{1}{2}}$

a. $(2x)^{-3} = (2)^{-3}(x)^{-3}$ **b.** $(27x^6)^{\frac{2}{3}} = (27)^{\frac{2}{3}}(x^6)^{\frac{2}{3}}$ **c.** $(81x)^{\frac{1}{2}} = (81)^{\frac{1}{2}}(x)^{\frac{1}{2}}$

$\qquad\qquad = \left(\frac{1}{8}\right)\left(\frac{1}{x^3}\right) \qquad\qquad\qquad = (3)^2 x^4 \qquad\qquad\qquad\qquad = 9\sqrt{x}$

$\qquad\qquad = \frac{1}{8x^3} \qquad\qquad\qquad\qquad\quad = 9x^4$

Simplify.

1. $(2x^3)^2$ **2.** $(3^2)(3^4)$ **3.** $(-32)^{\frac{2}{5}}$ **4.** $81^{-\frac{1}{4}}$ **5.** $(3x)^{-3}$ **6.** $4^{-2.5}$

The properties of rational exponents can also be applied to real number exponents.

EXAMPLE 1 **Simplify:** **a.** $(2^{\sqrt{30}})^{\sqrt{120}}$ **b.** $2^{-\sqrt{3}} \cdot 2^{\sqrt{48}}$

$\qquad\qquad$ **a.** $(2^{\sqrt{30}})^{\sqrt{120}} = 2^{\sqrt{3600}} = 2^{60}$

$\qquad\qquad$ **b.** $2^{-\sqrt{3}} \cdot 2^{\sqrt{48}} = 2^{-\sqrt{3}} \cdot 2^{4\sqrt{3}} = 2^{3\sqrt{3}}$

It is customary to say that an expression is in simplest form when every exponent is positive.

EXAMPLE 2 **Simplify and express with positive exponents:**

a. $\dfrac{-32x^{\frac{4}{5}}y^{\frac{1}{6}}}{4x^{\frac{1}{5}}y^{\frac{5}{6}}}$ **b.** $\dfrac{x^{3\sqrt{2}}y^{-2\sqrt{3}}}{x^{4\sqrt{2}}y^{-4\sqrt{3}}}$ **c.** $(x^{-\sqrt{8}}y^{\sqrt{50}})^{\sqrt{2}}$

a. $\dfrac{-32x^{\frac{4}{5}}y^{\frac{1}{6}}}{4x^{\frac{1}{5}}y^{\frac{5}{6}}} = -8(x^{\frac{4}{5}-\frac{1}{5}})(y^{\frac{1}{6}-\frac{5}{6}})$

$$= -8x^{\frac{3}{5}}y^{-\frac{4}{6}}$$

$$= -8x^{\frac{3}{5}}y^{-\frac{2}{3}}$$

$$= \dfrac{-8x^{\frac{3}{5}}}{y^{\frac{2}{3}}}$$

b. $\dfrac{x^{3\sqrt{2}}y^{-2\sqrt{3}}}{x^{4\sqrt{2}}y^{-4\sqrt{3}}} = (x^{3\sqrt{2}-4\sqrt{2}})(y^{-2\sqrt{3}-(-4\sqrt{3})})$

$$= x^{-\sqrt{2}}y^{2\sqrt{3}}$$

$$= \dfrac{y^{2\sqrt{3}}}{x^{\sqrt{2}}}$$

c. $(x^{-\sqrt{8}}y^{\sqrt{50}})^{\sqrt{2}} = x^{-\sqrt{16}}y^{\sqrt{100}}$

$$= x^{-4}y^{10}$$

$$= \dfrac{y^{10}}{x^4}$$

You can also use a scientific calculator to evaluate expressions containing real number exponents. The exponent key makes this possible. To enter the exponent, use the parentheses keys or the memory.

EXAMPLE 3 **Evaluate the following to the nearest thousandth:**

a. $248^{\frac{1}{5}}$ **b.** $3^{\frac{2}{7}}$ **c.** $4^{-\sqrt{6}}$ **d.** $2^{\sqrt{2}+\sqrt{3}}$

a. $248^{\frac{1}{5}} = 3.012$ *Use the exponent key. Enter the exponent in parentheses: (1 ÷ 5) or use the memory.*

b. $3^{\frac{2}{7}} = 1.369$ *Use the exponent key. Enter the exponent in parentheses or use the memory.*

c. $4^{-\sqrt{6}} = 0.034$ *Use the exponent, $\sqrt{}$, and negative keys.*

d. $2^{\sqrt{2}+\sqrt{3}} = 8.854$ *Use the exponent key. Enter the exponent in parentheses or use the memory.*

An equation in which the variable is in the exponent is called an **exponential equation.** Examples of exponential equations are $3^x = 81$ and $2^x = \frac{1}{16}$.

To solve exponential equations, recall that for $b > 0$, $b \neq 1$, $b^m = b^n$ if and only if $m = n$. You can use this fact to solve an exponential equation by first expressing both sides of the equation in terms of the same base.

EXAMPLE 4 **Solve for x:**

a. $81 = 27^x$ **b.** $25^{n-1} = \left(\frac{1}{5}\right)^{4n-1}$

a. $81 = 27^x$ **b.** $25^{n-1} = \left(\frac{1}{5}\right)^{4n-1}$

$\quad 3^4 = (3^3)^x$ $\quad (5^2)^{n-1} = (5^{-1})^{4n-1}$

$\quad 3^4 = 3^{3x}$ $\quad\quad 5^{2n-2} = 5^{-4n+1}$

$\quad\ \ 4 = 3x$ $\quad\quad 2n - 2 = -4n + 1$

$\quad \dfrac{4}{3} = x$ $\quad\quad\quad\ 6n = 3$

$\quad\quad\quad\quad\quad\quad\quad\quad\quad\quad n = \dfrac{1}{2}$

CLASS EXERCISES

Simplify.

1. $2^{\sqrt{7}} \cdot 2^{\sqrt{28}}$

2. $\left(3^{\sqrt{2}}\right)^{\sqrt{18}}$

3. $5^{-\sqrt{5}} \cdot 5^{\sqrt{45}}$

Simplify and express with positive exponents.

4. $(3x^{0.7})(81^{-0.5}x^{0.3})$

5. $\dfrac{x^{5\sqrt{2}}y^{-3\sqrt{5}}}{x^{2\sqrt{2}}y^{\sqrt{5}}}$

6. $(x^{-\sqrt{3}}y)^{\sqrt{12}}$

Evaluate the following to the nearest thousandth.

7. $5^{0.75}$

8. $6^{\sqrt{7}}$

9. $12^{\sqrt{14}}$

Solve for x.

10. $7^4 = 7^{2x+1}$

11. $8^{x-1} = 16^{3x}$

12. $3^{4x-3} = 27^{2x-4}$

PRACTICE EXERCISES   Use appropriate technology.

Simplify.

1. $5^{\sqrt{6}} \cdot 5^{\sqrt{216}}$ **2.** $6^{\sqrt{12}} \cdot 6^{\sqrt{243}}$ **3.** $\left(4^{\sqrt{5}}\right)^{\sqrt{180}}$ **4.** $\left(3^{-\sqrt{8}}\right)^{-\sqrt{800}}$

Simplify and express with positive exponents.

5. $2^{\sqrt{200}} \cdot 2^{-\sqrt{8}}$ **6.** $3^{\sqrt{5}} \cdot 3^{-\sqrt{500}}$ **7.** $(3^{\sqrt{8}})^{-\sqrt{800}}$ **8.** $(4^{-\sqrt{128}})^{\sqrt{5}}$

9. $\dfrac{15x^{5\sqrt{3}}y^{-\sqrt{5}}}{45x^{\sqrt{3}}y^{\sqrt{5}}}$ **10.** $\dfrac{25x^{2\sqrt{2}}y^{7\sqrt{7}}}{15x^{5\sqrt{2}}y^{2\sqrt{7}}}$

11. $(x^{-\sqrt{162}}y^{\sqrt{242}})^{\sqrt{2}}$ **12.** $(x^{\sqrt{32}}y^{-\sqrt{200}})^{\sqrt{8}}$

Evaluate each of the following to the nearest thousandth.

13. $6^{\frac{3}{7}}$ **14.** $5^{\frac{5}{9}}$ **15.** $7^{\frac{6}{11}}$ **16.** $3^{\frac{6}{7}}$

17. $6^{-\frac{6}{7}}$ **18.** $5^{-\frac{5}{9}}$ **19.** $7^{-\frac{6}{11}}$ **20.** $3^{-\frac{6}{7}}$

21. $3^{\sqrt{2}+\sqrt{3}}$ **22.** $4^{\sqrt{3}+\sqrt{2}}$ **23.** $2^{\sqrt{5}+\sqrt{6}}$ **24.** $6^{\sqrt{7}+\sqrt{3}}$

Solve for x.

25. $3^x = 3^{2x-3}$ **26.** $5^4 = 25^{2x}$

27. $6^{x+4} = 6^{2x-1}$ **28.** $9^x = 9^{2x-1}$

29. $5^{x-2} = 5^{3x+1}$ **30.** $3^{3x} = 3^{5x+2}$

31. $2^{x+3} = 4^{x-5}$ **32.** $25^{x-1} = 125^x$

33. $4^{2x+1} = 8^{x+5}$ **34.** $8^{2x-1} = 16^x$

35. $\left(\dfrac{1}{2}\right)^{x+2} = 16^{-x}$ **36.** $\left(\dfrac{1}{5}\right)^{x+3} = 25^{-x}$

37. $(3a)^x = (9a^2)^{x+2}$ **38.** $(4a)^x = (16a^2)^{x+2}$

39. $3^{\sqrt{x}+4} = 27^{\sqrt{x}}$ **40.** $5^{\sqrt{x}+8} = 125^{\sqrt{x}}$

Applications

41. Biology   The number N of bacteria in a particular solution can be determined by the formula $N = N_0 2^t$, where t is the time measured in hours, and N_0 is the number of bacteria present at time $t = 0$. Find N if $N_0 = 20{,}000$ and $t = \frac{3}{7}$ h.

42. Archaeology   The number A of milligrams (mg) of carbon-14 in a fossil can be determined by the formula $A = A_0 2^{-0.75t}$, where t is the time in years, and A_0 is the number of milligrams of carbon-14 present at time $t = 0$. Find the amount of carbon-14 left after 5 years if $A_0 = 150$.

CHALLENGE

Which is greater, $16^{\sqrt{75}}$ or $32^{\sqrt{48}}$?

8.2

Exponential Functions

Objectives: To define exponential functions
To graph exponential functions

Exponential functions are very useful for expressing growth and decay in mathematical terms.

Preview

If you invest an amount of money in an account in which interest is compounded, the total amount A in the account at the end of one year is

$$A = P\left(1 + \frac{r}{n}\right)^n$$

where n is the number of times per year that interest is compounded, r is the yearly rate of interest, and P is the principal invested.

EXAMPLE **Mary Ann invests $1000 at 8% in an account in which interest is compounded twice per year. How much money is in her account at the end of one year?**

$$A = P\left(1 + \frac{r}{n}\right)^n$$
$$= 1000\left(1 + \frac{0.08}{2}\right)^2$$
$$= 1000(1.04)^2 \qquad \textit{Calculation-ready form}$$
$$= 1081.60 \qquad \textit{To the nearest cent}$$

There is $1081.60 in her account at the end of one year.

Find A to the nearest cent under each of the following conditions.

1. $P = \$1000, r = 8\%, n = 3$
2. $P = \$1000, r = 5\%, n = 4$
3. $P = \$12,000, r = 6\%, n = 4$
4. $P = \$100,000, r = 7.5\%, n = 4$

5. How much more interest will Mary Ann earn on $1000 invested at 8% if the interest is compounded monthly rather than twice per year?

6. How much more interest will she earn if the interest in Exercise 5 is compounded daily (assume 365 days per year) rather than twice per year?

An **exponential function** is a function in which the variable is in the exponent. If x and b are real numbers such that $b > 0$ and $b \neq 1$, then $f(x) = b^x$ is an exponential function with base b. Examples of exponential functions include $y = 3^x$, $f(x) = 6^x$, and $f(x) = 2^x$.

EXAMPLE 1   Graph the function $y = 2^x$.

Begin by constructing a table of values. Write nonintegral powers of 2 in decimal form. Then plot the points and join them with a smooth curve.

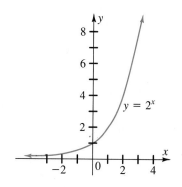

x	-3	-2	-1	0	1	2	3
y	0.125	0.25	0.5	1	2	4	8

Since the base of the function $y = 2^x$ is positive, the function is never negative. The x-axis is a horizontal asymptote. Any *horizontal line* intersects the graph in at most one point, so $y = 2^x$ is a *one-to-one function*.

Note that irrational numbers are included in the domain and range of $y = 2^x$. Its graph is a smooth curve including a point such as $(\sqrt{3}, 2^{\sqrt{3}})$. The approximate location of $(\sqrt{3}, 2^{\sqrt{3}})$ is shown on the graph at $(1.7, 3.3)$.

The value of y in Example 1 *increases* rapidly as x increases. Functions of this form are often said to increase *exponentially*. Some functions exhibit great *decreases* as x increases in value. The next example shows the graph of an exponential function in which the base is a positive number less than 1.

EXAMPLE 2   Graph the function $y = \left(\frac{1}{3}\right)^x$.

Write nonintegral powers of $\frac{1}{3}$ in decimal form.

Then plot the points and join them with a smooth curve.

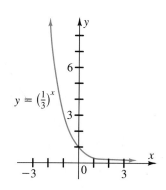

x	-3	-2	-1	0	1	2	3
y	27	9	3	1	0.333	0.111	0.037

The x-axis is again a horizontal asymptote. Since any horizontal line intersects the graph in at most one point, $y = \left(\frac{1}{3}\right)^x$ is a one-to-one function.

In Example 1, the graph approaches the negative x-axis. In Example 2, the graph approaches the positive x-axis. In general, the graph of $y = b^x$ rises to the right for $b > 1$, and to the left for $0 < b < 1$.

In advanced mathematics, science, and finance, the exponential function with base e is used frequently. The number e is irrational and can be approximated by 2.7182818. Many scientific calculators have an e^x key. To use most such calculators to evaluate e^x, press the e^x key and then enter the exponent.

EXAMPLE 3   **Find the value of:**   **a.** e^3 **b.** $e^{0.4}$ **c.** $e^{-1.2}$
Round your answers to three decimal places.

   a. $e^3 = 20.086$ **b.** $e^{0.4} = 1.492$ **c.** $e^{-1.2} = 0.301$

The function $f(x) = e^x$ is called the **natural exponential function.** These functions, like most exponential functions, can be graphed with a graphing utility.

EXAMPLE 4   **Use a graphing utility to graph the function $f(x) = e^x$.**

Use the e^x key to enter the function on your graphing utility.

x scl: 2   *y* scl: 2

CLASS EXERCISES

Evaluate each of the following for the function $f(x) = 3^x$.

1. $f(5)$ **2.** $f(-5)$ **3.** $f(0.5)$ **4.** $f(-0.5)$

5. Graph on the same set of axes: $f(x) = 2^x$, $f(x) = \left(\frac{1}{2}\right)^x$, $f(x) = 2^{-x}$.

6. How do the graphs of $f(x) = 2^x$ and $f(x) = \left(\frac{1}{2}\right)^x$ differ?

7. **Thinking Critically** The exponential function is defined for $b > 0$ and $b \neq 1$. How could you describe the graph if $b = 0$?   if $b = 1$?

PRACTICE EXERCISES Use appropriate technology.

Evaluate each of the following for the function $f(x) = 4^x$.

1. $f(2)$ 　　　　　　 2. $f(-3)$ 　　　　　　 3. $f(0.5)$ 　　　　　　 4. $f(-0.2)$

Evaluate each of the following for the function $f(x) = e^x$.

5. $f(4)$ 　　　　　　 6. $f(-4)$ 　　　　　　 7. $f(1.4)$ 　　　　　　 8. $f(-4.2)$

Graph each function.

9. $f(x) = 4^x$ 　　　 10. $f(x) = 7^x$ 　　　 11. $f(x) = 5^x$ 　　　 12. $f(x) = 6^x$

13. $f(x) = 5^{-x}$ 　　　 14. $f(x) = 10^{-x}$ 　　　 15. $f(x) = 2^{-x}$ 　　　 16. $f(x) = 4^{-x}$

17. $f(x) = \left(\frac{1}{2}\right)^x$ 　　 18. $f(x) = \left(\frac{1}{5}\right)^x$ 　　 19. $f(x) = \left(\frac{1}{4}\right)^x$ 　　 20. $f(x) = \left(\frac{1}{10}\right)^x$

21. $f(x) = 4^{2x}$ 　　 22. $f(x) = 3^{2x}$ 　　 23. $f(x) = 2^{2x+1}$ 　　 24. $f(x) = 3^{2x-1}$

25. $f(x) = \left(\frac{1}{2}\right)^{2x}$ 　 26. $f(x) = \left(\frac{1}{3}\right)^{2x}$ 　 27. $f(x) = e^{-x}$ 　　 28. $f(x) = e^{2x}$

29. $f(x) = 2^{x^2}$ 　　　　　　　　　　　　　 30. $f(x) = 2^{-x^2}$

Find the base of an exponential function, $y = b^x$, containing the points.

31. $\left(-2, \frac{1}{16}\right)$ 　　　　　　　　　 32. $\left(\frac{1}{2}, 2\right)$

Applications

33. **Algebra** If a number is squared and raised to a certain power, the result is the same number raised to a power that is one more than three times the power of the number being squared. What is the power?

34. **Optics** The optical intensity of objects can be found using the formula $I = 10^{-d}$. Graph the function and determine the intensity when $d = 0.8$.

EXTRA

One sheet of paper is 0.005 in. thick. Suppose you make stacks of computer paper such that each stack has twice as many sheets as the previous stack. How tall would the 30th stack be if the first stack has only 1 sheet?

8.3 Logarithmic Functions

Objectives: To define and graph logarithmic functions
To solve logarithmic equations

The concentration of hydrogen ions in a solution is called the pH of the solution. The pH is defined using a special function called a logarithm. The logarithmic function is the inverse of the exponential function.

Preview

Recall that the inverse of a function $y = f(x)$ can be found by interchanging the x- and y-coordinates and then solving for y. As you learned when studying the trigonometric functions, not all functions have inverses that are also functions.

EXAMPLE **Graph the function $y = x^2$ and its inverse on the same set of axes. Is the inverse also a function?**

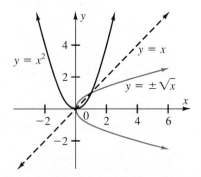

$y = x^2$
$x = y^2$ *Interchange x and y.*
$\pm\sqrt{x} = y$ *Solve for y.*

Since for each positive value of x there is more than one value of y, $y = \pm\sqrt{x}$ is *not* a function.

Note that the graph of $y = x^2$ is symmetric to the graph of its inverse, $y = \pm\sqrt{x}$, about the line $y = x$. Recall that the graphs of *all* functions are symmetric to their inverses about the line $y = x$.

Graph each function and its inverse on the same set of axes. Tell whether or not each inverse is also a function.

1. $y = 2x + 3$ **2.** $y = 3x + 6$ **3.** $y = x^3$

4. $y = -x^3$ **5.** $y = \sin x$ **6.** $y = \cos x$

The exponential function $y = b^x$ is a one-to-one function. Therefore, its inverse, $x = b^y$, is also a function. In order to write the inverse function $x = b^y$ in the form $y = f(x)$, the *logarithmic function* is defined.

For all positive real numbers x and b, $b \neq 1$, the inverse of the exponential function $y = b^x$ is the **logarithmic function**:

$$y = \log_b x$$ *Read: "y equals the logarithm to the base b of x" or "y equals log base b of x."*

$y = \log_b x$ if and only if $x = b^y$.

Recall that the graph of a function is symmetric to the graph of its inverse about $y = x$. Thus, the graph of $y = \log_b x$ is symmetric to that of $y = b^x$ about $y = x$.

 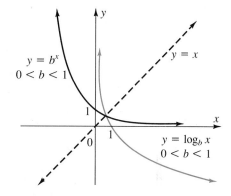

For the logarithmic function with $b > 1$, notice that as x increases, y increases. As x decreases, y decreases and the curve approaches the y-axis, its asymptote. For the logarithmic function with $0 < b < 1$, as x increases, y decreases. As x decreases, y increases and the curve approaches the y-axis, its asymptote.

Properties of Logarithmic Functions

- The domain is the set of positive real numbers.
- The range is the set of all real numbers.
- The x-intercept of the graph is 1.
- The y-axis is an asymptote of the graph.
- The function is one-to-one.

EXAMPLE 1   Graph $y = \log_3 x$.

Recall that $y = \log_3 x$ is equivalent to $x = 3^y$. Choose values for y and calculate values of x.

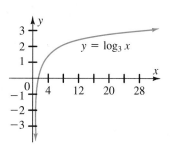

x	$\frac{1}{27}$	$\frac{1}{9}$	$\frac{1}{3}$	1	3	9	27
y	-3	-2	-1	0	1	2	3

You can express an equation given in logarithmic form in exponential form.

EXAMPLE 2   **Express in exponential form:**
   a. $\log_4 64 = 3$ **b.** $\log_2 32 = 5$ **c.** $\log_5 0.04 = -2$ **d.** $\log_{36} 6 = 0.5$

   a. $\log_4 64 = 3$ **b.** $\log_2 32 = 5$ **c.** $\log_5 0.04 = -2$ **d.** $\log_{36} 6 = 0.5$
          $4^3 = 64$ $2^5 = 32$ $5^{-2} = 0.04$ $36^{0.5} = 6$

You can also convert an expression given in exponential form to logarithmic form.

EXAMPLE 3   **Express in logarithmic form:**

   a. $7^2 = 49$ **b.** $5^4 = 625$ **c.** $16^{\frac{5}{4}} = 32$ **d.** $27^{-\frac{4}{3}} = \frac{1}{81}$

   a. $7^2 = 49$ **b.** $5^4 = 625$ **c.** $16^{\frac{5}{4}} = 32$ **d.** $27^{-\frac{4}{3}} = \frac{1}{81}$

        $\log_7 49 = 2$ $\log_5 625 = 4$ $\log_{16} 32 = \frac{5}{4}$ $\log_{27} \frac{1}{81} = -\frac{4}{3}$

Some logarithms can be evaluated by setting them equal to a variable and then rewriting the resulting equation in exponential form.

EXAMPLE 4   **Evaluate:**   **a.** $\log_3 81$ **b.** $\log_5 \frac{1}{25}$ **c.** $\log_2 2\sqrt{2}$

              a. $\log_3 81$ **b.** $\log_5 \frac{1}{25}$ **c.** $\log_2 2\sqrt{2}$

                 Let $x = \log_3 81$ Let $x = \log_5 \frac{1}{25}$ Let $x = \log_2 2\sqrt{2}$

                     $3^x = 81$ $5^x = \frac{1}{25}$ $2^x = 2\sqrt{2}$

                     $3^x = 3^4$ $5^x = 5^{-2}$ $2^x = 2^{\frac{3}{2}}$

                     $x = 4$ $x = -2$ $x = \frac{3}{2}$

Some logarithmic equations can be solved if they are converted to exponential form. If the base is not an integer, use a scientific calculator.

EXAMPLE 5   **Solve:**   **a.** $\log_{100} x = \frac{5}{2}$ **b.** $\log_5 125 = x$ **c.** $\log_x 32 = 3$

              a. $\log_{100} x = \frac{5}{2}$ **b.** $\log_5 125 = x$ **c.** $\log_x 32 = 3$

                     $100^{\frac{5}{2}} = x$ $5^x = 125$ $x^3 = 32$

                     $100{,}000 = x$ $5^x = 5^3$ $x = 32^{\frac{1}{3}}$

                                              $x = 3$ $x = 3.17$

CLASS EXERCISES

1. Thinking Critically How do the graphs of $y = \log_3 x$ and $y = \log_{10} x$ differ?

Express in logarithmic form.

2. $2^6 = 64$

3. $16^{0.5} = 4$

4. $10^{-3} = 0.001$

Express in exponential form.

5. $\log_6 36 = 2$

6. $\log_{25} 5 = 0.5$

7. $\log_{10} 0.01 = -2$

Evaluate.

8. $\log_{25} 5$

9. $\log_{10} 100$

10. $\log_{10} 0.001$

Solve.

11. $\log_x 625 = 4$

12. $\log_4 x = 3$

13. $\log_x 16 = 4$

PRACTICE EXERCISES   Use appropriate technology.

Express in exponential form.

1. $\log_8 64 = 2$

2. $\log_6 36 = 2$

3. $\log_3 27 = 3$

4. $\log_4 256 = 4$

5. $\log_{36} 6 = 0.5$

6. $\log_{81} 9 = 0.5$

7. $\log_2 \frac{1}{16} = -4$

8. $\log_3 \frac{1}{81} = -4$

Express in logarithmic form.

9. $4^3 = 64$

10. $6^3 = 216$

11. $49^{\frac{1}{2}} = 7$

12. $343^{\frac{1}{3}} = 7$

13. $27^{-\frac{2}{3}} = \frac{1}{9}$

14. $16^{-\frac{1}{4}} = \frac{1}{2}$

15. $7^{\frac{3}{2}} = 7\sqrt{7}$

16. $13^{\frac{3}{2}} = 13\sqrt{13}$

Evaluate.

17. $\log_{10} 1000$

18. $\log_{10} 1$

19. $\log_{10} 0.1$

20. $\log_{10} 0.0001$

21. $\log_4 0.5$

22. $\log_8 16$

23. $\log_9 27$

24. $\log_{16} 2$

Graph.

25. $y = \log_4 x$

26. $y = \log_3 x$

27. $y = \log_{\frac{1}{4}} x$

28. $y = \log_{\frac{1}{3}} x$

Solve.

29. $\log_4 64 = y$

30. $\log_x 16 = 4$

31. $\log_6 z = 3$

32. $\log_9 27 = y$

33. $\log_x 18 = 1$

34. $\log_3 z = -2$

35. $\log_3 243 = y$

36. $\log_x 10 = 0.5$

37. $\log_7 z = 0.5$　　**38.** $\log_{10} 10 = y$　　**39.** $\log_x 4 = \frac{1}{3}$　　**40.** $\log_{\frac{1}{2}} z = 8$

41. $\log_x 10 = 3.8$　　**42.** $\log_x 18 = 2.1$　　**43.** $\log_x 24 = 0.3$　　**44.** $\log_x 36 = 0.28$

45. $\log_x 9 = 14$　　**46.** $\log_x 17 = 5.4$　　**47.** $\log_x 4.2 = 0.7$　　**48.** $\log_x 12 = 0.19$

Graph and state the domain and range.

49. $y = 3 + \log_2 x$　　　　　　　　**50.** $y = -3 + \log_2 x$

51. $y = \log_{10}(x + 2)$　　　　　　　**52.** $y = \log_{10}(x + 4)$

Solve.

53. $\log_{\frac{2}{5}} \dfrac{4}{25} = \dfrac{x}{4}$　　　　**54.** $\log_{\frac{3}{5}} \dfrac{27}{125} = 2x$　　　　**55.** $\log_{\frac{3}{8}} \dfrac{9}{64} = 2x + 2$

56. $\log_{\frac{5}{8}} \dfrac{125}{512} = 2x - 7$　　　**57.** $\log_{\frac{2}{7}} \dfrac{8}{343} = 3x - 6$　　　**58.** $\log_{\frac{3}{7}} \dfrac{81}{2401} = 6x - 4$

Applications

59. Physics　The electric current I in a particular circuit is given by $\log_2 I = -t$. Write this formula in exponential form.

60. Technology　Use a graphing utility to graph the function $y = \log_{10} x$. Let the domain be the interval $0 \le x \le 20$ in increments of 1. Use the graph to evaluate $\log_{10} 1.9$ and $\log_{10} 5.3$.

TEST YOURSELF

Evaluate each of the following to the nearest thousandth.

1. $7^{\sqrt{11}}$　　　　　　　　　　　　**2.** $6^{-\frac{2}{9}}$　　　　　　　　　8.1

Solve each exponential equation.

3. $5^{2x+1} = 125^{x-1}$　　　　　　　**4.** $\left(\dfrac{1}{4}\right)^{x+3} = 2^{-x}$

5. Graph $f(x) = 3^{-x}$ and $f(x) = 3^x$ on the same set of axes.　　　8.2

6. Evaluate $6^{1.2}$ and $e^{-0.3}$.

7. Express $4^{-3} = \frac{1}{64}$ in logarithmic form and $\log_{49} 7 = \frac{1}{2}$ in exponential form.　　8.3

Solve.

8. $\log_4 1024 = x$　　**9.** $\log_5 \frac{1}{25} = x$　　**10.** $\log_x 11 = 0.5$　　**11.** $\log_{10} x = 6$

12. Graph $y = \log_5 x$ and $y = \log_{\frac{1}{5}} x$ on the same set of axes.

8.4

Properties of Logarithms

Objectives: To state the properties of logarithms
To use these properties to express logarithms in expanded form and to solve logarithmic equations

Loudness is a sensation that is perceived differently by different people. It is related to intensity, which is a physically measurable quantity. The human ear can detect sounds over a very wide range of intensities. However, what a human ear perceives as loudness is not directly proportional to the intensity of the sound. Therefore, a logarithmic scale is used to specify sound intensity.

Preview

The **intensity level** (β), measured in decibels (dB), of any sound is related to its **intensity** (I), measured in watts/meter^2 (W/m^2), using the following formula:

$$\beta = 10 \log_{10} \frac{I}{I_0}$$

where $I_0 = 1 \times 10^{-12}$ W/m^2, which is the minimum intensity audible to an average person and is called the *threshold of hearing*.

EXAMPLE   **Find the intensity level in decibels of the rustle of leaves, which has intensity 1×10^{-11} W/m^2.**

$$\beta = 10 \log_{10} \frac{10^{-11}}{10^{-12}}$$
$$= 10 \log_{10} 10$$
$$= 10(1)$$
$$= 10 \text{ dB}$$

1. Find the intensity level in dB of a whisper, which has intensity 1×10^{-10} W/m^2.

2. Find the intensity level of a quiet radio, which has intensity 1×10^{-8} W/m^2.

3. As the intensity is increased from 1×10^{-10} W/m^2 to 1×10^{-8} W/m^2 (by a factor of 100), the intensity level increases by how many decibels?

4. How much more intense would a 70-dB sound be than a 50-dB sound?

The properties of logarithms follow directly from the properties of exponents, because logarithmic functions and exponential functions are inverses. Recall that logarithms are defined only for positive real numbers and that the base of a logarithm can be any positive real number except 1.

Properties of Logarithms

If M, N, and b $(b \neq 1)$ are positive real numbers, and r is any real number, then

$$\log_b MN = \log_b M + \log_b N$$

$$\log_b \frac{M}{N} = \log_b M - \log_b N$$

$$\log_b N^r = r \log_b N$$

The expression on the right side of each of the above equations is called the *expanded form* of the logarithm on the left side.

Each of these properties can be derived from the properties of exponents.

To derive the first property, $\log_b MN = \log_b M + \log_b N$, let

$$\log_b M = u \quad \text{and} \quad \log_b N = v$$

Then $\quad b^u = M \quad$ and $\quad b^v = N$

$$MN = b^u b^v = b^{u+v}$$

$\log_b MN = u + v \qquad$ *Definition of logarithms*

Therefore $\quad \log_b MN = \log_b M + \log_b N \qquad$ *Substitute $\log_b M$ for u and $\log_b N$ for v.*

You will be asked to derive the other two properties in the exercises.

EXAMPLE 1 **Express in expanded form:**

   a. $\log_4 xy$ **b.** $\log_{10} \dfrac{I}{I_0}$ **c.** $\log_3 x^4$ **d.** $\log_5 \sqrt[4]{y}$

   a. $\log_4 xy = \log_4 x + \log_4 y$

   b. $\log_{10} \dfrac{I}{I_0} = \log_{10} I - \log_{10} I_0$

   c. $\log_3 x^4 = 4 \log_3 x$

   d. $\log_5 \sqrt[4]{y} = \log_5 y^{\frac{1}{4}} = \dfrac{1}{4} \log_5 y$

It is often necessary to use more than one property of logarithms when expanding an expression.

EXAMPLE 2 **Express in expanded form:** **a.** $\log_4 x^3 y^2$ **b.** $\log_5 \sqrt{\dfrac{x}{y^5}}$

a. $\log_4 x^3 y^2 = \log_4 x^3 + \log_4 y^2$ $log_b\ MN = log_b\ M + log_b\ N$
$\qquad\qquad\ = 3 \log_4 x + 2 \log_4 y$ $log_b\ N^r = r\ log_b\ N$

b. $\log_5 \sqrt{\dfrac{x}{y^5}} = \log_5 \left(\dfrac{x}{y^5}\right)^{\frac{1}{2}}$

$\qquad\qquad\ = \dfrac{1}{2} \log_5 \dfrac{x}{y^5}$ $log_b\ N^r = r\ log_b\ N$

$\qquad\qquad\ = \dfrac{1}{2} (\log_5 x - \log_5 y^5)$ $log_b\ \dfrac{M}{N} = log_b\ M - log_b\ N$

$\qquad\qquad\ = \dfrac{1}{2} (\log_5 x - 5 \log_5 y)$ $log_b\ N^r = r\ log_b\ N$

If the terms of a logarithmic expression in expanded form have the same base, you can rewrite the expression as a single logarithm.

EXAMPLE 3 **Express as a single logarithm:**
    a. $3 \log_3 x + \log_3 y$ **b.** $2 \log_4 x + \log_4 3 - \log_4 y$

a. $3 \log_3 x + \log_3 y = \log_3 x^3 + \log_3 y$ $r\ log_b\ N = log_b\ N^r$
$\qquad\qquad\qquad\ = \log_3 x^3 y$ $log_b\ M + log_b\ N = log_b\ MN$

b. $2 \log_4 x + \log_4 3 - \log_4 y = \log_4 x^2 + \log_4 3 - \log_4 y$ $r\ log_b\ N = log_b\ N^r$
$\qquad\qquad\qquad\qquad\ = \log_4 3x^2 - \log_4 y$ $log_b\ M + log_b\ N = log_b\ MN$

$\qquad\qquad\qquad\qquad\ = \log_4 \dfrac{3x^2}{y}$ $log_b\ M - log_b\ N = log_b\ \dfrac{M}{N}$

Since logarithmic functions and exponential functions are inverses, the following properties are true:

$$\log_b b^x = x \qquad \text{and} \qquad b^{\log_b x} = x$$

Two other important properties of logarithms are

$$\log_b b = 1 \qquad \textit{Recall that } b^1 = b.$$
$$\log_b 1 = 0 \qquad \textit{Recall that } b^0 = 1.$$

You can use these properties to evaluate logarithmic expressions.

EXAMPLE 4 **Evaluate the following expressions:**
    a. $3 \log_5 5 - \log_5 25$ **b.** $\log_4 16^2 + \log_4 8^3$

a. $3 \log_5 5 - \log_5 25 = 3(1) - \log_5 25$ $log_b\ b = 1$
$\qquad\qquad\qquad\qquad = 3 - \log_5 5^2$
$\qquad\qquad\qquad\qquad = 3 - 2 = 1$ $log_b\ b^x = x$

b. $\log_4 16^2 + \log_4 8^3 = 2 \log_4 16 + 3 \log_4 8$ $\quad log_b \ N^r = r \ log_b \ N$

$$= 2 \log_4 4^2 + 3 \log_4 4^{\frac{3}{2}}$$

$$= 2(2) + 3\left(\frac{3}{2}\right) \qquad\qquad log_b \ b^x = x$$

$$= 4 + \frac{9}{2}$$

$$= \frac{17}{2}$$

Since the logarithmic function is continuous and one-to-one, every positive real number has a unique logarithm to the base b. Therefore,

$$\log_b N = \log_b M \qquad \text{if and only if} \qquad N = M$$

This property can be used to solve *logarithmic equations*. It is important to check all possible solutions, since the logarithm of a variable expression is defined only when the expression is positive.

EXAMPLE 5   **Solve and check:**

a. $\log_3 4x + \log_3 x = \log_3 144$ **b.** $\log_5 (x^2 - 25) - \log_5 (x - 5) = 2$

a. $\log_3 4x + \log_3 x = \log_3 144$ **Check:** Reject -6 because logarithms are
$\qquad\quad \log_3 4x^2 = \log_3 144$ only defined for positive numbers.
$\qquad\qquad\quad 4x^2 = 144$ $\log_3 4x + \log_3 x = \log_3 144$
$\qquad\qquad\qquad x^2 = 36$ $\log_3 4(6) + \log_3 6 \overset{?}{=} \log_3 144$
$\qquad\qquad\qquad\quad x = \pm 6$ $\log_3 24 + \log_3 6 \overset{?}{=} \log_3 144$
$\qquad$ The solution is 6. $\log_3 144 \overset{?}{=} \log_3 144 \ \checkmark$

b. $\log_5 (x^2 - 25) - \log_5 (x - 5) = 2$ **Check:**
$$\log_5 \frac{x^2 - 25}{x - 5} = 2 \qquad \log_5 (x^2 - 25) - \log_5 (x - 5) = 2$$
$$\frac{(x + 5)(x - 5)}{x - 5} = 25 \qquad \log_5 [20^2 - 25] - \log_5 (20 - 5) \overset{?}{=} 2$$
$$x + 5 = 25 \qquad \log_5 (400 - 25) - \log_5 15 \overset{?}{=} 2$$
$$x = 20 \qquad \log_5 375 - \log_5 15 \overset{?}{=} 2$$
$$\log_5 25 \overset{?}{=} 2$$
The solution is 20. $\qquad\qquad\qquad\qquad\qquad\qquad 2 \overset{?}{=} 2 \ \checkmark$

CLASS EXERCISES

Thinking Critically   Tell whether each equation is true or false. If it is false, explain why.

1. $\log_5 \dfrac{x}{y} = \dfrac{\log_5 x}{\log_5 y}$ **2.** $\log_3 \sqrt{2x^2} = \dfrac{1}{2} \log_3 2 + \log_3 x^2$

3. $\log_4 \sqrt[n]{x^m} = \dfrac{m}{n} \log_4 x$

4. $\log_4 \sqrt[5]{5x} = \dfrac{1}{5} \log_4 5 + \dfrac{1}{5} \log_4 x$

5. $\log_2 x^3 y^4 = 3 \log_2 x + 4 \log_2 y$

6. $\dfrac{\log_3 x}{\log_3 y} = \log_3 x - \log_3 y$

7. $\log_6 \sqrt[5]{2y} = \dfrac{1}{5} \log_6 2y$

8. $\log_3 9 + \log_3 27 = 2 + 3$

PRACTICE EXERCISES

Express in expanded form.

1. $\log_3 xy$ **2.** $\log_4 xz$ **3.** $\log_5 2x$ **4.** $\log_2 3y$

5. $\log_2 \dfrac{x}{z}$ **6.** $\log_3 \dfrac{y}{x}$ **7.** $\log_3 \dfrac{x}{5}$ **8.** $\log_4 \dfrac{10}{y}$

9. $\log_3 x^2$ **10.** $\log_2 x^3$ **11.** $\log_6 \sqrt[5]{x}$ **12.** $\log_7 \sqrt[8]{y}$

13. $\log_2 x^4 y^3$ **14.** $\log_3 x^5 y^2$ **15.** $\log_2 (3x)^3$ **16.** $\log_3 (4x)^4$

17. $\log_3 \sqrt[3]{4}$ **18.** $\log_2 \sqrt[5]{6}$ **19.** $\log_5 xyz$ **20.** $\log_6 2xq$

Express as a single logarithm.

21. $\log_2 x + \log_2 y$

22. $\log_4 w + \log_4 z$

23. $\log_5 z + 2 \log_5 w - 3 \log_5 x$

24. $3 \log_n 4x - (\log_n 5y + \log_n 6z)$

Evaluate.

25. $\log_2 8 + \log_2 16$ **26.** $3 \log_3 81 - \log_3 27$ **27.** $\log_5 25 + \log_5 125$

28. $\log_8 16 + \log_8 256$ **29.** $\log_3 27 - 6 \log_3 9$ **30.** $\log_2 64 - 7 \log_2 4$

Solve and check.

31. $\log_{10} (x^2 + 36) = \log_{10} 100$

32. $\log_{10} (x^2 - 25) = \log_{10} 144$

Express in expanded form.

33. $\log_3 \sqrt{\dfrac{x^2}{y}}$ **34.** $\log_4 \sqrt{\dfrac{x}{y^2}}$ **35.** $\log_2 \sqrt{\dfrac{x^2}{y^3}}$ **36.** $\log_3 \sqrt{\dfrac{x^3}{y^2}}$

37. $\log_4 3x^2 y^4$ **38.** $\log_5 6x^3 y^3$ **39.** $\log_3 (2xy)^2$ **40.** $\log_4 (4yz)^3$

41. $\log_b \sqrt[4]{x^5}$ **42.** $\log_b \sqrt[5]{x^4}$ **43.** $\log_b \sqrt[3]{x^2 y^4}$ **44.** $\log_b \sqrt{7x^5}$

45. $\log_b \dfrac{3\sqrt{x}}{y^4}$ **46.** $\log_b \dfrac{2x^2}{\sqrt{y}}$ **47.** $\log_b \dfrac{6x^4 y}{\sqrt[3]{z}}$ **48.** $\log_b \dfrac{2x^2 y^2}{\sqrt[5]{z}}$

Express as a single logarithm.

49. $\frac{1}{2} \log_4 x^2 + \frac{1}{3} \log_4 y$

50. $\frac{1}{5} \log_4 x^2 + \frac{1}{2} \log_4 y^3$

51. $\frac{1}{3} \log_4 x - \frac{1}{2} \log_4 y$

52. $\log_3 x - \frac{1}{4} \log_3 y$

Solve and check.

53. $\log_6 n - \log_6 3 = 2$

54. $\log_5 n - \log_5 4 = 2$

55. $\log_5 4x + \log_5 x = \log_5 100$

56. $\log_2 9x + \log_2 x = \log_2 900$

Express in expanded form.

57. $\log_b \dfrac{(\sqrt[3]{x})(\sqrt{y^3})}{\sqrt[4]{z^3}}$

58. $\log_b \dfrac{(\sqrt{x^3})(y^4)}{\sqrt[3]{z^2}}$

59. $\log_2 \dfrac{\sqrt{x^2 y^5}}{zx^3}$

60. $\log_b \dfrac{\sqrt[3]{xyz}}{6yz}$

Solve and check.

61. $\log_3 (x + 1) - \log_3 x = \log_3 4$

62. $\log_4 x - \log_4 (x - 1) = \log_4 5$

63. $\log_7 (x - 5) + \log_7 (x + 1) = 1$

64. $\log_2 (9x + 5) - \log_2 (x^2 - 1) = 2$

65. Derive $\log_b \dfrac{M}{N} = \log_b M - \log_b N$ using properties of exponents.

66. Derive $\log_b N^r = r \log_b N$ using properties of exponents.

Applications

67. Physics Expand the formula for intensity level $\beta = 10 \log_{10} \dfrac{I}{I_0}$.

68. Physics Find the intensity level of an indoor rock concert with intensity 1.

TRIGONOMETRY IN CHEMISTRY

Chemists use the pH scale (values range from 0 to 14) to measure the acidity or alkalinity of a solution. The hydrogen ion concentration of a solution determines whether it is an acid (with a pH below 7), a base (with a pH above 7), or neutral (with a pH equal to 7). The pH of a solution is determined by the formula $pH = \log_{10} \dfrac{1}{H^+}$, where H^+ is the concentration of hydrogen ions in gram atoms per liter.

Find the pH of a solution if $H^+ = 10^{-5}$.

Evaluating Logarithms and Solving Exponential Equations

Objectives: To evaluate common and natural logarithms
To solve exponential equations
To change the base of a logarithmic expression

Logarithms were developed to simplify complicated calculations. Since our number system is based on powers of 10, base-10 logarithms, or **common logarithms,** were tabulated. In order to find the value of the common logarithm of a number using a table, you must first express the number in scientific notation. Calculators, which frequently use a form of scientific notation, considerably simplify work involving logarithms.

Preview

A positive number of the form $a \times 10^b$, where $1 \le a < 10$ and b is an integer, is in scientific notation.

Decimal form	Scientific notation
6,100,000	6.1×10^6
0.000123	1.23×10^{-4}

Express each number in scientific notation.

1. 6120 **2.** 160,000,000 **3.** 0.0525 **4.** 0.0000028

Express each number in decimal form.

5. 4.05×10^4 **6.** 2.4×10^{-2} **7.** 1.02×10^{11} **8.** 7.13×10^{-5}

It is customary to omit the base when writing a common logarithm.

$$\log N = \log_{10} N$$

 You can use the log key on a scientific calculator to find the common logarithm of a number. Although most logarithms displayed by a calculator are approximations, the $=$ symbol, rather than the $\approx$ symbol, is customarily used.

EXAMPLE 1 **Find each logarithm to four decimal places:**
a. log 340 **b.** log 34 **c.** log 3.4 **d.** log 0.34 **e.** log 0.034

a. $\log 340 = 2.5315$ **b.** $\log 34 = 1.5315$ **c.** $\log 3.4 = 0.5315$

d. $\log 0.34 = -0.4685$ **e.** $\log 0.034 = -1.4685$

Note that common logarithms are of base 10. Therefore, if numbers differ by a factor of 10, like those in Example 1, their common logarithms differ by 1. Thus, log 340 is between 2 and 3, log 34 is between 1 and 2, log 3.4 is between 0 and 1, log 0.34 is between -1 and 0, and log 0.034 is between -2 and -1. The common logarithm of a number can be expressed as the sum of an integer, called the *characteristic*, and a nonnegative number less than 1, called the *mantissa* (0.5315 in this example).

If you know log x, you can find x. The number x is called the *antilogarithm*, or *antilog*, of log x.

 Recall that 10^x is the inverse of $\log_{10} x$. If your calculator has a 10^x key, you can use it to find the antilog. If not, you can use the inverse key and the log key. Remember that with some calculators you must press the function key before the number; with others the order is reversed.

EXAMPLE 2   **Find x to four significant digits:**

     a. log x = 3.233   **b.** log x = 13.06

     a. log x = 3.233   *Enter 3.233. Use the inverse and log keys, or the 10^x key.*

           x = 1710   *To four significant digits*

     b. log x = 13.06   *Enter 13.06. Use the inverse and log keys, or the 10^x key.*

           x = 1.148 × 10^{13} *To four significant digits. Your calculator uses scientific notation to express this number.*

Natural logarithms are to the base e. They are of great importance in the study of calculus and in other areas of mathematics. Just as the symbol *log x* is used for common logarithms, the symbol *ln x* is used for natural logarithms.

$$\ln x = \log_e x \qquad \textit{ln is read ``el en.''}$$

 You can use the ln key on a scientific calculator to find the natural logarithm of a number.

EXAMPLE 3   **Find each logarithm to four decimal places:**

     a. ln 367   **b.** ln 0.023

     a. ln 367 = 5.9054   *Enter 367 and use the ln key.*

     b. ln 0.023 = -3.7723 *Enter 0.023 and use the ln key.*

 The inverse of $\log_e x$, or ln x, is e^x. If your calculator has an e^x key, you can use it to find the antilog of a natural logarithm. If your calculator does not have an e^x key, you can use the inverse key and the ln key.

EXAMPLE 4 **Find x to four significant digits:**
a. $\ln x = 3.2335$ **b.** $\ln x = -28.62$

a. $\ln x = 3.2335$ *Enter 3.2335. Use the inverse and*
In keys, or the e^x key.
$x = 25.37$ *To four significant digits*

b. $\ln x = -28.62$ *Enter 28.62. Use the $+/-$ key. Then use*
the inverse and In keys, or the e^x key.
$x = 3.720 \times 10^{-13}$ *To four significant digits. Your*
calculator uses scientific notation
to express this number.

A scientific calculator makes it easier to evaluate complicated logarithmic expressions. There is often more than one correct way to make entries. You may have to use the parentheses keys or the memory feature. Use your owner's manual and experiment with the keys to determine the sequences of keys that yield the correct answers in Example 5 below.

EXAMPLE 5 **Evaluate. Round each answer to four decimal places.**

a. $5(\ln 2.3 + \ln 18)$ **b.** $\dfrac{\log 34}{\log 8 + \log 3}$

a. $5(\ln 2.3 + \ln 18) = 18.6164$ *Use the parentheses keys:*
$5 \times$ (In 2.3 + In 18).

b. $\dfrac{\log 34}{\log 8 + \log 3} = 1.1096$ *Use the parentheses keys:*
log 34 $\div$ (log 8 + log 3).

Logarithms can be used to solve exponential equations in which both sides cannot be written in terms of the same base, as the following example illustrates.

EXAMPLE 6 **Find x to four significant digits:**
a. $3^x = 8$ **b.** $4^{x+2} = 15$

a. $3^x = 8$
$\log 3^x = \log 8$ *Take the log of both sides.*
$x \log 3 = \log 8$ *$\log_b N^r = r \log_b N$*
$x = \dfrac{\log 8}{\log 3}$ *Calculation-ready form*
$x = 1.893$

Check: Use the y^x key on your calculator to show that $3^{1.893}$ is approximately equal to 8.

b.
$$4^{x+2} = 15$$
$$\log 4^{x+2} = \log 15 \qquad \text{Take the log of both sides.}$$
$$(x + 2)\log 4 = \log 15 \qquad log_b \, N^r = r \, log_b \, N$$
$$x \log 4 + 2 \log 4 = \log 15 \qquad \text{Distributive property}$$
$$x \log 4 = \log 15 - 2 \log 4$$
$$x = \frac{\log 15 - 2 \log 4}{\log 4} \qquad \text{Calculation-ready form}$$
$$x = -0.0466$$

Check: Use the y^x key on your calculator to show that $4^{-0.0466+2}$ is approximately equal to 15.

The method demonstrated in Example 7 *changes* the base of an exponential expression. If you know the logarithm of a number to one base (the *reference base*), you can use the *change-of-base formula* to find the logarithm of that number to some other base.

Change-of-Base Formula

$$\log_b x = \frac{\log_a x}{\log_a b} \qquad \textit{The reference base is a.}$$

Let $\qquad n = \log_b x$

Then $\qquad b^n = x \qquad\qquad$ *Definition of logarithm*

$\qquad \log_a b^n = \log_a x \qquad$ *Take log_a of both sides.*

$\qquad n \log_a b = \log_a x \qquad log_b \, N^r = r \, log_b \, N$

$\qquad n = \dfrac{\log_a x}{\log_a b}$

$\qquad \log_b x = \dfrac{\log_a x}{\log_a b} \qquad$ *Replace n with log_b x.*

Since a scientific calculator can be used to approximate values of common logarithms, you can replace a with 10 in the change-of-base formula.

EXAMPLE 7 **Use common logarithms to find $\log_6 5$ to four decimal places.**

$$\log_6 5 = \frac{\log_{10} 5}{\log_{10} 6} \qquad log_b \, x = \frac{log_{10} \, x}{log_{10} \, b}$$

$$= \frac{\log 5}{\log 6}$$

$$= 0.8982$$

CLASS EXERCISES

1. **Thinking Critically** Would the results of Example 6 be different if you took the natural log of both sides of the equation instead of the common log? Explain.

Find each logarithm to four decimal places.

2. log 456

3. log 46.27

4. log 0.45

5. ln 32.1

6. ln 0.345

7. ln 68.234

8. $\log_3 7$

9. $\log_2 2.4$

10. $\log_5 11.1$

Find x to four significant digits.

11. log $x = 3.2341$

12. log $x = -1.7823$

13. log $x = 0.2891$

14. ln $x = 0.3679$

15. ln $x = -3.2738$

16. ln $x = 2.9845$

PRACTICE EXERCISES Use appropriate technology.

Find each logarithm to four decimal places.

1. log 792

2. log 26.36

3. log 0.37

4. log 312.4

5. log 0.0034

6. log 0.0000468

7. ln 56.2

8. ln 0.123

9. ln 32.567

10. ln 0.345

11. ln 0.000425

12. ln 1.239

Find x to four significant digits.

13. log $x = 2.3176$

14. log $x = -3.1842$

15. log $x = 0.3345$

16. log $x = 1.3924$

17. log $x = -1.3657$

18. log $x = -3.5743$

19. ln $x = 0.3679$

20. ln $x = 3.2738$

21. ln $x = 2.9845$

22. ln $x = -2.5691$

23. ln $x = 0.8734$

24. ln $x = -1.2363$

Evaluate. Round each answer to four decimal places.

25. log 3 + log 7

26. log 153 − log 34

27. 6(log 3.2 + log 4.5)

28. $\dfrac{\log 5}{\log 7}$

29. $\dfrac{5(\log 13)}{\log 4}$

30. $\dfrac{\log 45.1}{3(\log 8)}$

Find x to four significant digits.

31. log $x = 12.2481$

32. log $x = 21.2934$

33. log $x = -13.4218$

34. ln $x = -23.5347$

35. ln $x = 10.4321$

36. ln $x = 23.2634$

Solve and check. Round answers to four significant digits.

37. $2^x = 9$

38. $3^x = 10$

39. $4^x = 23$

40. $5^x = 27$

41. $2^{2x} = 19$

42. $3^{2x} = 20$

43. $3^{x+2} = 11$

44. $7^{x+3} = 17$

45. $7^{x-1} = 19$

Use common logarithms to find each logarithm to four decimal places.

46. $\log_3 11$

47. $\log_4 9$

48. $\log_6 86$

49. $\log_5 65$

50. $\log_{0.8} 70$

51. $\log_{2.8} 75$

52. $\log_{1.2} 7$

53. $\log_{\sqrt{3}} 17$

54. $\log_{\sqrt{2}} 38$

Solve and check. Round answers to four significant digits.

55. $6.2^{2x-1} = 74.1$

56. $2.9^{\sqrt{2x}\,+\,1} = 87.6$

57. $2.5^{\sqrt{2x}\,+\,2} = 90.5$

Evaluate. Give each answer to four decimal places.

58. $\dfrac{(\log 22)^2}{2 + \log 17}$

59. $\sqrt{\dfrac{2 + \log 6}{1 + \log 12}}$

60. $\dfrac{1 + 3\ln 3}{3 + \sqrt{\ln 3}}$

Applications

61. Astronomy The limiting magnitude L of an optical telescope with a lens of diameter d, in inches, is $L = 8.8 + 5.1 \log d$. Find the limiting magnitude of a 6-in. telescope.

62. Business Solve $1200 = 1000\left(1 + \dfrac{0.05}{4}\right)^{4t}$ for t to determine how many years it will take \$1000, deposited at 5% interest compounded quarterly, to grow to \$1200. (Assume no withdrawals or additional deposits are made.)

63. Business Solve $1200 = 1000\left(1 + \dfrac{0.05}{12}\right)^{12t}$ for t to determine how many years it will take \$1000, deposited at 5% interest compounded monthly, to grow to \$1200. (Assume no withdrawals or additional deposits are made.)

64. Business Solve $1200 = 1000e^{0.05t}$ for t to determine how many years it will take \$1000, deposited at 5% interest compounded continuously, to grow to \$1200.

MATH CLUB ACTIVITY

How can common logarithms be used to determine the number of digits in a number of the form n^{n^n}, where n is an integer? Use your answer to find the number of digits in 4^{4^4}.

Applications: Exponential and Logarithmic Equations

Objectives: To use exponential and logarithmic equations to solve real-world problems

Exponential and logarithmic equations can be used to describe exponential growth and decay, learning curves, and logistic growth. These equations are used by scientists, economists, and sociologists. Exponential equations also can be used to model the interest accumulated on bank deposits or loans.

Preview

Interest may be compounded periodically. For example, it may be compounded semiannually, quarterly, or daily. The following formula gives the total amount in a savings account after a given period of time.

Compound Interest Formula

$$A = P\left(1 + \frac{r}{n}\right)^{nt}$$

A = total amount (interest + principal), in dollars
P = principal invested, in dollars
r = rate of annual interest, expressed as a decimal
n = number of times per year the interest is compounded
t = number of years the principal is invested

If interest is compounded *continuously*, it can be shown that the formula becomes $A = Pe^{rt}$.

EXAMPLE **How much money will Wai Ming have at the end of 5 yr if he deposits $1000 at 9% interest compounded semiannually?**

$$A = P\left(1 + \frac{r}{n}\right)^{nt}$$

$$= 1000\left(1 + \frac{0.09}{2}\right)^{2(5)} = \$1552.97 \quad \textit{Calculation-ready form}$$

1. How much money will Wai Ming have if interest is compounded quarterly?
2. How much money will he have if interest is compounded monthly?
3. How much money will he have if interest is compounded daily?
4. How much money will he have if interest is compounded continuously?
5. Do your answers to questions 3 and 4 differ greatly? Why?
6. If interest is compounded monthly, approximately how many years will it take the investment to increase to $2500?

The number of bacteria in a culture increases exponentially.

EXAMPLE 1 **The growth rate for a certain bacterial culture can be calculated using the formula $B = 1000\,(2)^{\frac{t}{48}}$, where B is the number of bacteria and t is the elapsed time in hours.**

a. How many bacteria will be present after 4 hours?
b. How long will it take, to the nearest hour, for there to be 12,000 bacteria in the culture?

a. $B = 1000(2)^{\frac{t}{48}}$

$\qquad = 1000(2)^{\frac{4}{48}}$ *Calculation-ready form*

$\qquad = 1059$

There will be approximately 1059 bacteria present after 4 h.

b. $\qquad B = 1000(2)^{\frac{t}{48}}$

$\quad 12{,}000 = 1000(2)^{\frac{t}{48}}$

$\qquad 12 = 2^{\frac{t}{48}}$

$\qquad \log 12 = \log 2^{\frac{t}{48}}$ *Take the log of both sides.*

$\qquad \log 12 = \dfrac{t}{48}\,\log 2$ *$\log_b N^r = r\,\log_b N$*

$\qquad \dfrac{48\,\log 12}{\log 2} = t$ *Calculation-ready form*

$\qquad 172 = t$

It will take approximately 172 hours for there to be 12,000 bacteria in the culture.

It is possible to approximate the age of fossils by measuring the amount of carbon-14, a radioactive isotope, that remains inside them, and then applying the *exponential decay formula*.

Exponential Decay Formula

$$A = A_0 2^{-\frac{t}{k}}$$

A = present amount of the radioactive isotope
A_0 = original amount of the radioactive isotope, measured in the same units as A
t = time it takes to reduce original amount of the isotope to present amount
k = half-life of the isotope, measured in the same units as t

The half-life is the average time required for one-half of the atoms of a sample of a radioactive substance to decay.

EXAMPLE 2 A fossil that originally contained 100 mg of carbon-14 now contains 75 mg of the isotope. Determine the approximate age of the fossil, to the nearest 100 years, if the half-life of carbon-14 is 5570 yr.

$$A = A_0 2^{-\frac{t}{k}}$$

$$75 = 100(2)^{-\frac{t}{5570}}$$

$$0.75 = 2^{-\frac{t}{5570}}$$

$$\log 0.75 = \log 2^{-\frac{t}{5570}} \qquad \textit{Take the log of both sides.}$$

$$\log 0.75 = -\frac{t}{5570} \log 2 \qquad \textit{log}_b \; N^r = r \; \textit{log}_b \; N$$

$$\frac{5570 \log 0.75}{\log 2} = -t \qquad \textit{Calculation-ready form}$$

$$2311.76 = t$$

The fossil is approximately 2300 years old.

Sociologists use exponential equations to describe the spread of a rumor.

EXAMPLE 3 In a town of 15,000 people, the spread of a rumor that the local transit company would go on strike was such that t hours after the rumor started, $f(t)$ persons heard the rumor, where experience over time has shown that

$$f(t) = \frac{15,000}{1 + 7499e^{-0.8t}}$$

a. How many people started the rumor?
b. How many people heard the rumor after 5 hours?
c. How many hours does it take for 14,000 people to hear the rumor?

a. $f(t) = \dfrac{15,000}{1 + 7499e^{-0.8t}}$

$f(0) = \dfrac{15,000}{1 + 7499e^0}$

$= 2$

Two people started the rumor.

b. $f(t) = \dfrac{15,000}{1 + 7499e^{-0.8t}}$

$f(5) = \dfrac{15,000}{1 + 7499e^{-0.8(5)}}$

$= 108$

After 5 hours, 108 people heard the rumor.

c.

$$f(t) = \frac{15{,}000}{1 + 7499e^{-0.8t}}$$

$$14{,}000 = \frac{15{,}000}{1 + 7499e^{-0.8t}}$$

$$14{,}000 + 14{,}000(7499e^{-0.8t}) = 15{,}000$$

$$104{,}986{,}000e^{-0.8t} = 1000$$

$$e^{-0.8t} = \frac{1000}{104{,}986{,}000}$$

$$\ln e^{-0.8t} = \ln\left(\frac{1000}{104{,}986{,}000}\right) \quad \text{Take ln of}$$
$$\text{both sides.}$$

$$-0.8t \ln e = \ln\left(\frac{1000}{104{,}986{,}000}\right) \quad log_b \; N^r = r \; log_b \; N$$

$$-0.8t = \ln\left(\frac{1000}{104{,}986{,}000}\right) \quad ln \; e = 1$$

$$t = 14.45$$

It takes approximately 14 hours for 14,000 people to hear the rumor.

CLASS EXERCISES

1. **Thinking Critically** Why was the natural logarithm (ln) taken of both sides of the equation in Example 3(c) instead of the common logarithm (log)?

Solve.

2. If the bacteria culture in Example 1 is allowed to grow for 12 h, how many bacteria will be present?

3. A radioactive isotope has a half-life of 5 days. How many days will it take for a 10 g sample to decay to 4 g?

4. Jones Savings and Loan pays 7% interest compounded semiannually. If $2500 is deposited for 10 yr, how much money will be in the account at the end of that time period?

5. Approximately how many years will it take for $1600 to increase to $2100 if it is deposited in a bank that pays 7.5% interest compounded quarterly?

PRACTICE EXERCISES   Use appropriate technology.

Solve. Round answers to the nearest whole number unless otherwise told.

1. The growth rate for a particular bacterial culture can be calculated using the formula $B = 900(2)^{\frac{t}{50}}$, where B is the number of bacteria and t is the elapsed time in hours. How many bacteria will be present after 5 h?

2. Approximately how many hours will it take for there to be 9000 bacteria present in the culture in Exercise 1?

3. How many hours will it take for there to be 18,000 bacteria present in the culture in Exercise 1?

4. A dinosaur bone was found to have 80 g of carbon-14. If the bone originally contained 200 g of the isotope, determine the approximate age of the bone to the nearest 100 years. (Carbon-14 has a half-life of 5570 yr.)

5. A fossil that originally contained 150 mg of carbon-14 now contains 85 mg of the isotope. Determine the approximate age of the fossil to the nearest 100 years.

6. A shell fossil now contains 60 mg of carbon-14. If there were originally 150 mg of the isotope in the fossil, determine the approximate age of the fossil, to the nearest 100 years.

7. A radioactive isotope has a half-life of 4 days. How many days will it take for a 15-g sample of this isotope to decay to 3 g?

8. A particular isotope has a half-life of 25 days. How many days will it take for a 8-mg sample to decay to 2 mg?

In Exercises 9–12, round answers to the nearest tenth.

9. Dan invests $5000 at 8% annual interest compounded monthly. How many years will it take him to earn $2000 interest?

10. Debbie invests $4000 at 7% annual interest compounded semiannually. How many years will it take her to earn $3000 interest?

11. Allison invests $3400 at 8.5% annual interest compounded quarterly. How many years will it take her money to double in value?

12. Eric invests $2500 at 10.5% annual interest compounded monthly. How many years will it take his money to double in value?

13. If Allison invests the same amount of money at the same interest rate as in Exercise 11 compounded *continuously*, how many years will it take for her investment to double in value?

14. Mr. Kakutani found 12 mg of a radioactive isotope in a soil sample. After 5 h, only 8.2 mg of the isotope remained. Determine the approximate half-life of the isotope.

15. Ms. Carmenza discovered 15 mg of an isotope in a fossil. After 10 h, only 13.2 mg of the isotope remained. Determine the approximate half-life of the isotope.

16. In the town of Middlebury, the spread of a virus was such that t weeks after its outbreak, $f(t)$ persons had contracted it, where $f(t) = \dfrac{12{,}000}{1 + 599e^{-0.7t}}$. How many people had the virus at its outbreak?

17. In Exercise 16, how many weeks will it take for 3000 people to contract the virus?

18. In Exercise 16, how many weeks will it take for 6000 people to contract the virus?

19. As a person's experience increases, competence at completing a task increases rapidly at first, and then slows down as additional experience is gained. This phenomena can be represented by a *learning curve*. After t hours of typing, José can type $f(t) = 80(1 - e^{-0.025t})$ words per minute. How many words per minute can he be expected to type after 10 h of practice?

20. In Exercise 19, how many words per minute can José type after 25 h of practice?

TEST YOURSELF

Express as a single logarithm.

1. $\log_4 d + 3 \log_4 2e - 3 \log_4 f$ 8.4

2. $(\log_5 2x + 3 \log_5 y) - \log_5 3z$

Express in expanded form.

3. $\log_b 2xy^2 z^3$ **4.** $\log_3 \dfrac{8x^3}{y^2}$

Find each logarithm to four decimal places.

5. $\log 329$ **6.** $\log 0.042$ 8.5

7. $\ln 0.321$ **8.** $\ln 24.3$

Find x to four significant digits.

9. $\log x = 2.5642$ **10.** $\log x = 2.1226$

11. $\ln x = 0.2639$ **12.** $\ln x = -1.1728$

13. Use common logarithms to find $\log_5 8$ to four decimal places.

14. Evaluate $3(\log 60.2 + \log 7.5) - \log 21.3$ to four decimal places.   8.6

15. Sally invests $9000 at 7.5% annual interest compounded quarterly. How many years will it take for her investment to increase to $11,500?

16. Kareem found 25 mg of an isotope in a fossil. After 12 h, only 20 mg of the isotope remained. Find the approximate half-life of the isotope.

TECHNOLOGY:
The Richter Scale

Magnitude	Intensity Level	Probable Effect
1	10^1	Detectable only by
2	10^2	seismographic instruments
3	10^3	Slight damage
4	10^4	within a
5	10^5	limited area
6	10^6	Moderate damage
7	10^7	Major damage
8	10^8	Catastrophic damage

The Richter scale, created in 1935 by seismologist Charles Richter, is used to measure the intensity of an earthquake.

Because the Richter scale is logarithmic, the energy released increases by powers of 10 in relation to the Richter numbers. For example, an earthquake of magnitude 7 is 10 times more powerful than one of magnitude 6 and 10^2 or 100 times more powerful than one of magnitude 5. Thus, an earthquake of magnitude 8 is not twice as powerful as one of magnitude 4, but 10^4 or 10,000 times more powerful.

EXERCISES

1. Explain why $10^{R_2 - R_1}$ can be used to compare an earthquake of magnitude R_2 and one of magnitude R_1.

2. The highest Richter magnitude ever recorded was 8.9. How many times more powerful was this quake than the 1988 Armenian quake of magnitude 6.9?

3. Is it true that an earthquake of magnitude 2.5 is half as intense as an earthquake of magnitude 5?

4. Compare the difference between a quake which measures 3.4 and one which measures 3.9, and the difference between a quake which measures 7.4 and one which measures 7.9. Is this result surprising?

5. Research the history of earthquakes in the Western hemisphere to learn if seismologists use mathematical models to predict earthquakes.

Vocabulary

change-of-base formula (356)
common logarithm (353)
compound interest formula (359)
exponential decay formula (360)
exponential equation (336)

exponential function (339)
logarithmic function (343)
natural exponential function (340)
natural logarithm (354)

Real Exponents   Expressions containing real number exponents can be   **8.1**
simplified using the properties of exponents and evaluated using the y^x key on
a scientific calculator.

An equation in which the variable is in the exponent is called an exponential
equation. Some exponential equations can be solved by first expressing both
sides of the equation in terms of the same base.

Evaluate the following to the nearest thousandth.

1. $81^{\frac{1}{11}}$

2. $25^{\sqrt{3}}$

Solve for x.

3. $5^4 = 5^{2x+1}$

4. $4^{x-1} = 8^{3x}$

Exponential Functions   An exponential function is a function in which the   **8.2**
variable is in the exponent. If x and b are real numbers such that $b > 0$ and
$b \neq 1$, then $f(x) = b^x$ is an exponential function.

Evaluate each of the following for the function $f(x) = 3^x$.

5. $f(5)$ **6.** $f(-2)$ **7.** $f(0.5)$

8. Graph $f(x) = 3^x$.

Logarithmic Functions $y = \log_b x$ means y is the exponent to which b is   **8.3**
raised to produce x. Logarithmic functions and exponential functions are
inverses.

Express in logarithmic form.

9. $3^4 = 81$

10. $49^{0.5} = 7$

Express in exponential form.

11. $\log_4 64 = 3$

12. $\log_9 3 = 0.5$

Evaluate.

13. $\log_{81} 9$

14. $\log_{10} 0.01$

Solve for x.

15. $\log_x 125 = 3$

16. $\log_6 x = 3$

17. $\log_3 27 = x$

18. $\log_x 50 = 4$

Properties of Logarithms $\log_b MN = \log_b M + \log_b N$ 8.4

$$\log_b \frac{M}{N} = \log_b M - \log_b N$$

$$\log_b N^r = r \log_b N$$

19. Express $\log_4 \dfrac{x^5 y^2}{z}$ in expanded form.

20. Express $2 \log_n 3s - (\log_n 4t + 3 \log_n v)$ as a single logarithm.

Solve and check.

21. $\log_2 6 + \log_2 3 - \log_2 n = \log_2 9$ **22.** $\log_{10} (x - 2) + \log_{10} (x + 1) = 1$

Evaluating Logarithms and Solving Exponential Equations   8.5
Common (base 10) and natural (base e) logarithms can be computed using a scientific calculator. If you know the common or natural logarithm of a number, you can find the number. Logarithms can be used to solve exponential equations. The change-of-base formula is $\log_b x = \dfrac{\log_a x}{\log_a b}$.

Evaluate to four decimal places.

23. $\log 47.65$

24. $\ln 0.056$

Find x to four significant digits.

25. $\log x = -3.1252$

26. $\ln x = 1.5378$

27. Evaluate $2(\log 51.3 + \log 7.8) - \log 32$ to four decimal places.

28. Use common logarithms to evaluate $\log_3 8$ to four decimal places.

29. Use natural logarithms to evaluate $\log_4 7$ to four decimal places.

30. Joanie invests $8000 at 6.75% annual interest compounded semiannually.   8.6
How many years will it take her investment to double in value?

31. A radioactive isotope has a half-life of 4 days. How many days will it take
for a 12-g sample to decay to 3 g?

Evaluate to the nearest thousandth.

1. $(64)^{\frac{7}{9}}$

2. $6^{\sqrt{5}}$

3. $e^{1.2}$

Solve for x.

4. $7^{x+4} = 49^{2x-1}$

5. $8^{x+2} = 16^{2x}$

Evaluate.

6. $\log_5 125$

7. $\log_{27} 3$

8. $\log_8 16$

Solve for x.

9. $\log_3 x = 6$

10. $\log_2 x = 4$

11. $\log_x 21 = 3$

12. $\log_x 8 = 1.6$

13. Express $\log_2 4m + 3 \log_2 n - 3 \log_2 p^3$ as a single logarithm.

14. Express $\log_7 \dfrac{8x^3 y}{z}$ in expanded form.

Find each logarithm to four decimal places.

15. $\log 26.37$

16. $\ln 0.1572$

Find x to four significant digits.

17. $\log x = 2.6134$

18. $\ln x = -2.5124$

19. Use common logarithms to find $\log_5 9$ to four decimal places.

20. The growth rate for a particular bacterial culture can be calculated using the formula $B = 1000(2)^{\frac{t}{40}}$, where B is the number of bacteria and t is the elapsed time in hours. How many hours will it take for there to be 8000 bacteria present?

Challenge

A rare diamond was purchased in 1930 for $1500, and its value has doubled every 10 yr since that time. Let $f(t)$ represent its value t yr after its purchase. Find an equation for $f(t)$ and use it to find the value of the diamond in 1990.

In each item you are to compare a quantity in Column 1 with a quantity in Column 2. Write the letter of the correct answer from these choices:

A. The quantity in Column 1 is greater than the quantity in Column 2.
B. The quantity in Column 2 is greater than the quantity in Column 1.
C. The quantity in Column 1 is equal to the quantity in Column 2.
D. The relationship cannot be determined from the information given.

Notes: Information centered over both columns refers to one or both of the quantities being compared. A symbol that appears in both columns has the same meaning in each column. All variables represent real numbers. Most figures are not drawn to scale.

Column 1	Column 2
$f(x) = 2^x$	
1. $f(0)$	$f(0.5)$
2. $\log_5 1$	$\log_{64} 8$
3. $\log (MN)$	$\log M + \log N$
4. $\log (8^2)$	$\log \frac{8}{2}$
$\log_5 \sqrt[8]{5^7} = x$	
5. x	0.25
$-6x < 2x - 16$	
6. x	2
$5^x = 7$	
7. x	$\log 7$
$f(x) = \left(\frac{1}{3}\right)^{-x}$	
8. $f(1)$	$f(-1)$
$\log_x 0.04 = -2$	
9. 5	x

Column 1	Column 2
$a: 3x - 2y = 4$	
$b: 4x + y = -2$	
10. slope of line a	slope of line b
$x < y$	
11. $\frac{x^2}{y}$	$\frac{y}{x^2}$
$3^{x+2} = 27^x$	
12. x	2
$\log (x - 3) + \log x = 1$	
13. x	6
14. $\frac{1}{3} \log 27$	$\log \sqrt{9}$
15. $\ln e$	1
16. 10^x	e^x
17. $\frac{\log M}{\log N}$	$\log M - \log N$
$x > y$	
18. $\log x$	$\log \frac{1}{y}$

Evaluate each of the following for the function $f(x) = e^{x+2}$.

1. $f(3)$ **2.** $f(-3.1)$

3. Find the three cube roots of -8. Express your answers in rectangular form.

4. Prove $\dfrac{\cot^2\alpha + 1}{\sec^2\alpha} = \cot^2\alpha$.

5. Construct a table of values and graph the equation $r = 2 + 2\cos\theta$.

6. Evaluate $\arccos\left(-\dfrac{\sqrt{2}}{2}\right)$ in radians.

Find each sum or difference.

7. $(7 + 2i) - (5 + i)$ **8.** $(4 - i\sqrt{2}) + (-3 - 2i\sqrt{2})$

Find each logarithm to four decimal places.

9. $\ln 159.7$ **10.** $\log 0.0798$

Evaluate each of the following using a double-angle identity.

11. $\sin 2\left(\dfrac{\pi}{4}\right)$ **12.** $\tan 2(60°)$

Express in logarithmic form.

13. $6^4 = 1296$ **14.** $32^{-\frac{1}{5}} = \dfrac{1}{2}$

Find each product or quotient and simplify.

15. $5(\cos 42° + i\sin 42°) \cdot 2(\cos 78° + i\sin 78°)$

16. $10(\cos 90° + i\sin 90°) \div 5(\cos 45° + i\sin 45°)$

17. Find the length of the arc intercepted by a central angle of $\theta = \dfrac{\pi}{2}$ and a radius of 8 in. Express the answer in terms of π.

18. Solve and check $2^x = 7$.

19. Use the law of sines to find $\angle B$ given $\angle A = 112°$, $a = 23$, and $b = 13$.

20. Sketch the graph of the function $y = 2\sin\left(x - \dfrac{\pi}{4}\right) + 3$ over the interval $0 \le x < 2\pi$.

Sequences and Series

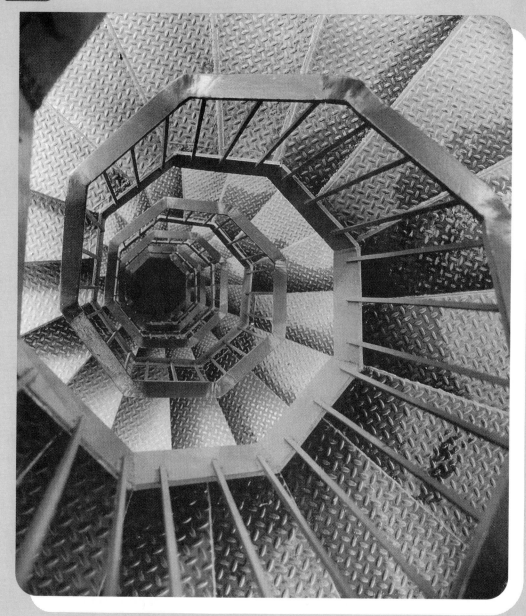

Spirals often occur in nature. They are used as a symbol or ornament in the world of art. The spiral staircase shown above is indeed a piece of artwork. Archimedean spirals are characterized by arithmetic sequences, and logarithmic spirals by geometric sequences.

9.1

Arithmetic Sequences

Objectives: To develop and use a formula to find specified terms of an arithmetic sequence

To find arithmetic means

In order to define a *sequence*, it is helpful to recall the definitions of a relation and a function.

Preview

A *relation* is a set of ordered pairs. Set T, shown below, is a relation.

$$T = \{(1, 2), (2, 4), (3, 6), (4, 8)\}$$

The domain of this relation is $\{1, 2, 3, 4\}$ and the range is $\{2, 4, 6, 8\}$. A relation is a *function* if for every element of the domain, there is one and only one corresponding element of the range. Thus, the relation T is also a function. T can be described by the rule $f(x) = 2x$, where x is a member of the set $\{1, 2, 3, 4\}$.

Find the domain and range of each relation and state whether or not the relation is a function.

1. $\{(2, 4), (3, 4), (5, 4), (6, 4)\}$
2. $\{(-2, 1), (1, 3), (4, 1), (5, -2), (-2, 0)\}$
3. $f(x) = \dfrac{1}{x}$
4. $f(x) = x^2$
5. $f(x) = \pm x$
6. $f(x) = -2x^2 + 1$

The following table shows the value, in pennies, of from 1 to 8 nickels.

number of nickels	1	2	3	4	5	6	7	8
value in pennies	5	10	15	20	25	30	35	40

The set of ordered pairs $\{(1, 5), (2, 10), \ldots, (8, 40)\}$ follows a pattern and is an example of a *sequence*. A **sequence** is a function whose domain is a set of consecutive positive integers. Each number in the range of a sequence is called a *term*. Using symbols, the first term is denoted by a_1, the second term by a_2, and the nth term, or **general term,** by a_n. The nickels-and-pennies sequence can also be written as 5, 10, 15, . . . , 40, where $a_1 = 5$, $a_2 = 10$, and so on. This is an example of a **finite sequence,** since it has a last term.

A sequence can be defined **explicitly** by expressing a_n as a function of n, where n is a positive integer. An explicit formula for the nickels-and-pennies sequence is $a_n = 5n$, where n is a member of the set $\{1, 2, 3, 4, 5, 6, 7, 8\}$. That is, the general term of the sequence is $5n$.

A sequence can also be defined *recursively* by stating the first term, a_1, and a rule for obtaining the $(n + 1)$th term from the preceding term, a_n. A **recursive formula** for the nickels-and-pennies sequence is $a_1 = 5$, $a_{n+1} = a_n + 5$, where n is a member of $\{1, 2, 3, 4, 5, 6, 7, 8\}$.

An **infinite sequence** does not have a last term. However, any desired number of consecutive terms can be found if the sequence is defined recursively.

EXAMPLE 1 **The following infinite sequence is defined recursively: $a_1 = 7$ and $a_{n+1} = a_n + 2$. Write the first five terms.**

$a_1 = 7$

$a_2 = 7 + 2 = 9$ $\qquad$ $a_2 = a_1 + 2$

$a_3 = 9 + 2 = 11$ $\qquad$ $a_3 = a_2 + 2$

$a_4 = 11 + 2 = 13$ $\qquad$ $a_4 = a_3 + 2$

$a_5 = 13 + 2 = 15$ $\qquad$ $a_5 = a_4 + 2$

The first five terms of the sequence are 7, 9, 11, 13, and 15.

An **arithmetic sequence,** or an *arithmetic progression*, is a sequence of numbers in which the difference between any two successive terms is a constant. This constant is called the **common difference.** You can find the common difference, d, by subtracting any two consecutive terms. That is,

$$d = a_{n+1} - a_n$$

This formula can be written in a different way to obtain a recursive formula for an arithmetic sequence. If the first term, a_1, is given, then

$$a_{n+1} = a_n + d$$

EXAMPLE 2 **Find the next three terms of the infinite arithmetic sequence 18, 24, 30, ...**

Since $24 - 18 = 6$, the common difference is 6.

Then, $\quad 30 + 6 = 36$ $\qquad$ $a_{n+1} = a_n + 6$

$36 + 6 = 42$

$42 + 6 = 48$ $\qquad$ The next three terms are 36, 42, and 48.

Consider the terms of the sequence in Example 2.

a_1	a_2	a_3	a_4	a_5	a_6
18	24	30	36	42	48
$18 + 0(6)$	$18 + 1(6)$	$18 + 2(6)$	$18 + 3(6)$	$18 + 4(6)$	$18 + 5(6)$
$a_1 + 0(d)$	$a_1 + 1(d)$	$a_1 + 2(d)$	$a_1 + 3(d)$	$a_1 + 4(d)$	$a_1 + 5(d)$

In general, an explicit formula for the general term, a_n, of an arithmetic sequence is

$$a_n = a_1 + (n - 1)d$$

where a_1 is the first term and d is the common difference.

EXAMPLE 3   **Find the general term a_n of the arithmetic sequence -8, -3, 2, ...**

$a_1 = -8$ and $d = -3 - (-8) = 5$

$a_n = -8 + (n - 1)(5)$ $\qquad\qquad a_n = a_1 + (n - 1)d$

$a_n = -8 + 5n - 5$

$a_n = 5n - 13$

The formula for the general term a_n can be used to find any specified term of a sequence if the first term and the common difference are known.

EXAMPLE 4   **Find the sixth term of the arithmetic sequence if $a_1 = 11$ and $d = 4$.**

$a_6 = 11 + (6 - 1)(4)$ *Substitute $a_1 = 11$, $n = 6$, and $d = 4$ in the formula $a_n = a_1 + (n - 1)d$.*

$= 11 + 20 = 31$ *The sixth term is 31.*

If two terms of an arithmetic sequence are known, a_1 and d can be found by using the formula for the general term twice and then solving the resulting system of equations.

EXAMPLE 5   **The tenth term of an arithmetic sequence is 77 and the sixteenth term is 119. Find a_1 and d.**

$a_n = a_1 + (n - 1)d$ *Use the formula for the nth term twice.*

$77 = a_1 + (10 - 1)d$ *Let $a_{10} = 77$ and $n = 10$.*

$119 = a_1 + (16 - 1)d$ *Let $a_{16} = 119$ and $n = 16$.*

$77 = a_1 + 9d$ *Simplify each equation and solve the system of equations.*

$119 = a_1 + 15d$

$-77 = -a_1 - 9d$ *Multiply both sides of the first equation by -1.*

$\underline{119 = a_1 + 15d}$

$42 = 6d$ *Add.*

$7 = d$ *Solve for d.*

$77 = a_1 + 9(7)$ *Substitute $d = 7$ in one of the original equations and solve for a_1.*

$77 = a_1 + 63$

$14 = a_1$ *Thus, $a_1 = 14$ and $d = 7$.*

The terms between any two nonconsecutive terms of an arithmetic sequence are called **arithmetic means.** In the sequence 5, 12, 19, 26, 33, . . . , the terms 12, 19, and 26 are the arithmetic means between 5 and 33. A single arithmetic mean between any two given numbers is called *the arithmetic mean* or the *average* of the two numbers. For all real numbers a and b, the arithmetic mean is $\dfrac{a + b}{2}$.

Any finite number of arithmetic means can be inserted between two numbers.

EXAMPLE 6 **Insert five arithmetic means between 13 and −11.**

Let $a_1 = 13$ and $a_7 = -11$. The sequence is 13, a_2, a_3, a_4, a_5, a_6, −11.

$a_7 = a_1 + (7 - 1)d$ *Use the formula $a_n = a_1 + (n - 1)d$, where $n = 7$.*
$-11 = 13 + 6d$ *Substitute: $a_7 = -11$, $a_1 = 13$.*
$-24 = 6d$
$-4 = d$ *Solve for d.*

Therefore, $a_2 = 13 + (-4) = 9$ $a_{n+1} = a_n + d$
$a_3 = 9 + (-4) = 5$
$a_4 = 5 + (-4) = 1$
$a_5 = 1 + (-4) = -3$
$a_6 = -3 + (-4) = -7$

The five arithmetic means between 13 and −11 are 9, 5, 1, −3, and −7.

> For any arithmetic sequence in which a_n is the general term, a_1 is the first term, d is the common difference, and n is a positive integer
> $$a_n = a_1 + (n - 1)d$$
> $$a_{n+1} = a_n + d \quad \text{or} \quad d = a_{n+1} - a_n$$

CLASS EXERCISES

Consider the arithmetic sequence 3, 8, 13, ... Find each value.

1. d **2.** a_1 **3.** a_3 **4.** a_4 **5.** a_5 **6.** a_6

Determine whether or not each sequence is arithmetic. If it is, find the common difference and write the general term, a_n.

7. 15, 18, 21   **8.** 14, 16, 19   **9.** 10, 5, 10   **10.** −5, −3, −1

11. 7, 11, 16   **12.** 1, 2, 3   **13.** −1, −2, −3   **14.** 4, 0, 4

PRACTICE EXERCISES

The following sequences are defined recursively. Give the first five terms of each sequence.

1. $a_{n+1} = a_n + 4$;
$a_1 = 5$

2. $a_{n+1} = a_n + 6$;
$a_1 = 3$

3. $a_{n+1} = a_n - 3$;
$a_1 = 4$

4. $a_{n+1} = a_n - 4$;
$a_1 = 6$

5. $a_{n+1} = a_n + 6$;
$a_1 = -17$

6. $a_{n+1} = a_n + 3$;
$a_1 = -1$

List the next four terms of each arithmetic sequence. Then, find the general term, a_n.

7. $7, 11, 15, \ldots$

8. $3, 6, 9, \ldots$

9. $-32, -28, -24, \ldots$

10. $\frac{1}{2}, \frac{3}{2}, \frac{5}{2}, \ldots$

11. $1, -4, -9, \ldots$

12. $-2.5, 1.5, 5.5, \ldots$

Find the given term of each arithmetic sequence.

13. $a_1 = 5, d = 5; a_7$

14. $a_1 = 4, d = 3; a_3$

15. $a_1 = -\frac{1}{3}, d = -\frac{2}{3}; a_5$

16. $a_1 = 15, d = -2; a_8$

17. $a_1 = 3, d = -2.5; a_{11}$

18. $a_1 = 0.8, d = 1; a_{15}$

Find the first term and the common difference for each arithmetic sequence, using the given terms.

19. $a_3 = 11, a_8 = 26$

20. $a_5 = 24, a_{12} = 45$

21. $a_6 = -19, a_8 = -29$

22. $a_7 = -23, a_{14} = -37$

23. $a_{21} = 13, a_{27} = 16$

24. $a_{50} = 5.4, a_{75} = 7.9$

25. Insert three arithmetic means between 6 and 46.

26. Insert six arithmetic means between 18 and 53.

27. Insert four arithmetic means between 6 and -6.

28. Insert three arithmetic means between -8 and 1.

29. Insert two arithmetic means between 6.2 and 7.8.

30. Insert two arithmetic means between -5.4 and -2.6.

Find the indicated term in each arithmetic sequence.

31. a_{12} of $-16, -13, -10, \ldots$

32. a_{21} of $10, 7, 4, \ldots$

33. a_{32} of $4, 7, 10, \ldots$

34. a_6 of $8, 3, -2, \ldots$

35. a_{12} of $\frac{3}{4}, \frac{3}{2}, \frac{9}{4}, \ldots$

36. a_{10} of $\frac{5}{3}, \frac{6}{3}, \frac{7}{3}, \ldots$

37. In the arithmetic sequence $-9, -2, 5, \ldots$, which term is 131?

38. In the arithmetic sequence $-3, 2, 7, \ldots$, which term is 142?

39. In the arithmetic sequence 7, 2, -3, . . . , which term is -28?

40. In the arithmetic sequence $\frac{9}{4}$, 2, $\frac{7}{4}$, . . . , which term is $-\frac{17}{4}$?

41. Find three arithmetic means between a and b.

42. The arithmetic mean of -22 and another number is -8. Find the other number.

43. The last term of an arithmetic sequence is 207, the common difference is 3, and the number of terms is 14. What is the first term of the sequence?

44. The fifth term of an arithmetic sequence is 19 and the eleventh term is 43. Find the first term and the sixty-seventh term of the sequence.

45. The fifth term of an arithmetic sequence is 3 and the fifteenth term is 8. What is the general term of the sequence?

46. The third term of an arithmetic sequence is 12 and the thirteenth term is 7. What is the general term of the sequence?

Find a_1 and d in terms of r and s for each arithmetic sequence.

47. $a_4 = r; a_9 = s$ **48.** $a_6 = r + s;\ a_{11} = 25$

Applications

49. Consumerism   In 1970, a family membership in a country club cost $650. This cost increased $35 each year since 1970. How much will a 1990 membership cost?

50. Business   A department store chain had 112 franchises in 1978 and opened 8 new franchises each year thereafter. How many franchises were there at the end of 1985?

EXTRA

Number theory is the study of relationships among numbers, particularly positive integers. Some number theory problems deal with sequences.

1. In an arithmetic sequence of three positive integers, can the sum of the terms be a prime number? *Hint*: Let a_1, $a_1 + d$, and $a_1 + 2d$ represent the three terms of the sequence, where d is the common difference. If the sum is divisible by an integer other than 1, it is not prime.

2. In an arithmetic sequence of five positive integers, can the sum of the terms be a prime number?

3. Is it possible to have an arithmetic sequence of six positive integers whose sum is not divisible by 6? If so, give an example.

Geometric Sequences

9.2

Objectives: To develop and use a formula to find specified terms of a
geometric sequence
To find geometric means

In many sequences, the terms increase or decrease by a constant factor rather
than by a constant addend.

Preview

The Singletons have just purchased a car for
$15,000. The car is expected to depreciate by
20 percent per year. Therefore, the value of the
car each year will be 0.8 times its value from
the previous year.

At the end of the first year, the car will be worth

$15,000(0.8) = $12,000

At the end of the second year, the car will
be worth

$12,000(0.8) = $15,000(0.8)^2 = $9600

Find the value of the car, to the nearest dollar, at the end of each year.

1. year 3   **2.** year 4   **3.** year 5   **4.** year 8

A **geometric sequence,** or *geometric progression*, is a sequence in which the
ratio of any two consecutive terms is a constant. This constant is called the
common ratio. To illustrate, in the geometric sequence 1, 3, 9, 27, . . . , each
term is 3 times the preceding term. That is, the common ratio is $\frac{3}{1} = \frac{9}{3} = \frac{27}{9} = 3$.

In general, the common ratio r can be found by dividing any term in the
sequence by the preceding term. Thus,

$$r = \frac{a_{n+1}}{a_n}$$

Note that this formula can be written in a different way to give a recursive
formula for a geometric sequence for which the first term, a_1, is known.

$$a_{n+1} = a_n r$$

EXAMPLE 1 **Write the first six terms of the geometric sequence with first term 2 and common ratio 5.**

Begin with 2 and multiply by 5 to get each succeeding term.

$a_1 = 2$

$a_2 = 2(5) = 10$ $\qquad$ $a_2 = a_1 r$

$a_3 = 10(5) = 50$ $\qquad$ $a_3 = a_2 r$

$a_4 = 50(5) = 250$ $\qquad$ $a_4 = a_3 r$

$a_5 = 250(5) = 1250$ $\qquad$ $a_5 = a_4 r$

$a_6 = 1250(5) = 6250$ $\qquad$ $a_6 = a_5 r$

Each term of a geometric sequence can be written as the product of the first term, a_1, and a power of r.

Term		
1	a_1	$= a_1$
2	a_2	$= a_1 r$
3	$a_3 = a_2 r = a_1 r \cdot r$	$= a_1 r^2$
4	$a_4 = a_3 r = a_1 r^2 \cdot r$	$= a_1 r^3$
5	$a_5 = a_4 r = a_1 r^3 \cdot r$	$= a_1 r^4$

An explicit formula for the nth term, or general term, of a geometric sequence, where a_1 is the first term and r is the common ratio, is

$$a_n = a_1 r^{n-1}$$

EXAMPLE 2 **Write the general term, a_n, of the geometric sequence 3, 12, 48, ...**

$a_1 = 3$ and $r = \dfrac{12}{3} = 4$

$a_n = 3(4)^{n-1}$ $\qquad\qquad$ $a_n = a_1 r^{n-1}$

You may wish to use a calculator when you work with geometric sequences.

EXAMPLE 3 **Find the sixth term of a geometric sequence with first term 3 and common ratio 4**

$a_6 = 3(4)^{6-1}$ $\qquad$ *Use the formula $a_n = a_1 r^{n-1}$, where $a_1 = 3$, $n = 6$, $r = 4$.*

$= 3(4)^5$ $\qquad\qquad$ *Calculation-ready form.*

$= 3072$

It is possible to find the first term and the common ratio of a geometric sequence if any two terms are given.

EXAMPLE 4   **The third term of a geometric sequence is 32 and the fifth term is 128. Find the first term and the common ratio.**

$$32 = a_1r^2$$ *Use the formula $a_n = a_1r^{n-1}$, where $a_3 = 32$ and $n = 3$.*
$$128 = a_1r^4$$ *Now let $a_5 = 128$ and $n = 5$.*

$$\frac{128}{32} = \frac{a_1r^4}{a_1r^2}$$ *Solve the system of equations by dividing.*

$$4 = r^2$$
$$\pm 2 = r$$

$$32 = a_1(2)^2 \qquad 32 = a_1(-2)^2$$ *Use one of the original equations*
$$32 = 4a_1 \qquad\quad 32 = 4a_1$$ *to solve for a_1.*
$$8 = a_1 \qquad\qquad 8 = a_1$$ *Thus $a_1 = 8$ and $r = 2$ or $r = -2$.*

The terms between any two nonconsecutive terms of a geometric sequence are called **geometric means.** Any finite number of geometric means can be inserted between two given numbers.

EXAMPLE 5   **Insert four geometric means between 972 and 4.**

Let $a_1 = 972$ and $a_6 = 4$. Then, the sequence is: $972, a_2, a_3, a_4, a_5, 4$.

$$a_6 = a_1r^{6-1}$$ *Use the formula $a_n = a_1r^{n-1}$, where $n = 6$.*
$$4 = 972r^5$$ *Substitute: $a_6 = 4$, $a_1 = 972$.*

$$\frac{4}{972} = r^5$$

$$\left(\frac{4}{972}\right)^{\frac{1}{5}} = r$$ *Solve for r.*

$$\frac{1}{3} = r$$

Thus, $a_2 = 972\left(\frac{1}{3}\right) = 324$ $a_3 = 324\left(\frac{1}{3}\right) = 108$ $a_{n+1} = a_nr$

$\qquad\quad a_4 = 108\left(\frac{1}{3}\right) = 36$ $a_5 = 36\left(\frac{1}{3}\right) = 12$

The geometric means are 324, 108, 36, and 12.

A single geometric mean between two numbers is called *the geometric mean,* or the **mean proportional,** between the two numbers. If $\dfrac{a}{m} = \dfrac{m}{b}$, where a and b are nonzero real numbers, then m is the mean proportional between a and b. Thus, $m^2 = ab$ and $m = \sqrt{ab}$ or $m = -\sqrt{ab}$. It is customary to let m equal $\sqrt{ab}$ when a and b are positive, and $-\sqrt{ab}$ when a and b are negative.

EXAMPLE 6 **Find the mean proportional between -12 and -300.**

a and b are negative, so $m = -\sqrt{ab}$.

$m = -\sqrt{(-12)(-300)}$ **Check:** $\dfrac{-12}{-60} \overset{?}{=} \dfrac{-60}{-300}$

$= -60$ $\dfrac{1}{5} = \dfrac{1}{5}$

For any geometric sequence in which a_n is the general term, a_1 is the first term, r is the common ratio, and n is a positive integer,

$$a_n = a_1 r^{n-1}$$

$$a_{n+1} = a_n r \quad \text{or} \quad r = \frac{a_{n+1}}{a_n}$$

CLASS EXERCISES

Determine whether or not each sequence is geometric. If it is, find the common ratio and write the general term a_n.

1. $-3, -12, -48$ **2.** $-2, 4, -6$ **3.** $20, 10, 5$ **4.** $18, 6, 2$

Find the specified term of each geometric sequence.

5. $a_1 = -1, r = 3; a_4$ **6.** $a_1 = 2, r = \frac{1}{2}; a_6$

7. $a_1 = 16, r = \frac{1}{4}; a_5$ **8.** $a_1 = 81, r = -\frac{1}{3}; a_8$

9. Insert three geometric means between 2 and 162.

10. Find the mean proportional of 16 and 36. Check your answer.

PRACTICE EXERCISES   Use appropriate technology.

Find the first four terms of each geometric sequence.

1. $a_1 = 3.5, r = 3$ **2.** $a_1 = 3, r = -4$ **3.** $a_1 = 1, r = -2$

4. $a_1 = 1.5, r = 2$ **5.** $a_1 = 8, r = -\frac{1}{2}$ **6.** $a_1 = 10, r = -\frac{1}{5}$

List the next three terms of each geometric sequence. Then, write the general term a_n.

7. $3, -9, 27, \ldots$ **8.** $64, -32, 16, \ldots$ **9.** $9, 3, 1, \ldots$

10. $-\frac{1}{2}, 1, -2, \ldots$ **11.** $-\frac{1}{9}, -\frac{1}{3}, -1, \ldots$ **12.** $\sqrt{3}, 3, 3\sqrt{3}, \ldots$

Find the specified terms of each geometric sequence.

13. $a_1 = 4, r = 6; a_3$ 14. $a_1 = 1, r = 5; a_4$ 15. $a_3 = 32, r = -\frac{1}{2}; a_6$

16. $a_4 = 16, r = \frac{1}{2}; a_8$ 17. $2x, 4x, 8x, \ldots; a_7$ 18. $y^3, y^4, y^5, \ldots; a_{10}$

19. Insert two geometric means between 1 and 8.

20. Insert five geometric means between 32 and $\frac{1}{2}$.

21. Insert four geometric means between $\frac{1}{3}$ and 81.

22. Insert three geometric means between 81 and 1.

23. Insert four geometric means between -3 and 96.

24. Insert three geometric means between 5 and $\frac{16}{125}$.

Find the mean proportional between each pair of numbers.

25. -12 and -108 26. -16 and -49 27. 15 and 60   28. 16 and 100

Find the common ratio and the first term for each geometric sequence.

29. $a_5 = 16, a_8 = 2$ 30. $a_8 = -21, a_{10} = -189$

31. $a_6 = 54, a_8 = 216$ 32. $a_4 = 100, a_6 = 10{,}000$

33. $a_7 = 6, a_{11} = \frac{3}{8}$ 34. $a_9 = 1.1, a_{13} = 0.00011$

35. The fourth term of a geometric sequence is -10 and the sixth term is -40. Find the ninth term of the sequence.

36. The second term of a geometric sequence is -2 and the fifth term is 54. Find the eighth term of the sequence.

37. Find the value of x so that $x - 1$, $x + 2$, and $x + 8$ are the first three terms of a geometric sequence.

38. Find the value of x so that $x - 4$, $x - 2$, and $x + 4$ are the first three terms of a geometric sequence.

39. If $a_1 = 100$ and $r = 0.01$, which term in the geometric sequence is 10^{-10}?

40. If $a_1 = x^2$ and $r = x^4$, which term in the geometric sequence is x^{18}?

Applications

41. **Business** Zoom Airlines' passenger load has been increasing by 12 percent annually. In 1980 they carried 20,500 passengers. How many passengers should they expect to carry in 1990?

42. **Business**   Monica invested $50,000 in a piece of equipment. The equipment depreciates at the rate of 20 percent per year. What will be the value of the equipment, to the nearest dollar, at the end of the fifth year?

43. **Investment**   Suppose an antique car you own appreciates (increases) in value at a rate of 8 percent annually. If the car was purchased for $12,000, what will it be worth during the eighth year?

44. **Geometry**   A 16-in. piece of wire is cut in half. Each piece is then cut in half, and the process is repeated 8 more times. What is the size of each final piece?

TRIGONOMETRY IN FINANCE

The Preview on page 359 introduced you to the compound interest formula. A somewhat different and equally useful version of the formula is $A_n = P(1 + r)^n$, where P is the original principal, n is the number of time periods for which the principal is invested, r is the interest rate per period (expressed as a decimal), and A_n is the amount in the account after n time periods. To understand how the formula is derived, study the pattern of the first few terms of the sequence. It is much like the pattern of the terms of a geometric sequence.

$$A_1 = P + rP = P(1 + r)^1$$
$$A_2 = P(1 + r) + r[P(1 + r)] = [P(1 + r)](1 + r) = P(1 + r)^2$$
$$A_3 = P(1 + r)^2 + r[P(1 + r)^2] = [P(1 + r)^2](1 + r) = P(1 + r)^3$$
$$A_n = P(1 + r)^n \qquad \textit{General term of the sequence}$$

To illustrate the use of this formula, consider the following problem:

If $1000 is invested at 8 percent per year, compounded quarterly, what will the value of the account be at the end of 2 years?

Since there are 4 quarters per year, $n = 2(4) = 8$.

Since the rate is 8% per year, $r = \frac{0.08}{4} = 0.02$.

$$
\begin{aligned}
A_8 &= 1000(1 + 0.02)^8 \\
&= 1000(1.02)^8 \\
&= 1171.659381
\end{aligned}
$$
   Use the formula $A_n = P(1 + r)^n$, where $n = 8$, $P = 1000$, and $r = 0.02$. The account will be worth $1171.66 to the nearest cent.

1. If $500 is invested at 7 percent per year, compounded annually, how much will the account be worth at the end of 3 years?

2. If $2000 is invested at 6 percent per year, compounded quarterly, how much will the account be worth at the end of 4 years?

3. If $45,000 is invested at 7.5 percent per year, compounded semiannually, how much will the account be worth at the end of 5 years?

Arithmetic Series

Objectives: To use sigma notation to represent the sum of a series
To develop and use formulas to find a partial sum of an arithmetic series

It is sometimes necessary to find the sum of the terms of a finite arithmetic sequence.

Preview

The bells in a tower chime as many times as the hour. From 1 AM to 12 noon, inclusive, how many times do they chime?

First, write the finite arithmetic sequence.

1, 2, 3, 4, 5, 6, 7, 8, 9, 10, 11, 12

Then add the terms of the sequence.

1 + 2 + 3 + 4 + 5 + 6 + 7 + 8 + 9 + 10 + 11 + 12 = 78

The bells chime 78 times from 1 AM to 12 noon, inclusive.

The front row of a triangular flower bed has 24 plants. Each successive row contains 3 more plants than the row in front of it. How many plants are there in each row?

1. 4th row   **2.** 6th row   **3.** 8th row   **4.** 10th row

5. What is the total number of plants in the first 8 rows?   in the first 10 rows?

A **series** is the indicated sum of the terms of a sequence. The indicated sum 2 + 4 + 6 + 8 is a finite series. The indicated sum 2 + 4 + 6 + 8 + · · · is an infinite series. An **arithmetic series** is the indicated sum of an arithmetic sequence. The sum of the first n terms of an arithmetic series, denoted S_n, is a **partial sum** of the series. The sum of the first five terms, S_5, of the series 2 + 4 + 6 + 8 + · · · is 30, since 2 + 4 + 6 + 8 + 10 = 30.

Sigma (Σ) notation, or **summation notation,** is a shorthand method for representing the sum of a series. The general term of a series is the same as the general term of the corresponding sequence. For the series 2 + 4 + 6 + 8 + · · ·, the general term, a_n, is $2n$. Therefore,

$$S_5 = \sum_{n=1}^{5} 2n \qquad Read \sum_{n=1}^{5} 2n \ as: \ the \ sum \ from \ n = 1 \ to \ 5 \ of \ 2n.$$

The variable n is called the **index of summation,** the numbers 1 and 5 are called the **limits of summation,** S_5 represents the sum of the first five terms, and $2n$ is the general term. Any letter can represent the index of summation.

EXAMPLE 1 Write $\displaystyle\sum_{j=3}^{6}(2j + 1)$ in expanded form and then find the sum.

Replace j successively with 3, 4, 5, and 6. Then add.

$$\sum_{j=3}^{6}(2j + 1) = [2(3) + 1] + [2(4) + 1] + [2(5) + 1] + [2(6) + 1]$$
$$= 7 + 9 + 11 + 13 = 40$$

To express in sigma notation a partial sum given in expanded form, first find the common difference and the general term.

EXAMPLE 2 **Express in sigma notation the sum of the first six terms of the arithmetic series $-2 + (-6) + (-10) + (-14) + \cdots$.**

First, find the common difference.

$d = -6 - (-2) = -4$ *Use the formula $d = a_{n+1} - a_n$, where $a_{n+1} = -6$ and $a_n = -2$.*

Then, find the general term.

$\begin{aligned} a_n &= -2 + (n - 1)(-4) \\ &= -2 - 4n + 4 \\ &= -2(2n - 1) \end{aligned}$ *Use the formula $a_n = a_1 + (n - 1)d$, where $a_1 = -2$ and $d = -4$.*

Then, $S_6 = \displaystyle\sum_{n=1}^{6} -2(2n - 1)$.

If a series has a large number of terms, it is not convenient to find the sum by addition. To find a general formula for the partial sum of an arithmetic series of n terms, write S_n in two ways, as shown below, and add.

$$S_n = a_1 \quad\quad + (a_1 + d) \ + (a_1 + 2d) + \cdots + (a_n - d) \ + a_n$$
$$S_n = a_n \quad\quad + (a_n - d) \ + (a_n - 2d) + \cdots + (a_1 + d) \ + a_1$$

$$2\,S_n = (a_1 + a_n) \ + (a_1 + a_n) \ + (a_1 + a_n) \ + \cdots + (a_1 + a_n) + (a_1 + a_n)$$
$2\,S_n = n(a_1 + a_n)$ *Since S_n has n terms, there are n terms*
$S_n = \dfrac{n(a_1 + a_n)}{2}$ *of the form $(a_1 + a_n)$ in $2S_n$.*

EXAMPLE 3 **Find S_{500}, the sum of the first 500 terms of an arithmetic series, if $a_1 = 2$ and $a_{500} = 1000$.**

$\begin{aligned} S_{500} &= \dfrac{500(2 + 1000)}{2} \\[2mm] &= 250,500 \end{aligned}$ *Use the formula $S_n = \dfrac{n(a_1 + a_n)}{2}$, where $n = 500$, $a_1 = 2$, and $a_n = 1000$.*

You can also use the formula $S_n = \dfrac{n(a_1 + a_n)}{2}$ to find the partial sum of a series expressed in sigma notation, without writing it in expanded form.

EXAMPLE 4   **Find the sum of the arithmetic series $\displaystyle\sum_{n=1}^{7} (5n - 2)$.**

$a_1 = 5(1) - 2 = 3$ and $a_7 = 5(7) - 2 = 33$ *Find a_1 and a_7.*

$S_7 = \dfrac{7(3 + 33)}{2} = 126$ $\qquad\qquad\qquad S_n = \dfrac{n(a_1 + a_n)}{2}$

An alternate formula for S_n can be obtained by substituting $a_1 + (n - 1)d$ for a_n. This alternate formula is useful when the nth term is not given.

$$S_n = \dfrac{n(a_1 + a_n)}{2} = \dfrac{n[a_1 + a_1 + (n - 1)d]}{2} = \dfrac{n[2a_1 + (n - 1)d]}{2}$$

EXAMPLE 5   **Find the sum of the first 50 terms of the arithmetic series**
$10 + 85 + 160 + 235 + \cdots$.

$d = a_{n+1} - a_n = 85 - 10 = 75$ *Find d.*

$S_{50} = \dfrac{50[2(10) + (50 - 1)(75)]}{2}$ $\qquad\qquad S_n = \dfrac{n[2a_1 + (n - 1)d]}{2}$

$= 92{,}375$ $\qquad\qquad\qquad\qquad\quad$ The sum of the first 50 terms is 92,375.

Specific terms of an arithmetic series can be found if the values of a_1, a_n, and S_n are known.

EXAMPLE 6   **Find the first three terms of the arithmetic series in which $a_1 = 5$,**
$a_n = 100$, and $S_n = 1050$.

$1050 = \dfrac{n(5 + 100)}{2}$ $\qquad\qquad$ *Solve $S_n = \dfrac{n(a_1 + a_n)}{2}$ for n.*

$n = \dfrac{2(1050)}{5 + 100} = 20$

Substitute $n = 20$ in $S_n = \dfrac{n[2a_1 + (n - 1)d]}{2}$ and solve for d.

$1050 = \dfrac{20[2(5) + (20 - 1)d]}{2}$

$1050 = 10(10 + 19d)$

$105 = 10 + 19d$ $\qquad\qquad$ *Divide both sides by 10.*

$5 = d$ $\qquad\qquad\qquad\quad$ The first three terms are 5, 10, and 15.

The formulas for the partial sum of an arithmetic series are summarized below.

> The partial sum, S_n, of an arithmetic series is given by either of the following formulas:
>
> $$S_n = \frac{n(a_1 + a_n)}{2} \quad \text{or} \quad S_n = \frac{n[2a_1 + (n-1)d]}{2}$$
>
> where n is the number of terms, a_1 is the first term, d is the common difference, and a_n is the nth term.

CLASS EXERCISES

Write in expanded form and then find the sum.

1. $\displaystyle\sum_{i=10}^{15} i$ **2.** $\displaystyle\sum_{i=6}^{12} i$ **3.** $\displaystyle\sum_{n=3}^{7} 2n$ **4.** $\displaystyle\sum_{n=1}^{5} 3n$

5. $\displaystyle\sum_{m=1}^{4} 2^m$ **6.** $\displaystyle\sum_{m=0}^{3} 3^m$ **7.** $\displaystyle\sum_{k=4}^{6} (k+3)$ **8.** $\displaystyle\sum_{k=4}^{7} (k-2)$

9. $\displaystyle\sum_{y=5}^{6} (y-4)$ **10.** $\displaystyle\sum_{y=3}^{4} (y+6)$ **11.** $\displaystyle\sum_{x=1}^{3} (2x-3)$ **12.** $\displaystyle\sum_{x=2}^{5} (-3x+9)$

13. Express in sigma notation the partial sum from $n = 1$ to $n = 5$ of the arithmetic series $7 + 10 + 13 + 16 + \cdots$.

Consider the series $\displaystyle\sum_{n=1}^{6} (2n-1)$. **Find the value of each partial sum.**

14. S_1 **15.** S_2 **16.** S_3

17. S_4 **18.** S_5 **19.** S_6

PRACTICE EXERCISES   Use appropriate technology.

Write each arithmetic series in expanded form, and then find its sum.

1. $\displaystyle\sum_{k=1}^{6} 2k$ **2.** $\displaystyle\sum_{k=1}^{4} 3k$ **3.** $\displaystyle\sum_{j=2}^{4} (j+3)$ **4.** $\displaystyle\sum_{j=3}^{7} (2j-1)$

5. $\displaystyle\sum_{k=7}^{9} (k+4)$ **6.** $\displaystyle\sum_{k=9}^{13} (2k-3)$ **7.** $\displaystyle\sum_{n=4}^{6} 2n + \frac{1}{2}$ **8.** $\displaystyle\sum_{n=5}^{8} 4n + \frac{1}{2}$

Find the indicated partial sum for each arithmetic series.

9. $a_1 = 2, a_{200} = 200; S_{200}$

10. $a_1 = 5, a_{30} = 100; S_{30}$

11. $a_1 = 1, a_{10} = -20; S_{10}$

12. $a_1 = 2, a_{20} = -40; S_{20}$

13. $a_1 = 73, d = -3; S_{15}$

14. $a_1 = 91, d = -4; S_{12}$

15. $a_1 = -6, d = 0.2; S_{10}$

16. $a_1 = 12, d = 0.4; S_6$

17. $9 + 11 + 13 + \cdots; S_{10}$

18. $-9 + (-6) + (-3) + \cdots; S_{16}$

19. $-11 + (-15) + (-19) + \cdots; S_{12}$

20. $3 + 9 + 15 + \cdots; S_{30}$

21. $\displaystyle\sum_{n=1}^{10} 4n$

22. $\displaystyle\sum_{n=1}^{10} (2n + 4)$

23. $\displaystyle\sum_{j=1}^{6} (j - 7)$

24. $\displaystyle\sum_{j=1}^{10} (30 - j)$

Express in sigma notation the sum of the first six terms of each arithmetic sequence.

25. $3 + 7 + 11 + 15 + \cdots$

26. $-5 + (-3) + (-1) + 1 + \cdots$

27. $8 + 3 + (-2) + (-7) + \cdots$

28. $-1 + (-8) + (-15) + \cdots$

Find the indicated term for each arithmetic sequence.

29. $a_1 = -2, S_{10} = 205; a_{10}$

30. $a_1 = -4, S_{15} = 885; a_{15}$

31. $a_1 = 10, S_{50} = -12{,}975; a_{50}$

32. $a_1 = -12, S_{60} = -7800; a_{60}$

33. $S_{20} = 40, d = 4; a_{20}$

34. $S_{30} = 120, d = 6; a_{30}$

35. $S_{100} = -200, d = -2; a_{100}$

36. $S_{200} = -400, d = -4; a_{200}$

Find the first three terms of each arithmetic series.

37. $a_1 = 6, a_n = 306, S_n = 1716$

38. $a_1 = 7, a_n = 139, S_n = 876$

39. $n = 25, a_{25} = 156, S_{25} = 2175$

40. $n = 14, a_{14} = 199, S_{14} = 1239$

41. For an arithmetic series with $a_n = -23, d = -4$, and $S_n = -72$, find a_1 and n.

42. For an arithmetic series with $a_n = 45, d = 5$, and $S_n = 210$, find a_1 and n.

43. For an arithmetic series with $a_1 = -11, S_n = -32$, and the fourth term is -5, find d and n.

44. For an arithmetic series with $a_1 = 6, S_n = -120$, and the third term is -6, find d and n.

45. Prove: $1 + 2 + 3 + \cdots + n = \dfrac{n(n + 1)}{2}$

46. Prove: $2 + 4 + 6 + \cdots + 2n = n(n + 1)$

47. Prove: $1 + 3 + 5 + \cdots + (2n - 1) = n^2$

48. Prove: $\displaystyle\sum_{j=1}^{n} 4j = 4 \sum_{j=1}^{n} j$

Applications

49. Building Design   The balcony of a theater has 12 rows of seats. The last row contains 8 seats, and each of the other rows contains one more seat than the row behind it. How many seats are there in the balcony?

50. Sports   During the first hour of her climb, a mountain climber ascends 675 ft. Each hour after that, she ascends 50 ft less than in the preceding hour. When will she be 4500 ft above her starting point?

51. Business   A contractor agreed that if a job was not done by a certain date, he would pay a $1000 penalty for the first day of delay, and for each day after that, he would pay $50 more than for the preceding day. How many penalty days did he use if his penalty was $10,800?

52. Number Theory   Find the sum of the integers from 1 through 100.

EXTRA

An expression given in sigma notation may often be written in a different form by changing the index. Consider the expression $\displaystyle\sum_{k=1}^{5} (3k + 7)$.

$$3k + 7 = 3k + 6 + 1 = 3(k + 2) + 1$$

Let $i = k + 2$. Then, the general term $3k + 7$ can be written $3i + 1$.
If the index is changed to i, the limits of summation must also be changed.

For $k = 1$,	$3(1) + 7 = 3i + 1$	For $k = 5$,	$3(5) + 7 = 3i + 1$
	$9 = 3i$		$22 = 3i + 1$
	$3 = i$		$7 = i$

Thus, the series $\displaystyle\sum_{k=1}^{5} (3k + 7)$ may be expressed as $\displaystyle\sum_{i=3}^{7} (3i + 1)$.

Write each expression in a different form, using the index j, where $j = k + 4$.

1. $\displaystyle\sum_{k=1}^{4} (2k + 8)$ **2.** $\displaystyle\sum_{k=1}^{6} (k + 5)$ **3.** $\displaystyle\sum_{k=3}^{8} (k^2 + 8k + 16)$

9.4
Geometric Series

Objectives: To develop and use formulas to find a partial sum of a
geometric series
To determine whether or not a series is arithmetic or
geometric

A geometric series can be formed from a geometric sequence, just as an
arithmetic series can be formed from an arithmetic sequence.

Preview

In a large corporation, there are five levels
of management: President, Vice President,
Regional Vice President, Regional Manager,
and Local Manager. There is 1 president,
and there are 4 vice presidents, 16 regional
vice presidents, and so on. The sequence
is geometric. How many people are on
the management team?

First, write the geometric sequence.

$$1, 4, 16, 64, 256$$

Then, add the terms of the sequence.

$$1 + 4 + 16 + 64 + 256 = 341$$

There are 341 people on the management team.

**Steven will receive a 5 percent salary increase at the end of each year. His
beginning salary is $10,000. How much will he earn during each year?**

1. 2nd year **2.** 3rd year **3.** 4th year **4.** 5th year **5.** 6th year

6. How much money will he earn in all during 6 years on the job?

A **geometric series** is the indicated sum of the terms of a geometric sequence.
The series $1 + 3 + 9 + 27 + 81 + 243$ is finite. The series
$1 + 4 + 16 + \cdots$ is infinite. A partial sum, S_n, of a geometric series is the
sum of the first n terms of the series.

$$S_n = a_1 + a_1r + a_1r^2 + \cdots + a_1r^{n-1}$$

To find a formula for the partial sum of a geometric series, multiply S_n by r
and subtract the resulting product from S_n.

$$S_n = a_1 + a_1r + a_1r^2 + \cdots + a_1r^{n-1}$$
$$\underline{rS_n = \qquad a_1r + a_1r^2 + \cdots + a_1r^{n-1} + a_1r^n}$$
$$S_n - rS_n = a_1 + 0 + 0 + \cdots + 0 - a_1r^n$$
$$S_n(1 - r) = a_1 - a_1r^n$$
$$S_n = \frac{a_1 - a_1r^n}{1 - r},\ r \neq 1 \qquad \textit{Divide each side by 1 − r.}$$
$$S_n = \frac{a_1(1 - r^n)}{1 - r},\ r \neq 1 \qquad \textit{Factor.}$$

EXAMPLE 1 **Find the sum of the first six terms of the geometric series**
$2 + 8 + 32 + 128 + \cdots$.

$$S_6 = \frac{2(1 - 4^6)}{1 - 4} = 2730 \qquad \textit{Use } S_n = \frac{a_1(1 - r^n)}{1 - r}, n = 6, a_1 = 2, \text{ and } r = 4.$$

You can find the first term if you know one partial sum and the common ratio.

EXAMPLE 2 **For a geometric series, find a_1 if $S_{10} = -2046$ and $r = 2$.**

$$-2046 = \frac{a_1(1 - 2^{10})}{1 - 2} \qquad S_n = \frac{a_1(1 - r^n)}{1 - r}$$
$$2046 = a_1(1 - 2^{10}) \qquad \textit{Multiply both sides by −1.}$$
$$\frac{2046}{1 - 2^{10}} = a_1 \qquad \textit{Calculation-ready form}$$
$$-2 = a_1$$

Recall that the general term of a geometric sequence is $a_n = a_1r^{n-1}$.
Multiplying both sides of this equation by r yields $ra_n = a_1r^n$. This new
equation can be used, along with algebraic manipulation, to find an alternate
formula for S_n. If $r \neq 1$, then

$$S_n = \frac{a_1(1 - r^n)}{1 - r} = \frac{a_1 - a_1r^n}{1 - r} = \frac{a_1 - ra_n}{1 - r} \qquad \textit{Substitute } ra_n \textit{ for } a_1r^n.$$

This formula is useful when the first and nth terms are known.

EXAMPLE 3 **Find the sum of the first four terms of a geometric series,**
where $a_1 = 343$, $a_4 = -1$, and $r = -\dfrac{1}{7}$.

$$S_4 = \frac{343 - \left(-\frac{1}{7}\right)(-1)}{1 - \left(-\frac{1}{7}\right)} = \frac{343 - \frac{1}{7}}{\frac{8}{7}} = \frac{2401 - 1}{8} = 300 \qquad \textit{Use } S_n = \frac{a_1 - ra_n}{1 - r}.$$

Geometric series can also be written using sigma notation. For example, the finite series $6 + 12 + 24 + 48 + 96 + 192 + 384 + 768$ can be written $\sum_{n=1}^{8} 6(2)^{n-1}$. In this series, $a_1 = 6$, $r = 2$, and $n = 8$. Before you can find the sum of a series given in sigma notation, you must first determine whether it is an arithmetic or a geometric series.

EXAMPLE 4 Find $\sum_{n=1}^{5} (4n - 1)$.

$$\sum_{n=1}^{5} (4n - 1) = [4(1) - 1] + [4(2) - 1] + [4(3) - 1] + [4(4) - 1] + [4(5) - 1]$$
$$= 3 + 7 + 11 + 15 + 19$$

This series is arithmetic, since the sequence 3, 7, 11, 15, 19 is arithmetic. Each succeeding term is found by adding 4 to the previous term. Because this is an arithmetic series, use $S_n = \dfrac{n(a_1 + a_n)}{2}$ to find the sum.

$$S_5 = \frac{5(3 + 19)}{2} \qquad n = 5, a_1 = 3, a_n = 19$$
$$= 55$$

If a geometric series is given in expanded form, it can be expressed in sigma notation. First, find the common ratio and the general term.

EXAMPLE 5 **Express in sigma notation the sum of the first eight terms of the geometric series $7 + 14 + 28 + 56 + \cdots$.**

First, find the common ratio.

$$r = \frac{14}{7} = 2 \qquad\qquad \text{Use } a_{n+1} = 14 \text{ and } a_n = 7.$$

Then, find the general term.

$$a_n = 7(2)^{n-1} \qquad\qquad a_1 = 7, r = 2$$

Then, $S_8 = \sum_{n=1}^{8} 7(2)^{n-1}$

It is possible to find one partial sum of a geometric series, given another partial sum and the common ratio. To do this, apply the formula for the partial sum twice.

EXAMPLE 6 Find S_{10} for the geometric sequence in which $S_6 = 728$ and $r = 3$.

$728 = \dfrac{a_1(1 - 3^6)}{1 - 3}$ Use $S_n = \dfrac{a_1(1 - r^n)}{1 - r}$ to find a_1;

$\dfrac{-1456}{1 - 3^6} = a_1$ $n = 6$, $S_n = 728$, and $r = 3$.

$2 = a_1$

$S_{10} = \dfrac{2(1 - 3^{10})}{1 - 3} = 59{,}048$ Use the same formula to find S_{10}; $n = 10$, $a_1 = 2$, and $r = 3$.

The partial sum, S_n, of a geometric series is given by either of the following formulas:

$$S_n = \dfrac{a_1(1 + r^n)}{1 - r} \quad r \neq 1 \qquad \text{or} \qquad S_n = \dfrac{a_1 - ra_n}{1 - r} \quad r \neq 1$$

where n is the number of terms, a_1 is the first term, r is the common ratio, and a_n is the nth term.

CLASS EXERCISES

Consider the geometric series $1 + 2 + 4 + 8 + \cdots$. Find each value.

1. r **2.** S_1 **3.** S_2 **4.** S_4 **5.** S_8 **6.** S_{12}

Find r for each geometric series. Then, find the indicated partial sum.

7. $3 + 6 + 12 + \cdots$; S_4 **8.** $10 + 100 + 1000 + \cdots$; S_5

9. $-32 + 16 + (-8) + \cdots$; S_6 **10.** $-81 + 27 + (-9) + \cdots$; S_4

PRACTICE EXERCISES   Use appropriate technology.

Find the indicated partial sum for each geometric series.

1. $a_1 = 6, r = 4$; S_4 **2.** $a_1 = 625, r = \frac{2}{5}$; S_5

3. $a_1 = 16, r = -\frac{1}{2}$; S_5 **4.** $a_1 = 243, r = -\frac{1}{3}$; S_6

5. $a_1 = 12, a_5 = 972, r = 3$; S_5 **6.** $a_1 = 4, a_6 = -972, r = -3$; S_6

7. $a_1 = -4, a_6 = \frac{1}{8}, r = -\frac{1}{2}$; S_6 **8.** $a_1 = 125, a_5 = \frac{1}{25}, r = -\frac{1}{5}$; S_5

9. $a_1 = 2, a_5 = \frac{1}{8}, r = -\frac{1}{2}$; S_5

10. $a_1 = -343, a_4 = 1, r = -\frac{1}{7}$; S_4

11. $1 + 2 + 4 + \cdots$; S_{10}

12. $1875 + 375 + 75 + \cdots$; S_5

13. $2 + (-6) + 18 + \cdots$; S_8

14. $3 + (-6) + 12 + \cdots$; S_7

15. $4 + 4 + 4 + \cdots$; S_{15}

16. $8 + 8 + 8 + \cdots$; S_{13}

17. $1024 + 256 + 64 + \cdots$; S_8

18. $2 + 8 + 32 + \cdots$; S_9

19. $\sum_{n=1}^{4} 2^n$

20. $\sum_{n=1}^{4} 2(3)^n$

21. $\sum_{n=1}^{5} 4^{n-1}$

22. $\sum_{n=1}^{5} 2^{n+2}$

For each geometric series, find the value of a_1.

23. $a_n = 420, r = \frac{3}{2}, S_n = 1270$

24. $a_n = 243, r = 3, S_n = 302$

25. $S_n = -172, r = -2, n = 7$

26. $S_n = -1785, r = -2, n = 8$

Determine whether each series is arithmetic or geometric and find its sum.

27. $\sum_{n=1}^{6} (3n + 5)$

28. $\sum_{n=1}^{4} (6n - 2)$

29. $\sum_{k=1}^{8} 4^k$

30. $\sum_{n=1}^{6} 2(2)^{n+2}$

Express in sigma notation the sum of the first 8 terms of each geometric series.

31. $3 + 1 + \frac{1}{3} + \frac{1}{9} + \cdots$

32. $\frac{1}{16} + \frac{1}{4} + 1 + 4 + \cdots$

33. $2 + (-10) + 50 + (-250) + \cdots$

34. $120 + (-60) + 30 + (-15) + \cdots$

Find the indicated partial sum for each geometric series.

35. $S_8 = 2040, r = 2$; S_6

36. $S_6 = 63, r = -\frac{1}{2}$; S_3

37. $S_4 = 240, r = 3$; S_7

38. $S_5 = -110, r = -2$; S_8

Find the missing values for each geometric series.

39. If $a_1 = 3, a_n = 192$, and $S_n = 129$, find r and n.

40. If $a_1 = 6, r = 2$, and $a_n = 3072$, find S_n and n.

Find the indicated partial sum.

41. $x + 3x^3 + 9x^5 + \cdots$; S_6

42. $xy + x^2y^4 + x^3y^7 + \cdots$; S_6

43. $\sum_{n=1}^{6} y(x^2)^{n+1}$

44. $\sum_{n=1}^{5} y^2(x^3)^n$

Applications

45. Construction A pile driver is used to drive a pipe into the ground. The first blow drives the pipe down 20 cm. If each succeeding blow drives the pipe down 0.4 times as far as the preceding blow, how far has the pipe been driven after 4 blows?

46. Entertainment A club sponsor needs to contact all of the club members by telephone to inform them of a change in plans. The sponsor calls 2 members, each of whom calls 2 other members not previously called, and so on. If all members have been contacted after this process has been repeated 6 times, how many members (excluding the sponsor) are in the club?

47. Finance An investment in a gas well earned a total of $54,450 during the first 5 yr. Each year after the first, the investment earned 3 times as much as during the preceding year. How much did the investment earn during the first year?   during the fifth year?

TEST YOURSELF

1. Find a_{16} for an arithmetic sequence in which $a_1 = 120$ and $d = -\frac{1}{3}$.   **9.1**

2. List the next five terms of the arithmetic sequence 100, 93, 86, ...

3. Insert four arithmetic means between -3 and 22.

4. Find a_7 for a geometric sequence in which $a_1 = 4$ and $r = -2$.   **9.2**

5. List the next five terms of the geometric sequence 1000, 200, 40, ...

6. Insert two geometric means between 81 and 3.

7. Find S_{15} for an arithmetic series in which $a_1 = 8$ and $d = 6$.   **9.3**

8. Find a_{12} for an arithmetic series in which $a_1 = -6$ and $S_{12} = 60$.

9. Express in sigma notation the sum of the first five terms of the arithmetic series $24 + 21 + 18 + \ldots$.

Find the indicated partial sum for each geometric series.

10. $a_1 = 10, r = 2; S_{20}$ **11.** $a_1 = 80, a_5 = 5; S_5$ **9.4**

Determine whether each series is arithmetic or geometric, and find its sum.

12. $\displaystyle\sum_{n=1}^{7} (4n - 2)$ **13.** $\displaystyle\sum_{n=1}^{6} 4^{n-2}$

14. Express in sigma notation the sum of the first nine terms of the geometric series $\frac{1}{2} + \left(-\frac{3}{2}\right) + \frac{9}{2} + \cdots$.

Infinite Geometric Series

Objectives: To develop and use a formula to find the sum of a convergent geometric series

To express repeating decimals as rational numbers in fraction form

You have learned how to compute the sum of a *finite* number of terms of a geometric series. However, there are many applications for which you will need to evaluate the sum of an *infinite* number of terms.

Preview

A tennis ball is tossed upward from the ground to a height of 25 ft. Each bounce is 0.8 as high as the previous bounce. The first four heights reached by the ball form the geometric sequence 25, 20, 16, 12.8. If the ball were to bounce indefinitely, the sequence would be infinite. If the sequence is infinite, it is not possible to find the total distance traveled by the ball simply by adding terms. However, you can look at some partial sums.

Before it bounces once, the ball travels up 25 ft and down 25 ft. Before it bounces a second time, it travels up 25(0.8) ft and down 25(0.8) ft. Let S_1 be the distance traveled before it bounces once, S_2 be the distance traveled before it bounces twice, and so on. Then,

$$S_1 = 2[25(0.8)^0] = 50$$

$$S_2 = 2[25(0.8)^0 + 25(0.8)^1] = 90$$

$$S_3 = 2[25(0.8)^0 + 25(0.8)^1 + 25(0.8)^2] = 122$$

How far has the ball traveled before it bounces the given number of times? *Hint*: **Use** $S_n = \dfrac{a_1(1 - r^n)}{1 - r}$ **and then multiply by 2.**

1. 5 times   **2.** 10 times   **3.** 15 times   **4.** 25 times   **5.** 50 times

6. Do you think the distance traveled will approach a certain number as the number of bounces increases indefinitely?

In this lesson, a formula will be developed that will enable you to find the sum of certain types of infinite geometric series.

EXAMPLE 1 Find the partial sums S_1, S_2, S_3, S_8, S_9, S_{10}, and S_n of the infinite geometric series $1 + \frac{1}{3} + \frac{1}{9} + \cdots$. Round to five decimal places.

$$r = \frac{a_{n+1}}{a_n} = \frac{\frac{1}{3}}{1} = \frac{1}{3} \qquad \textit{Find r.}$$

$$S_1 = 1$$

$$S_2 = 1 + \frac{1}{3} = 1.33333$$

$$S_3 = 1 + \frac{1}{3} + \frac{1}{9} = 1.44444$$

$$S_8 = 1 + \frac{1}{3} + \frac{1}{9} + \cdots + \frac{1}{2187} = 1.49977 \qquad a_n = a_1 r^{n-1}$$

$$S_9 = 1 + \frac{1}{3} + \frac{1}{9} + \cdots + \frac{1}{6561} = 1.49992 \qquad S_n = \frac{a_1(1 - r^n)}{1 - r}$$

$$S_{10} = 1 + \frac{1}{3} + \frac{1}{9} + \cdots + \frac{1}{19{,}683} = 1.49997$$

$$S_n = 1 + \frac{1}{3} + \frac{1}{9} + \cdots + \left(\frac{1}{3}\right)^{n-1} = \frac{1\left[1 - \left(\frac{1}{3}\right)^n\right]}{1 - \frac{1}{3}} = \frac{3}{2} - \frac{3}{2}\left(\frac{1}{3}\right)^n$$

This graph represents the partial sums in Example 1. The horizontal line at 1.5 is an *asymptote*. As n gets very large, S_n approaches 1.5 as a **limit.** Since the series has a limit, it is said to *converge*, or to be *convergent*. To see why S_n approaches the limit 1.5, consider the nth partial sum

$$S_n = \frac{3}{2} - \frac{3}{2}\left(\frac{1}{3}\right)^n$$

As n gets very large, $\frac{3}{2}\left(\frac{1}{3}\right)^n$ gets very close to zero, so S_n approaches $\frac{3}{2}$, or 1.5.

In general, an infinite geometric series **converges** if the sequence of partial sums approaches a limit as n increases without bound. It can be shown that an infinite geometric series converges if and only if $|r| < 1$, where r is the common ratio. This is true because if $|r| < 1$, then r^n approaches zero as n approaches infinity.

An infinite geometric series can be expressed in sigma notation, with the upper limit infinity (∞). The series in Example 1 can be written

$$\sum_{n=1}^{\infty} \left(\frac{1}{3}\right)^{n-1}$$

EXAMPLE 2 Find the partial sums S_1, S_2, S_3, S_4, S_5, and S_n for the infinite geometric series $\sum\limits_{n=1}^{\infty} 2^n$.

$$\sum_{n=1}^{\infty} 2^n = 2 + 4 + 8 + \cdots$$

$$r = \frac{a_{n+1}}{a_n} = \frac{4}{2} = 2$$

$S_1 = 2$

$S_2 = 2 + 4 = 6$

$S_3 = 2 + 4 + 8 = 14$

$S_4 = 2 + 4 + 8 + 16 = 30$

$S_5 = 2 + 4 + 8 + 16 + 32 = 62$

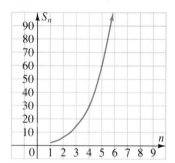

$$S_n = 2 + 4 + 8 + \cdots + 2(2)^{n-1} = \frac{2(1 - 2^n)}{1 - 2} = 2(2^n - 1)$$

As n gets very large, S_n also gets very large. Thus, this series has no finite sum. It is said to *diverge* or to be *divergent*.

A series **diverges** if the sequence of partial sums increases or decreases without bound as n increases without bound. It can be shown that an infinite geometric series diverges if and only if $|r| \geq 1$, where r is the common ratio.

The **sum of an infinite geometric series** in which the sequence of partial sums converges to a limit S is equal to S. To develop a formula for the sum of a convergent infinite geometric series, look at the formula for the partial sum S_n.

$$S_n = \frac{a_1(1 - r^n)}{1 - r}$$

Since the series converges, $|r| < 1$. Thus, as n gets very large, r^n approaches zero as a limit. Thus,

$$S_n = \frac{a_1(1 - r^n)}{1 - r} \quad \text{approaches} \quad \frac{a_1(1 - 0)}{1 - r} = \frac{a_1}{1 - r}$$

The sum S of a convergent infinite geometric series is given by

$$S = \frac{a_1}{1 - r} \quad |r| < 1$$

where a_1 is the first term and r is the common ratio.

EXAMPLE 3 Find the sum of each infinite geometric series, if it exists:

a. $\displaystyle\sum_{n=1}^{\infty} 4^{3-n}$ b. $\displaystyle\sum_{n=1}^{\infty} 4^{n-2}$

a. $\displaystyle\sum_{n=1}^{\infty} 4^{3-n} = 16 + 4 + 1 + \frac{1}{4} + \ldots$ $a_1 = 16, r = \frac{1}{4}$

Since $|r| < 1$, the series converges and the sum exists.

$$S = \frac{16}{1 - \frac{1}{4}} = \frac{64}{3} \qquad \text{Use the formula } S = \frac{a_1}{1 - r}.$$

b. $\displaystyle\sum_{n=1}^{\infty} 4^{n-2} = \frac{1}{4} + 1 + 4 + 16 + \cdots$ $a_1 = \frac{1}{4}, r = 4$

Since $|r| > 1$, the series diverges and the sum does not exist.

A repeating decimal can be thought of as an infinite geometric series. Therefore, you can use the formula for the sum of an infinite geometric series to write repeating decimals as rational numbers in fraction form.

EXAMPLE 4 Write $0.\overline{6}$ (or $0.666\ldots$) as a rational number in fraction form.

$0.\overline{6} = 0.6 + 0.06 + 0.006 + \cdots$ Write $0.\overline{6}$ as an infinite geometric series.
$a_1 = 0.6$ $r = 0.1$ Since $|r| < 1$, the series converges.

$$S = \frac{0.6}{1 - 0.1} = \frac{0.6}{0.9} = \frac{2}{3} \qquad \text{Therefore, } 0.\overline{6} = \frac{2}{3}.$$

CLASS EXERCISES

Tell whether each geometric series converges or diverges.

1. $1 + 5 + 25 + \cdots$

2. $1 + 6 + 36 + \cdots$

3. $1 + \frac{1}{5} + \frac{1}{25} + \cdots$

4. $1 + \frac{1}{6} + \frac{1}{36} + \cdots$

5. $48 - 36 + 27 - \cdots$

6. $80 - 64 + 51.2 - \cdots$

7. $-1 - 1 - 1 - \cdots$

8. $1 - 1 + 1 - \cdots$

Express each repeating decimal as a rational number in fraction form.

9. $0.\overline{3}$ 10. $0.\overline{24}$ 11. $0.2\overline{4}$ 12. $0.\overline{032}$

13. Use the formula for the sum of an infinite geometric series to find the total distance traveled by the ball in the Preview problem on page 396.

〰 Use appropriate technology.

Determine the sum S, if it exists, for each geometric series. If the sum does not exist, so state.

1. $1 + \frac{1}{7} + \frac{1}{49} + \cdots$

2. $1 + \frac{2}{7} + \frac{4}{49} + \cdots$

3. $16 + 8 + 4 + \cdots$

4. $20 + 10 + 5 + \cdots$

5. $2 + 2 + 2 + \cdots$

6. $2 - 2 + 2 - \cdots$

7. $4 + 5 + \frac{25}{4} + \cdots$

8. $3 + 4 + \frac{16}{3} + \cdots$

9. $-7 + 1 - \frac{1}{7} + \cdots$

10. $-10 + 1 - 0.1 + \cdots$

11. $0.8 + 0.4 + 0.2 + \cdots$

12. $0.2 + 0.4 + 0.8 + \cdots$

Write each repeating decimal as a rational number in fraction form.

13. $0.\overline{4}$

14. $0.\overline{8}$

15. $0.\overline{21}$

16. $0.\overline{12}$

17. $0.\overline{123}$

18. $0.\overline{456}$

Determine the sum S, if it exists, for each geometric series. If the sum does not exist, so state.

19. $\sum_{n=1}^{\infty} \left(\frac{1}{2}\right)^{n-1}$

20. $\sum_{n=1}^{\infty} \left(\frac{1}{3}\right)^{n}$

21. $\sum_{n=1}^{\infty} \frac{3}{4}(3)^{n-1}$

22. $\sum_{n=1}^{\infty} \frac{1}{2}(2)^{n+1}$

Write each repeating decimal as a rational number in fraction form.

23. $0.4\overline{1}$

24. $0.4\overline{23}$

25. $5.\overline{27}$

26. $6.\overline{37}$

Find the specified value for each infinite geometric series.

27. $a_1 = 12, S = 30; \ r$

28. $a_1 = 40, S = 200; \ r$

29. $r = -0.2, S = -50; \ a_1$

30. $r = -0.5, S = -28; \ a_1$

Determine whether each statement is true or false. If it is false, give a counterexample (a specific case in which the statement does not hold).

31. If the terms of an infinite geometric series alternate between positive and negative, then the series must be divergent.

32. If the terms of an infinite geometric series are all positive, then the series must be convergent.

33. If the sum of an infinite geometric series is twice the first term, then the common ratio is $\frac{1}{2}$.

34. If the common ratio of an infinite geometric series is 0.9999, then the series converges.

35. If the sum of an infinite geometric series is positive, all the terms must be positive.

36. If the sum of an infinite geometric series is negative, all the terms must be negative.

Applications

37. Physics The end of the pendulum of a clock travels 40 cm on its first swing. On each succeeding swing, it travels 0.8 as far as it did on the preceding one. If it continues in this manner, how far will the end of the pendulum travel?

38. Physics The pointer of a broken metronome travels 20 in. on its first swing. On each succeeding swing it travels $\frac{7}{8}$ as far as it did on the preceding one. If it continues in this manner, how far will the pointer travel in all?

39. Physics A ball tossed up 22 ft rebounds to $\frac{9}{10}$ of the height from which it falls on each bounce. What is the total distance (up and down) it will travel?

40. Physics After its release, a weather balloon rises 100 ft in the first minute. Each minute thereafter, the balloon rises only 80 percent as far as in the previous minute. What is the maximum height the balloon will reach?

41. Geometry Triangle *ABC* is equilateral with each side of length 40 cm. The midpoints of the sides are joined to form a second equilateral triangle *DEF*, and so on. If this midpoint-joining process continues without end, find the sum of the perimeters of all the triangles formed.

HISTORICAL NOTE

Limits It took several centuries to formulate precisely the concept of a limit, upon which calculus is based. A number of the best mathematical minds of each century worked with the idea before an acceptable definition was devised.

Century	Mathematicians	Contributions
17th	Isaac Newton	Developed and used infinite series
	Gottfried Wilhelm Leibniz	Invented convenient symbolism and developed rules of procedure
18th	Leonhard Euler	Tried to clarify basic limit ideas
	Joseph Louis Lagrange	
19th	Augustin-Louis Cauchy	Developed somewhat satisfactory definitions of limits
	Karl Weierstrass	Gave more precise definitions of the limit of a function and the limit of a series

Read about one of the men listed in the table and report on his additional accomplishments in mathematics and other fields.

Power Series and Trigonometric Functions

Objectives: To define power series

To estimate values of trigonometric functions using power series

Work with power series requires the simplification of expressions written in factorial notation.

Preview

Recall that $n!$ is read *n factorial* and means $n(n - 1) \ldots 3 \cdot 2 \cdot 1$.
Therefore, $6! = 6 \cdot 5 \cdot 4 \cdot 3 \cdot 2 \cdot 1 = 720$. Note that $0! = 1$ by definition.

EXAMPLE **Simplify:** $\dfrac{10!}{3! \, 5!}$

$$\frac{10!}{3! \, 5!} = \frac{10 \cdot 9 \cdot 8 \cdot 7 \cdot 6 \cdot 5!}{3! \, 5!} = \frac{10 \cdot 9 \cdot 8 \cdot 7 \cdot 6}{3 \cdot 2} = 10 \cdot 9 \cdot 8 \cdot 7 = 5040$$

Simplify each expression.

1. $\dfrac{8!}{4!}$ 2. $\dfrac{12!}{6!}$ 3. $\dfrac{5! \, 6!}{4!}$ 4. $\dfrac{n!}{(n - 1)!}$ 5. $\dfrac{(n + 1)!}{n!}$

The series previously considered had constant terms. Now consider series with variable terms, such as $1 + x + x^2 + \cdots + x^n + \cdots$. By substituting different values for x, a whole family of series can be generated. This series is an example of a *power series*. Expressions of the following form are called **power series** in x, where x is a variable and $a_0, a_1, a_2, \ldots, a_n, \ldots$ are constants.

$$\sum_{n=0}^{\infty} a_n x^n = a_0 x^0 + a_1 x^1 + a_2 x^2 + \cdots + a_{n-1} x^{n-1} + \cdots$$

 Power series can be used to represent certain functions, such as $\sin x$ and $\cos x$, where x is in *radians*. In fact, calculators and computers use power series to evaluate such functions.

$$\sin x = x - \frac{x^3}{3!} + \frac{x^5}{5!} - \frac{x^7}{7!} + \cdots \qquad \cos x = 1 - \frac{x^2}{2!} + \frac{x^4}{4!} - \frac{x^6}{6!} + \cdots$$

The sum of the first six terms of either expansion will provide a reasonable approximation, and it is helpful to use a calculator to find that sum. However, there are times when six terms are not enough to give a sufficiently precise approximation, and then more terms must be used.

EXAMPLE 1 **Approximate each value using the first six terms of the appropriate power series expansion. Round each answer to four decimal places.** **a.** $\cos \dfrac{\pi}{6}$ **b.** $\sin 23°$

a. $\cos \dfrac{\pi}{6} = 1 - \dfrac{\left(\dfrac{\pi}{6}\right)^2}{2!} + \dfrac{\left(\dfrac{\pi}{6}\right)^4}{4!} - \dfrac{\left(\dfrac{\pi}{6}\right)^6}{6!} + \dfrac{\left(\dfrac{\pi}{6}\right)^8}{8!} - \dfrac{\left(\dfrac{\pi}{6}\right)^{10}}{10!}$ $\qquad \cos x = 1 - \dfrac{x^2}{2!} + \dfrac{x^4}{4!} - \cdots$

$= 1 - \dfrac{\left(\dfrac{\pi}{6}\right)^2}{2} + \dfrac{\left(\dfrac{\pi}{6}\right)^4}{24} - \dfrac{\left(\dfrac{\pi}{6}\right)^6}{720} + \dfrac{\left(\dfrac{\pi}{6}\right)^8}{40320} - \dfrac{\left(\dfrac{\pi}{6}\right)^{10}}{3628800}$ $\qquad$ *The factorial function on your calculator will simplify these operations.*

$= 0.8660 \qquad$ Use the cosine key on your calculator to verify that

$\qquad\qquad \cos \dfrac{\pi}{6} = 0.8660$, correct to four decimal places.

b. $23° = 0.4014$ radians, to four decimal places $\qquad$ *Express 23° in radians.*

$\sin 0.4014 = 0.4014 - \dfrac{(0.4014)^3}{3!} + \dfrac{(0.4014)^5}{5!} - \dfrac{(0.4014)^7}{7!} + \dfrac{(0.4014)^9}{9!} - \dfrac{(0.4014)^{11}}{11!}$

$\qquad\qquad\qquad\qquad\qquad\qquad\qquad \sin x = x - \dfrac{x^3}{3!} + \dfrac{x^5}{5!} - \dfrac{x^7}{7!} + \cdots$

$= 0.4014 - \dfrac{(0.4014)^3}{6} + \dfrac{(0.4014)^5}{120} - \dfrac{(0.4014)^7}{5040} + \dfrac{(0.4014)^9}{362880} - \dfrac{(0.4014)^{11}}{39916800}$

$= 0.3907 \qquad$ Use the sine key on your calculator to verify that

$\qquad\qquad \sin 23° = 0.3907$, correct to four decimal places.

An important function in mathematics is the *exponential function, e^x*. Recall that e is the irrational number that is used as the base for natural logarithms. It can be shown that for all x

$$e^x = 1 + \frac{x}{1!} + \frac{x^2}{2!} + \frac{x^3}{3!} + \cdots = \sum_{n=0}^{\infty} \frac{x^n}{n!}$$

EXAMPLE 2 **Using $x = 1$, approximate the value of e using the first six terms of the power series for e^x. Round your answer to three decimal places.**

$e^1 = 1 + \dfrac{1}{1!} + \dfrac{1^2}{2!} + \dfrac{1^3}{3!} + \dfrac{1^4}{4!} + \dfrac{1^5}{5!} \qquad e^x = 1 + \dfrac{x}{1!} + \dfrac{x^2}{2!} + \dfrac{x^3}{3!} + \dfrac{x^4}{4!} + \dfrac{x^5}{5!}$

$e = 1 + 1 + \dfrac{1}{2} + \dfrac{1}{6} + \dfrac{1}{24} + \dfrac{1}{120} = 2.717$

Use a calculator to show that the value of e to three decimal places is 2.718. The difference is due to the fact that only six terms are used.

The Swiss mathematician Leonhard Euler (1707–1783) discovered a formula relating the sine, cosine, and exponential functions. **Euler's formula** is derived by replacing x with ix in the series for e^x, where $i = \sqrt{-1}$.

$$e^{ix} = 1 + ix + \frac{(ix)^2}{2!} + \frac{(ix)^3}{3!} + \frac{(ix)^4}{4!} + \frac{(ix)^5}{5!} + \cdots + \frac{(ix)^{n-1}}{(n-1)!} + \cdots$$

$$= 1 + ix + \frac{i^2 x^2}{2!} + \frac{i^3 x^3}{3!} + \frac{i^4 x^4}{4!} + \frac{i^5 x^5}{5!} + \cdots + \frac{i^{n-1} x^{n-1}}{(n-1)!} + \cdots$$

$$= 1 + ix - \frac{x^2}{2!} - i\frac{x^3}{3!} + \frac{x^4}{4!} + i\frac{x^5}{5!} - \frac{x^6}{6!} - i\frac{x^7}{7!} + \cdots \qquad i^2 = -1,\ i^3 = -i,\ i^4 = 1, \cdots$$

$$= \left(1 - \frac{x^2}{2!} + \frac{x^4}{4!} - \frac{x^6}{6!} + \cdots\right) + i\left(x - \frac{x^3}{3!} + \frac{x^5}{5!} - \frac{x^7}{7!} + \cdots\right) \qquad \text{Group real and imaginary terms.}$$

$$= \cos x + i \sin x \qquad 1 - \frac{x^2}{2!} + \frac{x^4}{4!} - \cdots = \cos x;\ x - \frac{x^3}{3!} + \frac{x^5}{5!} - \cdots = \sin x$$

Thus, Euler's formula is $e^{ix} = \cos x + i \sin x$.

EXAMPLE 3 **Find the exact value of $e^{\frac{3\pi i}{2}}$ using Euler's formula.**

$$e^{\frac{3\pi i}{2}} = \cos \frac{3\pi}{2} + i \sin \frac{3\pi}{2} = 0 + i(-1) = -i \qquad e^{ix} = \cos x + i \sin x$$

Another identity can be derived by substituting $-x$ for x in Euler's formula.

$$e^{i(-x)} = \cos(-x) + i \sin(-x) = \cos x + (-i \sin x) \qquad \cos(-x) = \cos x;$$
$$e^{-ix} = \cos x - i \sin x \qquad\qquad\qquad\qquad\qquad\quad \sin(-x) = -\sin x$$

Euler's formula and this identity can be used to prove other identities.

EXAMPLE 4 **Prove:** $\sin x = \dfrac{e^{ix} - e^{-ix}}{2i}$

$$\sin x \quad \left| \quad \begin{array}{l} \dfrac{e^{ix} - e^{-ix}}{2i} \\[2ex] \dfrac{\cos x + i \sin x - (\cos x - i \sin x)}{2i} \\[2ex] \dfrac{2i \sin x}{2i} \end{array} \right.$$

$$= \sin x \qquad \text{Thus, } \sin x = \frac{e^{ix} - e^{-ix}}{2i}$$

CLASS EXERCISES

Use the appropriate power series to write the first six terms needed to estimate each value. Do not evaluate the sum.

1. $\cos \dfrac{\pi}{7}$ 2. $\cos 3.6$ 3. $\sin 2.4$ 4. $\sin \dfrac{2\pi}{5}$ 5. $e^{1.3}$ 6. e^{π}

Find the exact value of each expression using Euler's formula.

7. $e^{2\pi i}$ **8.** $e^{-\pi i}$ **9.** $e^{\frac{\pi i}{4}}$

PRACTICE EXERCISES   Use appropriate technology.

Approximate each value using the first six terms of the appropriate power series. Round each answer to four decimal places.

1. $\sin \dfrac{\pi}{4}$ **2.** $\cos \dfrac{\pi}{4}$ **3.** $\cos \dfrac{\pi}{6}$ **4.** $\sin \pi$

5. $\cos \dfrac{2\pi}{3}$ **6.** $\sin \dfrac{3\pi}{4}$ **7.** $\sin 1.5$ **8.** $\cos 0.5$

Approximate each value using the first six terms of the power series for e^x. Round each answer to three decimal places.

9. $e^{0.1}$ **10.** $e^{0.2}$ **11.** $e^{\frac{1}{3}}$ **12.** $e^{\frac{1}{2}}$ **13.** $e^{-\frac{1}{2}}$ **14.** e^{-1}

Find the exact value of each expression using Euler's formula.

15. $e^{\pi i}$ **16.** $e^{\frac{\pi i}{2}}$ **17.** $e^{\frac{3\pi i}{4}}$ **18.** $e^{3\pi i}$ **19.** $e^{-4\pi i}$ **20.** $e^{\frac{7\pi i}{4}}$

Approximate each value using the first six terms of the appropriate power series. Remember to change degree measure to radian measure. Round each answer to four decimal places.

21. $\cos 180°$ **22.** $\sin 180°$ **23.** $\sin 112°$ **24.** $\cos 124°$

25. $\cos 246°$ **26.** $\sin 240°$ **27.** $\sin 162°$ **28.** $\cos 176°$

Prove each identity.

29. $\cos x = \dfrac{e^{ix} + e^{-ix}}{2}$ **30.** $\tan x = \dfrac{e^{ix} - e^{-ix}}{i(e^{ix} + e^{-ix})}$ **31.** $\csc x = \dfrac{2i}{e^{ix} - e^{-ix}}$

32. $\sec x = \dfrac{2}{e^{ix} + e^{-ix}}$ **33.** $\cot x = \dfrac{i(e^{ix} + e^{-ix})}{e^{ix} - e^{-ix}}$ **34.** $e^{(\pi + x)i} = -e^{ix}$

35. Prove: $\dfrac{e^{ix}}{e^{iy}} = e^{(x - y)i}$

36. Use Euler's formula and double-angle formulas to show that $(e^{ix})^2 = e^{2ix}$.

37. Justify the following statement using your calculator to substitute values for n: $\left(1 + \dfrac{1}{n}\right)^n$ approaches e as n gets larger and larger.

38. Show that $\sin^2 x + \cos^2 x = 1$, using $\sin x = \dfrac{e^{ix} - e^{-ix}}{2i}$ and $\cos x = \dfrac{e^{ix} + e^{-ix}}{2}$.

39. Show that $1 + \cot^2 x = \csc^2 x$, using the expressions for $\cot x$ and $\csc x$ given in Exercises 31 and 33.

40. Show that $1 + \tan^2 x = \sec^2 x$, using the expressions for $\tan x$ and $\sec x$ given in Exercises 30 and 32.

Statistics The *standard normal probability curve* $y = \dfrac{1}{\sqrt{2\pi}} e^{-\frac{x^2}{2}}$ is of great importance in the field of statistics.

41. Show that $y = \dfrac{1}{\sqrt{2\pi}} e^{-\frac{x^2}{2}}$ can be approximated by this equation

$$y = \frac{1}{\sqrt{2\pi}}\left(1 - \frac{x^2}{2} + \frac{x^4}{8} - \frac{x^6}{48} + \frac{x^8}{384} - \frac{x^{10}}{3840}\right)$$

Use the equation in Exercise 41 to find the y-coordinate of each point. Round to the nearest hundredth.

42. $A(0, \ \underline{?}\ )$ **43.** $B(0.25, \ \underline{?}\ )$

44. $C(-0.5, \ \underline{?}\ )$ **45.** $D(0.75, \ \underline{?}\ )$

46. $E(-1, \ \underline{?}\ )$ **47.** $F(1.25, \ \underline{?}\ )$

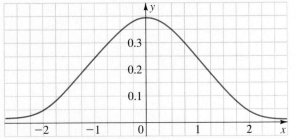

CHALLENGE

The letter e is used to represent the irrational number 2.7182818 (to seven decimal places). This unique number was discovered by the mathematician Leonhard Euler, who found that when n is replaced with successively greater values in the expression $\left(1 + \dfrac{1}{n}\right)^n$, the value of the expression approaches e as its limit. Euler also showed that e can be written as a continued fraction.

$$e = 2 + \cfrac{1}{1 + \cfrac{1}{2 + \cfrac{2}{3 + \cfrac{3}{4 + \cfrac{4}{\ddots}}}}}$$

1. Show that if the continued fraction above is evaluated through the "$3 + \frac{3}{4}$" row (that is, all rows below $3 + \frac{3}{4}$ are eliminated), the value obtained gives e correct to one decimal place. *Hint*: Start from the bottom and work up.

2. Through which row must the continued fraction be evaluated in order to obtain e correct to three decimal places?

9.7 Hyperbolic Functions

Objective: To define the hyperbolic functions in terms of e^x and e^{-x} and to express them as power series

Hyperbolic functions are used in physics and engineering. In order to understand them, it is important to be familiar with the fundamental identities, the double-angle identities, and the sum and difference identities.

In the last lesson, you learned that $e^x = 1 + x + \dfrac{x^2}{2!} + \dfrac{x^3}{3!} + \dfrac{x^4}{4!} + \cdots$. If you substitute $-x$ for x in this series, the following new series is generated

$$e^{-x} = 1 + (-x) + \frac{(-x)^2}{2!} + \frac{(-x)^3}{3!} + \frac{(-x)^4}{4!} + \cdots$$

$$e^{-x} = 1 - x + \frac{x^2}{2!} - \frac{x^3}{3!} + \frac{x^4}{4!} - \cdots$$

Two new functions can be generated by adding and subtracting the power series for e^x and e^{-x}. First add.

$$e^x = 1 + x + \frac{x^2}{2!} + \frac{x^3}{3!} + \frac{x^4}{4!} + \frac{x^5}{5!} + \frac{x^6}{6!} + \cdots$$

$$e^{-x} = 1 - x + \frac{x^2}{2!} - \frac{x^3}{3!} + \frac{x^4}{4!} - \frac{x^5}{5!} + \frac{x^6}{6!} - \cdots$$

$$e^x + e^{-x} = 2 + 0 + 2\left(\frac{x^2}{2!}\right) + 0 + 2\left(\frac{x^4}{4!}\right) + 0 + 2\left(\frac{x^6}{6!}\right) + \cdots$$

$$e^x + e^{-x} = 2\left(1 + \frac{x^2}{2!} + \frac{x^4}{4!} + \frac{x^6}{6!} + \cdots\right)$$

Thus, $\dfrac{e^x + e^{-x}}{2} = 1 + \dfrac{x^2}{2!} + \dfrac{x^4}{4!} + \dfrac{x^6}{6!} + \cdots$.

Recall that $\cos x = 1 - \dfrac{x^2}{2!} + \dfrac{x^4}{4!} - \dfrac{x^6}{6!} + \cdots$. Note that the power series

representation of $\dfrac{e^x + e^{-x}}{2}$ is the same as that of $\cos x$, except for the fact

that it does not have alternating signs. A new function, cosh x, is defined as

$$\cosh x = \frac{e^x + e^{-x}}{2} \qquad \textit{Read: hyperbolic cosine of x}$$

Next, subtract e^{-x} from e^x.

$$e^x = 1 + x + \frac{x^2}{2!} + \frac{x^3}{3!} + \frac{x^4}{4!} + \frac{x^5}{5!} + \cdots$$

$$e^{-x} = 1 - x + \frac{x^2}{2!} - \frac{x^3}{3!} + \frac{x^4}{4!} - \frac{x^5}{5!} + \cdots$$

$$e^x - e^{-x} = 0 + 2x + 0 + 2\left(\frac{x^3}{3!}\right) + 0 + 2\left(\frac{x^5}{5!}\right) + \cdots$$

$$e^x - e^{-x} = 2\left(x + \frac{x^3}{3!} + \frac{x^5}{5!} + \frac{x^7}{7!} + \cdots\right)$$

Thus, $\quad \dfrac{e^x - e^{-x}}{2} = x + \dfrac{x^3}{3!} + \dfrac{x^5}{5!} + \dfrac{x^7}{7!} + \cdots$.

Recall that $\sin x = x - \dfrac{x^3}{3!} + \dfrac{x^5}{5!} - \dfrac{x^7}{7!} + \cdots$. Note that the power series

representation of $\dfrac{e^x - e^{-x}}{2}$ is the same as that of $\sin x$, except for the fact that

it does not have alternating signs. A new function, sinh x, is defined as follows:

$$\sinh x = \frac{e^x - e^{-x}}{2} \qquad \textit{Read: hyperbolic sine of x}$$

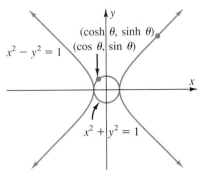

These new functions are associated with the unit
hyperbola, $x^2 - y^2 = 1$, graphed on the xy-plane
at the right. For each value of θ, the point (cosh θ,
sinh θ) lies on the unit hyperbola, just as the point
(cos θ, sin θ) lies on the unit circle $x^2 + y^2 = 1$.

EXAMPLE 1 **Evaluate cosh x and sinh x for $x = 1$. Round to four decimal places.**

$$\cosh x = \frac{e^x + e^{-x}}{2} \qquad\qquad\qquad \sinh x = \frac{e^x - e^{-x}}{2}$$

$$\cosh 1 = \frac{e^1 + e^{-1}}{2} \qquad\qquad\qquad \sinh 1 = \frac{e^1 - e^{-1}}{2}$$

$$\cosh 1 = 1.5431 \quad \textit{To four decimal places} \qquad \sinh 1 = 1.1752$$

Other similarities exist between the trigonometric functions and the **hyperbolic functions.** For example, the hyperbolic tangent, hyperbolic cotangent, hyperbolic secant, and hyperbolic cosecant functions can be defined in terms of the hyperbolic sine and cosine, just as the other trigonometric functions can be defined in terms of $\sin x$ and $\cos x$.

Trigonometric functions

$$\tan x = \frac{\sin x}{\cos x}$$

$$\cot x = \frac{\cos x}{\sin x}$$

$$\sec x = \frac{1}{\cos x}$$

$$\csc x = \frac{1}{\sin x}$$

Hyperbolic functions

$$\tanh x = \frac{\sinh x}{\cosh x} = \frac{e^x - e^{-x}}{e^x + e^{-x}}$$

$$\coth x = \frac{\cosh x}{\sinh x} = \frac{e^x + e^{-x}}{e^x - e^{-x}}$$

$$\text{sech } x = \frac{1}{\cosh x} = \frac{2}{e^x + e^{-x}}$$

$$\text{csch } x = \frac{1}{\sinh x} = \frac{2}{e^x - e^{-x}}$$

In addition, there are a number of identities for hyperbolic functions that are similar to those for the trigonometric functions.

Trigonometric identities

$\cos^2 x + \sin^2 x = 1$

$\sin (x + y) = \sin x \cos y + \cos x \sin y$

$\cos (x + y) = \cos x \cos y - \sin x \sin y$

$\sin 2x = 2 \sin x \cos x$

$\cos 2x = 2 \cos^2 x - 1$

Hyperbolic identities

$\cosh^2 x - \sinh^2 x = 1 \text{ (since } w^2 - z^2 = 1)$

$\sinh (x + y) = \sinh x \cosh y + \cosh x \sinh y$

$\cosh (x + y) = \cosh x \cosh y - \sinh x \sinh y$

$\sinh 2x = 2 \sinh x \cosh x$

$\cosh 2x = 2 \cosh^2 x - 1$

These identities can be used to prove other identities.

EXAMPLE 2   **Prove:** $-\sinh x = \sinh (-x)$

$-\sinh x$	$\sinh (-x)$
	$\dfrac{e^{-x} - e^{-(-x)}}{2}$
	$\dfrac{e^{-x} - e^{x}}{2}$
	$-\left(\dfrac{e^{x} - e^{-x}}{2}\right)$
$= -\sinh x$	

Thus, $-\sinh x = \sinh (-x)$.

CLASS EXERCISES

Match each expression in Exercises 1–6 with its value listed in the second row.

1. $\sinh 0$ **2.** $\cosh 0$ **3.** $\operatorname{sech} -1$ **4.** $\operatorname{csch} -1$ **5.** $\coth 1$ **6.** $\tanh 1$
a -0.8509 **b.** 0.7616 **c.** 0 **d.** 0.6481 **e.** 1.3130 **f.** 1

7. Prove $\cosh 2x = 2 \cosh^2 x - 1$

PRACTICE EXERCISES   Use appropriate technology.

Evaluate each function for $x = -1$. Round each answer to four decimal places.

1. $\sinh x$ **2.** $\cosh x$ **3.** $\tanh x$ **4.** $\coth x$

Prove each identity.

5. $\cosh x = \dfrac{1}{\operatorname{sech} x}$

6. $\sinh x = \dfrac{1}{\operatorname{csch} x}$

7. $\sinh x = \dfrac{\cosh x}{\coth x}$

8. $\tanh x = \dfrac{1}{\coth x}$

9. $\cosh x = \dfrac{\coth x}{\operatorname{csch} x}$

10. $\sinh x = \dfrac{\tanh x}{\operatorname{sech} x}$

11. $\operatorname{sech} x = \dfrac{\operatorname{csch} x}{\coth x}$

12. $\cosh(-x) = \cosh x$

13. $\operatorname{sech}^2 x = 1 - \tanh^2 x$

14. $\operatorname{csch}^2 x = \coth^2 x - 1$

15. $\sinh 2x = 2 \sinh x \cosh x$

16. $\cosh 2x = 1 + 2 \sinh^2 x$

17. $\sinh^2 x = \cosh^2 x - 1$

18. $\cosh 2x = \cosh^2 x + \sinh^2 x$

19. $e^x = \cosh x + \sinh x$

20. $e^{-2x} = (\sinh x - \cosh x)^2$

21. $\sinh(x - y) = \sinh x \cosh y - \cosh x \sinh y$

22. $\cosh(x - y) = \cosh x \cosh y - \sinh x \sinh y$

23. $\cosh \dfrac{x}{2} = \sqrt{\dfrac{1 + \cosh x}{2}}$

24. $\sinh^2\left(\dfrac{x}{2}\right) = \dfrac{\cosh x - 1}{2}$

Applications

Civil Engineering   Some bridge cables have the shape of a rope supported at both ends and hanging of its own weight. This curve is called a *catenary*. The catenary at the right is described by the equation $y = a \cosh x + b - a$, where b is the y-intercept and a is a quantity that depends upon forces acting on the cable.

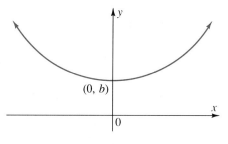

(0, b)

25. For simplicity, the x-axis in the figure can be chosen so that $b = a$. In that case, what is the simplified form of the equation of the catenary?

26. In Exercise 25, suppose $x = 3$. What is the value of y as a function of a?

TEST YOURSELF

Determine the sum S, if it exists, for each infinite geometric series.

1. $1 + \frac{1}{8} + \frac{1}{64} + \cdots$ **2.** $\frac{1}{64} + \frac{1}{8} + 1 + \cdots$ **9.5**

3. $\sum_{n=1}^{\infty} (-2)^n$ **4.** $\sum_{n=1}^{\infty} \left(-\frac{1}{2}\right)^n$

Write each repeating decimal as a fraction.

5. $0.\overline{52}$ **6.** $0.7\overline{4}$

Approximate each value using the first six terms of the appropriate power series. Round each answer to four decimal places.

7. $\sin \frac{5\pi}{4}$ **8.** $\cos 311°$ **9.** e^4 **10.** $e^{\frac{1}{4}}$ **9.6**

Find the exact value of each expression using Euler's formula.

11. $e^{\frac{-\pi i}{4}}$ **12.** $e^{\frac{-\pi i}{2}}$

Evaluate each function for $x = 2$. Round to four decimal places.

13. $\sinh x$ **14.** $\cosh x$ **9.7**

Prove each identity.

15. $\coth x = \frac{\operatorname{csch} x}{\operatorname{sech} x}$ **16.** $\cosh x - \sinh x = e^{-x}$

9.7 Hyperbolic Functions   **411**

TECHNOLOGY:
Spirals

Did you know that some spiders spin their webs in the shape of an Archimedean spiral? The grooves in a phonograph record, the coil of a spool of video or audio tape, and a coiled rope also take the form of an Archimedean spiral. In addition, these spirals are found in the plant kingdom in tomato plants, fiddlehead ferns, and sago palms.

What is a spiral, and in particular, what is an Archimedean spiral? A spiral is a curve traced by a point constantly moving around and away from a fixed point called a pole. The particular type of spiral is determined by the manner in which the curve moves away from the pole. An Archimedean spiral is a curve in which the distance from the origin of a point on the curve is directly proportional to the measure of the angle turned to reach that point. For example, if the distance from the pole is 1 when the angle is 60°, then the distance must be 2 when the angle is 120°, 3 when the angle is 180°, and so on. Thus, an Archimedean spiral increases in length at a constant rate per constant turn.

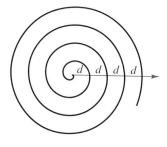

If a radius vector is superimposed on an Archimedean spiral, the spiral will intersect it at equally spaced intervals. The distances from the pole of the successive turns of the spiral along this radius vector form an arithmetic sequence: $d, 2d, 3d, \ldots$.

When spinning their webs, spiders are efficient at conserving energy and material while expanding over a maximum area. Spiders that spin their webs in the shape of an Archimedean spiral first create a framework of spokes that correspond to radius vectors. They then progress outward from the center, forming a spiral by guiding themselves by the previous turn and ensuring that the distance between each turn is constant.

Archimedean spirals can be represented by polar equations in the form $r = a\theta$. Recall from Chapter 7 that such equations can be graphed in parametric form on a graphing utility.

EXAMPLE **Graph $r = 0.1\theta$ over the interval $0° \leq \theta \leq 720°$.**

Remember to use the polar-rectangular conversion formulas:

$$x = r \cos \theta \text{ and } y = r \sin \theta$$

x scl: 30 *y* scl: 30

The loops of the spiral do appear to be equally spaced, as predicted. You can verify this by examining the values of successive intercepts.

When you trace values on a parametric graph, the trace follows θ, rather than *x* or *y*. The traced values of all of the *x*- and *y*-intercepts are shown here.

A careful examination will show that consecutive turns of the spiral are equidistant, 36 units apart.

θ	x	y
0°	0	0
90°	0	9
180°	−18	0
270°	0	−27
360°	36	0
450°	0	45
540°	−54	0
630°	0	−63
720°	72	0

EXERCISES

1. When differences in sign are ignored, the *x*- and *y*-intercepts of the graph above can be seen as successive distances from the pole (origin), at increments of 90°. What is the arithmetic sequence described? Write an explicit and a recursive rule for the sequence.

2. Graph $r = 0.10\theta$ and $r = 0.2\theta$ on the same screen, over $0° \leq \theta \leq 1440°$.
   a. Find the coordinates of one intersection of the two spirals.
   b. In what sense is this not an intersection?

3. Adjust the range of your graphing utility so that the angle is calculated in steps of 20°. Graph $r = 0.2\theta$. How does the graph differ when the step is changed to 30°? 60°? 65°? Explain.

4. Try to predict the result when $r = 0.1\theta$ and $r = -0.1\theta$ are graphed on the same screen. Graph and describe the results.

5. How could you graph a clockwise spiral?

Vocabulary

arithmetic means (375)

arithmetic sequence (373)

arithmetic series (384)

common difference (373)

common ratio (378)

convergent series (397)

divergent series (398)

Euler's formula (404)

explicit formula (372)

finite sequence (372)

general term (372)

geometric means (380)

geometric sequence (378)

geometric series (390)

hyperbolic functions (409)

index of summation (385)

infinite sequence (373)

limit (397)

limits of summation (385)

mean proportional (380)

partial sum (384)

power series (402)

recursive formula (373)

sequence (372)

series (384)

sigma notation (384)

sum of an infinite

geometric series (398)

Arithmetic Sequences   If a_1 is the first term, n is the number of   **9.1**
terms, d is the common difference, and a_n is the nth term of an arithmetic
sequence, then $a_n = a_1 + (n - 1)d$.

Find the next three terms and the general term of each arithmetic sequence.

  1. 10, 7, 4, . . .

  2. $-0.5, 0, 0.5, . . .$

Find the indicated term of each arithmetic sequence.

  3. a_{10} of $-20, -11, -2, . . .$

  4. a_8 of 18, 6, $-6, . . .$

  5. Find two arithmetic means between 6 and 36.

Geometric Sequences   If a_1 is the first term, n is the number of terms, r is   **9.2**
the common ratio, and a_n is the nth term, then $a_n = a_1 r^{n-1}$.

Find the next three terms and the general term of each geometric sequence.

  6. 36, 12, 4, . . .

  7. 1, 3, 9, . . .

Find the indicated term of each geometric sequence.

  8. a_6 of 4, $-8, 16, . . .$

  9. a_7 of 128, 64, 32, . . .

10. Find the mean proportional between -9 and -81.

Arithmetic Series   If a_1 is the first term, n is the number of terms, d is   **9.3**
the common difference, and a_n is the nth term, then the partial sum S_n of an
arithmetic series is given by

$$S_n = \frac{n(a_1 + a_n)}{2} \quad \text{or} \quad S_n = \frac{n[2a_1 + (n - 1)d]}{2}$$

Find the indicated partial sum of each arithmetic series.

11. $-15, -10, -5, \ldots ; S_{13}$ **12.** $\displaystyle\sum_{m=1}^{15} (2m + 1)$

13. Express in sigma notation the sum of the first six terms of the arithmetic series $-7 + (-3) + 1 + \cdots$.

14. If $a_8 = 100$ and $S_8 = 440$, and the series is arithmetic, find a_1.

Geometric Series If a_1 is the first term, n is the number of terms, r is the common ratio, and a_n is the nth term of a geometric series, then   **9.4**

$$S_n = \frac{a_1(1 - r^n)}{1 - r}, r \neq 1 \quad \text{or} \quad S_n = \frac{a_1 - ra_n}{1 - r}, r \neq 1$$

Find the indicated partial sum of each geometric series.

15. $135 + 45 + 15 + \cdots ; S_7$ **16.** $\displaystyle\sum_{j=1}^{6} (-3)^j$ **17.** $a_1 = 2, a_7 = 1458; S_7$

Infinite Geometric Series An infinite geometric series converges if and   **9.5**
only if $|r| < 1$. If S is the sum of a convergent geometric series, then $S = \dfrac{a_1}{1 - r}$.

Determine the sum S, if it exists, for each infinite geometric series.

18. $100 - 10 + 1 - \cdots$ **19.** $\frac{1}{2} + \frac{3}{2} + \frac{9}{2} + \cdots$ **20.** $\displaystyle\sum_{n=1}^{\infty} 2\left(\frac{1}{2}\right)^{n+5}$

Power Series and Trigonometric Functions   **9.6**

$$\sin x = x - \frac{x^3}{3!} + \frac{x^5}{5!} - \frac{x^7}{7!} + \cdots \qquad \cos x = 1 - \frac{x^2}{2!} + \frac{x^4}{4!} - \frac{x^6}{6!} + \cdots$$

$$e^x = 1 + \frac{x}{1!} + \frac{x^2}{2!} + \frac{x^3}{3!} + \cdots \qquad e^{ix} = \cos x + i \sin x$$

Approximate each value using the first six terms of the appropriate power series. Round each answer to four decimal places.

21. $\cos \dfrac{3\pi}{2}$ **22.** $\sin 1.7$ **23.** $e^{0.7}$

24. Use Euler's formula to find the exact value of $e^{\frac{\pi i}{6}}$.

Hyperbolic Functions $\cosh x = \dfrac{e^x + e^{-x}}{2}$ and $\sinh x = \dfrac{e^x - e^{-x}}{2}$ **9.7**

25. Prove: $\operatorname{sech} x = \dfrac{\tanh x}{\sinh x}$ **26.** Prove: $\coth^2 x = 1 + \operatorname{csch}^2 x$

1. In an arithmetic sequence, $a_1 = 24$ and $d = -0.5$. Find a_6.

2. Insert two arithmetic means between -10 and 8.

3. In a geometric sequence, $a_1 = 7$ and $r = 4$. Find a_9.

4. Find the mean proportional between 2 and 18.

5. The sixth term of an arithmetic sequence is 14 and the 12th term is 38. Find the first term and the common difference.

6. The fourth term of a geometric sequence is -27 and the seventh term is 729. Find the first term and the common ratio.

Find the indicated partial sum for each arithmetic series.

7. $a_1 = 10, d = 7$; S_{12}

8. $a_1 = 14, d = 12$; S_9

Find the indicated partial sum for each geometric series.

9. $4 + (-12) + 36 + \cdots$; S_{12}

10. $-7 + 14 + (-28) + \cdots$; S_9

Find the sum of each infinite geometric series, if it exists. If the sum does not exist, say so.

11. $\displaystyle\sum_{k=1}^{\infty} \left(\frac{4}{3}\right)^{-k}$

12. $\displaystyle\sum_{k=1}^{\infty} \left(\frac{3}{4}\right)^{-k}$

Approximate each value using the first six terms of the appropriate power series. Round each answer to four decimal places.

13. $\sin 2.8$

14. $\cos 60°$

15. $e^{0.3}$

16. Use Euler's formula to find the exact value of $e^{-6\pi i}$.

17. Prove: $\cosh x = \dfrac{\sinh x}{\tanh x}$.

Challenge

Prove: $\coth (x - y) = \dfrac{1 + \tanh x \tanh y}{\tanh x - \tanh y}$

Solve. Grid in your response on the answer sheet.

1. If $a_n = n^2 + 2$, what is the fourth term of the sequence?

2. $\sum\limits_{j=2}^{5} 3j^2 - 1$ equals what value?

3. If $f(t) = \dfrac{2t^2 + 1}{-t}$, find $f(-2)$.

4. Find the 22nd term of the arithmetic sequence 8, 11, 14, 17, ...

5. Evaluate: $\begin{vmatrix} 8 & 6 \\ 4 & 9 \end{vmatrix}$

6. Find the sum of the first 12 terms of the arithmetic sequence 1, 3, 5, ...

7. Solve $\sqrt{2x - 1} - x = -2$.

8. Find the 8th term of the geometric sequence 4, 8, 16, 32, ...

9. The perimeter of a rectangle is 48 ft. If the width of the rectangle is one half its length, find the width in feet.

10. A motorboat can travel 24 mi downstream in 2 h. It takes 3 h to make the return trip. Find the rate in mi/h of the boat in still water.

11. Find the sum of the geometric sequence 2, 6, 18, 54, 162. . . .

12. Find $\sum\limits_{n=1}^{\infty} 2^{3-n}$.

13. A closed rectangular tank 1 yd by 2 yd by 4 yd contains 4 cubic yards of water. When the tank is placed on its various sides, the water depth changes. What is the possible difference in water depths in yards?

14. Find the value of $\sin^2 (30°) + \cos^2 (30°)$.

15. Evaluate $\log_4 256 + \log_4 64 - \log_4 4$.

16. The degree measures of the angles of a triangle are a, b, and c. If a, b, and c are integers and $a < b < c$, what is the least possible value of c?

17. Wanda earned $127.50 in annual simple interest on an investment of $1500. What percent did she earn annually?

18. The cost of renting a car is $28 a day plus 12 cents a mile. How much in dollars will it cost Sara to rent the car for 2 days and drive a total distance of 200 mi?

Express in exponential form.

1. $\log_{125} 5 = \frac{1}{3}$

2. $\log_3 \frac{1}{243} = -5$

Find the approximate value in degrees.

3. $\text{Cos}^{-1}(0.3040)$

4. $\text{Arctan}\,(-0.4286)$

5. Construct a table of values and then graph the equation $r = 5 \cos 2\theta$.

Use half-angle identities to find the exact value of each trigonometric function.

6. $\tan \frac{5\pi}{8}$

7. $\sin 105°$

8. Apply De Moivre's theorem to evaluate $(2 + 2i)^4$. Express your answer in rectangular form.

Verify each identity for the given angle measure.

9. $\cot 135° \cos 135° = \csc 135° - \sin 135°$

10. $\dfrac{2 \sin \frac{\pi}{6} \cos \frac{\pi}{6}}{1 + \cos 2 \left(\frac{\pi}{6} \right)} = \tan \frac{\pi}{6}$

Simplify.

11. $\dfrac{-27\, r^{\frac{3}{4}} s^{\frac{2}{5}}}{-9\, r^{\frac{1}{4}} s^{\frac{4}{5}}}$

12. $(P^{\sqrt{16}} Q^{-\sqrt{12}})^{\sqrt{3}}$

13. Determine how many solutions exist to $\triangle ABC$, where $\angle C = 47°$, $a = 45$, and $c = 41$. Solve the triangle or triangles.

14. Find the sum of the odd integers from 1 to 49, inclusive.

15. Graph $y = \tan (x + \pi)$ over the interval $-2\pi < x < 2\pi$.

16. Prove $\dfrac{\cos^2 \alpha - 1}{\sec \alpha} = \dfrac{-\sin \alpha \sin 2\alpha}{2}$.

17. Find the length of the altitude from $\angle A$ to side a for a triangle whose sides are $a = 16$, $b = 14$, and $c = 20$.

Find the degree measure of the angle for each rotation.

18. $\frac{25}{3}$ clockwise rotation

19. $\frac{3}{4}$ counterclockwise rotation

Express each complex number in rectangular form.

20. $6 (\cos 330° + i \sin 330°)$

21. $5 (\cos 135° + i \sin 135°)$

22. Use the fact that the cosine function is an even function to find $\cos (-709°)$ if $\cos 709° = 0.9816$.

Determine whether each series is arithmetic or geometric, and then find its sum.

23. $\displaystyle\sum_{n=1}^{6} 5^{n-1}$

24. $\displaystyle\sum_{k=1}^{10} (2k + 7)$

Evaluate.

25. $\cos \left(\text{Arccos} \left(-\frac{\sqrt{2}}{2} \right) \right)$

26. $\text{Arccos} \left(\cos \left(\frac{5\pi}{4} \right) \right)$

27. Use the law of cosines to find the length of side c for $\triangle ABC$ if $\angle C = 58°$, $a = 12$, and $b = 21$.

Determine whether each function is even, odd, or neither even nor odd.

28. $h(x) = 3x - 4$

29. $g(x) = 2x^3 - x^2$

30. The entrance to a hardware store is 3.7 ft above ground level. A ramp for the handicapped from the ground level to the entrance is to be built at a 12° angle of elevation. Find the length of the ramp.

31. Let θ be an angle in standard position. Evaluate $\cos \theta$, $\tan \theta$, $\cot \theta$, $\sec \theta$, and $\csc \theta$ if θ lies in quadrant IV and $\sin \theta = -\frac{12}{13}$.

32. Given $\angle A = 59°$, $\angle B = 76°$, and $c = 33$ cm, compute the area of $\triangle ABC$ to the nearest square centimeter.

33. A ladder rests against a house at a point that is 18 ft from the ground. If the ladder makes a 45° angle with the ground, what is the length of the ladder to the nearest foot?

34. Express in sigma notation the sum of the first 8 terms of the geometric series $3 + (-6) + 12 + (-24) + \cdots$.

35. Use the sum identity for cosine to prove $\cos (2\pi + \alpha) = \cos \alpha$.

Graph each function.

36. $f(x) = e^{-x}$

37. $g(x) = 3^x$

38. Express $5 \cos 3x \sin x$ as a sum.

Calculate each product or quotient in polar form.

39. $6(\cos 15° + i \sin 15°) \cdot 3(\cos 30° + i \sin 30°)$

40. $\dfrac{12(\cos 250° + i \sin 250°)}{2(\cos 70° + i \sin 70°)}$

41. Insert four arithmetic means between -9 and 11.

42. Use the law of sines to solve $\triangle ABC$ with $\angle B = 105°$, $\angle A = 32°$, and $a = 47.2$. Express the lengths of the sides to three significant digits and angle measures to the nearest tenth of a degree.

Check graphically to see whether or not each equation is an identity.

43. $\dfrac{1 - \tan \alpha}{1 + \tan \alpha} = \dfrac{\cos 2\alpha}{1 + \sin 2\alpha}$

44. $\dfrac{\sin \beta}{1 - \cos \beta} = \dfrac{1 + \cos \beta}{\sin \beta}$

45. Approximate the solutions to $12 \sin^2 x - 8 \sin x + 1 = 0$ in degrees, for the interval $0° \le x \le 360°$.

46. Find the length of the arc intercepted by a central angle of $\dfrac{3\pi}{4}$ if the radius of the circle is 120 ft.

47. For the functions $f(x) = \sqrt{x} - 1$ and $g(x) = 2\sqrt{x}$, find $f(4) + g(9)$.

48. Express in sigma notation the sum of the first 8 terms of the arithmetic sequence $-6 + (-4) + (-2) + \cdots$.

49. The xy-axes are rotated about the origin through an angle of $60°$. Find the equation of the curve $x^2 - 2\sqrt{3}\,xy - y^2 = 8$ in the new coordinate system.

50. Write the expression $\dfrac{1}{\tan \beta + \sec \beta} + \dfrac{1}{\sec \beta - \tan \beta}$ in terms of $\cos \beta$.

51. How many years will it take for $1800 to increase to $2500, if it is deposited in an account that pays 7.25% interest, compounded monthly?

52. Find the exact values of the sine and cosine of an angle θ in standard position, if the point with coordinates $(4, -4)$ lies on its terminal side.

If $\sin \alpha = \dfrac{\sqrt{2}}{2}$, $\tan \beta = -1$, $\dfrac{\pi}{2} < \alpha < \pi$, and $\dfrac{3\pi}{2} < \beta < 2\pi$, find the exact values for each of the following.

53. $\sin(\alpha + \beta)$

54. $\sin(\alpha - \beta)$

55. Solve $5^x = 37$.

56. Find the distance between the points $(2, 3)$ and $(-7, -10)$.

Table of Squares, Cubes, Square and Cube Roots

N	N²	N³	√N	³√N	N	N²	N³	√N	³√N
1	1	1	1.000	1.000	51	2,601	132,651	7.141	3.708
2	4	8	1.414	1.260	52	2,704	140,608	7.211	3.733
3	9	27	1.732	1.442	53	2,809	148,877	7.280	3.756
4	16	64	2.000	1.587	54	2,916	157,464	7.348	3.780
5	25	125	2.236	1.710	55	3,025	166,375	7.416	3.803
6	36	216	2.449	1.817	56	3,136	175,616	7.483	3.826
7	49	343	2.646	1.913	57	3,249	185,193	7.550	3.849
8	64	512	2.828	2.000	58	3,364	195,112	7.616	3.871
9	81	729	3.000	2.080	59	3,481	205,379	7.681	3.893
10	100	1,000	3.162	2.154	60	3,600	216,000	7.746	3.915
11	121	1,331	3.317	2.224	61	3,721	226,981	7.810	3.936
12	144	1,728	3.464	2.289	62	3,844	238,328	7.874	3.958
13	169	2,197	3.606	2.351	63	3,969	250,047	7.937	3.979
14	196	2,744	3.742	2.410	64	4,096	262,144	8.000	4.000
15	225	3,375	3.873	2.466	65	4,225	274,625	8.062	4.021
16	256	4,096	4.000	2.520	66	4,356	287,496	8.124	4.041
17	289	4,913	4.123	2.571	67	4,489	300,763	8.185	4.062
18	324	5,832	4.243	2.621	68	4,624	314,432	8.246	4.082
19	361	6,859	4.359	2.668	69	4,761	328,509	8.307	4.102
20	400	8,000	4.472	2.714	70	4,900	343,000	8.367	4.121
21	441	9,261	4.583	2.759	71	5,041	357,911	8.426	4.141
22	484	10,648	4.690	2.802	72	5,184	373,248	8.485	4.160
23	529	12,167	4.796	2.844	73	5,329	389,017	8.544	4.179
24	576	13,824	4.899	2.884	74	5,476	405,224	8.602	4.198
25	625	15,625	5.000	2.924	75	5,625	421,875	8.660	4.217
26	676	17,576	5.099	2.962	76	5,776	438,976	8.718	4.236
27	729	19,683	5.196	3.000	77	5,929	456,533	8.775	4.254
28	784	21,952	5.292	3.037	78	6,084	474,552	8.832	4.273
29	841	24,389	5.385	3.072	79	6,241	493,039	8.888	4.291
30	900	27,000	5.477	3.107	80	6,400	512,000	8.944	4.309
31	961	29,791	5.568	3.141	81	6,561	531,441	9.000	4.327
32	1,024	32,768	5.657	3.175	82	6,724	551,368	9.055	4.344
33	1,089	35,937	5.745	3.208	83	6,889	571,787	9.110	4.362
34	1,156	39,304	5.831	3.240	84	7,056	592,704	9.165	4.380
35	1,225	42,875	5.916	3.271	85	7,225	614,125	9.220	4.397
36	1,296	46,656	6.000	3.302	86	7,396	636,056	9.274	4.414
37	1,369	50,653	6.083	3.332	87	7,569	658,503	9.327	4.431
38	1,444	54,872	6.164	3.362	88	7,744	681,472	9.381	4.448
39	1,521	59,319	6.245	3.391	89	7,921	704,969	9.434	4.465
40	1,600	64,000	6.325	3.420	90	8,100	729,000	9.487	4.481
41	1,681	68,921	6.403	3.448	91	8,281	753,571	9.539	4.498
42	1,764	74,088	6.481	3.476	92	8,464	778,688	9.592	4.514
43	1,849	79,507	6.557	3.503	93	8,649	804,357	9.644	4.531
44	1,936	85,184	6.633	3.530	94	8,836	830,584	9.695	4.547
45	2,025	91,125	6.708	3.557	95	9,025	857,375	9.747	4.563
46	2,116	97,336	6.782	3.583	96	9,216	884,736	9.798	4.579
47	2,209	103,823	6.856	3.609	97	9,409	912,673	9.849	4.595
48	2,304	110,592	6.928	3.634	98	9,604	941,192	9.899	4.610
49	2,401	117,649	7.000	3.659	99	9,801	970,299	9.950	4.626
50	2,500	125,000	7.071	3.684	100	10,000	1,000,000	10.000	4.642

Using a Table of Trigonometric Values

The table on pages 434 to 438 contains the values of the six trigonometric functions from $0°00'$ to $90°00'$ in 10-minute increments. The radian measure of each angle value is also listed. A portion of the table is shown below.

θ Deg.	θ Rad.	Sin θ	Cos θ	Tan θ	Cot θ	Sec θ	Csc θ		
19°00'	0.3316	0.3256	0.9455	0.3443	2.9042	1.0576	3.0716	1.2392	**71°00'**
10'	0.3345	0.3283	0.9446	0.3476	2.8770	1.0587	3.0458	1.2363	**50'**
20'	0.3374	0.3311	0.9436	0.3508	2.8502	1.0598	3.0206	1.2334	**40'**
30'	0.3403	0.3338	0.9426	0.3541	2.8239	1.0608	2.9957	1.2305	**30'**
40'	0.3432	0.3365	0.9417	0.3574	2.7980	1.0619	2.9713	1.2275	**20'**
50'	0.3462	0.3393	0.9407	0.3607	2.7725	1.0631	2.9474	1.2246	**10'**
20°00'	0.3491	0.3420	0.9397	0.3640	2.7475	1.0642	2.9238	1.2217	**70°00'**
10'	0.3520	0.3448	0.9387	0.3673	2.7228	1.0653	2.9006	1.2188	**50'**
		Cos θ	Sin θ	Cot θ	Tan θ	Csc θ	Sec θ	θ Rad.	θ Deg.

To find the value of a function for an angle between $0°00'$ and $45°00'$, read *down* the first column to locate the angle and across the *top* row to locate the function. The value of the function is at the intersection of that row and column. To illustrate, sin $19°30' = 0.3338$. To find the value of a function for an angle between $45°00'$ and $90°00'$, read *up* the last column to locate the angle and across the *bottom* row to locate the function. For instance, tan $70°50' = 2.8770$.

To find the value for a trigonometric function of an angle expressed in tenths of degrees, convert the angle measure to degrees and minutes first. To find the value for a trigonometric function of an angle expressed in radians, convert to decimal radian measure and then use the second and next to last columns of the table.

EXAMPLE 1 Use a table to find the following:

$$\textbf{a. } \cos 42°20' \qquad \textbf{b. } \cot 54.5° \qquad \textbf{c. } \csc \frac{\pi}{8}$$

a. $\cos 42°20' = 0.7392$ *Use the table.*

b. $\cot 54.5° = \cot 54°30'$ *Convert to degrees and minutes.*
$\qquad\quad = 0.7133$ *Use the table.*

c. $\csc \dfrac{\pi}{8} = \csc 0.3927$ *Convert to decimal radian measure.*

$\qquad\quad = 2.613$ *Use second column of table for radians.*

To find the value for a trigonometric function of an angle greater than $90°$, use a reference triangle in the appropriate quadrant. Find the value of the *reference angle,* which is the acute angle of the reference triangle that the terminal side makes with the *x*-axis.

EXAMPLE 2   **Use a table to find the following.   a.** $\sin 245°40'$ **b.** $\sec \dfrac{23\pi}{12}$

a. $\sin 245°40' = -\sin 65°40'$ *Reference triangle is in quadrant III.*
$\qquad\qquad\quad\;\; = -0.9112$ *Use the table.*

b. $\sec \dfrac{23\pi}{12} = \sec \dfrac{\pi}{12}$ *Reference triangle is in quadrant IV.*

$\qquad\qquad = \sec 0.2618$ *Convert to decimal radian measure.*
$\qquad\qquad = 1.0353$ *Use second column of table for radians.*

The table of trigonometric function values can also be used to solve trigonometric equations. The acute reference angle θ of a reference triangle can be found directly if the value of a function is known.

EXAMPLE 3   **Solve for θ, $0° \le \theta < 360°$, to the nearest minute:**
$\qquad\qquad$ **a.** $\tan \theta = 0.8642$ **b.** $\sin \theta = -0.5628$

a. $\tan \theta = 0.8642$
$\quad$ reference angle $= 40°50'$ *Use the table.*
$\quad \theta = 40°50', 220°50'$ *Tangent is positive in quadrants I and III.*

b. $\sin \theta = -0.5616$ *Ignore negative to find reference angle.*
$\quad$ reference angle $= 34°10'$ *Use the table.*
$\quad \theta = 214°10', 325°50'$ *Sine is negative in quadrants III and IV.*

EXERCISES

Use a table to find the following.

1. $\cos 9°20'$
2. $\sin 74°10'$
3. $\tan 43°50'$
4. $\csc 88°40'$
5. $\sec \dfrac{\pi}{18}$
6. $\cot \dfrac{5\pi}{24}$
7. $\sin \dfrac{17\pi}{9}$
8. $\csc \dfrac{23\pi}{30}$
9. $\tan 223°10'$
10. $\cos 147°20'$
11. $\csc 289°50'$
12. $\sin 315°30'$
13. $\sec 12°50'$
14. $\cot 53°40'$
15. $\csc 237°40'$
16. $\sec 302°10'$

Solve for θ, $0° \le \theta < 360°$, to the nearest ten minutes.

17. $\sin \theta = 0.9013$
18. $\cot \theta = 14.301$
19. $\sec \theta = -5.4026$
20. $\tan \theta = -2.5605$
21. $\csc \theta = -1.4572$
22. $\cos \theta = 0.8843$

Linear Interpolation

The table on pages 434–438 includes angles expressed in 10-minute increments. If the measure of an angle falls between two entries, you can use a method called **linear interpolation** to find the values of trigonometric functions of that angle.

Over very small intervals, the graph of a trigonometric function is nearly a straight line. Consider the graph of $y = \sin x$ between the points $A(32°10', 0.5324)$ and $C(32°20', 0.5348)$. In order to estimate the value of $\sin x$ at a point between A and C where $x = 32°17'$, assume that the graph between points A and C is a straight line. Then, since triangles ABD and ACE are similar, corresponding sides are proportional. That is,

$$\frac{AD}{AE} = \frac{BD}{CE}$$

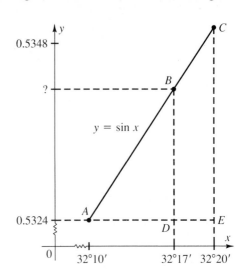

EXAMPLE 1   Use linear interpolation to approximate $\sin 32°17'$.

First set up a table.

x	$\sin x$
32°10′	0.5324
32°17′	?
32°20′	0.5348

10′   7′ d 0.0024

Then write and solve a proportion.

$$\frac{7}{10} = \frac{d}{0.0024}$$

$$\frac{AD}{AE} = \frac{BD}{CE}$$

$$d = \frac{7}{10}(0.0024) \approx 0.0017$$

Round to four decimal places, as in the sin table.

So, $\sin 32°17' = 0.5324 + d = 0.5324 + 0.0017 = 0.5341$

Linear interpolation can also be used to approximate the measure of an angle when the value of one of its trigonometric functions falls between entries in the table.

EXAMPLE 2 **Use linear interpolation to approximate θ if tan θ = 0.4000, $0° \leq \theta \leq 360°$.**

Use a trigonometric table to find the values of tangent just less than and just greater than 0.4000.

θ	tan θ
21°40'	0.3973
?	0.4000
21°50'	0.4006

$10'$ ⎡ d ⎣ ... ⎤ ... 0.0027 ... 0.0033

Write and solve a proportion.

$$\frac{d}{10'} = \frac{0.0027}{0.0033}$$

$$d = \frac{27}{33}(10') \approx 8 \qquad \textit{Round.}$$

So, $\theta = 21°40' + d = 21°40' + 8' = 21°48'$
Note that tan θ is also positive in Quadrant III, so θ can also equal $180° + 21°48' = 201°48'$ if tan θ = 0.4000.

EXERCISES

Use linear interpolation to approximate each function.

1. sin 27°23'
2. cos 31°42'
3. tan 16°17'

4. sec 87°21'
5. csc 12°52'
6. cot 33°47'

7. cos −15°53'
8. tan −25°34'
9. sin −52°08'

10. sec −12°05'
11. cot −82°14'
12. csc −55°22'

13. sin 0.4111
14. cos 0.4660
15. tan 0.3210

16. cos −0.6000
17. sin −0.5302
18. tan −0.7777

Use linear interpolation to approximate θ, $0° \leq \theta \leq 360°$, to the nearest minute.

19. sin θ = 0.3205
20. cos θ = 0.8111
21. tan θ = 1.7032

22. sin θ = −0.5107
23. cos θ = 0.2215
24. tan θ = 6.0721

25. cos θ = 0.1155
26. tan θ = 1.1617
27. sin θ = 0.5152

28. tan θ = −2.5487
29. sec θ = −5.381
30. csc θ = −1.458

31. cos θ = −0.1205
32. cot θ = −7.0123
33. sin θ = −0.2222

34. csc θ = −5.562
35. sec θ = −1.018
36. cot θ = −4.4163

Using a Table of Common Logarithms

The table on pages 430 and 431 contains the common logarithms of numbers from 1 to 9.99. A portion of the table is shown below. The logarithm values are rounded to four decimal places and the decimal points are omitted.

N	0	1	2	3	4	5	6	7	8	9
4.1	6128	6138	6149	6160	6170	6180	6191	6201	6212	6222

To find the logarithm of a number N between 1 and 10, find the first two digits of N in the column on the left and the third digit in the top row. The logarithm of N is shown at the intersection of that row and column. To illustrate, the logarithm of 4.17 is found at the intersection of the row for 41 and the column for 7. To four decimal places, log 4.17 = 0.6201.

To find the logarithm of a number greater than 10 or between 0 and 1, write the number in scientific notation and then use the properties of logarithms.

EXAMPLE 1 Use a table to find each logarithm:
    a. log 4170   **b.** log 0.0417

a. log 4170 $= \log (4.17 \times 10^3)$
        $= \log 4.17 + \log 10^3$ $\log_b MN = \log_b M + \log_b N$
        $= 0.6201 + 3$ $\log_b b^x = x$
        $= 3.6201$ *To four decimal places*

b. log 0.0417 $= \log (4.17 \times 10^{-2})$
        $= \log 4.17 + \log 10^{-2}$
        $= 0.6201 + (-2)$
        $= 0.6201 - 2$, or -1.3799 *To four decimal places*

A common logarithm can be written as the sum of a *nonnegative* number less than 1, called the **mantissa,** and an integer, called the **characteristic.**

$$\begin{array}{cc} & \textit{mantissa} \quad \textit{characteristic} \\ \log 4170 = & 0.6201 \;+\; 3 \\ \log 0.0417 = & 0.6201 \;+\; (-2) \end{array}$$

The logarithm of a number such as 0.0417 can be written in many ways. Sometimes it is more convenient to use one form than another.

log 0.0417 = −1.3799   *Not* in mantissa/characteristic form
log 0.0417 = 0.6201 − 2   Mantissa: 0.6201   Characteristic: −2
log 0.0417 = 2.6201 − 4   Mantissa: 0.6201   Characteristic: 2 − 4 = −2
log 0.0417 = 8.6201 − 10   Mantissa: 0.6201   Characteristic: 8 − 10 = −2

The table of common logarithms can also be used to find a number N if log N is known. If log N is between 0 and 1, then N is between 1 and 10 and its value can be read directly from the table. For instance, if log $N = 0.7932$, the table on page 431 shows that the logarithm closest to this value is found at the intersection of the row for 62 and the column for 1. That is, N is approximately 6.21, to three significant digits.

EXAMPLE 2   Use a table to find N:
   a. log $N = 5.1904$ **b.** log $N = -3.2165$.

First write each number in mantissa/characteristic form.

a. log $N = 5.1905$
   $= 0.1905 + 5$ *Mantissa/characteristic form*
   $= $ log $1.55 + 5$ *Use the table.*
   $= $ log $1.55 + $ log 10^5
   $= $ log (1.55×10^5)
   $= $ log $155,000$
   So, $N = 155,000$ *Three significant digits*

b. log $N = -3.2165$
   $= 4 + -3.2165 - 4$
   $= 0.7835 - 4$
   $= $ log $6.07 - 4$
   $= $ log $6.07 + $ log 10^{-4}
   $= $ log (6.07×10^{-4})
   $= $ log 0.000607
   So, $N = 0.000607$

EXERCISES

Use a table to find each common logarithm to four decimal places.

1. log 2.58   **2.** log 9.04   **3.** log 16.8   **4.** log 340

5. log 72,500   **6.** log 0.143   **7.** log 0.00329   **8.** log 0.0605

Use a table to find N to three significant digits.

9. log $N = 0.8982$ **10.** log $N = 0.5528$ **11.** log $N = 1.9805$

12. log $N = 3.6944$ **13.** log $N = 0.9842 - 2$ **14.** log $N = 0.1732 - 1$

Use a table to find each common logarithm to four decimal places.

15. log 300,000   **16.** log 75,000,000   **17.** log 0.000086   **18.** 0.0000006

Use a table to find N to three significant digits.

19. log $N = 8.8131$ **20.** log $N = 9.5265$ **21.** log $N = 4.7378 - 10$

22. log $N = -1.7714$ **23.** log $N = -2.3100$ **24.** log $N = 6.9272 - 10$

25. log $N = 11.9559$ **26.** log $N = 13.0042$

27. log $N = -12.5406$ **28.** log $N = -10.7393$

Computation Using Common Logarithms

Common logarithms can be used to simplify computations. In this process, addition and subtraction replace the operations of multiplication and division, while multiplication replaces the operation of exponentiation.

EXAMPLE 1 **Use common logarithms from a table to evaluate:**

$$\textbf{a. } 0.00496 \times 23{,}600 \qquad \textbf{b. } \frac{0.000408}{7.73} \qquad \textbf{c. } \sqrt[3]{0.425}$$

a. Let $\qquad N = 0.00496 \times 23{,}600$

Then $\log N = \log (0.00496 \times 23{,}600)$

$\qquad\qquad = \log 0.00496 + \log 23{,}600 \qquad$ *log MN = log M + log N*

$\qquad\qquad = \log (4.96 \times 10^{-3}) + \log (2.36 \times 10^{4})$

$\qquad\qquad = (0.6955 - 3) + (4.3729) \qquad$ *Use the table.*

$\qquad\qquad = 2.0684$

$\qquad\qquad = 0.0684 + 2 \qquad$ *Mantissa/characteristic form*

So, $\qquad N = 1.17 \times 10^{2} \qquad$ *Use the table again. Find the closest entry.*

$\qquad\qquad N = 117 \qquad$ *Three significant digits*

b. Let $\qquad N = \dfrac{0.000408}{7.73}$

Then $\log N = \log \left(\dfrac{0.000408}{7.73} \right)$

$\qquad\qquad = \log 0.000408 - \log 7.73 \qquad \log \dfrac{M}{N} = log\ M - log\ N$

$\qquad\qquad = \log (4.08 \times 10^{-4}) - \log 7.73$

$\qquad\qquad = (0.6107 - 4) - 0.8882 \qquad$ *Use the table.*

$\qquad\qquad = (1.6107 - 5) - 0.8882 \qquad$ *To obtain a positive mantissa,*
$\qquad\qquad\qquad\qquad\qquad\qquad\qquad\qquad$ *write 0.6107 − 4 as 1.6107 − 5.*

$\qquad\qquad = 0.7225 - 5 \qquad$ *Mantissa/characteristic form*

So, $\qquad N = 5.28 \times 10^{-5} \qquad$ *Use the table again. Find the closest entry.*

$\qquad\qquad N = 0.0000528 \qquad$ *Three significant digits*

c. Let $\qquad N = \sqrt[3]{0.425} = 0.425^{\frac{1}{3}}$

Then $\log N = \log 0.425^{\frac{1}{3}}$

$\qquad\qquad = \frac{1}{3} \log 0.425 \qquad$ *log N^r = r log N*

$\qquad\qquad = \frac{1}{3} (0.6284 - 1) \qquad$ *Use the table.*

$\qquad\qquad = \frac{1}{3} (2.6284 - 3) \qquad$ *To get an integer for the negative part of the*
$\qquad\qquad\qquad\qquad\qquad\qquad\qquad$ *characteristic, write 0.6284 − 1 as 2.6284 − 3.*

$\qquad\qquad = 0.8761 - 1 \qquad$ *Mantissa/characteristic form*

So, $\qquad N = 7.52 \times 10^{-1} \qquad$ *Use the table again. Find the closest entry.*

$\qquad\qquad N = 0.752 \qquad$ *Three significant digits*

In the next example, several operations are performed using common logarithms.

EXAMPLE 2 Use common logarithms from a table to evaluate: $\dfrac{0.0493(3.71)^4}{52.6}$

Let $\qquad N = \dfrac{0.0493(3.71)^4}{52.6}$

Then $\quad \log N = \log \left(\dfrac{0.0493(3.71)^4}{52.6} \right)$

$\qquad\qquad\quad = \log 0.0493 + \log 3.71^4 - \log 52.6$

$\qquad\qquad\quad = \log 0.0493 + 4 \log 3.71 - \log 52.6$

$\qquad\qquad\quad = (0.6928 - 2) + 4(0.5694) - 1.7210$

$\qquad\qquad\quad = 0.6928 - 2 + 2.2776 - 1.7210$

$\qquad\qquad\quad = -0.7506, \quad \text{or } 0.2494 - 1$

So, $\qquad\quad N = 0.178 \qquad$ *Three significant digits*

EXERCISES

Use common logarithms from a table to evaluate to three significant digits.

1. 2.96×9.74

2. 7.45×8.93

3. 0.0309×768

4. 0.00265×37.6

5. 3.49^3

6. 6.27^4

7. $\sqrt[5]{925,000}$

8. $\sqrt[4]{34,800}$

9. $256(4.95)^2$

10. $0.747(8.71)^2$

11. $3760\sqrt[4]{46,800}$

12. $0.0129\sqrt[3]{5080}$

13. $\dfrac{5620}{3.85}$

14. $\dfrac{8.38}{0.125}$

15. $\dfrac{0.714(7.36)}{247}$

16. $\dfrac{0.00368(2.54)}{1.13}$

17. $\dfrac{56.5(5.66)^4}{363,000}$

18. $\dfrac{0.296(4.87)^3}{1370}$

19. $0.0165\sqrt[6]{7280}$

20. $45,200\sqrt[8]{0.968}$

21. $\sqrt{26.8(127)^3}$

22. $\sqrt[3]{0.0561}(4.99)^2$

23. $56.3(4.26)^{-6}$

24. $7250(36.2)^{-5}$

25. $\dfrac{4.72}{0.928(42.6)^2}$

26. $\dfrac{38.3}{2.09(0.0256)^3}$

27. $\dfrac{7.66(0.0478)}{0.395(26.7)}$

28. $\dfrac{686(0.00261)}{0.985(49.7)}$

29. $\dfrac{\sqrt{21.3}}{\sqrt[3]{26,700}}$

30. $\dfrac{\sqrt[5]{0.796}}{\sqrt[3]{545}}$

31. $\left(\dfrac{\sqrt[3]{751}}{275} \right)^4$

32. $\left(\dfrac{368}{\sqrt[4]{911}} \right)^3$

33. $\sqrt{\dfrac{92.6(1.49)^3}{0.368}}$

34. $\sqrt[3]{\dfrac{476(0.155)^2}{76.7}}$

35. $\sqrt{\dfrac{529(1.48)^5}{1.37\sqrt[3]{0.0245}}}$

36. $\sqrt[4]{\dfrac{29.4\sqrt[3]{0.766}}{301(0.135)^3}}$

Table of Common Logarithms

N	0	1	2	3	4	5	6	7	8	9
1.0	0000	0043	0086	0128	0170	0212	0253	0294	0334	0374
1.1	0414	0453	0492	0531	0569	0607	0645	0682	0719	0755
1.2	0792	0828	0864	0899	0934	0969	1004	1038	1072	1106
1.3	1139	1173	1206	1239	1271	1303	1335	1367	1399	1430
1.4	1461	1492	1523	1553	1584	1614	1644	1673	1703	1732
1.5	1761	1790	1818	1847	1875	1903	1931	1959	1987	2014
1.6	2041	2068	2095	2122	2148	2175	2201	2227	2253	2279
1.7	2304	2330	2355	2380	2405	2430	2455	2480	2504	2529
1.8	2553	2577	2601	2625	2648	2672	2695	2718	2742	2765
1.9	2788	2810	2833	2856	2878	2900	2923	2945	2967	2989
2.0	3010	3032	3054	3075	3096	3118	3139	3160	3181	3201
2.1	3222	3243	3263	3284	3304	3324	3345	3365	3385	3404
2.2	3424	3444	3464	3483	3502	3522	3541	3560	3579	3598
2.3	3617	3636	3655	3674	3692	3711	3729	3747	3766	3784
2.4	3802	3820	3838	3856	3874	3892	3909	3927	3945	3962
2.5	3979	3997	4014	4031	4048	4065	4082	4099	4116	4133
2.6	4150	4166	4183	4200	4216	4232	4249	4265	4281	4298
2.7	4314	4330	4346	4362	4378	4393	4409	4425	4440	4456
2.8	4472	4487	4502	4518	4533	4548	4564	4579	4594	4609
2.9	4624	4639	4654	4669	4683	4698	4713	4728	4742	4757
3.0	4771	4786	4800	4814	4829	4843	4857	4871	4886	4900
3.1	4914	4928	4942	4955	4969	4983	4997	5011	5024	5038
3.2	5051	5065	5079	5092	5105	5119	5132	5145	5159	5172
3.3	5185	5198	5211	5224	5237	5250	5263	5276	5289	5302
3.4	5315	5328	5340	5353	5366	5378	5391	5403	5416	5428
3.5	5441	5453	5465	5478	5490	5502	5514	5527	5539	5551
3.6	5563	5575	5587	5599	5611	5623	5635	5647	5658	5670
3.7	5682	5694	5705	5717	5729	5740	5752	5763	5775	5786
3.8	5798	5809	5821	5832	5843	5855	5866	5877	5888	5899
3.9	5911	5922	5933	5944	5955	5966	5977	5988	5999	6010
4.0	6021	6031	6042	6053	6064	6075	6085	6096	6107	6117
4.1	6128	6138	6149	6160	6170	6180	6191	6201	6212	6222
4.2	6232	6243	6253	6263	6274	6284	6294	6304	6314	6325
4.3	6335	6345	6355	6365	6375	6385	6395	6405	6415	6425
4.4	6435	6444	6454	6464	6474	6484	6493	6503	6513	6522
4.5	6532	6542	6551	6561	6571	6580	6590	6599	6609	6618
4.6	6628	6637	6646	6656	6665	6675	6684	6693	6702	6712
4.7	6721	6730	6739	6749	6758	6767	6776	6785	6794	6803
4.8	6812	6821	6830	6839	6848	6857	6866	6875	6884	6893
4.9	6902	6911	6920	6928	6937	6946	6955	6964	6972	6981
5.0	6990	6998	7007	7016	7024	7033	7042	7050	7059	7067
5.1	7076	7084	7093	7101	7110	7118	7126	7135	7143	7152
5.2	7160	7168	7177	7185	7193	7202	7210	7218	7226	7235
5.3	7243	7251	7259	7267	7275	7284	7292	7300	7308	7316
5.4	7324	7332	7340	7348	7356	7364	7372	7380	7388	7396

Table of Common Logarithms

N	0	1	2	3	4	5	6	7	8	9
5.5	7404	7412	7419	7427	7435	7443	7451	7459	7466	7474
5.6	7482	7490	7497	7505	7513	7520	7528	7536	7543	7551
5.7	7559	7566	7574	7582	7589	7597	7604	7612	7619	7627
5.8	7634	7642	7649	7657	7664	7672	7679	7686	7694	7701
5.9	7709	7716	7723	7731	7738	7745	7752	7760	7767	7774
6.0	7782	7789	7796	7803	7810	7818	7825	7832	7839	7846
6.1	7853	7860	7868	7875	7882	7889	7896	7903	7910	7917
6.2	7924	7931	7938	7945	7952	7959	7966	7973	7980	7987
6.3	7993	8000	8007	8014	8021	8028	8035	8041	8048	8055
6.4	8062	8069	8075	8082	8089	8096	8102	8109	8116	8122
6.5	8129	8136	8142	8149	8156	8162	8169	8176	8182	8189
6.6	8195	8202	8209	8215	8222	8228	8235	8241	8248	8254
6.7	8261	8267	8274	8280	8287	8293	8299	8306	8312	8319
6.8	8325	8331	8338	8344	8351	8357	8363	8370	8376	8382
6.9	8388	8395	8401	8407	8414	8420	8426	8432	8439	8445
7.0	8451	8457	8463	8470	8476	8482	8488	8494	8500	8506
7.1	8513	8519	8525	8531	8537	8543	8549	8555	8561	8567
7.2	8573	8579	8585	8591	8597	8603	8609	8615	8621	8627
7.3	8633	8639	8645	8651	8657	8663	8669	8675	8681	8686
7.4	8692	8698	8704	8710	8716	8722	8727	8733	8739	8745
7.5	8751	8756	8762	8768	8774	8779	8785	8791	8797	8802
7.6	8808	8814	8820	8825	8831	8837	8842	8848	8854	8859
7.7	8865	8871	8876	8882	8887	8893	8899	8904	8910	8915
7.8	8921	8927	8932	8938	8943	8949	8954	8960	8965	8971
7.9	8976	8982	8987	8993	8998	9004	9009	9015	9020	9025
8.0	9031	9036	9042	9047	9053	9058	9063	9069	9074	9079
8.1	9085	9090	9096	9101	9106	9112	9117	9122	9128	9133
8.2	9138	9143	9149	9154	9159	9165	9170	9175	9180	9186
8.3	9191	9196	9201	9206	9212	9217	9222	9227	9232	9238
8.4	9243	9248	9253	9258	9263	9269	9274	9279	9284	9289
8.5	9294	9299	9304	9309	9315	9320	9325	9330	9335	9340
8.6	9345	9350	9355	9360	9365	9370	9375	9380	9385	9390
8.7	9395	9400	9405	9410	9415	9420	9425	9430	9435	9440
8.8	9445	9450	9455	9460	9465	9469	9474	9479	9484	9489
8.9	9494	9499	9504	9509	9513	9518	9523	9528	9533	9538
9.0	9542	9547	9552	9557	9562	9566	9571	9576	9581	9586
9.1	9590	9595	9600	9605	9609	9614	9619	9624	9628	9633
9.2	9638	9643	9647	9652	9657	9661	9666	9671	9675	9680
9.3	9685	9689	9694	9699	9703	9708	9713	9717	9722	9727
9.4	9731	9736	9741	9745	9750	9754	9759	9763	9768	9773
9.5	9777	9782	9786	9791	9795	9800	9805	9809	9814	9818
9.6	9823	9827	9832	9836	9841	9845	9850	9854	9859	9863
9.7	9868	9872	9877	9881	9886	9890	9894	9899	9903	9908
9.8	9912	9917	9921	9926	9930	9934	9939	9943	9948	9952
9.9	9956	9961	9965	9969	9974	9978	9983	9987	9991	9996

Table of Natural Logarithms (ln x)

x	0.00	0.01	0.02	0.03	0.04	0.05	0.06	0.07	0.08	0.09
1.0	0.0000	0.0100	0.0198	0.0296	0.0392	0.0488	0.0583	0.0677	0.0770	0.0862
1.1	0.0953	0.1044	0.1133	0.1222	0.1310	0.1398	0.1484	0.1570	0.1655	0.1740
1.2	0.1823	0.1906	0.1989	0.2070	0.2151	0.2231	0.2311	0.2390	0.2469	0.2546
1.3	0.2624	0.2700	0.2776	0.2852	0.2927	0.3001	0.3075	0.3148	0.3221	0.3293
1.4	0.3365	0.3436	0.3507	0.3577	0.3646	0.3716	0.3784	0.3853	0.3920	0.3988
1.5	0.4055	0.4121	0.4187	0.4253	0.4318	0.4383	0.4447	0.4511	0.4574	0.4637
1.6	0.4700	0.4762	0.4824	0.4886	0.4947	0.5008	0.5068	0.5128	0.5188	0.5247
1.7	0.5306	0.5365	0.5423	0.5481	0.5539	0.5596	0.5653	0.5710	0.5766	0.5822
1.8	0.5878	0.5933	0.5988	0.6043	0.6098	0.6152	0.6206	0.6259	0.6313	0.6166
1.9	0.6419	0.6471	0.6523	0.6575	0.6627	0.6678	0.6729	0.6780	0.6831	0.6881
2.0	0.6931	0.6981	0.7031	0.7080	0.7130	0.7178	0.7227	0.7275	0.7324	0.7372
2.1	0.7419	0.7467	0.7514	0.7561	0.7608	0.7655	0.7701	0.7747	0.7793	0.7839
2.2	0.7885	0.7930	0.7975	0.8020	0.8065	0.8109	0.8154	0.8198	0.8242	0.8286
2.3	0.8329	0.8372	0.8416	0.8459	0.8502	0.8544	0.8587	0.8629	0.8671	0.8713
2.4	0.8755	0.8796	0.8838	0.8879	0.8920	0.8961	0.9002	0.9042	0.9083	0.9123
2.5	0.9163	0.9203	0.9243	0.9282	0.9322	0.9361	0.9400	0.9439	0.9478	0.9517
2.6	0.9555	0.9594	0.9632	0.9670	0.9708	0.9746	0.9783	0.9821	0.9858	0.9895
2.7	0.9933	0.9969	1.0006	1.0043	1.0080	1.0116	1.0152	1.0188	1.0225	1.0260
2.8	1.0296	1.0332	1.0367	1.0403	1.0438	1.0473	1.0508	1.0543	1.0578	1.0613
2.9	1.0647	1.0682	1.0716	1.0750	1.0784	1.0818	1.0852	1.0886	1.0919	1.0953
3.0	1.0986	1.1019	1.1053	1.1086	1.1119	1.1151	1.1184	1.1217	1.1249	1.1282
3.1	1.1314	1.1346	1.1378	1.1410	1.1442	1.1474	1.1506	1.1537	1.1569	1.1600
3.2	1.1632	1.1663	1.1694	1.1725	1.1756	1.1787	1.1817	1.1848	1.1878	1.1909
3.3	1.1939	1.1970	1.2000	1.2030	1.2060	1.2090	1.2119	1.2149	1.2179	1.2208
3.4	1.2238	1.2267	1.2296	1.2326	1.2355	1.2384	1.2413	1.2442	1.2470	1.2499
3.5	1.2528	1.2556	1.2585	1.2613	1.2641	1.2669	1.2698	1.2726	1.2754	1.2782
3.6	1.2809	1.2837	1.2865	1.2892	1.2920	1.2947	1.2975	1.3002	1.3029	1.3056
3.7	1.3083	1.3110	1.3137	1.3164	1.3191	1.3218	1.3244	1.3271	1.3297	1.3324
3.8	1.3350	1.3376	1.3403	1.3429	1.3455	1.3481	1.3507	1.3533	1.3558	1.3584
3.9	1.3610	1.3635	1.3661	1.3686	1.3712	1.3737	1.3762	1.3788	1.3813	1.3838
4.0	1.3863	1.3888	1.3913	1.3938	1.3962	1.3987	1.4012	1.4036	1.4061	1.4085
4.1	1.4110	1.4134	1.4159	1.4183	1.4207	1.4231	1.4255	1.4279	1.4303	1.4327
4.2	1.4351	1.4375	1.4398	1.4422	1.4446	1.4469	1.4493	1.4516	1.4540	1.4563
4.3	1.4586	1.4609	1.4633	1.4656	1.4679	1.4702	1.4725	1.4748	1.4770	1.4793
4.4	1.4816	1.4839	1.4861	1.4884	1.4907	1.4929	1.4952	1.4974	1.4996	1.5019
4.5	1.5041	1.5063	1.5085	1.5107	1.5129	1.5151	1.5173	1.5195	1.5217	1.5239
4.6	1.5261	1.5282	1.5304	1.5326	1.5347	1.5369	1.5390	1.5412	1.5433	1.5454
4.7	1.5476	1.5497	1.5518	1.5539	1.5560	1.5581	1.5602	1.5623	1.5644	1.5665
4.8	1.5686	1.5707	1.5728	1.5748	1.5769	1.5790	1.5810	1.5831	1.5851	1.5872
4.9	1.5892	1.5913	1.5933	1.5953	1.5974	1.5994	1.6014	1.6034	1.6054	1.6074
5.0	1.6094	1.6114	1.6134	1.6154	1.6174	1.6194	1.6214	1.6233	1.6253	1.6273
5.1	1.6292	1.6312	1.6332	1.6351	1.6371	1.6390	1.6409	1.6429	1.6448	1.6467
5.2	1.6487	1.6506	1.6525	1.6544	1.6563	1.6582	1.6601	1.6620	1.6639	1.6658
5.3	1.6677	1.6696	1.6715	1.6734	1.6752	1.6771	1.6790	1.6808	1.6827	1.6845
5.4	1.6864	1.6882	1.6901	1.6919	1.6938	1.6956	1.6974	1.6993	1.7001	1.7029

Table of Natural Logarithms (ln x)

x	0.00	0.01	0.02	0.03	0.04	0.05	0.06	0.07	0.08	0.09
5.5	1.7047	1.7066	1.7084	1.7102	1.7120	1.7138	1.7156	1.7174	1.7192	1.7210
5.6	1.7228	1.7246	1.7263	1.7281	1.7299	1.7317	1.7334	1.7352	1.7370	1.7387
5.7	1.7405	1.7422	1.7440	1.7457	1.7475	1.7492	1.7509	1.7527	1.7544	1.7561
5.8	1.7579	1.7596	1.7613	1.7630	1.7647	1.7664	1.7682	1.7699	1.7716	1.7733
5.9	1.7750	1.7766	1.7783	1.7800	1.7817	1.7834	1.7851	1.7867	1.7884	1.7901
6.0	1.7918	1.7934	1.7951	1.7967	1.7984	1.8001	1.8017	1.8034	1.8050	1.8066
6.1	1.8083	1.8099	1.8116	1.8132	1.8148	1.8165	1.8181	1.8197	1.8213	1.8229
6.2	1.8245	1.8262	1.8278	1.8294	1.8310	1.8326	1.8342	1.8358	1.8374	1.8390
6.3	1.8406	1.8421	1.8437	1.8453	1.8469	1.8485	1.8500	1.8516	1.8532	1.8547
6.4	1.8563	1.8579	1.8594	1.8610	1.8625	1.8641	1.8656	1.8672	1.8687	1.8703
6.5	1.8718	1.8733	1.8749	1.8764	1.8779	1.8795	1.8810	1.8825	1.8840	1.8856
6.6	1.8871	1.8886	1.8901	1.8916	1.8931	1.8946	1.8961	1.8976	1.8991	1.9006
6.7	1.9021	1.9036	1.9051	1.9066	1.9081	1.9095	1.9110	1.9125	1.9140	1.9155
6.8	1.9169	1.9184	1.9199	1.9213	1.9228	1.9242	1.9257	1.9272	1.9286	1.9301
6.9	1.9315	1.9330	1.9344	1.9359	1.9373	1.9387	1.9402	1.9416	1.9430	1.9445
7.0	1.9459	1.9473	1.9488	1.9502	1.9516	1.9530	1.9544	1.9559	1.9573	1.9587
7.1	1.9601	1.9615	1.9629	1.9643	1.9657	1.9671	1.9685	1.9699	1.9713	1.9727
7.2	1.9741	1.9755	1.9769	1.9782	1.9796	1.9810	1.9824	1.9838	1.9851	1.9865
7.3	1.9879	1.9892	1.9906	1.9920	1.9933	1.9947	1.9961	1.9974	1.9988	2.0001
7.4	2.0015	2.0028	2.0042	2.0055	2.0069	2.0082	2.0096	2.0109	2.0122	2.0136
7.5	2.0149	2.0162	2.0176	2.0189	2.0202	2.0215	2.0229	2.0242	2.0255	2.0268
7.6	2.0282	2.0295	2.0308	2.0321	2.0334	2.0347	2.0360	2.0373	2.0386	2.0399
7.7	2.0412	2.0425	2.0438	2.0451	2.0464	2.0477	2.0490	2.0503	2.0516	2.0528
7.8	2.0541	2.0554	2.0567	2.0580	2.0592	2.0605	2.0618	2.0631	2.0643	2.0665
7.9	2.0669	2.0681	2.0694	2.0707	2.0719	2.0732	2.0744	2.0757	2.0769	2.0782
8.0	2.0794	2.0807	2.0819	2.0832	2.0844	2.0857	2.0869	2.0882	2.0894	2.0906
8.1	2.0919	2.0931	2.0943	2.0956	2.0968	2.0980	2.0992	2.1005	2.1017	2.1029
8.2	2.1041	2.1054	2.1066	2.1078	2.1090	2.1102	2.1114	2.1126	2.1138	2.1150
8.3	2.1163	2.1175	2.1187	2.1199	2.1211	2.1223	2.1235	2.1247	2.1258	2.1270
8.4	2.1282	2.1294	2.1306	2.1318	2.1330	2.1342	2.1353	2.1365	2.1377	2.1389
8.5	2.1401	2.1412	2.1424	2.1436	2.1448	2.1459	2.1471	2.1483	2.1494	2.1506
8.6	2.1518	2.1529	2.1541	2.1552	2.1564	2.1576	2.1587	2.1599	2.1610	2.1622
8.7	2.1633	2.1645	2.1656	2.1668	2.1679	2.1691	2.1702	2.1713	2.1725	2.1736
8.8	2.1748	2.1759	2.1770	2.1782	2.1793	2.1804	2.1815	2.1827	2.1838	2.1849
8.9	2.1861	2.1872	2.1883	2.1894	2 1905	2.1917	2.1928	2.1939	2.1950	2.1961
9.0	2.1972	2.1983	2.1994	2.2006	2.2017	2.2028	2.2039	2.2050	2.2061	2.2072
9.1	2.2083	2.2094	2.2105	2.2116	2.2127	2.2138	2.2148	2.2159	2.2170	2.2181
9.2	2.2192	2.2203	2.2214	2.2225	2.2235	2.2246	2.2257	2.2268	2.2279	2.2289
9.3	2.2300	2.2311	2.2322	2.2332	2.2343	2.2354	2.2364	2.2375	2.2386	2.2396
9.4	2.2407	2.2418	2.2428	2.2439	2.2450	2.2460	2.2471	2.2481	2.2492	2.2502
9.5	2.2513	2.2523	2.2534	2.2544	2.2555	2.2565	2.2576	2.2586	2.2597	2.2607
9.6	2.2618	2.2628	2.2638	2.2649	2.2659	2.2670	2.2680	2.2690	2.2701	2.2711
9.7	2.2721	2.2732	2.2742	2.2752	2.2762	2.2773	2.2783	2.2793	2.2803	2.2814
9.8	2.2824	2.2834	2.2844	2.2854	2.2865	2.2875	2.2885	2.2895	2.2905	2.2915
9.9	2.2925	2.2935	2.2946	2.2956	2.2966	2.2976	2.2986	2.2996	2.3006	2.3016

Table of Values of the Trigonometric Functions

θ Deg.	θ Rad.	Sin θ	Cos θ	Tan θ	Cot θ	Sec θ	Csc θ		
0°00′	0.0000	0.0000	1.0000	0.0000		1.000		1.5708	90°00′
10′	0.0029	0.0029	1.0000	0.0029	343.77	1.000	343.8	1.5679	50′
20′	0.0058	0.0058	1.0000	0.0058	171.89	1.000	171.9	1.5650	40′
30′	0.0087	0.0087	1.0000	0.0087	114.59	1.000	114.6	1.5621	30′
40′	0.0116	0.0116	0.9999	0.0116	85.940	1.000	85.95	1.5592	20′
50′	0.0145	0.0145	0.9999	0.0145	68.750	1.000	68.76	1.5563	10′
1°00′	0.0175	0.0175	0.9998	0.0175	57.290	1.000	57.30	1.5533	89°00′
10′	0.0204	0.0204	0.9998	0.0204	49.104	1.000	49.11	1.5504	50′
20′	0.0233	0.0233	0.9997	0.0233	42.964	1.000	42.98	1.5475	40′
30′	0.0262	0.0262	0.9997	0.0262	38.188	1.000	38.20	1.5446	30′
40′	0.0291	0.0291	0.9996	0.0291	34.368	1.000	34.38	1.5417	20′
50′	0.0320	0.0320	0.9995	0.0320	31.242	1.001	31.26	1.5388	10′
2°00′	0.0349	0.0349	0.9994	0.0349	28.636	1.001	28.65	1.5359	88°00′
10′	0.0378	0.0378	0.9993	0.0378	26.432	1.001	26.45	1.5330	50′
20′	0.0407	0.0407	0.9992	0.0407	24.542	1.001	24.56	1.5301	40′
30′	0.0436	0.0436	0.9990	0.0437	22.904	1.001	22.93	1.5272	30′
40′	0.0465	0.0465	0.9989	0.0466	21.470	1.001	21.49	1.5243	20′
50′	0.0495	0.0494	0.9988	0.0495	20.206	1.001	20.23	1.5213	10′
3°00′	0.0524	0.0523	0.9986	0.0524	19.081	1.001	19.11	1.5184	87°00′
10′	0.0553	0.0552	0.9985	0.0553	18.075	1.002	18.10	1.5155	50′
20′	0.0582	0.0581	0.9983	0.0582	17.169	1.002	17.20	1.5126	40′
30′	0.0611	0.0610	0.9981	0.0612	16.350	1.002	16.38	1.5097	30′
40′	0.0640	0.0640	0.9980	0.0641	15.605	1.002	15.64	1.5068	20′
50′	0.0669	0.0669	0.9978	0.0670	14.924	1.002	14.96	1.5039	10′
4°00′	0.0698	0.0698	0.9976	0.0699	14.301	1.002	14.34	1.5010	86°00′
10′	0.0727	0.0727	0.9974	0.0729	13.727	1.003	13.76	1.4981	50′
20′	0.0756	0.0756	0.9971	0.0758	13.197	1.003	13.23	1.4952	40′
30′	0.0785	0.0785	0.9969	0.0787	12.706	1.003	12.75	1.4923	30′
40′	0.0814	0.0814	0.9967	0.0816	12.251	1.003	12.29	1.4893	20′
50′	0.0844	0.0843	0.9964	0.0846	11.826	1.004	11.87	1.4864	10′
5°00′	0.0873	0.0872	0.9962	0.0875	11.430	1.004	11.47	1.4835	85°00′
10′	0.0902	0.0901	0.9959	0.0904	11.059	1.004	11.10	1.4806	50′
20′	0.0931	0.0929	0.9957	0.0934	10.712	1.004	10.76	1.4777	40′
30′	0.0960	0.0958	0.9954	0.0963	10.385	1.005	10.43	1.4748	30′
40′	0.0989	0.0987	0.9951	0.0992	10.078	1.005	10.13	1.4719	20′
50′	0.1018	0.1016	0.9948	0.1022	9.7882	1.005	9.839	1.4690	10′
6°00′	0.1047	0.1045	0.9945	0.1051	9.5144	1.006	9.567	1.4661	84°00′
10′	0.1076	0.1074	0.9942	0.1080	9.2553	1.006	9.309	1.4632	50′
20′	0.1105	0.1103	0.9939	0.1110	9.0098	1.006	9.065	1.4603	40′
30′	0.1134	0.1132	0.9936	0.1139	8.7769	1.006	8.834	1.4573	30′
40′	0.1164	0.1161	0.9932	0.1169	8.5555	1.007	8.614	1.4544	20′
50′	0.1193	0.1190	0.9929	0.1198	8.3450	1.007	8.405	1.4515	10′
7°00′	0.1222	0.1219	0.9925	0.1228	8.1443	1.008	8.206	1.4486	83°00′
10′	0.1251	0.1248	0.9922	0.1257	7.9530	1.008	8.016	1.4457	50′
20′	0.1280	0.1276	0.9918	0.1287	7.7704	1.008	7.834	1.4428	40′
30′	0.1309	0.1305	0.9914	0.1317	7.5958	1.009	7.661	1.4399	30′
40′	0.1338	0.1334	0.9911	0.1346	7.4287	1.009	7.496	1.4370	20′
50′	0.1367	0.1363	0.9907	0.1376	7.2687	1.009	7.337	1.4341	10′
8°00′	0.1396	0.1392	0.9903	0.1405	7.1154	1.010	7.185	1.4312	82°00′
10′	0.1425	0.1421	0.9899	0.1435	6.9682	1.010	7.040	1.4283	50′
20′	0.1454	0.1449	0.9894	0.1465	6.8269	1.011	6.900	1.4254	40′
30′	0.1484	0.1478	0.9890	0.1495	6.6912	1.011	6.765	1.4224	30′
40′	0.1513	0.1507	0.9886	0.1524	6.6506	1.012	6.636	1.4195	20′
50′	0.1542	0.1536	0.9881	0.1554	6.4348	1.012	6.512	1.4166	10′
9°00′	0.1571	0.1564	0.9877	0.1584	6.3138	1.012	6.392	1.4137	81°00′
		Cos θ	Sin θ	Cot θ	Tan θ	Csc θ	Sec θ	θ Rad.	θ Deg.

Table of Values of the Trigonometric Functions

θ Deg.	θ Rad.	Sin θ	Cos θ	Tan θ	Cot θ	Sec θ	Csc θ		
9°00′	0.1571	0.1564	0.9877	0.1584	6.3138	1.012	6.392	1.4137	**81°00′**
10′	0.1600	0.1593	0.9872	0.1614	6.1970	1.013	6.277	1.4108	50′
20′	0.1629	0.1622	0.9868	0.1644	6.0844	1.013	6.166	1.4079	40′
30′	0.1658	0.1650	0.9863	0.1673	5.9758	1.014	6.059	1.4050	30′
40′	0.1687	0.1679	0.9858	0.1703	5.8708	1.014	5.955	1.4021	20′
50′	0.1716	0.1708	0.9853	0.1733	5.7694	1.015	5.855	1.3992	10′
10°00′	0.1745	0.1736	0.9848	0.1763	5.6713	1.015	5.759	1.3963	**80°00′**
10′	0.1774	0.1765	0.9843	0.1793	5.5764	1.016	5.665	1.3934	50′
20′	0.1804	0.1794	0.9838	0.1823	5.4845	1.016	5.575	1.3904	40′
30′	0.1833	0.1822	0.9833	0.1853	5.3955	1.017	5.487	1.3875	30′
40′	0.1862	0.1851	0.9827	0.1883	5.3093	1.018	5.403	1.3846	20′
50′	0.1891	0.1880	0.9822	0.1914	5.2257	1.018	5.320	1.3817	10′
11°00′	0.1920	0.1908	0.9816	0.1944	5.1446	1.019	5.241	1.3788	**79°00′**
10′	0.1949	0.1937	0.9811	0.1974	5.0658	1.019	5.164	1.3759	50′
20′	0.1978	0.1965	0.9805	0.2004	4.9894	1.020	5.089	1.3730	40′
30′	0.2007	0.1994	0.9799	0.2035	4.9152	1.020	5.016	1.3701	30′
40′	0.2036	0.2022	0.9793	0.2065	4.8430	1.021	4.945	1.3672	20′
50′	0.2065	0.2051	0.9787	0.2095	4.7729	1.022	4.876	1.3643	10′
12°00′	0.2094	0.2079	0.9781	0.2126	4.7046	1.022	4.810	1.3614	**78°00′**
10′	0.2123	0.2108	0.9775	0.2156	4.6382	1.023	4.745	1.3584	50′
20′	0.2153	0.2136	0.9769	0.2186	4.5736	1.024	4.682	1.3555	40′
30′	0.2182	0.2164	0.9763	0.2217	4.5107	1.024	4.620	1.3526	30′
40′	0.2211	0.2193	0.9757	0.2247	4.4494	1.025	4.560	1.3497	20′
50′	0.2240	0.2221	0.9750	0.2278	4.3897	1.026	4.502	1.3468	10′
13°00′	0.2269	0.2250	0.9744	0.2309	4.3315	1.026	4.445	1.3439	**77°00′**
10′	0.2298	0.2278	0.9737	0.2339	4.2747	1.027	4.390	1.3410	50′
20′	0.2327	0.2306	0.9730	0.2370	4.2193	1.028	4.336	1.3381	40′
30′	0.2356	0.2334	0.9724	0.2401	4.1653	1.028	4.284	1.3352	30′
40′	0.2385	0.2363	0.9717	0.2432	4.1126	1.029	4.232	1.3323	20′
50′	0.2414	0.2391	0.9710	0.2462	4.0611	1.030	4.182	1.3294	10′
14°00′	0.2443	0.2419	0.9703	0.2493	4.0108	1.031	4.134	1.3265	**76°00′**
10′	0.2473	0.2447	0.9696	0.2524	3.9617	1.031	4.086	1.3235	50′
20′	0.2502	0.2476	0.9689	0.2555	3.9136	1.032	4.039	1.3206	40′
30′	0.2531	0.2504	0.9681	0.2586	3.8667	1.033	3.994	1.3177	30′
40′	0.2560	0.2532	0.9674	0.2617	3.8208	1.034	3.950	1.3148	20′
50′	0.2589	0.2560	0.9667	0.2648	3.7760	1.034	3.906	1.3119	10′
15°00′	0.2618	0.2588	0.9659	0.2679	3.7321	1.035	3.864	1.3090	**75°00′**
10′	0.2647	0.2616	0.9652	0.2711	3.6891	1.036	3.822	1.3061	50′
20′	0.2676	0.2644	0.9644	0.2742	3.6470	1.037	3.782	1.3032	40′
30′	0.2705	0.2672	0.9636	0.2773	3.6059	1.038	3.742	1.3003	30′
40′	0.2734	0.2700	0.9628	0.2805	3.5656	1.039	3.703	1.2974	20′
50′	0.2763	0.2728	0.9621	0.2836	3.5261	1.039	3.665	1.2945	10′
16°00′	0.2793	0.2756	0.9613	0.2867	3.4874	1.040	3.628	1.2915	**74°00′**
10′	0.2822	0.2784	0.9605	0.2899	3.4495	1.041	3.592	1.2886	50′
20′	0.2851	0.2812	0.9596	0.2931	3.4124	1.042	3.556	1.2857	40′
30′	0.2880	0.2840	0.9588	0.2962	3.3759	1.043	3.521	1.2828	30′
40′	0.2909	0.2868	0.9580	0.2994	3.3402	1.044	3.487	1.2799	20′
50′	0.2938	0.2896	0.9572	0.3026	3.3052	1.045	3.453	1.2770	10′
17°00′	0.2967	0.2924	0.9563	0.3057	3.2709	1.046	3.420	1.2741	**73°00′**
10′	0.2996	0.2952	0.9555	0.3089	3.2371	1.047	3.388	1.2712	50′
20′	0.3025	0.2979	0.9546	0.3121	3.2041	1.048	3.356	1.2683	40′
30′	0.3054	0.3007	0.9537	0.3153	3.1716	1.049	3.326	1.2654	30′
40′	0.3083	0.3035	0.9528	0.3185	3.1397	1.049	3.295	1.2625	20′
50′	0.3113	0.3062	0.9520	0.3217	3.1084	1.050	3.265	1.2595	10′
18°00′	0.3142	0.3090	0.9511	0.3249	3.0777	1.051	3.236	1.2566	**72°00′**
		Cos θ	Sin θ	Cot θ	Tan θ	Csc θ	Sec θ	θ Rad.	θ Deg.

Table of Values of the Trigonometric Functions

θ Deg.	θ Rad.	Sin θ	Cos θ	Tan θ	Cot θ	Sec θ	Csc θ		
18°00′	0.3142	0.3090	0.9511	0.3249	3.0777	1.051	3.236	1.2566	72°00′
10′	0.3171	0.3118	0.9502	0.3281	3.0475	1.052	3.207	1.2537	50′
20′	0.3200	0.3145	0.9492	0.3314	3.0178	1.053	3.179	1.2508	40′
30′	0.3229	0.3173	0.9483	0.3346	2.9887	1.054	3.152	1.2479	30′
40′	0.3258	0.3201	0.9474	0.3378	2.9600	1.056	3.124	1.2450	20′
50′	0.3287	0.3228	0.9465	0.3411	2.9319	1.057	3.098	1.2421	10′
19°00′	0.3316	0.3256	0.9455	0.3443	2.9042	1.058	3.072	1.2392	71°00′
10′	0.3345	0.3283	0.9446	0.3476	2.8770	1.059	3.046	1.2363	50′
20′	0.3374	0.3311	0.9436	0.3508	2.8502	1.060	3.021	1.2334	40′
30′	0.3403	0.3338	0.9426	0.3541	2.8239	1.061	2.996	1.2305	30′
40′	0.3432	0.3365	0.9417	0.3574	2.7980	1.062	2.971	1.2275	20′
50′	0.3462	0.3393	0.9407	0.3607	2.7725	1.063	2.947	1.2246	10′
20°00′	0.3491	0.3420	0.9397	0.3640	2.7475	1.064	2.924	1.2217	70°00′
10′	0.3520	0.3448	0.9387	0.3673	2.7228	1.065	2.901	1.2188	50′
20′	0.3549	0.3475	0.9377	0.3706	2.6985	1.066	2.878	1.2159	40′
30′	0.3578	0.3502	0.9367	0.3739	2.6746	1.068	2.855	1.2130	30′
40′	0.3607	0.3529	0.9356	0.3772	2.6511	1.069	2.833	1.2101	20′
50′	0.3636	0.3557	0.9346	0.3805	2.6279	1.070	2.812	1.2072	10′
21°00′	0.3665	0.3584	0.9336	0.3839	2.6051	1.071	2.790	1.2043	69°00′
10′	0.3694	0.3611	0.9325	0.3872	2.5826	1.072	2.769	1.2014	50′
20′	0.3723	0.3638	0.9315	0.3906	2.5605	1.074	2.749	1.1985	40′
30′	0.3752	0.3665	0.9304	0.3939	2.5386	1.075	2.729	1.1956	30′
40′	0.3782	0.3692	0.9293	0.3973	2.5172	1.076	2.709	1.1926	20′
50′	0.3811	0.3719	0.9283	0.4006	2.4960	1.077	2.689	1.1897	10′
22°00′	0.3840	0.3746	0.9272	0.4040	2.4751	1.079	2.669	1.1868	68°00′
10′	0.3869	0.3773	0.9261	0.4074	2.4545	1.080	2.650	1.1839	50′
20′	0.3898	0.3800	0.9250	0.4108	2.4342	1.081	2.632	1.1810	40′
30′	0.3927	0.3827	0.9239	0.4142	2.4142	1.082	2.613	1.1781	30′
40′	0.3956	0.3854	0.9228	0.4176	2.3945	1.084	2.595	1.1752	20′
50′	0.3985	0.3881	0.9216	0.4210	2.3750	1.085	2.577	1.1723	10′
23°00′	0.4014	0.3907	0.9215	0.4245	2.3559	1.086	2.559	1.1694	67°00′
10′	0.4043	0.3934	0.9194	0.4279	2.3369	1.088	2.542	1.1665	50′
20′	0.4072	0.3961	0.9182	0.4314	2.3183	1.089	2.525	1.1636	40′
30′	0.4102	0.3987	0.9171	0.4348	2.2998	1.090	2.508	1.1606	30′
40′	0.4131	0.4014	0.9159	0.4383	2.2817	1.092	2.491	1.1577	20′
50′	0.4160	0.4041	0.9147	0.4417	2.2637	1.093	2.475	1.1548	10′
24°00′	0.4189	0.4067	0.9135	0.4452	2.2460	1.095	2.459	1.1519	66°00′
10′	0.4218	0.4094	0.9124	0.4487	2.2286	1.096	2.443	1.1490	50′
20′	0.4247	0.4120	0.9112	0.4522	2.2113	1.097	2.427	1.1461	40′
30′	0.4276	0.4147	0.9100	0.4557	2.1943	1.099	2.411	1.1432	30′
40′	0.4305	0.4173	0.9088	0.4592	2.1775	1.100	2.396	1.1403	20′
50′	0.4334	0.4200	0.9075	0.4628	2.1609	1.102	2.381	1.1374	10′
25°00′	0.4363	0.4226	0.9063	0.4663	2.1445	1.103	2.366	1.1345	65°00′
10′	0.4392	0.4253	0.9051	0.4699	2.1283	1.105	2.352	1.1316	50′
20′	0.4422	0.4279	0.9038	0.4734	2.1123	1.106	2.337	1.1286	40′
30′	0.4451	0.4305	0.9026	0.4770	2.0965	1.108	2.323	1.1257	30′
40′	0.4480	0.4331	0.9013	0.4806	2.0809	1.109	2.309	1.1228	20′
50′	0.4509	0.4358	0.9001	0.4841	2.0655	1.111	2.295	1.1199	10′
26°00′	0.4538	0.4384	0.8988	0.4877	2.0503	1.113	2.281	1.1170	64°00′
10′	0.4567	0.4410	0.8975	0.4913	2.0353	1.114	2.268	1.1141	50′
20′	0.4596	0.4436	0.8962	0.4950	2.0204	1.116	2.254	1.1112	40′
30′	0.4625	0.4462	0.8949	0.4986	2.0057	1.117	2.241	1.1083	30′
40′	0.4654	0.4488	0.8936	0.5022	1.9912	1.119	2.228	1.1054	20′
50′	0.4683	0.4514	0.8923	0.5059	1.9768	1.121	2.215	1.1025	10′
27°00′	0.4712	0.4540	0.8910	0.5095	1.9626	1.122	2.203	1.0996	63°00′
		Cos θ	Sin θ	Cot θ	Tan θ	Csc θ	Sec θ	θ Rad.	θ Deg.

Table of Values of the Trigonometric Functions

Tables

θ Deg.	θ Rad.	Sin θ	Cos θ	Tan θ	Cot θ	Sec θ	Csc θ		
27°00′	0.4712	0.4540	0.8910	0.5095	1.9626	1.122	2.203	1.0996	63°00′
10′	0.4741	0.4566	0.8897	0.5132	1.9486	1.124	2.190	1.0966	50′
20′	0.4771	0.4592	0.8884	0.5169	1.9347	1.126	2.178	1.0937	40′
30′	0.4800	0.4617	0.8870	0.5206	1.9210	1.127	2.166	1.0908	30′
40′	0.4829	0.4643	0.8857	0.5243	1.9074	1.129	2.154	1.0879	20′
50′	0.4858	0.4669	0.8843	0.5280	1.8940	1.131	2.142	1.0850	10′
28°00′	0.4887	0.4695	0.8829	0.5317	1.8807	1.133	2.130	1.0821	62°00′
10′	0.4916	0.4720	0.8816	0.5354	1.8676	1.134	2.118	1.0792	50′
20′	0.4945	0.4746	0.8802	0.5392	1.8546	1.136	2.107	1.0763	40′
30′	0.4974	0.4772	0.8788	0.5430	1.8418	1.138	2.096	1.0734	30′
40′	0.5003	0.4797	0.8774	0.5467	1.8291	1.140	2.085	1.0705	20′
50′	0.5032	0.4823	0.8760	0.5505	1.8165	1.142	2.074	1.0676	10′
29°00′	0.5061	0.4848	0.8746	0.5543	1.8040	1.143	2.063	1.0647	61°00′
10′	0.5091	0.4874	0.8732	0.5581	1.7917	1.145	2.052	1.0617	50′
20′	0.5120	0.4899	0.8718	0.5619	1.7796	1.147	2.041	1.0588	40′
30′	0.5149	0.4924	0.8704	0.5658	1.7675	1.149	2.031	1.0559	30′
40′	0.5178	0.4950	0.8689	0.5696	1.7556	1.151	2.020	1.0530	20′
50′	0.5207	0.4975	0.8675	0.5735	1.7437	1.153	2.010	1.0501	10′
30°00′	0.5236	0.5000	0.8660	0.5774	1.7321	1.155	2.000	1.0472	60°00′
10′	0.5265	0.5025	0.8646	0.5812	1.7205	1.157	1.990	1.0443	50′
20′	0.5294	0.5050	0.8631	0.5851	1.7090	1.159	1.980	1.0414	40′
30′	0.5323	0.5075	0.8616	0.5890	1.6977	1.161	1.970	1.0385	30′
40′	0.5352	0.5100	0.8601	0.5930	1.6864	1.163	1.961	1.0356	20′
50′	0.5381	0.5125	0.8587	0.5969	1.6753	1.165	1.951	1.0327	10′
31°00′	0.5411	0.5150	0.8572	0.6009	1.6643	1.167	1.942	1.0297	59°00′
10′	0.5440	0.5175	0.8557	0.6048	1.6534	1.169	1.932	1.0268	50′
20′	0.5469	0.5200	0.8542	0.6088	1.6426	1.171	1.923	1.0239	40′
30′	0.5498	0.5225	0.8526	0.6128	1.6319	1.173	1.914	1.0210	30′
40′	0.5527	0.5250	0.8511	0.6168	1.6212	1.175	1.905	1.0181	20′
50′	0.5556	0.5275	0.8496	0.6208	1.6107	1.177	1.896	1.0152	10′
32°00′	0.5585	0.5299	0.8480	0.6249	1.6003	1.179	1.887	1.0123	58°00′
10′	0.5614	0.5324	0.8465	0.6289	1.5900	1.181	1.878	1.0094	50′
20′	0.5643	0.5348	0.8450	0.6330	1.5798	1.184	1.870	1.0065	40′
30′	0.5672	0.5373	0.8434	0.6371	1.5697	1.186	1.861	1.0036	30′
40′	0.5701	0.5398	0.8418	0.6412	1.5597	1.188	1.853	1.0007	20′
50′	0.5730	0.5422	0.8403	0.6453	1.5497	1.190	1.844	0.9977	10′
33°00′	0.5760	0.5446	0.8387	0.6494	1.5399	1.192	1.836	0.9948	57°00′
10′	0.5789	0.5471	0.8371	0.6536	1.5301	1.195	1.828	0.9919	50′
20′	0.5818	0.5495	0.8355	0.6577	1.5204	1.197	1.820	0.9890	40′
30′	0.5847	0.5519	0.8339	0.6619	1.5108	1.199	1.812	0.9861	30′
40′	0.5876	0.5544	0.8323	0.6661	1.5013	1.202	1.804	0.9832	20′
50′	0.5905	0.5568	0.8307	0.6703	1.4919	1.204	1.796	0.9803	10′
34°00′	0.5934	0.5592	0.8290	0.6745	1.4826	1.206	1.788	0.9774	56°00′
10′	0.5963	0.5616	0.8274	0.6787	1.4733	1.209	1.781	0.9745	50′
20′	0.5992	0.5640	0.8258	0.6830	1.4641	1.211	1.773	0.9716	40′
30′	0.6021	0.5664	0.8241	0.6873	1.4550	1.213	1.766	0.9687	30′
40′	0.6050	0.5688	0.8225	0.6916	1.4460	1.216	1.758	0.9657	20′
50′	0.6080	0.5712	0.8208	0.6959	1.4370	1.218	1.751	0.9628	10′
35°00′	0.6109	0.5736	0.8192	0.7002	1.4281	1.221	1.743	0.9599	55°00′
10′	0.6138	0.5760	0.8175	0.7046	1.4193	1.223	1.736	0.9570	50′
20′	0.6167	0.5783	0.8158	0.7089	1.4106	1.226	1.729	0.9541	40′
30′	0.6196	0.5807	0.8141	0.7133	1.4019	1.228	1.722	0.9512	30′
40′	0.6225	0.5831	0.8124	0.7177	1.3934	1.231	1.715	0.9483	20′
50′	0.6254	0.5854	0.8107	0.7221	1.3848	1.233	1.708	0.9454	10′
36°00′	0.6283	0.5878	0.8090	0.7265	1.3764	1.236	1.701	0.9425	54°00′
		Cos θ	Sin θ	Cot θ	Tan θ	Csc θ	Sec θ	θ Rad.	θ Deg.

Table of Values of the Trigonometric Functions **437**

Table of Values of the Trigonometric Functions

θ Deg.	θ Rad.	Sin θ	Cos θ	Tan θ	Cot θ	Sec θ	Csc θ		
36°00′	0.6283	0.5878	0.8090	0.7265	1.3764	1.236	1.701	0.9425	**54°00′**
10′	0.6312	0.5901	0.8073	0.7310	1.3680	1.239	1.695	0.9396	50′
20′	0.6341	0.5925	0.8056	0.7355	1.3597	1.241	1.688	0.9367	40′
30′	0.6370	0.5948	0.8039	0.7400	1.3514	1.244	1.681	0.9338	30′
40′	0.6400	0.5972	0.8021	0.7445	1.3432	1.247	1.675	0.9308	20′
50′	0.6429	0.5995	0.8004	0.7490	1.3351	1.249	1.668	0.9279	10′
37°00′	0.6458	0.6018	0.7986	0.7536	1.3270	1.252	1.662	0.9250	**53°00′**
10′	0.6487	0.6041	0.7969	0.7581	1.3190	1.255	1.655	0.9221	50′
20′	0.6516	0.6065	0.7951	0.7627	1.3111	1.258	1.649	0.9192	40′
30′	0.6545	0.6088	0.7934	0.7673	1.3032	1.260	1.643	0.9163	30′
40′	0.6574	0.6111	0.7916	0.7720	1.2954	1.263	1.636	0.9134	20′
50′	0.6603	0.6134	0.7898	0.7766	1.2876	1.266	1.630	0.9105	10′
38°00′	0.6632	0.6157	0.7880	0.7813	1.2799	1.269	1.624	0.9076	**52°00′**
10′	0.6661	0.6180	0.7862	0.7860	1.2723	1.272	1.618	0.9047	50′
20′	0.6690	0.6202	0.7844	0.7907	1.2647	1.275	1.612	0.9018	40′
30′	0.6720	0.6225	0.7826	0.7954	1.2572	1.278	1.606	0.8988	30′
40′	0.6749	0.6248	0.7808	0.8002	1.2497	1.281	1.601	0.8959	20′
50′	0.6778	0.6271	0.7790	0.8050	1.2423	1.284	1.595	0.8930	10′
39°00′	0.6807	0.6293	0.7771	0.8098	1.2349	1.287	1.589	0.8901	**51°00′**
10′	0.6836	0.6316	0.7753	0.8146	1.2276	1.290	1.583	0.8872	50′
20′	0.6865	0.6338	0.7735	0.8195	1.2203	1.293	1.578	0.8843	40′
30′	0.6894	0.6361	0.7716	0.8243	1.2131	1.296	1.572	0.8814	30′
40′	0.6923	0.6383	0.7698	0.8292	1.2059	1.299	1.567	0.8785	20′
50′	0.6952	0.6406	0.7679	0.8342	1.1988	1.302	1.561	0.8756	10′
40°00′	0.6981	0.6428	0.7660	0.8391	1.1918	1.305	1.556	0.8727	**50°00′**
10′	0.7010	0.6450	0.7642	0.8441	1.1847	1.309	1.550	0.8698	50′
20′	0.7039	0.6472	0.7623	0.8491	1.1778	1.312	1.545	0.8668	40′
30′	0.7069	0.6494	0.7604	0.8541	1.1708	1.315	1.540	0.8639	30′
40′	0.7098	0.6517	0.7585	0.8591	1.1640	1.318	1.535	0.8610	20′
50′	0.7127	0.6539	0.7566	0.8642	1.1571	1.322	1.529	0.8581	10′
41°00′	0.7156	0.6561	0.7547	0.8693	1.1504	1.325	1.524	0.8552	**49°00′**
10′	0.7185	0.6583	0.7528	0.8744	1.1436	1.328	1.519	0.8523	50′
20′	0.7214	0.6604	0.7509	0.8796	1.1369	1.332	1.514	0.8494	40′
30′	0.7243	0.6626	0.7490	0.8847	1.1303	1.335	1.509	0.8465	30′
40′	0.7272	0.6648	0.7470	0.8899	1.1237	1.339	1.504	0.8436	20′
50′	0.7301	0.6670	0.7451	0.8952	1.1171	1.342	1.499	0.8407	10′
42°00′	0.7330	0.6691	0.7431	0.9004	1.1106	1.346	1.494	0.8378	**48°00′**
10′	0.7359	0.6713	0.7412	0.9057	1.1041	1.349	1.490	0.8348	50′
20′	0.7389	0.6734	0.7392	0.9110	1.0977	1.353	1.485	0.8319	40′
30′	0.7418	0.6756	0.7373	0.9163	1.0913	1.356	1.480	0.8290	30′
40′	0.7447	0.6777	0.7353	0.9217	1.0850	1.360	1.476	0.8261	20′
50′	0.7476	0.6799	0.7333	0.9271	1.0786	1.364	1.471	0.8232	10′
43°00′	0.7505	0.6820	0.7314	0.9325	1.0724	1.367	1.466	0.8203	**47°00′**
10′	0.7534	0.6841	0.7294	0.9380	1.0661	1.371	1.462	0.8174	50′
20′	0.7563	0.6862	0.7274	0.9435	1.0599	1.375	1.457	0.8145	40′
30′	0.7592	0.6884	0.7254	0.9490	1.0538	1.379	1.453	0.8116	30′
40′	0.7621	0.6905	0.7234	0.9545	1.0477	1.382	1.448	0.8087	20′
50′	0.7650	0.6926	0.7214	0.9601	1.0416	1.386	1.444	0.8058	10′
44°00′	0.7679	0.6947	0.7193	0.9657	1.0355	1.390	1.440	0.8029	**46°00′**
10′	0.7709	0.6967	0.7173	0.9713	1.0295	1.394	1.435	0.7999	50′
20′	0.7738	0.6988	0.7153	0.9770	1.0235	1.398	1.431	0.7970	40′
30′	0.7767	0.7009	0.7133	0.9827	1.0176	1.402	1.427	0.7941	30′
40′	0.7796	0.7030	0.7112	0.9884	1.0117	1.406	1.423	0.7912	20′
50′	0.7825	0.7050	0.7092	0.9942	1.0058	1.410	1.418	0.7883	10′
45°00′	0.7854	0.7071	0.7071	1.0000	1.0000	1.414	1.414	0.7854	**45°00′**
		Cos θ	**Sin θ**	**Cot θ**	**Tan θ**	**Csc θ**	**Sec θ**	**θ Rad.**	**θ Deg.**

Tables

Chapter 1 Trigonometric Functions

Practice Exercises, pages 5–7 1. $D = \{1, 2, 5, 7, 16\}; R = \{0, 1, 3, 5, 9\}$ **3.** $D = \{-2, -1, 0, 1\}; R = \{-10, -1, 8, 11\}$ **5.** $D = \{-3, 2, 4\}; R = \{1, 2, 3\}$ **7.** $D = \{4, 6, 8, 10\}; R = \{25\}$
9. not a function **11.** function **13.** function
15. not a function **17.** function **19.** not a function **21.** 25 **23.** -2 **25.** -4 **27.** -1.6
29. $D = \{-1, 5, 8\}; R = \{-10, -1, 3, 8\}$; not a function **31.** $D = \{x: x \text{ is a real number}\}; R = \{f(x): f(x) \geq 0\}$; function **33.** $D = \{x: x \text{ is a real number}\}; R = \{f(x): f(x) \text{ is real number}\}$; function
35. $D = \{x: x \leq 0\}; R = \{y: y \text{ is a real number}\}$; not a function **37.** $D = \{x: x \leq 0, x \neq -1\}; R = \{y: y \geq 0, y \neq 1\}$; function **39.** 3 **41.** 24 **43.** 4
45. $x \neq 0$ **47.** $x \neq \pm 7$ **49.** $-\dfrac{1}{9}$ **51.** -81
53. 3 **55.** -26 **57.** 95°F **59.** 54 in.^{2}

Practice Exercises, pages 10–11 1. 5 **3.** $7\sqrt{2}$
5. $6\sqrt{2}$ **7.** $\sqrt{37}$ **9.** 5 **11.** $\sqrt{10}$ **13.** 7 **15.** 11
17. $2\sqrt{15}$ **19.** $\sqrt{2.5}$ **21.** $3\sqrt{2}$ **23.** $3\sqrt{3}$
25. $2\sqrt{a^2 + b^2}$ **27.** $x^2 + y^2 = 25$
29. $x^2 + y^2 = 45$ **31.** $x^2 + y^2 = 27$
33. $x = 9$ or $x = -7$ **35.** 24 units^2
37. $(-1, -6)$ **39.** $3\sqrt{5}$ mi **41.** 13 ft **43.** 74 ft

Practice Exercises, pages 15–16
1.

3.

5.

7.

9.
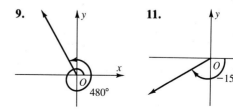
11.

21. 835° **23.** 115° **25.** $-2909°$ **27.** 763°
29. $-150°$ **31.** 378° **33.** 702° **35.** 570°
37. $-410°$ **39.** 260° **41.** 174° **43.** 180°
45. 350° **47.** 3 **49.** 18 **51.** 12,000°

Practice Exercises, pages 21–22 1. II **3.** III
5. IV **7.** 27°30′ **9.** 20°12′ **11.** 134°18′
13. 135.5° **15.** 220.3° **17.** 25.2° **19.** $\dfrac{\pi}{3}$
21. $\dfrac{2\pi}{3}$ **23.** $\dfrac{5\pi}{6}$ **25.** $\dfrac{3\pi}{10}$ **27.** $\dfrac{67\pi}{36}$ **29.** $-\dfrac{23\pi}{36}$
31. 240° **33.** 75° **35.** $-130°$ **37.** 210°
39. $-135°$ **41.** 300° **43.** 21°33′36″
45. 160°6′36″ **47.** $-313°31′48″$
49. $-381°52′12″$ **51.** 18.25° **53.** 315.80°
55. $-102.03°$ **57.** $-515.25°$ **59.** $\dfrac{5\pi}{2}$ **61.** $-\dfrac{9\pi}{4}$
63. $-\dfrac{23\pi}{4}$ **65.** 945° **67.** $-1080°$ **69.** $-1200°$
71. 576° **73.** $-\dfrac{5\pi}{3}$ **75.** $\dfrac{5\pi}{3}$ **77.** $-\dfrac{19\pi}{6}$
79. $\dfrac{11\pi}{2}$ **81.** $-\dfrac{17\pi}{4}$ **83.** $\dfrac{11\pi}{6}$ **85.** $-30°; -540°$
87. $-90°; -\dfrac{\pi}{2}$

Practice Exercises, pages 26–28 1. 5π in.
3. $\dfrac{25\pi}{2}$ in. **5.** 45π in. **7.** π rad **9.** $\dfrac{2\pi}{3}$ rad
11. $\dfrac{6\pi}{5}$ rad **13.** $\dfrac{\pi}{2}$ rad/min **15.** $\dfrac{2\pi}{15}$ rad/min
17. $\dfrac{\pi}{10}$ rad/min **19.** 200π cm/s **21.** 75π cm/s

23. $\dfrac{\pi}{30}$ rad/s 25. $\dfrac{8\pi}{3}$ rad/min 27. $\dfrac{24\pi}{5}$ rad/min

29. $1033\dfrac{1}{3}$ mph 31. 890 rpm 33. 8π ft/s; 8π ft/s

Test Yourself, page 28 1. $D = \{1, 2, 4, 6\}$; $R = \{3, 5, 7\}$; function 3. $\dfrac{1}{7}$ 5. undefined 7. II

9. I 11. $\dfrac{5\pi}{12}$ 13. 12π m 15. 84π in./s

Practice Exercises, pages 32–33

1. $\dfrac{4}{5}; \dfrac{3}{5}$ 3. $-\dfrac{12}{13}; -\dfrac{5}{13}$

5. $0; -1$ 7. $\dfrac{\sqrt{5}}{5}; -\dfrac{2\sqrt{5}}{5}$

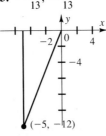

9. $\dfrac{3}{5}$ 11. $\dfrac{5}{13}$ 13. $-\dfrac{1}{2}$ 15. $-\dfrac{4}{5}$

17. $\dfrac{7\sqrt{74}}{74}; \dfrac{5\sqrt{74}}{74}$ 19. $-\dfrac{\sqrt{2}}{2}; \dfrac{\sqrt{2}}{2}$

21. $\dfrac{5\sqrt{34}}{34}; -\dfrac{3\sqrt{34}}{34}$

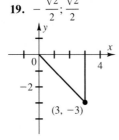

23. $\dfrac{\sqrt{3}}{2}$ 25. $-\dfrac{1}{2}$ 27. $-\dfrac{4\sqrt{6}}{11}$

29. I; II 31. I; III

Practice Exercises, pages 37–38 Answers are in the order: sin; cos; tan; csc; sec; cot.

1. $\dfrac{4}{5}; \dfrac{3}{5}; \dfrac{4}{3}; \dfrac{5}{4}; \dfrac{5}{3}; \dfrac{3}{4}$

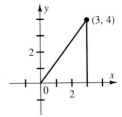

3. $-\dfrac{5}{13}; \dfrac{12}{13}; -\dfrac{5}{12};$
$-\dfrac{13}{5}; \dfrac{13}{12}; -\dfrac{12}{5}$

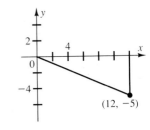

5. $-\dfrac{\sqrt{2}}{2}; -\dfrac{\sqrt{2}}{2}; 1;$
$-\sqrt{2}; -\sqrt{2}; 1$

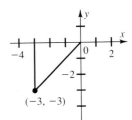

7. $-\dfrac{\sqrt{21}}{7}; \dfrac{2\sqrt{7}}{7}; -\dfrac{\sqrt{3}}{2};$
$-\dfrac{\sqrt{21}}{3}; \dfrac{\sqrt{7}}{2}; -\dfrac{2\sqrt{3}}{3}$

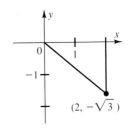

9. $-\dfrac{24}{25}; \dfrac{7}{25}; -\dfrac{24}{7};$
$-\dfrac{25}{24}; \dfrac{25}{7}; -\dfrac{7}{24}$

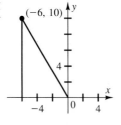

11. $\dfrac{\sqrt{3}}{2}; -\dfrac{1}{2}; -\sqrt{3};$
$\dfrac{2\sqrt{3}}{3}; -2; -\dfrac{\sqrt{3}}{3}$

$(-2, 2\sqrt{3})$

13. I, IV **15.** IV **17.** II, IV **19.** III, IV
21. $-\dfrac{3}{5}; \dfrac{4}{3}; -\dfrac{5}{4}; -\dfrac{5}{3}; \dfrac{3}{4}$ **23.** $\dfrac{4}{5}; \dfrac{3}{5}; \dfrac{5}{4}; \dfrac{5}{3}; \dfrac{3}{4}$
25. $-\dfrac{\sqrt{3}}{2}; -\dfrac{1}{2}; \sqrt{3}; -\dfrac{2\sqrt{3}}{3}; \dfrac{\sqrt{3}}{3}$ **27.** $-\dfrac{7}{25};$
$-\dfrac{24}{25}; \dfrac{7}{24}; -\dfrac{25}{7}; -\dfrac{25}{24}$ **29.** $\dfrac{\sqrt{7}}{4}; -\dfrac{3}{4}; -\dfrac{\sqrt{7}}{3}$
31. $-\dfrac{\sqrt{3}}{2}; \dfrac{1}{2}; -\sqrt{3}$ **33.** $-\dfrac{12}{13}; -\dfrac{5}{13}; \dfrac{12}{5}$
35. $(-8, 6)$ **37.** $(1, -1)$ **39.** $(4\sqrt{2}, -4\sqrt{2})$
41. tangent **43.** $\dfrac{2}{5}$

Practice Exercises, pages 44–45 **1.** $\dfrac{1}{2}$ **3.** $\dfrac{\sqrt{3}}{3}$
5. $\dfrac{2\sqrt{3}}{3}$ **7.** $\dfrac{\sqrt{2}}{2}$ **9.** 1 **11.** $\sqrt{2}$ **13.** $\dfrac{\sqrt{3}}{2}$ **15.** $\sqrt{3}$
17. 2 **19.** 9° **21.** 55° **23.** 52° **25.** 70°
Answers to 27–45 are in the order: sin; cos; tan; csc;
sec; cot.
27. $\dfrac{1}{2}; -\dfrac{\sqrt{3}}{2}; -\dfrac{\sqrt{3}}{3}; 2; -\dfrac{2\sqrt{3}}{2}; -\sqrt{3}$
29. $\dfrac{\sqrt{2}}{2}; -\dfrac{\sqrt{2}}{2}; -1; \sqrt{2}; -\sqrt{2}; -1$ **31.** $-\dfrac{\sqrt{2}}{2};$
$-\dfrac{\sqrt{2}}{2}; 1; -\sqrt{2}; -\sqrt{2}; 1$ **33.** $-\dfrac{1}{2}; -\dfrac{\sqrt{3}}{2}; \dfrac{\sqrt{3}}{3}; -2;$
$-\dfrac{2\sqrt{3}}{3}; \sqrt{3}$ **35.** $\dfrac{1}{2}; \dfrac{\sqrt{3}}{2}; \dfrac{\sqrt{3}}{3}; 2; \dfrac{2\sqrt{3}}{3}; \sqrt{3}$
37. $\dfrac{\sqrt{2}}{2}; -\dfrac{\sqrt{2}}{2}; -1; \sqrt{2}; -\sqrt{2}; -1$ **39.** $-\dfrac{1}{2};$
$\dfrac{\sqrt{3}}{2}; -\dfrac{\sqrt{3}}{3}; -2; \dfrac{2\sqrt{3}}{3}; -\sqrt{3}$ **41.** $\dfrac{\sqrt{3}}{2}; -\dfrac{1}{2}; -\sqrt{3};$
$\dfrac{2\sqrt{3}}{3}, -2; -\dfrac{\sqrt{3}}{3}$ **43.** $-\dfrac{\sqrt{3}}{2}; \dfrac{1}{2}; -\sqrt{3}; -\dfrac{2\sqrt{3}}{3};$
$2; -\dfrac{\sqrt{3}}{3}$ **45.** $-\dfrac{\sqrt{3}}{2}; -\dfrac{1}{2}; \sqrt{3}; -\dfrac{2\sqrt{3}}{3}; -2; \dfrac{\sqrt{3}}{3}$
47. $\dfrac{\pi}{3}; \dfrac{5\pi}{3}$ **49.** $\dfrac{3\pi}{4}; \dfrac{7\pi}{4}$ **51.** $\dfrac{\pi}{4}; \dfrac{7\pi}{4}$ **53.** $\dfrac{5\pi}{6}; \dfrac{11\pi}{6}$
55. 78° **57.** 7° **59.** 45° **61.** 150° **63.** 240°
65. 270° **67.** 315° **69.** 26 ft

Practice Exercises, pages 50–51 **1.** 0.3190
3. 9.0579 **5.** 0.9755 **7.** 0.3179 **9.** 21.2049
11. 0.8090 **13.** 0.8660 **15.** -1 **17.** 2.8291
19. 1.1343 **21.** -1.2134 **23.** 1.0122
25. 51.1°, 128.9° **27.** 34.1°, 214.1° **29.** 52.0°,
308.0° **31.** -0.7630 **33.** 2.8556 **35.** -0.3714
37. -1.5959 **39.** 0.9749 **41.** 3.4057
43. -0.2260 **45.** 1.4142 **47.** 23.6°, 156.4°
49. 12.7°, 192.7° **51.** 351.4°, 171.4° **53.** 48.3°
55. 235.5° **57.** 328.2° **59.** 102.0° **61.** $\pi \div 180$

Test Yourself, page 51 **1.** $\cos\theta = -\dfrac{1}{2}$
Answers to 3–5 are in the order: sin; cos; tan; csc;
sec; cot.
3. $-\dfrac{12}{13}; -\dfrac{5}{13}; \dfrac{12}{5}; -\dfrac{13}{12}; -\dfrac{13}{5}; \dfrac{5}{12}$ **5.** $\dfrac{\sqrt{95}}{12};$
$-\dfrac{7}{12}; -\dfrac{\sqrt{95}}{7}; \dfrac{12\sqrt{95}}{95}; -\dfrac{12}{7}; -\dfrac{7\sqrt{95}}{95}$ **7.** 48°
9. $\dfrac{\pi}{4}$ **11.** -0.5329 **13.** 2.3048

Summary and Review, pages 54–55 **1.** $D = \{1,$
$5\}; R = \{3, 7, 9\};$ no **3.** 10 **7.** 495° **9.** 4π in.
11. $-\dfrac{8}{17}$ **13.** 0; -1; 0; undefined; -1; undefined
15. 0.7071; -1.2314

Maintaining Skills, page 58 **1.** -10 **3.** $-\dfrac{7}{4}$
5. $-\dfrac{5}{2}$ **7.** $(2x + 3)(2x - 3)$ **9.** $(2 - x)$
$(4 + 2x + x^2)$ **11.** $(x - 3y)(x^2 + 3xy + 9y^2)$
13. $0, \sqrt{2}, -\sqrt{2}$ **15.** $0, 2, -1$ **17.** $0, 5, -1$
19.

21.

23.

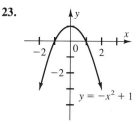

Answers

Chapter 2   Graphing Trigonometric Functions

Practice Exercises, pages 64–66   1. yes; yes; 2
3. yes; no   **5.** yes; yes; 4   **7.** period 2; $0 \le x \le 8$

9. period 1; $-2 \le x \le 3$

11. period 2; $-4 \le x \le 6$

13. even   **15.** odd   **17.** even   **19.** odd   **21.** even
23. odd   **25.** y-axis   **27.** none   **29.** origin
31.

35.

37. 31–2nd, 3rd; 33–2nd, 3rd; 35–2nd, 3rd
39. $x^2 + y^2 = 36$; y-axis, x-axis; origin

41.

Practice Exercises, pages 71–72   1. $\cos 62°$
3. $\sec 5°$ **5.** $\csc 63°$ **7.** $\tan 37°$ **9.** -0.5299
11. 0.6691   **13.** 0.8480   **15.** 0.8090   **17.** 0.4695
19. -0.6691

Answers

21.

23.

25. 0.3746 **27.** 0.4540 **29.** 0.6293
31. −0.7314 **33.** 0.6561
35.

37.

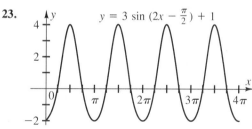

39.

41. For $0 \le x \le 0.6$, $x \approx \sin x$, to the nearest tenth

Practice Exercises, pages 82–83 **1.** $2; \frac{2\pi}{3}$; none;
none **3.** $3; \frac{\pi}{2}$; none; none **5.** 1; π; none; down 5
7. $\frac{1}{4}$; π; none; none **9.** 4; $\frac{2\pi}{3}$; right $\frac{\pi}{9}$; up 2

11.

13.

15.

17.

19.

21.

23.

25. $y = 4 \sin 2x - 3$ **27.** $y = \frac{3}{5} \cos \frac{4}{3}x +$
$\frac{22}{5}$ **29.** $y = \frac{1}{25} \sin 2(196\pi x)$ **31.** $0.004; \frac{1}{150}; \frac{\pi}{2}$

Test Yourself, page 83 **1.** even **3.** odd
5. x-axis **7.** −0.8572 **9.** 5; 2π **11.** 4; 6π
13. 2; 2π; $\frac{\pi}{2}$ right; 5 up **15.** 1; π; $\frac{3\pi}{2}$ left; 1 down

17.

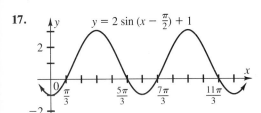

$y = 2 \sin (x - \frac{\pi}{2}) + 1$

11.

$y = \sin 2x - \cos 2x$

Practice Exercises, pages 86–87

1.

$y = 2 \sin x + \sin x$

13.

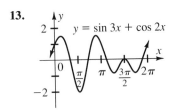

$y = \sin 3x + \cos 2x$

3.

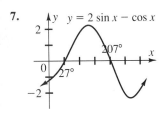

$y = \sin x + 2 \cos x$

117°

297°

15.

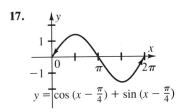

$y = -2 \sin x - \cos x$

5.

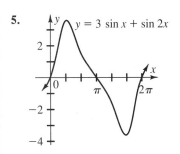

$y = 3 \sin x + \sin 2x$

17.

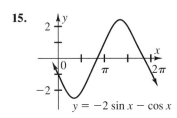

$y = \cos (x - \frac{\pi}{4}) + \sin (x - \frac{\pi}{4})$

7.

$y = 2 \sin x - \cos x$

207°

27°

19.

$y = \sin x + \cos x + \sin 2x$

9.

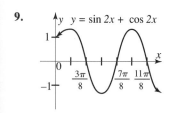

$y = \sin 2x + \cos 2x$

21.

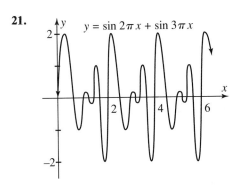

$y = \sin 2\pi x + \sin 3\pi x$

Practice Exercises, pages 92–93

1. $\dfrac{\pi}{3}$; π right   **3.** π; $\dfrac{\pi}{4}$ right   **5.** π; $\dfrac{2\pi}{3}$ left

7. $\dfrac{\pi}{2}$; π left; 3 up   **9.** π; $\dfrac{4\pi}{3}$ right; 1 down

11. π; $\dfrac{2\pi}{3}$ left; 2 down   **13.** π; $\dfrac{\pi}{6}$ right; 2 up

15.

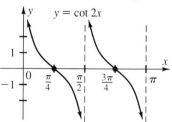

$y = \cot 2x$

17.

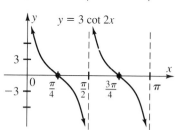

$y = 3 \cot 2x$

19.

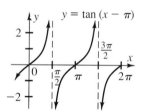

$y = \tan (x - \pi)$

21.

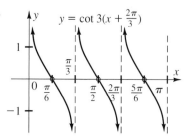

$y = \cot 3(x + \dfrac{2\pi}{3})$

23.

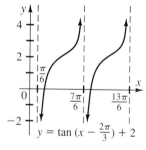

$y = \tan (x - \dfrac{2\pi}{3}) + 2$

25.

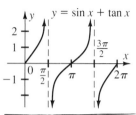

$y = \sin x + \tan x$

27.

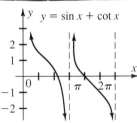

$y = \sin x + \cot x$

Practice Exercises, pages 98–99   **1.** $\dfrac{2\pi}{3}$; π right; none   **3.** 2π; $\dfrac{\pi}{4}$ right; none   **5.** 2π; $\dfrac{2\pi}{3}$ left; none

7. π; π left; 3 up   **9.** 2π; $\dfrac{4\pi}{3}$ right; 1 down

11. $\dfrac{2\pi}{3}$; $\dfrac{2\pi}{9}$ left; 2 down   **13.** π; $\dfrac{\pi}{4}$ left; 1 down

15.

$y = \csc 2x$

17.

$y = 3 \csc 2x$

19.

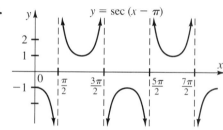

$y = \sec (x - \pi)$

Answers

21.

$y = \csc 2\left(x + \frac{2\pi}{3}\right)$

23. $y = \sec (2x - \pi)$

25. $y = \sin x + \sec x$

27. $y = \sin x + \csc x$

Practice Exercises, pages 103–104

1. $y = 3 \sin 2\pi\left(t - \frac{1}{4}\right)$ **3.** $y = 5 \sin \pi\left(t - \frac{1}{2}\right)$

5. $y = 5 \sin 4\pi\left(t - \frac{1}{8}\right)$ **7.** $y = 20 \sin \frac{\pi}{2} (t - 1)$

9. $y = 20 \sin 8\pi \left(t - \frac{1}{16}\right)$ **11.** $y = \sin (528 \,\pi t)$

13. $y = 3.2 \sin \frac{\pi}{6.2} (t + 3.1)$ **15.** $x = 6 \sin \frac{\pi}{2} t$

Test Yourself, page 104

1.

$y = \cos x + \cos 3x$

3. $\frac{\pi}{4}$; π left; none   **5.** π; $\frac{\pi}{4}$ left; 1 down

7. 2π; $\frac{\pi}{2}$ left; 1 up   **9.** π; $\frac{\pi}{4}$ left; 3 down

11.

$y = \tan \frac{1}{2}x$

13.

$y = \sec \frac{1}{2}x$

15. $y = 8 \sin \frac{\pi}{2} (t - 1)$

Summary and Review, pages 106–107   **1.** yes; period $= 6$, y-axis symmetry   **3.** -0.8192

5. $4; \dfrac{2\pi}{3}$ **7.** 3; π; π left; 4 down

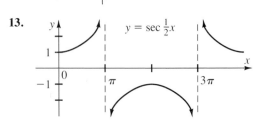

$y = 3 \cos 2(x + \pi) - 4$

9.

$y = 3 \cos x + \cos x$

11. $\dfrac{\pi}{3}$; none; none

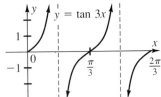

$y = \tan 3x$

446   Answers to Selected Exercises

13. $\dfrac{2\pi}{3}$; none; none

15. $y = \sin 264\pi t$

Cumulative Review, page 110
1.

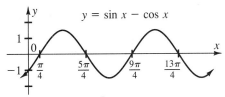

$y = \sin\left(x - \dfrac{\pi}{3}\right)$

3. even **5.** no **7.** $y = 6 \sin \pi \left(t - \dfrac{1}{2}\right)$

9. $\sin 41°$

11. $y = \sin x - \cos x$

$y = \sin x - \cos x$

13. 24π ft/s **15.** $\dfrac{7\pi}{4}$ **17.** $330°$ **19.** $48.8°$ and $311.2°$

Chapter 3 Solving Right Triangles

Practice Exercises, pages 116–117 **1.** $\angle A = 65°$; $a = 30$; $c = 33$ **3.** $\angle A = 73°$; $a = 21$; $b = 6.4$
5. $\angle A = 57°$; $\angle B = 33°$; $c = 6.6$ **7.** $\angle B = 32°$; $a = 23$; $b = 14$ **9.** $\angle A = 74.9°$; $a = 10.3$; $b = 2.79$ **11.** $\angle B = 27°$; $b = 5.6$; $c = 12$
13. $\angle A = 47.5°$; $b = 172$; $c = 255$ **15.** $\angle B = 53°19'$; $b = 25.94$; $c = 32.34$ **17.** $\angle A = 17°32'$; $b = 268.5$; $c = 281.6$ **19.** $\angle A = 42.53°$; $\angle B = 47.47°$; $a = 16.16$ **21.** $\angle A = 52.68°$; $\angle B = 37.32°$; $c = 23.45$ **23.** $\angle A = 54.09°$; $\angle B = 35.91°$; $b = 1182$ Answers to 25–27 are in the order: sin; cos; tan; csc; sec; cot.

25. $\dfrac{3}{t}$; $\dfrac{\sqrt{t^2-9}}{t}$; $\dfrac{3\sqrt{t^2-9}}{t^2-9}$; $\dfrac{t}{3}$; $\dfrac{t\sqrt{t^2-9}}{t^2-9}$; $\dfrac{\sqrt{t^2-9}}{3}$

27. $\dfrac{\sqrt{t^2-1}}{t}$; $\dfrac{1}{t}$; $\sqrt{t^2-1}$; $\dfrac{t\sqrt{t^2-1}}{t^2-1}$; t; $\dfrac{\sqrt{t^2-1}}{t^2-1}$

29. 86 ft **31.** 16 ft

Practice Exercises, pages 121–123 **1.** 22 ft
3. 89 ft **5.** 39° **7.** 4590 ft **9.** 1089 ft **11.** 147 ft **13.** 150 ft **15.** 55 m **17.** 13.91° **19.** 53 m

Practice Exercises, pages 126–129 **1.** 124°; 64 km **3.** 221 mi **5.** 916 ft **7.** 95 yd **9.** 61°
11. 19.8 mi **13.** 145 mi **15.** 216 ft
17. 198.0 mi

Test Yourself, page 129 **1.** $\angle B = 47°$; $a = 13$; $b = 14$ **3.** $\angle A = 38°$; $\angle B = 52°$; $c = 24$
5. 20° **7.** 149°; 37 mi **9.** 92 m

Practice Exercises, pages 133–135
9. $1 + (\sqrt{3})^2 = (2)^2$ **11.** $\left(\dfrac{\sqrt{2}}{2}\right)^2 + \left(\dfrac{\sqrt{2}}{2}\right)^2 = 1$

13. $-\sqrt{3} = \dfrac{-\dfrac{\sqrt{3}}{2}}{\dfrac{1}{2}}$ **15.** $-\dfrac{\sqrt{3}}{2} = -\left(\dfrac{\sqrt{3}}{2}\right)$

17. $1 - \left(\dfrac{\sqrt{2}}{2}\right)^2 = \left(-\dfrac{\sqrt{2}}{2}\right)^2$ **19.** $\dfrac{1}{\cos\theta} = \dfrac{1}{\dfrac{x}{r}} =$
$\dfrac{r}{x} = \sec\theta$ **21.** $\dfrac{\cos\theta}{\sin\theta} = \dfrac{\dfrac{x}{r}}{\dfrac{y}{r}} = \dfrac{x}{y} = \cot\theta$ **23.** $1 +$
$\tan^2\theta = 1 + \left(\dfrac{y}{x}\right)^2 = \dfrac{x^2+y^2}{x^2} = \dfrac{r^2}{x^2} = \sec^2\theta$

25. $\sin\theta \csc\theta = \left(\dfrac{y}{r}\right)\left(\dfrac{1}{\dfrac{y}{r}}\right) = \left(\dfrac{y}{r}\right)\left(\dfrac{r}{y}\right) = 1$

27. $1 - \sin^2\theta = 1 - \left(\dfrac{y^2}{r^2}\right) = \dfrac{r^2-y^2}{r^2} =$
$\dfrac{x^2}{r^2} = \cos^2\theta$ **29.** $\left(\dfrac{\sqrt{2}}{2}\right)(1) = \dfrac{\sqrt{2}}{2}$
31. $\left(\dfrac{\sqrt{2}}{2}\right)\left(-\dfrac{\sqrt{2}}{2}\right)(-1) = 1 - \left(-\dfrac{\sqrt{2}}{2}\right)^2$
33. $41^2 = 40^2 + 9^2$; $1681 = 1681$
35. $\left(\dfrac{41}{40}\right)\left(\dfrac{9}{41}\right)\left(\dfrac{40}{9}\right) = 1$; $1 = 1$
37. 0.9903; 20.0 m

Practice Exercises, pages 137–139 **1.** $1 - \sin^2\theta$
3. $\sec^2\beta - 1$ **5.** $\dfrac{(\sin\alpha - 5)(\sin\alpha + 5)}{(\sin\alpha + 5)(\sin\alpha + 5)}$
7. $\dfrac{(\cos\alpha - 7)(\cos\alpha - 7)}{(\cos\alpha - 7)(\cos\alpha + 7)}$

9. $\dfrac{(\tan \theta - 3)(\tan^2 \theta + 3 \tan \theta + 9)}{(\tan \theta - 3)(\tan \theta + 3)}$

11. $\dfrac{(\cot \theta + 2)(\cot^2 \theta - 2 \cot \theta + 4)}{(\cot \theta + 2)(\cot \theta - 2)}$

13. $\csc^2 \theta - \cot^2 \theta$ **15.** $\dfrac{\sin^2 \theta}{\cos^2 \theta}$ **17.** $\dfrac{\sin \theta}{1 - \sin^2 \theta}$

19. $\dfrac{2}{\sin \theta}$ **21.** $\cos^2 \theta$ **23.** $\dfrac{2}{\cos \theta}$ **25.** $\tan \theta$

27. $\dfrac{1}{\tan \theta}$ **29.** $\sin^2 \theta + \sin \theta \cos \theta + \cos^2 \theta = \sin \theta \cos \theta + 1$ **31.** $a^2 - b^2$

Practice Exercises, pages 143–144

1. $\dfrac{1 - \cos^2 \theta}{\cos^2 \theta} = \dfrac{\sin^2 \theta}{\cos^2 \theta} = \left(\dfrac{\sin \theta}{\cos \theta}\right)^2$

3. $\dfrac{\cos \alpha}{1 - \sin^2 \alpha} = \dfrac{\cos \alpha}{\cos^2 \alpha} = \dfrac{1}{\cos \alpha}$

5. $\dfrac{1}{\tan \alpha} = \dfrac{\cot \alpha}{1} = \dfrac{\cot \alpha}{\sin^2 \alpha + \cos^2 \alpha}$

7. $\dfrac{1 + \cos \mu}{\cos \mu} = \dfrac{1}{\cos \mu} + \dfrac{\cos \mu}{\cos \mu} = \sec \mu + 1$

9. $\dfrac{\sin^2 \alpha}{1 - \sin^2 \alpha} = \dfrac{\sin^2 \alpha}{\cos^2 \alpha} = \tan^2 \alpha = \sec^2 \alpha - 1$

11. $\sin^2 \theta \cos \theta \sec \theta = \sin^2 \theta (\cos \theta)\left(\dfrac{1}{\cos \theta}\right) =$

$\sin^2 \theta = 1 - \cos^2 \theta$ **13.** $1 - \dfrac{1}{\sec^2 \beta} =$

$1 - \cos^2 \beta = \sin^2 \beta$

Practice Exercises, pages 146–147 **1.** identity
3. identity **5.** identity **7.** identity **9.** identity
11. not an identity **13.** identity **15.** not an
identity **17.** identity **19.** not an identity
21. identity **23.** not an identity **25.** identity
27. not an identity **29.** $4 \cos \theta + 3 \sin \theta = 5 \cos (\theta - 36°52')$

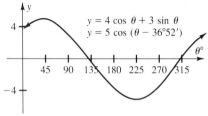

$y = 4 \cos \theta + 3 \sin \theta$
$y = 5 \cos (\theta - 36°52')$

Test Yourself, page 147

1. $\left(-\dfrac{\sqrt{3}}{3}\right)^2 = \left(\dfrac{2\sqrt{3}}{3}\right)^2 - 1$

3. $\dfrac{\cot \theta}{\sin \theta}(\sec \theta - \cos \theta) =$

$\dfrac{\cos \theta}{\sin \theta}$
$\dfrac{\cos \theta}{\sin \theta}\left(\dfrac{1}{\cos \theta} - \cos \theta\right) = \dfrac{\cos \theta}{\sin^2 \theta}\left(\dfrac{1 - \cos^2 \theta}{\cos \theta}\right) =$

$\left(\dfrac{\cos \theta}{\sin^2 \theta}\right)\left(\dfrac{\sin^2 \theta}{\cos \theta}\right) = 1$ **7.** identity

Summary and Review, pages 150–151 **1.** $\angle A =$
$69°$; $b = 65.3$; $c = 182$ **3.** $21°$ **5.** $213°$;

44.6 km **7.** 680 ft **9.** $\dfrac{\sqrt{3}}{3} = \dfrac{-\dfrac{1}{2}}{-\dfrac{\sqrt{3}}{2}}$ **11.** $\dfrac{\cos^2 \theta}{\sin^2 \theta}$

Maintaining Skills, page 154

1. $\dfrac{6 + q}{3q}$ **3.** $\dfrac{2q - p^2 - pq}{pq}$ **5.** 1 **7.** $\dfrac{15R + 12}{10}$

9. $\dfrac{abd - ac}{2d}$ **11.** $\dfrac{5RY - 5SY}{8}$ **13.** $x < -\dfrac{7}{2}$

15. $x > \dfrac{2}{3}$ **17.** $x > -7$ **19.** $80°$

Chapter 4 Oblique Triangles

Practice Exercises, pages 160–162 **1.** $b = 18$
3. $a = 170$ **5.** 30 **7.** 100 **9.** 79 **11.** 19
13. 30 **15.** $\angle P = 95°$, $q = 13$, $r = 11$
17. $\angle Q = 66°$, $p = 47$, $r = 41$ **19.** $\angle Q = 100.5°$, $q = 155$, $p = 84.9$ **21.** $\angle R = 79.9°$, $p = 62.7$, $q = 55.5$ **23.** $\angle R = 39.1°$, $r = 76.9$, $q = 112$ **25.** $\angle P = 14.8°$, $p = 19.9$, $r = 63.9$
27. $\angle R = 67.5°$, $r = 121$, $q = 119$ **29.** $\angle R = 19.2°$, $r = 47.1$, $q = 105$ **31.** yes

33. $\sin A = \dfrac{a}{c}$, $\sin B = \dfrac{b}{c}$, $\sin C = 1$; $\dfrac{a}{c} \div a =$

$\dfrac{b}{c} \div b = \dfrac{1}{c}$ **37.** $48.87°$; 256.0 ft; 320.8 ft

39. 930 ft

Practice Exercises, pages 166–167 **1.** 0 **3.** 2;
$\angle B = 131°$, $\angle C = 17°$, $c = 3.9$ or $\angle B = 49°$,
$\angle C = 99°$, $c = 13$ **5.** 0 **7.** 0 **9.** 2; $\angle B = 42°$,
$\angle C = 101°$, $c = 80$ or $\angle B = 138°$, $\angle C = 5°$,
$c = 7.1$ **11.** 1; $\angle B = 52.2°$, $\angle C = 68.0°$, $c = 86.7$ **13.** 2; $\angle B = 40.9°$, $\angle C = 107.2°$, $c = 55.3$ or $\angle B = 139.1°$, $\angle C = 9.0°$, $c = 9.1$
15. 0 **17.** 2; $\angle B = 70.4°$, $\angle A = 62.5°$, $a = 14.4$ or $\angle B = 109.6°$, $\angle A = 23.3°$, $a = 6.43$
19. 2; $\angle C = 40.9°$, $\angle A = 102.8°$, $a = 76.3$ or $\angle C = 139.1°$, $\angle A = 4.6°$, $a = 6.27$ **21.** 0
23. 1; $\angle A = 63.3°$, $\angle B = 8.0°$, $b = 7.98$ **25.** 0

Answers

27. 2; $\angle A = 31.0°$, $\angle B = 125.5°$, $b = 41.9$ or
$\angle A = 149.0°$, $\angle B = 7.5°$, $b = 6.75$ **29.** $\angle B =$
$56°$, $\angle C = 73°$ or $\angle B = 124°$, $\angle C = 5°$
31. $15°, 134°$ **37.** $112°$

Practice Exercises, pages 172–174 1. 10 **3.** 7.4
5. 28 **7.** 30 **9.** 8.8 **11.** 40° **13.** 105° **15.** 90°
17. $r = 9.0$, $\angle Q = 63°$, $\angle P = 87°$ **19.** $p =$
63, $\angle R = 43°$, $\angle Q = 54°$ **21.** $q = 59$, $\angle P =$
$25°$, $\angle R = 42°$ **23.** $\angle R = 84°$, $\angle P = 41°$,
$\angle Q = 55°$ **25.** $\angle R = 31°$, $\angle Q = 53°$, $\angle P =$
$96°$ **27.** $\angle P = 41.47°$, $\angle Q = 59.51°$, $\angle R =$
$79.02°$ **35.** $82.0°, 51.7°, 46.3°$ **37.** $106.0°$
39. 410 mi

Practice Exercises, pages 179–180 1. $\angle A =$
$110°$, $\angle B = 23°$, $c = 36$ **3.** $\angle A = 51°$, $\angle B =$
$25°$, $c = 34$ **5.** $\angle B = 87°$, $\angle C = 50°$, $a = 42$
7. $\angle B = 31°$, $\angle C = 26°$, $a = 25$ **9.** $\angle A =$
$108°$, $\angle C = 46°$, $b = 21$ **11.** $\angle A = 44°$, $\angle C =$
$25°$, $b = 120$ **13.** $\angle C = 104°$, $a = 29$, $b = 20$,
$c = 39$ **15.** $\angle B = 64°$, $a = 60$, $b = 54$, $c = 23$
17. $\angle C = 114.7°$, $a = 60.1$, $b = 32.2$, $c = 79.2$
19. $\angle A = 94.4$, $a = 77.0$, $b = 46.1$, $c = 58.2$
23. 84 m

Test Yourself, page 180 1. $\angle C = 91°$, $b = 23$,
$c = 28$ **3.** $\angle B = 87°$, $a = 11$, $b = 16$ **5.** 1;
$\angle B = 48°$, $\angle C = 37°$, $c = 19$ **7.** $\angle A = 42°$,
$\angle B = 54°$, $c = 62$ **9.** $\angle A = 97°$, $\angle B = 51°$,
$\angle C = 32°$ **11.** 0 **13.** $\angle A = 21°$, $\angle B = 39°$,
$c = 48$, **15.** $\angle C = 51°$, $a = 7.7$, $b = 36$,
$c = 31$ **17.** 11 cm

Practice Exercises, pages 184–185 1. 440
3. 380 **5.** 60 **7.** 730 **9.** 30 **11.** 860 **13.** 540
15. 40 **17.** 108 **19.** 1506 **21.** 201 **23.** 1039
25. 121 **27.** 120 **31.** 2100 cm^2 **33.** 4888 ft^2
35. 5.0 acres

Practice Exercises, pages 188–189 1. 6 **3.** 280
5. 150 **7.** 270 **9.** 60 **11.** 220 **13.** 15,500
15. 13,900 **17.** 12 **19.** 26 **21.** 1200 **23.** 2000
27. 170 ft^2

Practice Exercises, pages 195–197

1. $\|\vec{AB}\| = 13$

3. $\|\vec{AB}\| = \sqrt{290}$

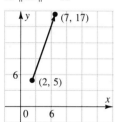

5. 13; 8 **7.** 16; 8 **9.** -4; 11

11.

13.

15. $\mathbf{v} - \mathbf{t}$

17.

19. 22; 44° **21.** 31; 19° **23.** 3.6; 64° **25.** 29.1;
18.5° **27.** 33.8; 61.4°

29.

31.

$\mathbf{u}$

$\mathbf{v}$

$\mathbf{u} + \mathbf{v}$

$2(\mathbf{u} + \mathbf{v})$

33. 280 mi/h; 86° **35.** 230 mi/h; 50° **37.** 590
mi/h; 27° E of N **39.** 320 mi/h; 33°

Test Yourself, page 197 1. 560 in.^2 **3.** 350 cm^2
5. 150 **7.** 140 **9.** $\sqrt{34}$ **11.** 18.9; 17.5°

Summary and Review, pages 200–201 1. 24.6
3. $c = 2$, $\angle B = 115°$, $\angle C = 9°$ or $c = 11$,
$\angle B = 65°$, $\angle C = 59°$ **5.** $\angle B = 36°$, $\angle C = 45°$,
$c = 18$ **7.** $\angle A = 47°$, $\angle B = 75°$,
$c = 14$ **9.** triangle is not determined **11.** 160
13. 334 **15.** $10\sqrt{2}$; $10\sqrt{2}$

Cumulative Review, pages 204–206 1. $\angle A =$
$52°$, $\angle B = 94°$, $c = 7.9$ **3.** 132.31° **5.** 0.6691
7. 18 cm **9.** 10π in./s **11.** yes **13.** 7; 54°
17. 180° **19.** $\left(\frac{1}{2}\right)^2 \left(\frac{\sqrt{3}}{2}\right) \left(\frac{2\sqrt{3}}{3}\right) = 1 - \left(\frac{\sqrt{3}}{2}\right)^2$;
$\left(\frac{1}{4}\right) 1 = 1 - \left(\frac{3}{4}\right); \frac{1}{4} = \frac{1}{4}$ **21.** y-axis

23. 500 in.² **25.** quadrant IV **27.** $\sin \theta = \frac{3}{5}$;

$\tan \theta = -\frac{3}{4}$; $\csc \theta = \frac{5}{3}$; $\sec \theta = -\frac{5}{4}$;

$\cot \theta = -\frac{4}{3}$ **29.** 790 ft **31.** $\angle A = 83°$;

$\angle C = 32°$ **33.** $x = 8$ or $x = -4$ **35.** neither

37. odd **39.** 303°; 663°; −417° **41.** 1.1303

43. 0.3907 **45.** 0.5446 **47.** 16° **49.** $\frac{1}{2}$ acre

Chapter 5 Trigonometric Identities

Practice Exercises, pages 212–213

1. $\dfrac{\sqrt{6} - \sqrt{2}}{4}$ **3.** $\dfrac{-\sqrt{6} - \sqrt{2}}{4}$ **5.** $\dfrac{\sqrt{2} + \sqrt{6}}{4}$

7. $\dfrac{\sqrt{2} + \sqrt{6}}{4}$ **9.** $\dfrac{33}{65}$ **11.** $\dfrac{3}{5}$ **13.** $-\dfrac{44}{125}$

15. $-\dfrac{77}{85}$ **17.** $\cos(360° + \alpha) = \cos 360° \cos \alpha - \sin 360° \sin \alpha = 1(\cos \alpha) - 0(\sin \alpha) = \cos \alpha$

19. $\cos(270° - \beta) = \cos 270° \cos \beta + \sin 270° \sin \beta = 0(\cos \beta) + (-1)(\sin \beta) = -\sin \beta$

21. $\cos\left(\dfrac{\pi}{3} - \beta\right) = \cos\dfrac{\pi}{3}\cos \beta + \sin\dfrac{\pi}{3}\sin \beta =$

$\dfrac{1}{2}\cos \beta + \dfrac{\sqrt{3}}{2}\sin \beta = \dfrac{\cos \beta + \sqrt{3}\sin \beta}{2}$

23. $\cos\left(\dfrac{3\pi}{4} + \alpha\right) = \cos\dfrac{3\pi}{4}\cos \alpha - \sin\dfrac{3\pi}{4} =$

$\sin \alpha = -\dfrac{\sqrt{2}}{2}\cos \alpha - \dfrac{\sqrt{2}}{2}\sin \alpha = -\dfrac{\sqrt{2}}{2}(\cos \alpha +$

$\sin \alpha)$ **25.** $\dfrac{\sqrt{6} - \sqrt{2}}{4}$ **27.** $\dfrac{\sqrt{2} - \sqrt{6}}{4}$

29. $\cos(\alpha - \beta) - \cos(\alpha + \beta) = \cos \alpha \cos \beta + \sin \alpha \sin \beta - [\cos \alpha \cos \beta - \sin \alpha \sin \beta] = \cos \alpha \cos \beta + \sin \alpha \sin \beta - \cos \alpha \cos \beta +$

$\sin \alpha \sin \beta = 2 \sin \alpha \sin \beta$ **31.** $\dfrac{\sqrt{2} - \sqrt{6}}{4}$

Practice Exercises, pages 218–219

1. $\dfrac{\sqrt{6} - \sqrt{2}}{4}$ **3.** $\dfrac{\sqrt{2} - \sqrt{6}}{4}$ **5.** $\dfrac{\sqrt{6} - \sqrt{2}}{4}$

7. $\dfrac{\sqrt{2} + \sqrt{6}}{4}$ **9.** $\dfrac{323}{325}$ **11.** $-\dfrac{36}{85}$ **13.** $-\dfrac{16}{65}$

15. $-\dfrac{323}{325}$ **17.** $\sin(180° - \alpha) = \sin 180° \cos \alpha -$

$\cos 180° \sin \alpha = 0 (\cos \alpha) - (-1)(\sin \alpha) = \sin \alpha$

19. $\sin(210° + \alpha) = \sin 210° \cos \alpha +$

$\cos 210° \sin \alpha = \left(-\dfrac{1}{2}\right)\cos \alpha +$

$\left(-\dfrac{\sqrt{3}}{2}\right)\sin \alpha = \dfrac{-\cos \alpha - \sqrt{3}\sin \alpha}{2}$

21. $\sin(45° + \alpha) = \sin 45° \cos \alpha + \cos 45°$

$\sin \alpha = \dfrac{\sqrt{2}}{2}\cos \alpha + \dfrac{\sqrt{2}}{2}\sin \alpha = \dfrac{\sqrt{2}}{2}(\cos \alpha + \sin \alpha)$

23. $\sin\left(\dfrac{\pi}{3} - \beta\right) = \sin\dfrac{\pi}{3}\cos \beta - \cos\dfrac{\pi}{3}\sin \beta =$

$\dfrac{\sqrt{3}}{2}\cos \beta - \dfrac{1}{2}\sin \beta = \dfrac{1}{2}(\sqrt{3}\cos \beta - \sin \beta)$

25. $\sin\left(\beta - \dfrac{7\pi}{6}\right) = \sin \beta \cos\dfrac{7\pi}{6} - \cos \beta \sin\dfrac{7\pi}{6} =$

$\sin \beta\left(-\dfrac{\sqrt{3}}{2}\right) - \cos \beta\left(-\dfrac{1}{2}\right) = -\dfrac{1}{2}(\sqrt{3}\sin \beta -$

$\cos \beta)$ **27.** $\dfrac{-\sqrt{2} - \sqrt{6}}{4}$ **29.** $\dfrac{-\sqrt{2} - \sqrt{6}}{4}$

31. $\sin(\alpha + \beta) + \sin(\alpha - \beta) = \sin \alpha \cos \beta + \cos \alpha \sin \beta + \sin \alpha \cos \beta - \cos \alpha \sin \beta =$

$2 \sin \alpha \sin \beta$ **33.** $y = -2.5 \sin 4t + \dfrac{5\sqrt{3}}{2}\cos 4t$

Practice Exercises, pages 223–224

1. $2 + \sqrt{3}$ **3.** $-2 - \sqrt{3}$ **5.** $\sqrt{3} - 2$ **7.** $\sqrt{3} - 2$

9. $-\dfrac{33}{56}$ **11.** $-\dfrac{13}{84}$ **13.** $-\dfrac{117}{44}$ **15.** $\dfrac{44}{117}$

17. $\tan(45° - \alpha) = \dfrac{\tan 45° - \tan \alpha}{1 + \tan 45° \tan \alpha} =$

$\dfrac{1 - \tan \alpha}{1 + \tan \alpha}$ **19.** $\tan(180° + \alpha) =$

$\dfrac{\tan 180° + \tan \alpha}{1 - \tan 180° \tan \alpha} = \dfrac{0 + \tan \alpha}{1 - 0(\tan \alpha)} = \tan \alpha$

21. $\tan(360° + \alpha) = \dfrac{\tan 360° + \tan \alpha}{1 - \tan 360° \tan \alpha} =$

$\dfrac{0 + \tan \alpha}{1 - 0(\tan \alpha)} = \tan \alpha$ **23.** $\tan\left(\dfrac{3\pi}{4} + \alpha\right) =$

$\dfrac{\tan\dfrac{3\pi}{4} + \tan \alpha}{1 - \tan\dfrac{3\pi}{4}\tan \alpha} = \dfrac{-1 + \tan \alpha}{1 - (-1)\tan \alpha} = \dfrac{\tan \alpha - 1}{\tan \alpha + 1}$

25. $\tan\left(\alpha - \dfrac{5\pi}{4}\right) = \dfrac{\tan \alpha - \tan\dfrac{5\pi}{4}}{1 + \tan \alpha \tan\dfrac{5\pi}{4}} =$

$\dfrac{\tan \alpha - 1}{1 + (\tan \alpha)(1)} = \dfrac{\tan \alpha - 1}{\tan \alpha + 1}$ **27.** $\sqrt{3} - 2$

29. $-2 - \sqrt{3}$

Test Yourself, page 225

1. $\dfrac{\sqrt{6} - \sqrt{2}}{4}$ **3.** $\dfrac{\sqrt{6} - \sqrt{2}}{4}$ **5.** $2 - \sqrt{3}$ **7.** $\dfrac{24}{25}$

9. $\dfrac{117}{44}$ **11.** $\cos\left(\dfrac{\pi}{2} + \alpha\right) = \cos\dfrac{\pi}{2}\cos\alpha -$

$\sin\dfrac{\pi}{2}\sin\alpha = 0(\cos\alpha) - 1(\sin\alpha) = -\sin\alpha$

13. $\tan(360° - \beta) = \dfrac{\tan 360° - \tan\beta}{1 + \tan 360° \tan\beta} =$

$\dfrac{0 - \tan\beta}{1 + 0(\tan\beta)} = -\tan\beta$

Practice Exercises, pages 228–230

1. $-\dfrac{7}{25}$ **3.** $-\dfrac{24}{7}$ **5.** $-\dfrac{24}{25}$ **7.** $-\dfrac{120}{169}$

9. $\dfrac{7}{25}$ **11.** $\dfrac{120}{119}$ **13.** $-\dfrac{336}{625}$ **15.** $-\dfrac{7}{25}$

17. $\sin 3\theta = 3\sin\theta - 4\sin^3\theta$

19. $\dfrac{1 - \tan^2\beta}{\sec^2\beta} = \dfrac{1 - \dfrac{\sin^2\beta}{\cos^2\beta}}{\dfrac{1}{\cos^2\beta}} \cdot \dfrac{\cos^2\beta}{\cos^2\beta}$

$= \cos^2\beta - \sin^2\beta = \cos 2\beta$

21. $\dfrac{1 + \cos 2\beta}{2\sin\beta\cos\beta} = \dfrac{2\cos^2\beta}{2\sin\beta\cos\beta} = \dfrac{\cos\beta}{\sin\beta} = \cot\beta$

33. 11 ft

Practice Exercises, pages 235–236 **1.** $-2 - \sqrt{3}$

3. $\dfrac{\sqrt{2 - \sqrt{2}}}{2}$ **5.** $\dfrac{\sqrt{2 - \sqrt{2}}}{2}$ **7.** $\dfrac{\sqrt{2 + \sqrt{2}}}{2}$

9. $\dfrac{-\sqrt{2 + \sqrt{2}}}{2}$ **11.** $2 - \sqrt{3}$ **13.** $\dfrac{3\sqrt{10}}{10}$

15. $\dfrac{\sqrt{3}}{2}$ **17.** $\dfrac{-\sqrt{5} - 1}{2}$ **19.** $-\dfrac{3}{4}$ **21.** $\dfrac{2\sqrt{13}}{13}$

33. $\left(s\sin\dfrac{\alpha}{2}\right)\left(s\cos\dfrac{\alpha}{2}\right) = s^2\sqrt{\dfrac{1 - \cos\alpha}{2}} \cdot$

$\sqrt{\dfrac{1 + \cos\alpha}{2}} = s^2\sqrt{\dfrac{1 - \cos^2\alpha}{4}} =$

$s^2\left(\dfrac{\sqrt{\sin^2\alpha}}{2}\right) = \dfrac{1}{2}s^2\sin\alpha$

Practice Exercises, pages 240–241 **1.** $\cos 24° -$

$\cos 60°$ **3.** $\sin 7x - \sin 3x$ **5.** $\sin\dfrac{\pi}{4} - \sin\dfrac{\pi}{12}$

7. $\dfrac{1}{2}\left(\cos\dfrac{\pi}{4} + \cos\pi\right)$

9. $2\cos 72° \cos 34°$ **11.** $2\sin\dfrac{15\theta}{2}\cos\dfrac{9\theta}{2}$

13. $2\sin\dfrac{4\pi}{7}\sin\dfrac{2\pi}{7}$

Test Yourself, pages 241–242 **1.** $\dfrac{119}{169}$

3. $\dfrac{\sqrt{2 + \sqrt{3}}}{2}$ **5.** $-\sqrt{5}$ **9.** $2\sin 5x\cos x$

Summary and Review, pages 244–245

1. $\dfrac{\sqrt{2} - \sqrt{6}}{4}$ **3.** $\dfrac{-3\sqrt{3} + 4}{10}$ **5.** $\dfrac{\sqrt{6} + \sqrt{2}}{4}$

7. $-\dfrac{4}{5}$ **9.** $2 + \sqrt{3}$ **11.** $\sqrt{3}$ **13.** $-\dfrac{24}{25}$ **15.** $-\dfrac{24}{7}$

17. $\dfrac{3\sqrt{13}}{13}$ **19.** $-\dfrac{3}{2}$ **21.** $\cos 5x - \cos 13x$

23. $2\sin 57° \sin 16°$

Maintaining Skills, page 248

1. $\dfrac{2\sqrt{3} + 3}{3}$ **3.** $\dfrac{2\sqrt{2} - \sqrt{14}}{4}$ **5.** $\dfrac{\sqrt[5]{6}}{2}$ **7.** 28

9. 127 **11.** 4 **13.** $-1, -2$ **15.** $-3, 1$

17. 0, 3 **19.** $y = \dfrac{1}{3}x + \dfrac{5}{3}$ **21.** $y = 2x + 1$

Chapter 6 Inverse Trigonometric Functions

Practice Exercises, pages 253–254

1.

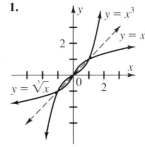

3. $y = \dfrac{1}{10}x$ **5.** $y = x + 8$ **7.** $y = -\dfrac{1}{7}x$

9. $y = \dfrac{1}{2}x - \dfrac{1}{2}$ **11.** $y = \dfrac{1}{5}x + \dfrac{4}{5}$

13. $y = \dfrac{3}{2}x - 9$ **15.** $90° + n(360°)$

17. $180° + n(360°)$ **19.** $\frac{\pi}{2} + \pi n$

21. $\frac{\pi}{2} + 2\pi n$ **23.** $\frac{2\pi}{3} + 2\pi n$ or $\frac{4\pi}{3} +$

$2\pi n$ **25.** $\frac{7\pi}{6} + 2\pi n$ or $\frac{11\pi}{6} + 2\pi n$

27. **29.**

33.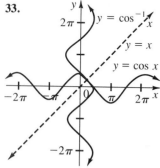

35. $f^{-1}(x) = \sqrt[3]{x-1}; f(f^{-1}(x)) = x; f^{-1}(f(x)) = x$

37. $F(x) = \frac{9}{5}x + 32$

Practice Exercises, pages 257–259

1. $-\frac{\pi}{3}$ **3.** $\frac{\pi}{3}$ **5.** $-\frac{\pi}{3}$ **7.** $-\frac{\pi}{2}$ **9.** 54° **11.** 19°

13. 38° **15.** 135° **17.** 0.4 **19.** $\frac{\sqrt{2}}{2}$ **21.** $\frac{1}{2}$

23. $\frac{\pi}{3}$ **25.** $\frac{\pi}{6}$ **27.** $\frac{\pi}{4}$ **29.** 0.9109 **31.** 0.7887

33. 0.5675 **35.** 0.8588

37. **39.**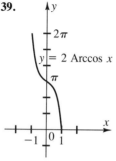

41. $-\frac{1}{2}$ **43.** $-\frac{\sqrt{2}}{2}$ **45.** yes **47.** $-1 \le x \le \frac{\sqrt{2}}{2}$

Practice Exercises, pages 263–264

1. $\frac{\pi}{4}$ **3.** $-\frac{\pi}{3}$ **5.** 66° **7.** 144° **9.** 0.6 **11.** 5

13. $-\frac{\sqrt{3}}{3}$ **15.** $\frac{\sqrt{2}}{2}$ **17.** $\frac{\sqrt{3}}{3}$ **19.** $\frac{4\sqrt{15}}{15}$

21. $\sec^{-1}2 + \tan^{-1}\frac{\sqrt{3}}{3} = \frac{\pi}{3} + \frac{\pi}{6} = \frac{\pi}{2}$

23. $\cos(\text{Arcsec } w) = \cos\left(\text{Arccos }\frac{1}{w}\right) = \frac{1}{w}$

25. $\frac{\sqrt{6} + \sqrt{2}}{4}$ **27.** **29.** 8.1°

Test Yourself, page 264

1. $\frac{\pi}{6} + 2\pi n$ or $\frac{5\pi}{6} + 2\pi n$ **3.** $\frac{\pi}{6} + 2\pi n$ or $-\frac{\pi}{6}$

$+ 2\pi n$ **5.** 14° **7.** 54° **9.** 0.7125 **11.** 2 **13.** $\frac{5}{3}$

Practice Exercises, pages 267–268

1. $\frac{\pi}{6}; \frac{5\pi}{6}$ **3.** $\frac{\pi}{4}; \frac{7\pi}{4}$ **5.** $\frac{7\pi}{6}; \frac{11\pi}{6}$ **7.** 180°

9. 150°; 330° **11.** 120°; 240° **13.** 0 **15.** $\frac{3\pi}{4}; \frac{7\pi}{4}$

17. $\frac{\pi}{3}; \frac{2\pi}{3}; \frac{4\pi}{3}; \frac{5\pi}{3}$ **19.** $\frac{\pi}{4}; \frac{3\pi}{4}; \frac{5\pi}{4}; \frac{7\pi}{4}$ **21.** $\frac{3\pi}{2};$

$\frac{7\pi}{6}; \frac{11\pi}{6}$ **23.** $0; \frac{\pi}{4}; \pi; \frac{5\pi}{4}$ **25.** $\frac{\pi}{4}; \frac{3\pi}{4}; \frac{5\pi}{4}; \frac{7\pi}{4}$

27. $0; \frac{\pi}{6}; \frac{5\pi}{6}; \pi$ **29.** $\frac{\pi}{4}; \frac{5\pi}{4}$ **31.** $\frac{\pi}{6}; \frac{\pi}{2}; \frac{5\pi}{6}; \frac{3\pi}{2}$

33. $\frac{\pi}{6}; \frac{5\pi}{6}; \frac{7\pi}{6}; \frac{11\pi}{6}$ **35.** $0; \frac{\pi}{2}; \pi; \frac{7\pi}{6}; \frac{11\pi}{6}$

37. $0; \frac{\pi}{3}; \frac{\pi}{2}; \frac{2\pi}{3}; \pi; \frac{4\pi}{3}; \frac{3\pi}{2}; \frac{5\pi}{3}$ **39.** $\frac{\pi}{4}; \frac{3\pi}{4}; \frac{5\pi}{4}; \frac{7\pi}{4}$

41. $0; \frac{\pi}{3}; \pi; \frac{5\pi}{3}$

Practice Exercises, pages 271–272 **1.** 42°; 138°
3. 141°; 219° **5.** 78°; 102°; 258°; 282° **7.** 90°;
190°; 350° **9.** 104°; 109°; 251°; 256° **11.** 9°;
171°; 232°; 308° **13.** 48°; 76°; 284°; 312°
15. 85°; 275° **17.** 184°; 356° **19.** 10°; 80°; 190°;
260° **21.** 38°; 142°; 218°; 322° **23.** 44°; 134°;
224°; 314° **25.** 3°; 31°; 149°; 177° **27.** 1.77;
2.09; 4.19; 4.51 **29.** 0.10; 3.04; 4.71 **33.** 0.09;
0.44; 0.98; 1.34; 2.19; 2.54; 3.08; 3.43; 4.28;
4.63; 5.17; 5.52 **35.** no solution **37.** 1.58 s

Practice Exercises, pages 276–277

1. $(\sqrt{2}, -4\sqrt{2})$ **3.** $\left(\dfrac{8 - 5\sqrt{3}}{2}, \dfrac{-8\sqrt{3} - 5}{2}\right)$

5. $\left(\dfrac{-6\sqrt{3} - 7}{2}, \dfrac{6 - 7\sqrt{3}}{2}\right)$ **7.** $(2\sqrt{2}, \sqrt{2})$

9. $3(x')^2 - (y')^2 = 4$ **11.** $7(x')^2 - 5(y')^2 = 70$
13. $4(x')^2 + (y')^2 = 4$
15. $11(x')^2 + 9(y')^2 = 8$

17.

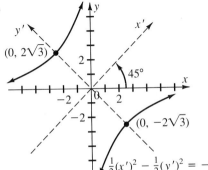

$\frac{1}{2}(x')^2 - \frac{1}{2}(y')^2 = -6$

19.

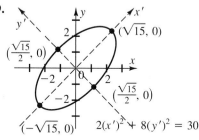

$2(x')^2 + 8(y')^2 = 30$

25. $1.35(x')^2 - 5.35(y')^2 = 8$

27. The new equation is equivalent to the original.
29. $\theta = 30°$; $4(x')^2 + (y')^2 = 8$

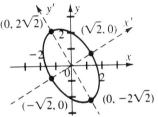

Test Yourself, page 277   **1.** $\dfrac{\pi}{4}; \dfrac{5\pi}{4}$ **3.** $x = 0$;

5. $45°$; $168.7°$; $225°$; $348.7°$ **7.** $30°$

Summary and Review, pages 280–281

1. $y = x + \dfrac{3}{4}$ **3.** $\dfrac{\pi}{2} + 2\pi k$, k is an integer

5. $\dfrac{\sqrt{3}}{2}$ **7.** $-\dfrac{\pi}{4}$ **9.** $-\sqrt{3}$ **11.** $\dfrac{\pi}{6}; \dfrac{5\pi}{6}; \dfrac{3\pi}{2}$

13. $\dfrac{\pi}{6}, \dfrac{\pi}{2}, \dfrac{5\pi}{6}, \dfrac{3\pi}{2}$ **15.** $56°$; $111°$; $249°$; $304°$

17. $\left(\dfrac{19\sqrt{3} - 15}{2}, \dfrac{-15\sqrt{3} - 19}{2}\right)$

19. $4(y')^2 - 5\,x'y' - 2(x')^2 = 9$
21. $\theta = 45°$;

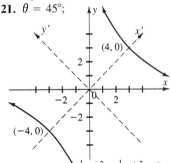

$\frac{1}{2}(x')^2 - \frac{1}{2}(y')^2 = 8$

Cumulative Review, page 284

1. $2\cos 70° \cos 18°$ **5.** $\dfrac{\pi}{2}$

9. $\|\overrightarrow{AB}\| = 13$

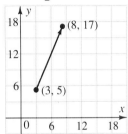

11. $\dfrac{1}{2}$ **13.** $10°$; $50°$; $130°$; $170°$; $250°$; $290°$
15. 3.36 **17.** $75°$

Chapter 7   Complex Numbers

Practice Exercises, pages 290–291

1.

3.

5. −420°   **7.**

9.

11.

13. $(3, 60°); (3, -300°); (-3, 240°); (-3, -120°)$
15. $(2, -210°); (2, 150°); (-2, 330°); (-2, -30°)$
17. $(0, 0)$ **19.** $(-1.29, 1.53)$ **21.** $(1.03, -2.82)$
23. $(1, 270°)$ **25.** $(2\sqrt{2}, 315°)$ **27.** $(2\sqrt{3}, 150°)$
29. $r = 4 \sec \theta$ **31.** $r = 6$ **33.** $x^2 + y^2 = 10y$
35. $y = 6$ **37.** $r = -\frac{3}{4} \sin \theta$ **39.** $r = 12 \cos \theta$
41. $4x^2 + 3y^2 + 2y = 1$ **43.** $x^2 = 6y$
45. $x^2 + y^2 = 50\sqrt{x^2 + y^2} + 50x$

Practice Exercises, pages 296–297

3.

5.

7.

9.

11.

13.

15.

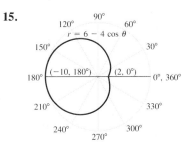

454   Answers to Selected Exercises

17.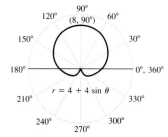

$r = 4 + 4 \sin \theta$

19.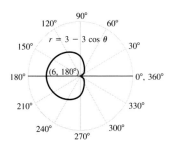

$r = 3 - 3 \cos \theta$
(6, 180°)

21.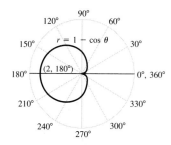

$r = 1 - \cos \theta$
(2, 180°)

23.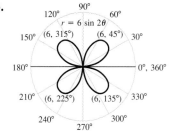

$r = 6 \sin 2\theta$
(6, 315°) (6, 45°)
(6, 225°) (6, 135°)

25.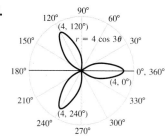

(4, 120°)
$r = 4 \cos 3\theta$
(4, 0°)
(4, 240°)

$r = 5 \cos 4\theta$
(5, 90°)
(5, 180°) (5, 0°)
(5, 270°)

$r^2 = \cos 2\theta$
(1, 180°) (1, 0°)

33. ellipse **35.** the pole; $\left(\dfrac{\sqrt{2}}{2}, 45°\right)$

Practice Exercises, pages 301–302

1. $5i$ **3.** $\dfrac{1}{7}i$ **5.** $2i\sqrt{7}$ **7.** $5i\sqrt{6}$ **9.** $6i$

11. $2i - 2i\sqrt{2}$ **13.** $11i$ **15.** i **17.** 1 **19.** -1

21. $14 + 14i$ **23.** $-3 - 10i$ **25.** $-7 + 16i$

27. $-14 - 2i$ **29.** -17 **31.** $i\sqrt{5} + i\sqrt{16}$

33. $-2 - 12i\sqrt{2}$ **35.** 1 **37.** $-i$ **39.** $-10 + 5i$

41. $3 - 9i$ **43.** $5 + 6i$ **45.** $x = -2; y = 7$

47. $x = 0; y = \dfrac{7}{3}$ **49.** no **51.** $15 + 4i$

53. $1 - 3i$ **55.** 5 **57.** 10 **59.** 10 ohms

Practice Exercises, pages 305–307 **1.** $2 + 4i$

3. $8 - i$ **5.** $9 + 33i$ **7.** -60 **9.** $-2\sqrt{6}$

11. $-9 + 40i$ **13.** $7 - 24i$ **15.** 13

17. $9 - 3i\sqrt{5}$ **19.** $\dfrac{5 - i}{26}$ **21.** $\dfrac{3 + 4i}{25}$ **23.** $\dfrac{-2 + i}{15}$

25. $\dfrac{i}{4}$ **27.** $-\dfrac{i}{3}$ **29.** i **31.** $\dfrac{-7 + 24i}{25}$

33. $\dfrac{-6 - 8i}{5}$ **35.** $\dfrac{-37 + 46i}{85}$ **37.** $\dfrac{-5 + i}{26}$

39. $\dfrac{-6 + 13i}{5}$ **41.** $3 + 2i$ **43.** $1 \pm 2i$

45. $\dfrac{1 \pm i\sqrt{6}}{2}$ **47.** $\pm i\sqrt{26}$ **49.** $-9 + 46i$

51. $-7 + 4i$ **53.** $x = 4; y = 0$ **55.** $(a - bi) + (c - di) = (a + c) + (-bi - di) = (a + c) - (b + d)i$ **57.** $28 + 24i$ volts **59.** $\dfrac{5 + 5i}{4}$

Test Yourself, page 307 **1.** $(2.6, -1.5)$

3. $(2, 330°)$ **5.** $r = 5 \csc \theta$ **7.** $x = 5$

Test Yourself, page 307 **1.** $(2.6, -1.5)$
3. $(2, 330°)$ **5.** $r = 5 \csc \theta$ **7.** $x = 5$
9.

$r = 3 + 4 \cos \theta$

$(7, 0°)$

11. $2i\sqrt{10}$ **13.** $-i$ **15.** $22 + i$ **17.** $-15 + 29i$

Practice Exercises, pages 311–312

1.

$-1 - i$

3.

$-2 + 2i$

5.

$\sqrt{3} - \sqrt{3}i$

7.

$4i$

9.

-5

11.

$-5 + 5\sqrt{5}i$

13. $\sqrt{2}(\cos 315° + i \sin 315°)$
15. $2(\cos 30° + i \sin 30°)$ **17.** $4(\cos 0° + i \sin 0°)$ **19.** $4(\cos 330° + i \sin 330°)$

21. $\frac{3\sqrt{2}}{2} + \frac{3\sqrt{2}}{2}i$ **23.** $\frac{1}{2} - \frac{\sqrt{3}}{2}i$ **25.** $\frac{5\sqrt{3}}{2} + \frac{5}{2}i$

27. $\sqrt{2} - \sqrt{2}i$ **29.** $1 + 0i$ **31.** $-\frac{3}{2} + \frac{3\sqrt{3}}{2}i$

33. $-2\sqrt{2} - 2\sqrt{2}i$ **41.**

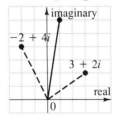

$-2 + 4i$

$3 + 2i$

Practice Exercises, pages 315–317

1. $6(\cos 130° + i \sin 130°)$ **3.** $24\left(\cos \frac{11\pi}{12} + \right.$

$\left. i \sin \frac{11\pi}{12} \right)$ **5.** $84(\cos 152° + i \sin 152°)$

7. $50(\cos 32° + i \sin 32°)$ **9.** $56\left(\cos \frac{5\pi}{4} + \right.$

$\left. i \sin \frac{5\pi}{4} \right)$ **11.** $4(\cos 20° + i \sin 20°)$

13. $4(\cos 25° + i \sin 25°)$ **15.** $\frac{7}{2}\left(\cos \frac{\pi}{2} + i \sin \frac{\pi}{2}\right)$

17. $16(\cos 45° + i \sin 45°)$ **19.** $\frac{10}{7}(\cos 150° +$

$i \sin 150°)$ **21.** $10(\cos 23° + i \sin 23°)$

23. $2\sqrt{10}(\cos 108° + i \sin 108°)$

25. $\frac{\sqrt{10}}{2}(\cos 288° + i \sin 288°)$ **27.** $\frac{5}{2}(\cos 83° +$

$i \sin 83°)$ **35.** $r = 8 \operatorname{cis} \theta$ or $r = 8$

Practice Exercises, pages 320–321

1. $-\frac{27}{2} + \frac{27\sqrt{3}}{2}i$ **3.** $\frac{125\sqrt{2}}{2} + \frac{125\sqrt{2}}{2}i$

5. $4 - 4\sqrt{3}i$ **7.** i **9.** $-108 - 108i\sqrt{3}$
11. -64 **13.** $8i$ **15.** $8i$ **17.** $-54 - 54i$

19. $-i$ **21.** $-\frac{1}{8}$ **23.** $4 \cos^3 \theta - 3 \cos \theta$

25. $4 \cos^3 \theta \sin \theta - 4 \cos \theta \sin^3 \theta$

27. $\frac{-1 + \sqrt{3}}{2} + \frac{1 + \sqrt{3}}{2}i$

Practice Exercises, pages 325–326

1. $\frac{3\sqrt{2}}{2} + \frac{3\sqrt{2}}{2}i; -\frac{3\sqrt{2}}{2} - \frac{3\sqrt{2}}{2}i$ **3.** $3; -\frac{3}{2} +$

$\frac{3\sqrt{3}}{2}i; -\frac{3}{2} - \frac{3\sqrt{3}}{2}i$ **5.** $2\sqrt{3} - 2i; 4i; -2\sqrt{3} - 2i$

7. $\frac{\sqrt[4]{18}}{2} - \frac{\sqrt[4]{2}}{2}i; \frac{\sqrt[4]{2}}{2} + \frac{\sqrt[4]{18}}{2}i; -\frac{\sqrt[4]{18}}{2} + \frac{\sqrt[4]{2}}{2}i; -\frac{\sqrt[4]{2}}{2} -$

$\frac{\sqrt[4]{18}}{2}i$ **9.** $2 + 0i; 0 + 2i; -2 + 0i; 0 - 2i$

11. $1; -1$ **13.** $1 + 0i; \frac{1}{2} + \frac{\sqrt{3}}{2}i; -\frac{1}{2} + \frac{\sqrt{3}}{2}i;$

$-1 + 0i; -\frac{1}{2} - \frac{\sqrt{3}}{2}i; \frac{1}{2} - \frac{\sqrt{3}}{2}i$
15. $1 + 0i; 0 + i;$
$-1 + 0i; 0 - i$

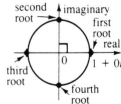

second root, first root, third root, fourth root, $1 + 0i$

17. $-\dfrac{3}{2} + \dfrac{3\sqrt{3}}{2}\,i$;

$3 + 0i$;

$-\dfrac{3}{2} - \dfrac{3\sqrt{3}}{2}\,i$

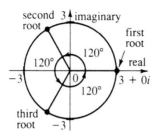

19. $3\cos 36° + 3i\sin 36°$; $3\cos 108° +$
$3i\sin 108°$; $-3 + 0i$; $3\cos 252° + 3i\sin 252°$;
$3\cos 324° + 3i\sin 324°$

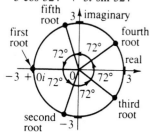

21. $-\dfrac{3}{2} + \dfrac{3\sqrt{3}}{2}\,i$;

$3 + 0i$;

$-\dfrac{3}{2} - \dfrac{3\sqrt{3}}{2}\,i$

23. $\cos 36° + i\sin 36°$, $\cos 108° + i\sin 108°$;
$\cos 180° + i\sin 180°$; $\cos 252° + i\sin 252°$;
$\cos 324° + i\sin 324°$

25. $1 + i\sqrt{3}$; $-2 + 0i$; $1 - i\sqrt{3}$

27. $\dfrac{\sqrt{2+\sqrt{2}}}{4} + \dfrac{\sqrt{2-\sqrt{2}}}{4}\,i$; $-\dfrac{\sqrt{2-\sqrt{2}}}{4}$

$+ \dfrac{\sqrt{2+\sqrt{2}}}{4}\,i$; $-\dfrac{\sqrt{2+\sqrt{2}}}{4} - \dfrac{\sqrt{2-\sqrt{2}}}{4}\,i$;

$\dfrac{\sqrt{2-\sqrt{2}}}{4} - \dfrac{\sqrt{2+\sqrt{2}}}{4}\,i$

29. $4\cos 20° + 4i\sin 20°$;
$4\cos 140° + 4i\sin 140°$;
$4\cos 260° + 4i\sin 260°$

31. $2 + 0i$; $2\cos 72° +$
$2i\sin 72°$; $2\cos 144° +$
$2i\sin 144°$; $2\cos 216° +$
$2i\sin 216°$; $2\cos 288° +$
$2i\sin 288°$

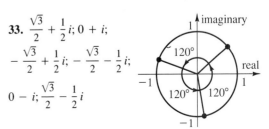

33. $\dfrac{\sqrt{3}}{2} + \dfrac{1}{2}\,i$; $0 + i$;

$-\dfrac{\sqrt{3}}{2} + \dfrac{1}{2}\,i$; $-\dfrac{\sqrt{3}}{2} - \dfrac{1}{2}\,i$;

$0 - i$; $\dfrac{\sqrt{3}}{2} - \dfrac{1}{2}\,i$

35. $\cos 40° + i\sin 40°$;
$\cos 160° + i\sin 160°$;
$\cos 280° + i\sin 280°$

37. $\dfrac{\sqrt[3]{4}}{2} + \dfrac{\sqrt[6]{432}}{2}\,i$; $-\sqrt[3]{4}$; $\dfrac{\sqrt[3]{4}}{2} - \dfrac{\sqrt[6]{432}}{2}\,i$

39. $2\sqrt{3} - 2i; 4i; -2\sqrt{3} - 2i$

41. $27^{\frac{1}{3}} \text{ cis} \left(\dfrac{120}{3} + k\dfrac{360}{3} \right)$ **43.** 2 cis 20; 2 cis 80;

2 cis 140; 2 cis 200; 2 cis 260; 2 cis 320

Test Yourself, page 326   1. $3\sqrt{2} (\cos 315° +$

$i \sin 315°)$ **3.** $-2\sqrt{2} + 2\sqrt{2}\, i$ **5.** 40 (cos 354°

$+ i \sin 354°)$ **7.** $-8,388,608 + 14,529,495i$

9. $\dfrac{3}{2} + \dfrac{3\sqrt{3}}{2}\, i; -3 + 0i; \dfrac{3}{2} - \dfrac{3\sqrt{3}}{2}\, i$

Summary and Review, pages 328–329

1.

3. (2, 60°)

5.

7. $2i\sqrt{5}$ **9.** $-8 + 7i$ **11.** $\dfrac{-27 - 8i}{13}$

13. $-3\sqrt{3} + 3i$ **15.** $2 (\cos 60° + i \sin 60°)$

17. $\dfrac{15625}{2} + \dfrac{15625\sqrt{3}}{2}\, i$ **19.** 5 + 0i; −5 + 0i;

$0 - 5i; 0 + 5i$

Maintaining Skills, page 332   1. -243 **3.** $-\dfrac{1}{16}$

5. $\dfrac{1}{16}$ **7.** $3x^3$ **9.** x^3y^6 **11.** a^2b **13.** 35

15. -7 **17.** 6   **19.** 1   **21.** $\dfrac{5}{6}$ **23.** 26

25. 6.25×10^5 **27.** 6.38×10^7 **29.** 3.72×10^7

Chapter 8   Exponential and Logarithmic Functions

Practice Exercises, pages 336–337

1. $5^{7\sqrt{6}}$ **3.** 4^{30} **5.** $2^{8\sqrt{2}}$ **7.** $\dfrac{1}{3^{80}}$ **9.** $\dfrac{x^{4\sqrt{3}}}{3y^{2\sqrt{5}}}$

11. $\dfrac{y^{22}}{x^{18}}$ **13.** 2.155   **15.** 2.890   **17.** 0.215

19. 0.346   **21.** 31.707   **23.** 25.733   **25.** 3

27. 5   **29.** $-\dfrac{3}{2}$ **31.** 13   **33.** 13   **35.** $\dfrac{2}{3}$

37. -4 **39.** 4   **41.** 26,918

Practice Exercises, page 341

1. 16   **3.** 2   **5.** 54.598   **7.** 4.055

9.

$f(x) = 4^x$

11.

$f(x) = 5^x$

13.

$f(x) = 5^{-x}$

15.

$f(x) = 2^{-x}$

17.

$f(x) = \left(\tfrac{1}{2}\right)^x$

19.

$f(x) = \left(\tfrac{1}{4}\right)^x$

21.

$f(x) = 4^{2x}$

23.

$f(x) = 2^{2x+1}$

25.

$f(x) = \left(\frac{1}{2}\right)^{2x}$

27.

$f(x) = e^{-x}$

29.

$f(x) = 2^{x^2}$

31. 4   **33.** −1

Practice Exercises, pages 345–346 **1.** $8^2 = 64$
3. $3^3 = 27$ **5.** $36^{0.5} = 6$ **7.** $2^{-4} = \frac{1}{16}$ **9.** $\log_4$
$64 = 3$ **11.** $\log_{49} 7 = \frac{1}{2}$ **13.** $\log_{27} \frac{1}{9} = -\frac{2}{3}$
15. $\log_7 7\sqrt{7} = \frac{3}{2}$ **17.** 3 **19.** −1 **21.** $-\frac{1}{2}$ **23.** $\frac{3}{2}$
25.

$y = \log_4 x$

27.

$y = \log_{\frac{1}{4}} x$

29. 3 **31.** 216 **33.** 18 **35.** 5 **37.** $\sqrt{7}$ **39.** 64
41. 1.83 **43.** 39,875.32 **45.** 1.17 **47.** 7.77
49. $D = \{x: x > 0\}$; $R =$ **51.** $D = \{x: x > -2\}$;
$\{y: y$ is a real number$\}$ $R = \{y: y$ is a real
number$\}$

$y = 3 + \log_2 x$

$y = \log_{10} (x + 2)$

53. 8 **55.** 0 **57.** 3 **59.** $2^{-t} = I$

Test Yourself, page 346 **1.** 635.145 **3.** 4
7. $\log_4 \frac{1}{64} = -3$; $49^{\frac{1}{2}} = 7$ **9.** −2
11. 1,000,000

Practice Exercises, pages 351–352 **1.** $\log_3 x +$
$\log_3 y$ **3.** $\log_5 2 + \log_5 x$ **5.** $\log_2 x - \log_2 z$
7. $\log_3 x - \log_3 5$ **9.** $2 \log_3 x$ **11.** $\frac{1}{5} \log_6 x$
13. $4 \log_2 x + 3 \log_2 y$ **15.** $3 \log_2 3 + 3 \log_2 x$
17. $\frac{1}{3} \log_3 4$ **19.** $\log_5 x + \log_5 y + \log_5 z$
21. $\log_2 xy$ **23.** $\log_5 \frac{zw^2}{x^3}$ **25.** 7 **27.** 5 **29.** −9
31. ±8 **33.** $\log_3 x - \frac{1}{2} \log_3 y$ **35.** $\log_2 x - \frac{3}{2}$
$\log_2 y$ **37.** $\log_4 3 + 2 \log_4 x + 4 \log_4 y$
39. $2 \log_3 2 + 2 \log_3 x + 2 \log_3 y$ **41.** $\frac{5}{4} \log_b x$
43. $\frac{2}{3} \log_b x + \frac{4}{3} \log_b y$ **45.** $\log_b 3 + \frac{1}{2} \log_b x -$
$4 \log_b y$ **47.** $\log_b 6 + 4 \log_b x + \log_b y - \frac{1}{3} \log_b z$
49. $\log_4 x\sqrt[3]{y}$ **51.** $\log_4 \frac{\sqrt[3]{x}}{\sqrt{y}}$ **53.** 108
55. 5 **57.** $\frac{1}{3} \log_b x + \frac{3}{2} \log_b y - \frac{3}{4} \log_b z$
59. $-2 \log_2 x + \frac{5}{2} \log_2 y - \log_2 z$
61. $\frac{1}{3}$ **63.** 6 **67.** $\beta = 10(\log_{10} I - \log_{10} I_0)$

Practice Exercises, pages 357–358 **1.** 2.8987
3. −0.4318 **5.** −2.4685 **7.** 4.0289 **9.** 3.4833
11. −7.7634 **13.** 207.8 **15.** 2.160 **17.** 0.0431
19. 1.445 **21.** 19.78 **23.** 2.395 **25.** 1.3222
27. 6.9502 **29.** 9.2511 **31.** 1.771×10^{12}
33. 3.786×10^{-14} **35.** 3.393×10^4 **37.** 3.170
39. 2.262 **41.** 2.124 **43.** 0.1827 **45.** 2.513
47. 1.5850 **49.** 2.5937 **51.** 4.1933 **53.** 5.1578
55. 1.680 **57.** 4.254 **59.** 1.1559 **61.** 12.77
63. 3.654 yr

Practice Exercises, pages 362–364 **1.** 965
3. 216 h **5.** 4600 yr **7.** 9 d **9.** 4.2 yr **11.** 8.2
yr **13.** 8.2 yr **15.** 54 h **17.** 8 wk **19.** 18

Test Yourself, page 364
1. $\log_4 \frac{8de^3}{f^3}$ **3.** $\log_b 2x + 2 \log_b y + 3 \log_b z$
5. 2.5172 **7.** −1.1363 **9.** 366.6 **11.** 1.302
13. 1.2920 **15.** 3.3 yr

Answers

Answers to Selected Exercises   **459**

Summary and Review, pages 366–367 **1.** 1.491

3. $\frac{3}{2}$ **5.** 243 **7.** 1.732 **9.** $\log_3 81 = 4$

11. $4^3 = 64$ **13.** $\frac{1}{2}$ **15.** 5 **17.** 3 **19.** $5\log_4 x$
$+ 2\log_4 y - \log_4 z$ **21.** 2 **23.** 1.6781
25. 0.0007495 **27.** 3.6993 **29.** 1.4037

Cumulative Review, page 370 **1.** 148.413
3. $-2 + 0i, 1 + i\sqrt{3}, 1 - i\sqrt{3}$ **5.**

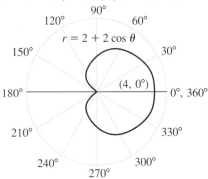

$r = 2 + 2\cos\theta$

(4, 0°)

7. $2 + i$ **9.** 5.0733 **11.** 1 **13.** $\log_6 1296 = 4$
15. $-5 + 5i\sqrt{3}$ **17.** 4π in. **19.** 31.6°

Chapter 9 Sequences and Series

Practice Exercises, pages 376–377 **1.** 5, 9, 13,
17, 21 **3.** 4, 1, -2, -5, -8 **5.** -17, -11,
-5, 1, 7 **7.** 19, 23, 27, 31; $a_n = 4n + 3$
9. -20, -16, -12, -8; $a_n = 4n - 36$
11. -14, -19, -24, -29; $a_n = -5n + 6$
13. 35 **15.** -3 **17.** -22 **19.** $a_1 = 5$; $d = 3$

21. $a_1 = 6$; $d = -5$ **23.** $a_1 = 3$; $d = \frac{1}{2}$

25. 16, 26, 36 **27.** 3.6, 1.2, -1.2, -3.6

29. $\frac{101}{15}$, $\frac{109}{15}$ **31.** 17 **33.** 97 **35.** 9 **37.** $n = 21$

39. $n = 8$ **41.** $\frac{3a+b}{4}$, $\frac{a+b}{2}$, $\frac{a+3b}{4}$

43. 168 **45.** $a_n = \frac{1}{2}n + \frac{1}{2}$ **47.** $a_1 = \frac{8r-3s}{5}$;

$d = \frac{s-r}{5}$ **49.** \$1350

Practice Exercises, pages 381–383 **1.** 3.5, 10.5,
31.5, 94.5 **3.** 1, -2, 4, -8 **5.** 8, -4, 2, -1

7. -81, 243, -729; $a_n = 3(-3)^{n-1}$ **9.** $\frac{1}{3}$, $\frac{1}{9}$, $\frac{1}{27}$;

$a_n = 9\left(\frac{1}{3}\right)^{n-1}$ **11.** -3, -9, -27; $a_n = -\frac{1}{9}(3)^{n-1}$

13. 144 **15.** -4 **17.** $128x$ **19.** 2, 4 **21.** 1, 3,
9, 27 **23.** 6, -12, 24, -48 **25.** -36 **27.** 30

29. $r = \frac{1}{2}$; $a_1 = 256$ **31.** $r = 2$; $a_1 = \frac{27}{16}$ or $r = -2$;

$a_1 = -\frac{27}{16}$ **33.** $r = \pm\frac{1}{2}$; $a_1 = 384$ **35.** -320

37. 4 **39.** seventh **41.** 63,670 **43.** \$22,211

Practice Exercises, pages 387–389 **1.** 42 **3.** 18
5. 36 **7.** 31.5 **9.** 20,200 **11.** -95 **13.** 780
15. -51 **17.** 180 **19.** -396 **21.** 220

23. -21 **25.** $\sum_{n=1}^{6}(4n - 1)$ **27.** $\sum_{n=1}^{6}(-5n + 13)$

29. 43 **31.** -529 **33.** 40 **35.** -101 **37.** 6,
36, 66 **39.** 18, 23.75, 29.5 **41.** $a_1 = 5$; $n = 8$
43. $d = 2$; $n = 4$ **49.** 162 **51.** $9d$

Practice Exercises, pages 393–395 **1.** 510 **3.** 11
5. 1452 **7.** -2.625 **9.** 1.375 **11.** 1023
13. -3280 **15.** 60 **17.** 1365.3125 **19.** 30
21. 341 **23.** -5 **25.** -4 **27.** arithmetic; 93

29. geometric; 87,380 **31.** $\sum_{k=1}^{8} 3\left(\frac{1}{3}\right)^{k-1}$

33. $\sum_{k=1}^{8} 2(-5)^{k-1}$ **35.** 504 **37.** 6558 **39.** $r =$

-2; $n = 7$ **41.** $\frac{x - 729x^{13}}{1 - 3x^2}$ **43.** $\frac{y(x^4 - x^{16})}{1 - x^2}$

45. 32.48 cm **47.** \$450; \$36,450

Test Yourself, page 395 **1.** 115 **3.** 2, 7, 12, 17

5. 8, $\frac{8}{5}$, $\frac{8}{25}$, $\frac{8}{125}$, $\frac{8}{625}$ **7.** 750 **9.** $\sum_{n=1}^{5}(-3n + 27)$

11. 155 or 55 **13.** geometric; 341.25

Practice Exercises, pages 400–401 **1.** $\frac{7}{6}$ **3.** 32

5. does not exist **7.** does not exist **9.** $-\frac{49}{8}$

11. $\frac{5}{4}$ **13.** $\frac{4}{9}$ **15.** $\frac{7}{33}$ **17.** $\frac{41}{333}$ **19.** 2 **21.** does

not exist **23.** $\frac{37}{90}$ **25.** $\frac{58}{11}$ **27.** $\frac{3}{5}$ **29.** -60

31. false **33.** true **35.** false **37.** 200 cm
39. 440 ft **41.** 240 cm

Practice Exercises, pages 405–406 **1.** 0.7072
3. 0.8660 **5.** -0.5184 **7.** 0.9975 **9.** 1.105

11. 1.396 **13.** 0.607 **15.** -1 **17.** $-\frac{\sqrt{2}}{2} + \frac{\sqrt{2}}{2}i$

19. 1 **21.** -1.0018 **23.** 0.9272 **25.** -0.4810
27. 0.3089 **43.** 0.39 **45.** 0.30 **47.** 0.18

Practice Exercises, pages 410–411 **1.** -1.1752

3. -0.7616 **5.** $\dfrac{1}{\operatorname{sech} x} = \dfrac{1}{\dfrac{1}{\cosh x}} = \cosh x$

7. $\dfrac{\cosh x}{\coth x} = \dfrac{\cosh x}{\dfrac{\cosh x}{\sinh x}} = \sinh x$ **9.** $\dfrac{\coth x}{\operatorname{csch} x} =$

$\dfrac{\dfrac{\cosh x}{\sinh x}}{\dfrac{1}{\sinh x}} = \cosh x$ **11.** $\dfrac{\operatorname{csch} x}{\coth x} = \dfrac{\dfrac{1}{\sinh x}}{\dfrac{\cosh x}{\sinh x}} = \dfrac{1}{\cosh x} =$

$\operatorname{sech} x$ **13.** $1 - \tanh^2 x = 1 - \dfrac{e^{2x} - 2 + e^{-2x}}{e^{2x} + 2 + e^{-2x}} =$

$\dfrac{e^{2x} + 2 + e^{-2x} - e^{2x} + 2 - e^{-2x}}{e^{2x} + 2 + e^{-2x}} =$

$\dfrac{4}{e^{2x} + 2 + e^{-2x}} = \left(\dfrac{2}{e^x + e^{-x}}\right)^2 = \operatorname{sech}^2 x$

15. $2 \sinh x \cosh x = 2 \left(\dfrac{e^x - e^{-x}}{2}\right)\left(\dfrac{e^x + e^{-x}}{2}\right) =$

$2\left(\dfrac{e^{2x} - e^{-2x}}{4}\right) = \dfrac{e^{2x} - e^{-2x}}{2} = \sinh 2x$

17. $\cosh^2 x - 1 = \left(\dfrac{e^x + e^{-x}}{2}\right)^2 - 1 =$

$\dfrac{e^{2x} - 2 + e^{-2x}}{4} = \left(\dfrac{e^x - e^{-x}}{2}\right)^2 = \sinh^2 x$

19. $\cosh x + \sinh x = \dfrac{e^x + e^{-x}}{2} + \dfrac{e^x - e^{-x}}{2} =$

$\dfrac{2e^x}{2} = e^x$ **25.** $y = a \cosh x$

Test Yourself, page 411 **1.** $\dfrac{8}{7}$ **3.** does not exist

5. $\dfrac{52}{99}$ **7.** -0.715 **9.** 42.8667 **11.** $\dfrac{\sqrt{2}}{2} - \dfrac{\sqrt{2}}{2} i$
13. 3.6269

Summary and Review, pages 414–415
1. $1, -2, -5; a_n = -3n + 13$ **3.** 61 **5.** $16,$
26 **7.** $27, 81, 243; a_n = 3^{n-1}$ **9.** 2 **11.** 195

13. $\displaystyle\sum_{k=1}^{6} (4k - 11)$ **15.** 202.4 **17.** 2186 or 1094

19. does not exist **21.** -0.2224 **23.** 2.0136

25. $\dfrac{\tanh x}{\sinh x} = \dfrac{\dfrac{\sinh x}{\cosh x}}{\sinh x} = \dfrac{1}{\cosh x} = \operatorname{sech} x$

Cumulative Review, pages 418–420
1. $125^{\frac{1}{3}} = 5$ **3.** $72°$
5.

7. $\dfrac{\sqrt{6} + \sqrt{2}}{4}$ **9.** $(-1)\left(-\dfrac{\sqrt{2}}{2}\right) = \sqrt{2} - \dfrac{\sqrt{2}}{2};$

$\dfrac{\sqrt{2}}{2} = \dfrac{\sqrt{2}}{2}$ **11.** $\dfrac{3r^{\frac{1}{2}}}{s^{\frac{2}{5}}}$ **13.** $2; \angle A = 53°, \angle B = 80°,$

$b = 55$ or $\angle A = 6°, \angle B = 127°, b = 6$
15. $y = \tan (x + \pi)$

17. 14 **19.** $270°$ **21.** $-\dfrac{5\sqrt{2}}{2} + \dfrac{5\sqrt{2}}{2} i$

23. geometric; 3906 **25.** $-\dfrac{\sqrt{2}}{2}$ **27.** 18

29. neither **31.** $\dfrac{5}{13}; -\dfrac{12}{5}; -\dfrac{5}{12}; \dfrac{13}{5}; -\dfrac{13}{12}$

33. 25 ft **35.** $\cos (2\pi + \alpha) = \cos 2\pi \cos \alpha -$
$\sin 2\pi \sin \alpha = (1) \cos \alpha - (0) \sin \alpha = \cos \alpha$

37.

g(x)

$g(x) = 3^x$

39. $9\sqrt{2} + 9\sqrt{2}\, i$ **41.** $-5, -1, 3, 7$
43. identity **45.** $10°, 30°, 150°, 170°$ **47.** 7
49. $\dfrac{(y')^2}{4} - \dfrac{(x')^2}{4} = 1$ **51.** 4.5 yr **53.** 1 **55.** 2.2

GLOSSARY

The explanations given in this glossary include definitions and brief descriptions of the key terms used in this book.

addition of ordinates (p. 85) A method for sketching the graph of a function that is the sum or difference of two trigonometric functions which involve adding or subtracting the two y-values for each x-value.

ambiguous case (p. 164) When the measures of two sides and the angle opposite one of them are given, it is not always possible to determine a unique triangle using the law of sines and a given value of the sine ratio.

amplitude (p. 73) Half the difference of the maximum and minimum values of a function.

angle of depression (p. 118) The angle formed between a horizontal line (the horizon) and an observer's line of sight to an object below the horizon.

angle of elevation (p. 118) The angle formed between a horizontal line (the horizon) and an observer's line of sight to an object above the horizon.

angular displacement (p. 24) The angle θ through which $\overrightarrow{OA}$ rotates where O is the axis of an object and A is a point on the edge of the object.

angular velocity (p. 24) The angular displacement per unit of time: $w = \dfrac{\theta}{t}$, where w represents angular velocity.

angle (p. 273) Formed by rotating a ray about its vertex from one position, called the initial side, to another, called its terminal side.

argument (p. 309) The measure of the angle formed by $\overrightarrow{OP}$ with the real axis in a complex plane.

arithmetic means (p. 377) The terms between any two nonconsecutive terms of an arithmetic sequence.

arithmetic sequence (p. 375) An arithmetic progression; a sequence of numbers in which the difference between any two successive terms is a constant.

arithmetic series (p. 386) The indicated sum of an arithmetic sequence.

asymptote (p. 88) A line that a graph approaches but never crosses.

cardioid (p. 294) A heart-shaped curve represented by polar equations of the form $r = a \pm b \cos \theta$ or $r = a \pm b \sin \theta$ in which $a = b$.

central angle (p. 17) An angle whose vertex is the center of a circle.

change-of-base formula (p. 356) $\log_b x = \dfrac{\log_a x}{\log_a b}$, where a is the reference base.

circular function (p. 31) The trigonometric functions are sometimes called circular functions.

cofunctions (p. 67) The sine and cosine, tangent and cotangent, and secant and cosecant, are pairs of cofunctions.

common difference (p. 375) The constant, d, between any two successive terms in an arithmetic sequence; $d = a_{n-1} - a_n$.

common logarithm (p. 353) A base-10 logarithm denoted $\log x$.

common ratio (p. 380) The constant, r, which is the ratio of any two consecutive terms in a geometric sequence; $r = \dfrac{a_n + 1}{a_n}$.

complementary angles (p. 67) Two acute angles the sum of whose measures is 90°.

complex number (p. 299) A number that can be written in the form $a + bi$, where a and b are real numbers and $i = \sqrt{-1}$.

complex plane (p. 308) A coordinate plane for plotting complex numbers of the form $a + bi$, where the horizontal axis is associated with the real part and the vertical axis is associated with the imaginary part.

components of a vector (p. 191) The x- and y-component of a vector in standard position such that the magnitude of x equals the product of the magnitude of the vector and the cosine of its directional angle and y equals the product of the magnitude of the vector and the sine of its directional angle.

conjugates (p. 304) Two complex numbers, such as $3 + 5i$ and $3 - 5i$, that differ only in the sign of the imaginary part.

convergent series (p. 399) An infinite geometric series whose sequence of partial sums approaches a limit as n increases without bound.

coordinate plane (p. 3) A rectangular coordinate system in which each point is associated with exactly one ordered pair in a relation.

cosecant (p. 34) If $P(x, y)$ is any point other than the origin on the terminal side of an angle θ in standard position, $\csc \theta = \dfrac{r}{y}$, $y \neq 0$.

cosine (p. 29, 34) The function defined by the set of ordered pairs (θ, x); if $P(x, y)$ is any point other than the origin on the terminal side of an angle θ in standard position, $\cos \theta = \dfrac{x}{r}$, where $r = \sqrt{x^2 + y^2}$.

cotangent (p. 34) If $P(x, y)$ is any point other than the origin on the terminal side of an angle θ in standard position, $\cot \theta = \dfrac{x}{y}$, $y \neq 0$.

coterminal angles (p. 13) Angles in standard position, whose terminal sides coincide.

cycle (p. 60) The part of the graph of a periodic function that is completed in each period.

degree (p. 13) A unit of angle measure equal to $\dfrac{1}{360}$ revolution.

direction angle of a vector (p. 191) The angle between a vector in standard position and the positive x-axis.

divergent series (p. 400) An infinite geometric series whose sequence of partial sums increase or decrease without bound as n increases without bound.

domain (p. 2) The set of replacements for the first variable in a relation or function.

equivalent vectors (p. 190) Vectors that have the same magnitude and the same direction.

even function (p. 62) A function in which $f(-x) = f(x)$ for every value in the domain of the function.

explicit formula (p. 374) A formula that defines a sequence by expressing the nth term, a_n, as a function of n.

exponent function (p. 339) A function in which the variable is in the exponent as in $f(x) = b^x$, $y = 3^x$, and $f(x) = 4^x$.

finite sequence (p. 374) A sequence that has a last term.

frequency (p. 100) For a sinusoidal graph the reciprocal of the period, which is $\dfrac{|b|}{2\pi}$.

function (p. 3) A relation in which each member of the domain is paired with exactly one member of the range.

fundamental identities (p. 133) The reciprocal, Pythagorean, ratio, and odd-even identities.

geometric means (p. 382) The terms between any two nonconsecutive terms of a geometric sequence.

geometric sequence (p. 380) A geometric progression; a sequence in which the ratio of any two consecutive terms is a constant.

geometric series (p. 392) The indicated sum of the terms of a geometric sequence.

general term (p. 374) The nth term, denoted a^n, in a sequence.

hyperbolic functions (p. 410–411) The trigonometric functions defined in terms of the exponential function and associated with the unit hyperbola, $\sinh x = \dfrac{e^x - e^{-x}}{2}$, $\cosh x = \dfrac{e^x + e^{-x}}{2}$, and $\tanh x = \dfrac{e^x - e^{-x}}{e^x + e^{-x}}$.

imaginary number i (p. 298) The square root of negative one, that is, $i = \sqrt{-1}$.

infinite sequence (p. 375) A sequence that does not have a last term.

initial point of a vector (p. 190) The point A in $\overrightarrow{AB}$; the tail of a vector.

initial side (p. 12) The position from which a ray is rotated about the vertex of an angle.

inverse function (p. 251) If the inverse of a function is also a function, it is called the inverse function.

law of cosines (p. 168) For any triangle ABC, where a, b, and c are the lengths of the sides opposite the angles with measures A, B, and C, respectively, $a^2 = b^2 + c^2 - 2bc \cos A$, $b^2 = a^2 + c^2 - 2ac \cos B$, and $c^2 = a^2 + b^2 - 2ab \cos C$.

law of sines (p. 156) For any triangle ABC in which a, b, and c are the lengths of the sides opposite the angles with measures A, B, and C, respectively, $\dfrac{\sin A}{a} = \dfrac{\sin B}{b} = \dfrac{\sin C}{c}$.

law of tangents (p. 175) For any triangle ABC in which a, b, and c are the lengths of the sides opposite angles with measures A, B, and C, respectively, $\dfrac{a - b}{a + b} = \dfrac{\tan \frac{1}{2}(A - B)}{\tan \frac{1}{2}(A + B)}$,

$\dfrac{c - a}{c + a} = \dfrac{\tan \frac{1}{2}(C - A)}{\tan \frac{1}{2}(C + A)}$, and

$\dfrac{b - c}{b + c} = \dfrac{\tan \frac{1}{2}(B - C)}{\tan \frac{1}{2}(B + C)}$.

limaçon (p. 294) A curve represented by polar equations of the form $r = a + b \cos \theta$ or $r = a + b \sin \theta$.

limit (p. 399) A number that the partial sums of an infinite geometric series approach but never reach as the number of terms summed increases without bound.

linear velocity (p. 24) The linear distance traveled by a point on a revolving ray per unit of time.

logarithmic function (p. 343) The function $y = \log_b x$ which is the inverse of the exponential function $y = b^x$.

magnitude of a vector (p. 190) The magnitude, or norm, of V is represented as $\|V\|$ and is found by using the distance formula.

mean proportional (p. 382) A single geometric mean between two numbers in a geometric sequence; $m = \pm\sqrt{ab}$, where a and b are real numbers and m is the mean proportional.

minute (p. 17) A subdivision of a degree, where $1° = 60'$ (60 min).

modulus (p. 310) The quantity r in $r(\cos \theta + i \sin \theta)$.

natural exponential function (p. 340) The function $f(x) = e^x$, where $e \approx 2.7182818$.

natural logarithm (p. 354) A base e logarithm denoted $\ln x$.

negative vector (p. 192) A vector of equal magnitude but opposite direction from another vector.

odd-even identities (p. 132) Identities derived from the definitions of odd and even functions: $\sin(-\theta) = -\sin \theta$, $\cos(-\theta) = \cos \theta$, and $\tan(-\theta) = -\tan \theta$.

odd function (p. 62) A function in which $f(-x) = -f(x)$ for every value in the domain of the function.

origin (p. 3) The point of intersection of the axes in a coordinate plane.

period (p. 60) For a periodic function $f(x)$, the smallest positive number p, such that $f(x + p) = f(x)$ whenever $f(x + p)$ and $f(x)$ are defined, is the period of the function.

phase shift (p. 78) The distance that a graph is shifted right or left.

polar coordinates (p. 286) The coordinates (r, θ) of a point P in the polar coordinate system, where $|r|$ is the distance from the pole and θ is the measure of the angle from the polar axis to $\overrightarrow{OP}$.

polar form of a complex number (p. 309) $a + bi = r(\cos \theta + i \sin \theta)$, where $r = \sqrt{a^2 + b^2}$ and $\tan \theta = \dfrac{b}{a}$, $a \neq 0$.

power series (p. 404) An expression of the form $\displaystyle\sum_{n=0}^{\infty} a_n x^n$, where x is a variable and a_0, a_1, a_2, . . . , a_n, . . . are constants.

principal values (p. 255) The values in the range of an inverse function.

Pythagorean theorem (p. 8) a and b are the lengths of the legs of a right triangle and c is the length of its hypotenuse if and only if $c^2 = a^2 + b^2$.

quadrantal angle (p. 39) An angle whose terminal side is on the x- or y-axis.

radian (p. 18) A unit of angle measure; a central angle of 1 radian intercepts an arc that has the same length as the radius of the circle.

range (p. 2) The set of replacements for the second variable in a relation or function.

reciprocal functions (p. 35) Sine and cosecant, cosine and secant, and tangent and cotangent are pairs of reciprocal functions.

recursive formula (p. 375) A formula that defines a sequence by stating the first term, a_1, and a rule for obtaining the $(n + 1)$th term from the preceding term, a_n.

reference angle (p. 42) The smallest positive acute angle determined by the x-axis and the terminal side of θ.

reference triangle (p. 30) A right triangle formed when a perpendicular is drawn from any point on the terminal side of an angle to the x-axis.

resultant vector (p. 191) A vector that is the sum of two vectors.

rose (p. 295) A curve $r = a \cos n\theta$ or $r = a \sin n\theta$, where n is a positive integer, consisting of n leaves if n is odd and $2n$ leaves if n is even.

rotation of axes (p. 273) A new $x'y'$-coordinate system formed by rotating the xy-coordinate axes about the origin.

scalar (p. 192) A real number used to multiply a vector.

secant (p. 34) If $P(x, y)$ is any point other than the origin on the terminal side of an angle θ in standard position, $\sec \theta = \dfrac{r}{x}$, $x \neq 0$.

second (p. 17) A subdivision of a minute, where $1' = 60''$ (60 s).

sequence (p. 372) A function whose domain is a set of consecutive positive integers.

series (p. 386) The indicated sum of the terms of a sequence.

sine (p. 29, 34) The function defined by the set of ordered pairs (θ, y); if $P(x, y)$ is any point other than the origin on the terminal side of an angle θ in standard position, $\sin \theta = \dfrac{y}{r}$, where $r = \sqrt{x^2 + y^2}$.

standard position (p. 12) When the vertex of an angle is at the origin of the coordinate plane and its initial side coincides with the positive x-axis, the angle is in standard position.

sum of an infinite geometric series (p. 400) The limit approached by the sequence of partial sums of an infinite geometric series; $S = \dfrac{a_1}{1 - r}$, $|r| < 1$ where a_1 is the first term and r is the common ratio.

tangent (p. 34) If $P(x, y)$ is any point other than the origin on the terminal side of an angle θ in standard position, $\tan \theta = \dfrac{y}{x}$, $x \neq 0$.

terminal side (p. 12) The side of an angle that determines the amount of rotation of the initial side of the angle.

trigonometric functions (p. 34) Sine, cosine, tangent, cosecant, secant, and cotangent are trigonometric functions.

unit circle (p. 29) A circle with a radius of 1.

vector (p. 190) A directed line segment that represents a vector quantity.

vertical shift (p. 80) The distance that a graph is shifted up or down.

INDEX

Index

solving, 156-178
Odd-even identities, 132
Odd function, 61-62
Omega (ω), 24
Ordered pair, 2-3
Ordered triplet, 11, 198
Origin, 3
Origin symmetry, 62-63
Oscilloscope, 67, 100

Parabola, 52
Parallelogram addition, 191-192
Parallels, 236
Partial sum(s)
    of arithmetic series, 384-387
    of geometric series, 390-393
Pascal's triangle, trigonometry
    and, 291
Period, 60
Periodic function(s), 60-61
    amplitude of, 73-75
pH, 342, 352
Phase shift, 78-81
Physics, applications, 77, 147,
    197, 264, 268, 272, 291, 296-
    297, 306-307, 346, 352, 401
Pi (π), 19
Pitch, 73
Plane
    complex, 308
    coordinate, 3
    slice, 105
    vectors in, 190-194
    wz-, 408
Polar axis, 286
Polar coordinate system, 286-290
Polar coordinates, 286-290
    converting to rectangular coor-
        dinates, 289
Polar equations, 292-295
Polar form
    of complex numbers, 309-310
    division in, 314-315
    multiplication in, 313-314
Polar graphs, 292-295
    rotations of, 297
Pole, 286
Power series, 402-404
    and hyperbolic functions, 407-
        409
Powers
    of complex numbers, 318-320
    logarithms of, 348-349
Principal values, 255-256, 260
Product(s)
    of complex numbers, 303-304,
        313-314

of conjugates, 304
logarithms of, 348-349
of reciprocal trigonometric
    functions, 35
of vector by scalar, 192
Product/sum identities, 237-239,
    241
Ptolemy, Claudius, 66
Pulleys, 25-26
Pure imaginary number, 299
Pythagoras, 17
Pythagorean identities, 131
Pythagorean theorem, 8, 117

Quadrantal angle, 39
Quadrants, 3
Quotient(s)
    of complex numbers, 304-305
    logarithms of, 348-349

Radian, 18
Radian measure(s), 18-20
    converting to degree measure,
        19
    equivalent degree measures, 20
Radius vector, 30
Rainbow, 278-279
Range, 2
Ratio
    common, 378
    identities, 131
    right triangle, 113
    trigonometric, 112
Real axis, 308
Real number exponents, 334-336
    properties of, 334
Reciprocal(s)
    of complex numbers, 304
    identities, 130
    properties, 35
    trigonometric functions, 35-36
Rectangular coordinate system, 3
    three-dimensional, 198-199
Rectangular coordinates, 286
    converting to polar coordinates,
        288
Rectangular form of complex
    numbers, 309
Regiomontanus (Johann Müller),
    66
Reference angle, 42-43
Reference triangle, 30-31
Refraction, 278-279
Relation, 2-3
    domain of, 2
    graphs of, 3

inverse of, 250-251
range of, 2
Repeating decimals, 399
Resultant of a vector, 191
Reviews
    Chapter Summary and Review,
        54-55, 106-107, 150-151,
        200-201, 244-245, 280-281,
        328-329, 366-367, 414-415
    College Entrance Exam, 57,
        109, 153, 203, 247, 283,
        331, 369, 417
    Cumulative, 110, 204-206, 284,
        370, 418-420
    Mixed Review, 58, 154, 248,
        332
Richter scale, 365
Right triangle(s)
    applications, 118-120, 124-125
    isosceles, 39
    ratios, 113
    solving, 112-115
Roots of complex numbers, 322-
    324
Rose, 295
Rotary motion, 24, 198
Rotation
    angles of, 273
    of axes, 273-276
    of polar graphs, 297

Scalar, 192
Scalene triangle, 156
Scientific notation, 353
Sec θ, 34
Sec x, 67
Sec^{-1} x, 261
Secant function(s), 34-36
    graphs of, 94-97
    hyperbolic, 409
    inverse, 261-263
    properties of, 94-97
    range of, 95
Sech x, 409
Seconds, 17
Semiperimeter, 186
Sequence(s), 372
    arithmetic, 373-375, 412-413
    defined explicitly, 372
    defined recursively, 373
    finite, 372
    general term of, 372
    geometric, 378-381
    infinite, 373
    number theory and, 377
    term of, 372
Self-test (*see* Tests, Test Yourself)

Photo Credits